Volume 1

Mechanisms of Inorganic and Organometallic Reactions

Volume 1

Mechanisms of Inorganic and Organometallic Reactions

Edited by

M. V. Twigg
Imperial Chemical Industries P. L. C.
Billingham, United Kingdom

SPRINGER SCIENCE+BUSINESS MEDIA, LLC

Library of Congress Cataloging in Publication Data

Main entry under title:

Mechanisms of inorganic and organometallic reactions.
  Includes bibliographical references and index.
  1. Chemical reactions. 2. Chemistry, Inorganic. 2. Organometallic compounds. I.
Twigg, M. V.
QD501.M426 1983   541.3/9   83-2140
ISBN 978-1-4615-7412-5   ISBN 978-1-4615-7410-1 (eBook)
DOI 10.1007/978-1-4615-7410-1

Contributors

Dr. J. Burgess   Chemistry Department, The University of Leicester, Leicester, U.K.

Dr. J. Coe   Chemistry Department, Kings College, London University, The Strand, London WC2, U.K.

Dr. J.M. Davidson   Dept. of Chemical Engineering, University of Edinburgh, Kings Building, Edinburgh EH9 3JL, Scotland, U.K.

Dr. A.J. Deeming   Chemistry Department, University College London, 20 Gordon Street, London WC1H OAJ, U.K.

Dr. M. Green   Chemistry Department, The University of York, York, North Yorkshire, U.K.

Dr. D. Hague   Chemical Laboratory, The University of Kent, Canterbury, Kent, U.K.

Dr. M.N. Hughes Chemistry Department, Queen Elizabeth College, University of London, London W8 7AH, U.K.

Dr. L.A.P. Kane-Maguire Chemistry Department, University College, Cardiff CF1 1XL, Wales, U.K.

* *Dr. A.G. Lappin* Chemistry Department, University of Glasgow, Glasgow, Scotland, U.K.

Professor A. McAuley Department of Chemistry, University of Victoria, British Columbia, Canada

Dr. P. Moore Department of Chemistry and Molecular Sciences, University of Warwick, Coventry CV4 7AL, U.K.

*Presently: Department of Chemistry, University of Notre Dame, Notre Dame, Indiana 46556, U.S.A.

Preface

During recent years a high level of interest has been maintained in the kinetics and mechanisms of inorganic compounds in solution, and there has also been a notable upsurge of literature concerned with reaction mechanisms of organotransition metal compounds. The reviews of the primary literature previously provided by "Inorganic Reaction Mechanisms" (Royal Society of Chemistry) and "Reaction Mechanisms in Inorganic Chemistry" in "MTP International Reviews of Science" (Butterworths) continue to be of considerable value to those concerned with mechanistic studies, and it is unfortunate they are no longer published.

The objective of the present series is to provide a continuing critical review of literature dealing with mechanisms of inorganic and organometallic reactions in solution. The scope of potentially relevant work is very large, particularly in the field of organotransition metal chemistry, and papers for inclusion have been chosen that specifically probe mechanistic aspects, rather than those of a preparative nature. This volume covers the literature published during the period July 1979 to December 1980 inclusive. Material is arranged basically by type of reaction and type of compound along generally accepted lines. Numerical data are usually reported in the units used by the original authors, though the units of some results have been converted in order to make comparisons.

Many people, most of whom are members of the Inorganic Mechanisms Discussion Group (UK), were involved in establishing this series. Their help is gratefully acknowledged, as is the enthusiastic and prompt way in which contributions were prepared. Comments on this and future volumes will be welcomed.

Contents

Part 1. Electron Transfer Reactions

Chapter 1. General Redox Processes and Reactions between Two Complexes

A.G. Lappin

Chapter 2. Metal-Ion–Ligand Redox Reactions

A. McAuley

Part 2.   Substitution and Related Reactions

Chapter 3.   Reactions of Compounds of the Nonmetallic Elements

M.N. Hughes

Chapter 6. Substitution Reactions of Labile Metal Complexes

D.N. Hague

Part 3. Reactions of Organometallic Compounds

Chapter 7. Metal-Alkyl Bond Fission and Formation

M. Green

Chapter 8. Substitution, Oxidative Addition–Reductive Elimination, and Migration–Insertion Reactions

M. Green

Chapter 11. Homogeneous Catalysis of Organic Reactions by Complexes of Metal Ions

J.M. Davidson

Part 1
Electron Transfer Reactions

Chapter 1

General Redox Processes and Reactions between Two Complexes

1.1. Introduction

In this section I have tried to follow a pattern comparable to the one used in the relevant chapter of the "Specialist Periodical Reports" series[1] though minor adjustments have been made where necessary. The major developments in theory are followed by an outline of the outer-sphere and inner-sphere electron transfer literature. Photoinduced electron transfer, which remains fashionable, and bioinorganic studies covering mainly the interactions of metalloproteins with small inorganic reagents complete the chapter. Coverage is as comprehensive as space will allow and I have continued the practice of tabulating all the relevant rate data at the end of the chapter.

1.2. General Background

1.2.1. General and Theoretical

The appearance of a book[2] devoted to the subject of electron transfer between metal ion complexes is to be welcomed. A book chapter[3] on redox mechanisms has also been published.

1.2.2. Theoretical Developments

Comparison of the self-exchange rates for $[Co(NH_3)_6]^{3+/2+}$ and $[Ru(NH_3)_6]^{3+/2+}$ has been made[4] using multiphonon theory for nonadiabatic electron transfer in an attempt to elucidate the very slow rate for the former process. The high-spin–low-spin change on going from cobalt(II) to cobalt(III) accounts for a reduction of four orders of magnitude in the self-exchange rate compared with the ruthenium system, but Franck–Condon factors have a much larger effect.

A thermodynamic derivation[5] of the Marcus cross relationship [equation (1)], where k_{11} and k_{22} are the self-exchange

$$k_{12} = (k_{11}k_{22}K_{12}f)^{\frac{1}{2}} \tag{1}$$

rate constants for the reactants, k_{12} the cross-reaction rate constant, and K_{12} the corresponding equilibrium constant, suggests that the equation has much wider applicability than to adiabatic outer-sphere electron transfer processes. Breakdown of the behavior occurs only when activation processes for the reactants are not independent of their reaction partners and when the activated states for the symmetric electron exchange reactions (k_{11}, k_{22}) and the asymmetric electron transfer process (k_{12}) differ. It is noteworthy that the relationship should hold in some cases where electron transfer is nonadiabatic and where there is a change in spin multiplicity.

The activation free energy for an outer-sphere electron transfer process, $\Delta G^{\ddagger}$, consists of an internal contribution, $\Delta G_{IN}^{\ddagger}$, associated with rearrangement of the inner-coordination spheres and an external contribution, $\Delta G_{OUT}^{\ddagger}$, due to solvent rearrangement [equation (2)]. Reaction entropies, ΔS_{RC}°, for

$$\Delta G^{\ddagger} = \Delta G_{IN}^{\ddagger} + \Delta G_{OUT}^{\ddagger} \tag{2}$$

the redox couple [equation (3)] have been measured[6] for a

$$ML_6^{n+} + e^- \rightleftharpoons ML_6^{(n-1)+} \tag{3}$$

series of aquo and amino complexes and show evidence for specific ligand contributions though the effects of charge $ML_6^{(n-1)+}$ and electronic structure are also important. These entropy changes reflect the difference in radii of the ions ML_6^{n+}, $ML_6^{(n-1)+}$ and can be used[7] to correlate with $\Delta G_{IN}^{\ddagger}$ as a measure of complex rearrangement, whereas $\Delta G_{OUT}^{\ddagger}$ is mainly determined by the absolute size of the reactants. There is a general tendency for $\Delta G^{\ddagger}$ for the self-exchange reactions to increase with ΔS°_{RC} for aquo and amino complexes, indicating that $\Delta G^{\ddagger}_{IN}$ has a substantial effect. However, spin multiplicity and nonadiabaticity factors complicate matters.

Rate constants, k_{12}, and activation parameters for a number of outer-sphere cross reactions of aquo and amino complexes have been compared[8] in detail with values calculated using Marcus theory. Activation free energies, corrected for charge effects involved in collision complex formation using a Debye–Hückel approach showed satisfactory agreement at low overall driving force, but calculated values were consistently smaller than experimental results for large driving forces. This reflects a change in the rate-determining step from electron transfer to a process independent of redox driving force, most likely collision complex formation. Reinforcement of this notion comes from examination of the activation parameters where the disparity in rate constants is shown to be primarily an entropic factor. That there is significant solvent redistribution on collision complex formation is revealed by the large negative values for $\Delta S^{\ddagger}_{12}$ for bimolecular electron transfer since reorganizational entropy within the binuclear complex is close to zero.[9]

Attempts[10] to compare homogeneous and electrochemical rate constants using the Marcus-derived expression (4) where k^{e}_{12}

$$k^{e}_{12}/Z_e = (k_{12}/Z_h)^{1/2} \tag{4}$$

is the heterogeneous rate constant at the intersection of rate–potential plots for the respective half-reactions and Z_e and Z_h are the heterogeneous and homogeneous collision frequencies, show significantly faster rates for homogeneous electron transfer involving aquo ions than are predicted by theory. It is suggested that these ions have a greater solvent structuring effect compared with amino complexes and that they approach the electrode surface less closely.

1.2.3. Optical Electron Transfer

Calculations[11] of electron transfer rate constants from the energies of intervalence transfer (IT) bands in mixed-valence complexes have provided some interest (Table 1.1). The ion pair formed from paraquat (1,1'-dimethyl-4,4'-bipyridine^{2+}) PQ^{2+}, and ferrocyanide, $[PQ^{2+}, Fe(CN)_6^{4-}]$, shows an IT band[12] from which the activation energy E_a, for thermal electron transfer within the ion pair can be derived (Figure 1.1) using Hush's theory to compare spectroscopic and kinetic data [equation (5)].

$$E_a \approx E^2_{op}/4(E_{op} - \Delta E) \tag{5}$$

TABLE 1.1. Intramolecular Electron Transfer Rate Constants at 25°C

Reaction	Medium	$(\mathbf{k}, s^{-1})$ $(K, mol^{-1} liter)$	$\Delta H^{\ddagger}$, kcal mol^{-1}	$\Delta S^{\ddagger}$, cal K^{-1} mol^{-1}	Ref.
Nonlabile system					
[(bipy)$_2$ClRu(III)N◯O NRu(II)Cl(bipy)$_2$]$^{3+}$	CH$_3$CN (IT)	3–9×10^9 (s^{-1})			11
[(bipy)$_2$ClRu(III)N◯O◯ NRu(II)Cl(bipy)$_2$]$^{3+}$	CH$_3$CN (IT)	1.0–2.8×10^8 (s^{-1})			11
[(bipy)$_2$ClRu(III)N◯ ═ ◯ O NRu(II)Cl(bipy)$_2$]$^{3+}$	CH$_3$CN (IT)	5.1–10×10^7 (s^{-1})			11
[(NH$_3$)$_5$Ru(III)N◯O Ru(II)Cl(bipy)$_2$]$^{4+}$	CH$_3$CN (IT)	5.7×10^7 (s^{-1})			11
[(bipy)$_2$ClRu(III)Ph$_2$PCH$_2$PPh$_2$Ru(II)Cl(bipy)$_2$]$^{3+}$	CH$_3$CN (IT)	8.8×10^8 (s^{-1})			13
[Co(III)(NH$_3$)$_5$-o-O$_2$CC$_6$H$_4$NO$_2$]$^+$ ^{a}	pH 7.0	4.0×10^5 (s^{-1})			27
[Co(III)(NH$_3$)$_5$-m-O$_2$CC$_6$H$_4$NO$_2$]$^+$ ^{a}	pH 7.0	1.5×10^2 (s^{-1})			27
[Co(III)(NH$_3$)$_5$-p-O$_2$CC$_6$H$_4$NO$_2$]$^+$ ^{a}	pH 7.0	2.6×10^3 (s^{-1})			27
[Co(III)(NH$_3$)$_5$-o-O$_2$CCH$_2$C$_6$H$_4$NO$_2$]$^+$ ^{a}	pH 7.0	3.5×10^4 (s^{-1})			27
[Co(III)(NH$_3$)$_5$-m-O$_2$CCH$_2$C$_6$H$_4$NO$_2$]$^+$ ^{a}	pH 7.0	1.0×10^2 (s^{-1})			27

Compound	Conditions	Rate constant			Ref.
$[Co(III)(NH_3)_5\text{-}p\text{-}O_2CCH_2C_6H_4NO_2]^+$ ^{a}	pH 7.0	3.9×10^2 (s^{-1})			27
$[Co(III)(NH_3)_5\text{-}o\text{-}O_2CCH{=}CHC_6H_4NO_2]^+$ ^{a}	pH 7.0	1.7×10^3 (s^{-1})			27
$[Co(III)(NH_3)_5\text{-}m\text{-}O_2CCH{=}CHC_6H_4NO_2]^+$ ^{a}	pH 7.0	3.1 (s^{-1})			27
$[Co(III)(NH_3)_5\text{-}p\text{-}O_2CCH{=}CHC_6H_4NO_2]^+$ ^{a}	pH 7.0	4.8×10^2 (s^{-1})			27
$[Co(III)NH_3)_5\text{-}p\text{-}O_2CC(CH_3)_2C_6H_4NO_2]^+$ ^{a}	pH 7	1.5×10^2 (s^{-1})			27
$[Co(III)(NH_3)_5\text{-}p\text{-}O_2CCH_2NHCOC_6H_4NO_2]^+$ ^{a}	pH 7	5.8 (s^{-1})			27
$[C(III)(NH_3)_5\text{-}p\text{-}O_2C(CH_2NHCO_2C_6H_4NO_2]^+$ ^{a}	pH 7	1.5×10^3 (s^{-1})			27
$[Co(III)(NH_3)_5\text{-}p\text{-}O_2CC_6H_4NO_2]^+$ ^{a}	1% 2-propanol	1.3×10^4 (s^{-1})	12.5	2.9	25
$[\mathbf{2},\ \mathbf{R} = p\text{-}C_6H_4NO_2]^{2+}$	1% 2-propanol	2.3×10^3 (s^{-1})	18.4	19.4	25
Labile systems					
$\{[\mathbf{2},\ \mathbf{R} = \text{dipic}]_2Fe(II)]^{4+}$	0.1	3.7×10^{-3} (s^{-1})	24.0	11	24
$\{[\mathbf{2},\ \mathbf{R} = \text{dipic}]Ti(III)]^{4+}$	0.1	0.25 (s^{-1})	19	3	24
$\{[\mathbf{2},\ (NH_3)_3 = A_3,\ \mathbf{R} = \text{dipic}]Ti(III)]^{4+}$	0.1	0.023 (s^{-1})	18.5	-4	24
$\{[\mathbf{2},\ \mathbf{R} = \text{dipic}]_2Ti(III)]^{5+}$	0.1	32 (s^{-1})	14	-6	24
$\{[\mathbf{2},\ (NH_3)_3 = A_3,\ \mathbf{R} = \text{dipic}]_2Ti(III)]^{5+}$	0.1	15.6 (s^{-1})	15	-2	24
$[\mathbf{2},\ \mathbf{R} = \text{napthalene}] + Fe^{2+}$	0.16	3.94×10^{-3} (s^{-1}); 64 (mol^{-1} liter)	20.2	-1.6	23
$[\mathbf{2},\ \mathbf{R} = \text{quinoline}] + Fe^{2+}$	0.16	7.73×10^{-3}; 54 (mol^{-1} liter)	19.1	-4	23
$[\mathbf{2},\ \mathbf{R} = \text{pyridine}] + Fe^{2+}$	0.16	1.61×10^{-3}; 50 (mol^{-1} liter)	21.4	0.3	23
$PQ^{2+} + [Fe(CN)_6]^{4-}$ ^{b}	0.1 (IT)	7.12×10^{-7} (s^{-1}); 220 (mol^{-1} liter)			12
$[Co(NH_3)_5(4\text{-phenylpyridine})]^{3+} + [Fe(CN)_5(4,4'\text{-bipy})]^{3-}$	0.0769	3.12×10^{-3} (s^{-1}); 1047 (mol^{-1} liter)			59
$[Co(NH_3)_5(4\text{-benzoylpyridine})]^{3+} + [Fe(CN)_5(4,4'\text{-bipy})]^{3-}$	0.0769	1.15×10^{-2} (s^{-1}); 904 (mol^{-1} liter)			59
$[Co(NH_3)_5(4\text{-phenylpyridine})]^{3+}$ $+ [Fe(CN)_5(1,2\text{-bis}(4\text{-pyridyl})\text{ethane})]^{3-}$	0.0769	8.99×10^{-3} (s^{-1}); 761 (mol^{-1} liter)			59
$[Co(NH_3)_5(4\text{-benzoylpyridine})]^{3+}$ $+ [Fe(CN)_5(1,2\text{-bis}(4\text{-pyridyl})\text{ethane})]^{3-}$	0.0769	2.75×10^{-2} (s^{-1}); 904 (mol^{-1} liter)			59

Continued

TABLE 1.1.   *(Continued)*

Reaction	Medium	$(\mathbf{k}, s^{-1})$ $(K, mol^{-1}\ liter)$	$\Delta H^{\ddagger}$, kcal mol^{-1}	$\Delta S^{\ddagger}$, cal K^{-1} mol^{-1}	Ref.
$[Co(NH_3)_5Cl]^{2+} + Fe^{2+}$	0.34 (DMF)	$1.16 \times 10^{-3}\ (s^{-1})$ $16.1\ (mol^{-1}\ liter)$	25.6 -4.8	13.6 -10.5	94
$[Co(NH_3)_5N_3]^{2+} + Fe^{2+}$	0.34 (DMF)	$12\ (s^{-1})$ $1.5\ (mol^{-1}\ liter)$			95
$cis\text{-}[Co(en)_2Cl_2]^+ + Fe^{2+}$	0.35 (DMSO)	$1.19 \times 10^{-2}\ (s^{-1})$ $2.61\ (mol^{-1}\ liter)$	24.9 -7.9	16.2 -24.4	96
$[Fe(CN)_6]^{3-} + cyt\ c(II)$	0.02 (pH 7.8)	$4.6 \times 10^4\ (s^{-1})$ $870\ (mol^{-1}\ liter)$			139
$cyt\ c(III) + [Fe(CN)_6]^{4-}$	0.02 (pH 7.8)	$3.3 \times 10^2\ (s^{-1})$ $400\ (mol^{-1}\ liter)$			130
$[Co(NH_3)_6]^{3+} + Fd_{red,red}$	0.1 (pH 8)	$98\ (s^{-1})$ $466\ (mol^{-1}\ liter)$	15.3 0.3	1.9 13.4	139
$[Co(en)_3]^{3+} + Fd_{red,red}$	0.1 (pH 8)	$11.8\ (s^{-1})$ $261\ (mol^{-1}\ liter)$	16.1 0.9	0.5 13.9	139
$[Pt(NH_3)_6]^{4+} + Fd_{red,red}$	0.1 (pH 8)	$111\ (s^{-1})$ $2400\ (mol^{-1}\ liter)$			139

^{a} 22°C.
^{b} 23°C.

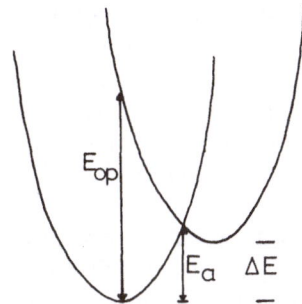

Figure 1.1. Optical and thermal electron transfer for a thermodynamically unfavorable change.

The electron transfer rate constant, obtained using

$$k_{et} = \nu_{et} \exp(-E_a/RT) \tag{6}$$

$$\nu_{et} = (2\pi V_{ab}^2/\hbar)[\pi/kT(E_{op} - \Delta E)]^{1/2} \tag{7}$$

where the resonance energy V_{ab} can be deduced from the IT band intensity, when combined with a statistical mechanical preassociation constant yields a second-order rate constant for reaction (8) of $2\text{--}3 \times 10^{-4}$ mol liter^{-1} s^{-1}

$$[Fe(CN)_6]^{4-} + PQ^{2+} \rightarrow [Fe(CN)_6]^{3-} + PQ^+ \tag{8}$$

($I = 0.10$ mol liter^{-1}, 23°C) (I is ionic strength). This compares favorably with a Marcus value for the reaction of 10^{-3} mol liter^{-1} s^{-1} calculated from self-exchange data.

The mixed-valence complex [(bipy)$_2$ClRu(III)Ph$_2$PCH$_2$PPh$_2$Ru(II)Cl(bipy)$_2$]$^{3+}$ is formed[13] in mixtures of the corresponding [Ru(III),(III)]$^{4+}$ and [Ru(II),(II)]$^{2+}$ species and the rate of intramolecular electron transfer, 8.8×10^8 s^{-1}, has been calculated from an IT band at 1245 nm in acetonitrile solution. Comparison with the [Ru(bipy)$_2$pyCl]$^{+/2+}$ self-exchange rate in acetonitrile is made possible by use of an estimated precursor complex formation constant of 0.6 mol liter^{-1} from which the outer-sphere intramolecular rate constant of 8×10^7 s^{-1} was evaluated. There is reasonable agreement especially when differences in the distance between redox sites (6.8 and 13.2 Å) are taken into account.

One-electron oxidation of complexes [(bipy)$_2$ClRu(II)LRu(II)Cl(bipy)$_2$]$^{2+}$ where L is an unsaturated N-donor bridge yields mixed-valence complexes with IT bands.[11,14] Effects of changing solvent polarity on the bands can be explained[14] using a dielectric continuum model which, with modification, can account for an increase in the band energies with increasing distance between the redox sites. A comparable treatment when applied to outer-sphere electron transfer reactions leads to the important conclusion that even between like charged ions, close contact in the precursor complex is energetically favored in polar solvents and there is no basis for expecting that long-range electron transfer should occur. The redox activity of ruthenium complexes bound on chemically modified electrodes has also been considered.[15]

Intervalence transfer in the solid phase with ion pair systems of the type $[Ru(III)(NH_3)_6][Fe(II)(CN)_5L]$ where L is CO, CN, DMSO, pyrazine, pyridine, or imidazole is strongly dependent[16] on the nature of L. A linear correlation between E_{op} and the difference in redox potential $\Delta E°$ of the component ions

$$E_{op} = 1.39\Delta E° + 108 \text{ (kJ mol}^{-1}) \tag{9}$$

is observed and the greater-than-unit slope, 1.39, suggests that the intrinsic barrier to electron transfer is not constant for the series but is proportional to the reduction free energy. Calculations of thermal electron transfer barriers are in reasonable agreement with experimental values.[17]

In acetonitrile solution, substituted derivatives of $[Fe(phen)_3]^{3+}$ and $[IrCl_6]^{2-}$ oxidize metal alkyls,[18,19] R_4Sn, R_4Pb, R_2Hg, and metal hydrides,[20] R_3GeH, R_3SnH, to give cation radical species.

$$SnR_4 + [Fe(phen)_3]^{3+} \rightarrow SnR_4^{\overset{+}{\bullet}} + [Fe(phen)_3]^{2+} \tag{10}$$

$$SnR_4^{\overset{+}{\bullet}} \xrightarrow{\text{fast}} SnR_3^+ + R^{\bullet} \tag{11}$$

Rate constants for the $[Fe(phen)_3]^{3+}$ reactions show Marcus behavior indicative of an outer-sphere process in which reorganization of the metal alkyl is the same regardless of the steric bulk of the R groups. In contrast a marked rate dependence on steric bulk is found for $[IrCl_6]^{2-}$ reactions and the presence of an alternative inner-sphere mechanism is confirmed by the fate of the radical cation [reaction (11)]. In general the rates of these reactions are considerably in excess of Marcus values and a chloride-bridged intermediate (**1**) is proposed. Detection of such

$$\left[>\!\!\underset{|}{\overset{|}{Sn}}\!\!-Cl\!\!-IrCl_5 \right]$$

1

an intermediate with tetracyanoethene as oxidant allows application of Mulliken theory of charge transfer. The energy, $h\nu_{CT}$, of the intermediate charge transfer bands can be related to the ionization potential of the alkyl metal, I_D, and the electron affinity of tetracyanoethene, E_A, as in equation (12),

$$hv_{CT} = I_D + E_A + W \tag{12}$$

where W is a measure of the interaction in the excited state and can be used to quantify the steric interaction in free energy correlations. Heterogeneous reactions of the alkyl metals have also been investigated.[21]

1.3. Reaction Mechanisms

1.3.1. Inner-Sphere and Outer-Sphere Reactions

Intrinsic electron transfer rate constants, k_{et}, devoid of the complications of precursor complex formation (K_0), have provided an important area of study [reaction (13)].

$$M(III) + N(II) \overset{K_0}{\rightleftharpoons} [M(III),N(II)] \overset{k_{et}}{\to} [M(II),N(III)] \to \text{products} \tag{13}$$

In some cases it has proved possible to obtain directly the K_0 and k_{et} terms from the rate law (14) with,

$$\text{rate} = \frac{k_{et}K_0[M(III)]\,[N(II)]}{1 + K_0[N(II)]} \tag{14}$$

in this instance, reductant N(II) in excess. However, where K_0 is small, it may not be possible to obtain k_{et} (Table 1.2). By measuring the rate of reduction of $[Co(NH_3)_6]^{3+}$ and $[Co(NH_3)_5OH_2]^{3+}$, both ions of similar charge and size so that K_0 will be similar, one can define[22] the selectivity, S, of a reductant

$$S = \log (k)_{[Co(NH_3)_6]^{3+}} - \log (k)_{[Co(NH_3)_5OH_2]^{3+}}$$
$$= \log (k_{et})_{[Co(NH_3)_6]^{3+}} - \log (k_{et})_{[Co(NH_3)_5OH_2]^{3+}} \tag{15}$$

A correlation where E_R° is the reduction potential of the reducing agent has been found to be as follows:

$$S = 1.3E_R^\circ + 2.3 \tag{16}$$

Table 1.2. Rate Constants and Activation Parameters for Reactions between Metal Ion Complexes at 25°C

Oxidant	Reductant	Medium, mol liter^{-1}	Rate law	Rate constant mol^{-1} liter s^{-1}	$\Delta H^{\ddagger}$, kcal mol^{-1}	$\Delta S^{\ddagger}$, cal K^{-1} mol^{-1}	Ref.
Reductant Ti^{3+}							
[Os(4,4'-Me$_2$bipy)$_3$]$^{3+}$	Ti^{3+}	3.0	k	$1.4 \times 10^{-}$			40
[Os(bipy)$_3$]$^{3+}$	Ti^{3+}	3.0	k	3.4×10^5			40
[Os(5-Cl phen)$_3$]$^{3+}$	Ti^{3+}	3.0	k	2.2×10^6			40
[Os(4,4'-Me$_2$bipy)$_3$]$^{3+}$	TiOH^{2+}	3.0	k	2.0×10^6		40	40
[Os(bipy)$_3$]$^{3+}$	TiOH^{2+}	3.0	k	1.7×10^7			40
[(NH$_3$)$_5$CoO$_2$CCH$_3$]$^{2+}$	Ti^{3+}	1.0	k[H$^+$]$^{-1}$	4.9×10^{-3} (s^{-1})			41
[(NH$_3$)$_5$CoO$_2$CCCl$_3$]$^{2+}$	Ti^{3+}	1.0	k[H$^+$]$^{-1}$	1.8×10^{-5} (s^{-1})			41
[(NH$_3$)$_5$CoO$_2$CCF$_3$]$^{2+}$	Ti^{3+}	1.0	k[H$^+$]$^{-1}$	1.6×10^{-5} (s^{-1})			41
[Co(NH$_3$)$_6$]$^{3+}$	[Ti(edtaH)OH$_2$]	0.10	k	0.165			43
[Co(NH$_3$)$_5$Cl]$^{2+}$	[Ti(edtaH)OH$_2$]	0.10	k	9.6			43
[Co(NH$_3$)$_5$OH$_2$]$^{3+}$	[Ti(edtaH)OH$_2$]	0.10	$k_1 + k_2$[H$^+$]$^{-1}$	22, 5.3 (s^{-1})			43
2, R = CH$_3$	[Ti(dipic)$_2$]$^{2-}$	0.16	k	0.64			23
2, R = CH$_2$F	[Ti(dipic)$_2$]$^{2-}$	0.16	k	2.13			23
2, R = CHF$_2$	[Ti(dipic)$_2$]$^{2-}$	0.16	k	4.14			23
2, R = CF$_3$	[Ti(dipic)$_2$]$^{2-}$	0.16	k	8.66			23
2, R = H	[Ti(dipic)$_2$]$^{2-}$	0.16	k	1.27			23
2, R = C$_6$H$_5$	[Ti(dipic)$_2$]$^{2-}$	0.16	k	1.77			24
2, R = C$_5$H$_4$N	[Ti(dipic)$_2$]$^{2-}$	0.16	k	4.15	14		24
2, R = C$_4$H$_3$N$_2$	[Ti(dipic)$_2$]$^{2-}$	0.16	k	5.17		-8	24
2, (NH$_3$)$_3$ = A$_3$, R = C$_5$H$_4$N	[Ti(dipic)$_2$]$^{2-}$	0.16	k	0.77	17.7	-4	24

Reductant V²⁺

Oxidant	Reductant						Ref.
[2-CO₂Co(NH₃)₅, 4-CO₂H-py]²⁺	V²⁺	1.0	k	8.0			44
[2-CO₂Co(NH₃)₅, 4-CONH₃-py]²⁺	V²⁺	1.0	k	e7.8			44
[2-CO₂Co(NH₃)₅, 3-CO₂H-py]²⁺	V²⁺	1.0	k	3.3			44
[2-CO₂Co(NH₃)₅, 3-CONH₂-py]²⁺	V²⁺	1.0	k	1.22			44
[2-CO₂Co(NH₃)₅, 5-CO₂H-py]²⁺	V²⁺	1.0	k	2.9			44
[2-CO₂Co(NH₃)₅, 5-CONH₂-py]²⁺	V²⁺	1.0	k	4.4			44
[2-CO₂Co(NH₃)₅, 6-CO₂H-py]²⁺	V²⁺	1.0	k	6.0			44
[2-CO₂Co(NH₃)₅, 6-CONH₂-py]²⁺	V²⁺	1.0	k	4.6			44
[2-CO₂Co(NH₃)₅, 6-CO₂Me-py]²⁺	V²⁺	1.0	k	3.9			44
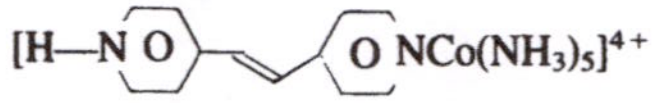 [H—N⟨O⟩NCo(NH₃)₅]⁴⁺	V²⁺	1.0	k	0.66			44
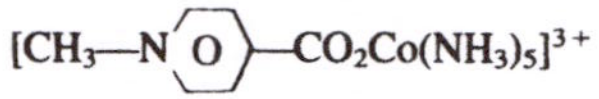 [CH₃—N⟨O⟩—CO₂Co(NH₃)₅]³⁺	V²⁺	1.0	k	0.82			44
2, R = CH₃	V²⁺	1.0	k	0.062			23
2, R = CH₂F	V²⁺	1.0	k	0.13			23
2, R = CHF₂	V²⁺	1.0	k	0.21			23
2, R = CF₃	V²⁺	1.0	k	0.38			23
2, R = H	V²⁺	1.0	k	0.113			23
trans-[Pt(NH₃)₄Cl₂]²⁺	V²⁺	1.0	k	28.3	10.7	−14.5	47
trans-[Pt(NH₂Me)₄Cl₂]²⁺	V²⁺	1.0	k	134	5.1	−31.0	47
[Pt(NH₃)₅Cl]³⁺	V²⁺	1.0	k	0.61	9.9	−25.1	47
[Pt(NH₃)₅Br]³⁺	V²⁺	1.0	k	6.0	7.5	−28.8	47
mer-[Pt(NH₃)₃Cl₃]⁺	V²⁺	1.0	k	159			47
[PtCl₆]²⁻	V²⁺	1.0	k	1.23×10^4	4.6	−23.9	47

continued

Table 1.2. (continued)

Oxidant	Reductant	Medium, mol liter^{-1}	Rate law	Rate constant mol^{-1} liter s^{-1}	$\Delta H^{\ddagger}$, kcal mol^{-1}	$\Delta S^{\ddagger}$, cal K^{-1} mol^{-1}	Ref.
Reductant Cr^{2+}							
[(NH$_3$)$_5$Co(H$_2$ida)]$^{3+}$	Cr^{2+}	1.0	k	5.0×10^{-2}			55
[(NH$_3$)$_5$Co(Hida)]$^{2+}$	Cr^{2+}	1.0	k	4.3×10^{-2}			55
[(NH$_3$)$_5$Co(ida)]$^{+}$	Cr^{2+}	1.0	k	9.2×10^{6}			55
[(NH$_3$)$_5$Co(H$_2$edda)]$^{3+}$	Cr^{2+}	1.0	k	4.1×10^{-2}			55
(en)$_2$Co...	Cr^{2+}	1.0	k	4.5×10^{-4}	13.0	-30	54
(en)$_2$Co...	Cr^{2+}	1.0	k	5.6×10^{-3}	11.0	-30	54
2, R = CH$_3$	Cr^{2+}	1.0	k	1.46×10^{-3}			23
2, R = CH$_2$F	Cr^{2+}	1.0	k	2.79×10^{-3}			23
2, R = CHF$_2$	Cr^{2+}	1.0	k	3.97×10^{-3}			23
2, R = CF$_3$	Cr^{2+}	1.0	k	6.30×10^{-3}			23
2, R = H	Cr^{2+}	1.0	k	2.72×10^{-3}			23
[Cr(OH$_2$)$_4$FBr]$^{+}$	CrCl^{+}	1.0	k	1.4			53
[Cr(OH$_2$)$_5$Br]$^{2+}$	[Cr(bipy)$_3$]$^{2+}$	1.0	k	110			53
[Co(NH$_3$)$_6$]$^{3+}$	[Cr(Me$_2$phen)$_3$]$^{2+}$	0.2	k	187			52
[Ru(NH$_3$)$_6$]$^{3+}$	[Cr(bipy)$_3$]$^{2+}$	0.1	k	1.4×10^{9}			51
[Cr(Me$_2$phen)$_3$]$^{3+}$		0.1 (95% MeOH)	k	9.6×10^{8}			51
[Fe$_4$S$_4$(SCH$_2$CH$_2$CO$_2$)$_4$]$^{7-}$	[Cr(edta)]$^{2-\ a}$	0.1	k	2.2×10^{5}	10.3	0.2	132

Reductant Fe²⁺

Oxidant	Reductant	Conditions		Rate const.	$\Delta V^{\ddagger}$	Ref.
trans-[Pt(NH₃)₄Br₂]²⁺	Fe²⁺ [b]	1.0	k	2.43×10^{-4}		48
trans-[Pt(NH₂Me)₄Br₂]²⁺	Fe²⁺ [b]	1.0	k	6.03×10^{-4}		48
trans-[Pt(NH₂Et)₄Br₂]²⁺	Fe²⁺ [b]	1.0	k	2.20×10^{-3}		48
trans-[Pt(NH₂Pr)₄Br₂]²⁺	Fe²⁺ [b]	1.0	k	8.75×10^{-4}		48
trans-[Pt(en)₂Br₂]²⁺	Fe²⁺ [b]	1.0	k	5.1×10^{-5}		48
[Co(DH)(DH₂)pyN₃]	Fe²⁺	1.0	k	0.196		65
[Co(DH)₂pyN₃]	Fe²⁺	1.0	k	2.6	17.1	65
[Co(DH)(DH₂)NH₃N₃]	Fe²⁺	1.0	k	7.97×10^{-3}	−10.3	65
[Co(DH)₂NH₃N₃]	Fe²⁺	1.0	k	0.166	19.6	65
Cu³⁺	Fe²⁺	HClO₄ (pH 2.1)	k	1.3×10^{8}	−7.7	81
2, R = CH₃	[Fe(dipic)₂]²⁻	0.16	k	1.43×10^{-2}		23
2, R = CH₂F	[Fe(dipic)₂]²⁻	0.16	k	5.08×10^{-2}		23
2, R = CHF₂	[Fe(dipic)₂]²⁻	0.16	k	11.9×10^{-2}		23
2, R = CF₃	[Fe(dipic)₂]²⁻	0.16	k	28.6×10^{-2}		23
2, R = H	[Fe(dipic)₂]²⁻	0.16	k	4.1×10^{-2}		23
[Co(edta)]⁻	[Fe(CN)₆]⁴⁻	0.0268	k	0.12		56
[Co(edta)]⁻	[NaFe(CN)₆]³⁻	0.0268	k	0.42		56
Rps Gelatinosa HiPIP	[Fe(CN)₆]⁴⁻	0.12 (pH 7.0)	k	5.3×10^{4}	0	135
Thiocapsa HiPIP	[Fe(CN)₆]⁴⁻	0.08 (pH 7.0)	k	2.8×10^{3}	−40.7	135
Paracoccus HiPIP	[Fe(CN)₆]⁴⁻	0.12 (pH 7.0)	k	2.3×10^{3}		135
[Fe(CN)₆]³⁻	*Chromatium* HiPIP	0 (pH 7.0)	k	5.1×10^{3}		133
[Fe(CN)₆]³⁻	*Rps Gelatinosa* HiPIP	0.12 (pH 7.0)	k	1×10^{3}	3.7	135
[Fe(CN)₆]³⁻	*Thiocapsa* HiPIP	0.08 (pH 7.0)	k	63	−32.7	135
[Fe(CN)₆]³⁻	*Paracoccus* HiPIP	0.12 (pH 7.0)	k	9		135
[Co(phen)₃]³⁺	*Chromatium* HiPIP	0.5 (pH 7.0)	k	2.73×10^{3}		123
[Co(ox)₃]³⁻	*Chromatium* HiPIP	0.5 (pH 7.0)	k	8×10^{-3}		123
[Co(NH₃)₅Cl]²⁺	*C. pasteurianum* Fd_{ox} [c]	0.1 (pH 7.0)	k	5.1×10^{5}		129
[Co(NH₃)₅Cl]²⁺	*C. pasteurianum* Fd_{red} [c]	0.1 (pH 7.0)	k	4.1×10^{5}		129
[Co(acac)₃]	*C. pasteurianum* Fd_{red} [c]	0.1 (pH 8.0)	k	3.1×10^{4}	7.6	130
[Co(edta)]⁻	*C. pasteurianum* Fd_{red} [c]	0.1 (pH 8.0)	k	1.1×10^{4}	−14.4	130
[Co(ox)₃]³⁻	*C. pasteurianum* Fd_{red} [c]	0.1 (pH 8.0)	k	4.8×10^{3}		130

continued

Table 1.2. (continued)

Oxidant	Reductant	Medium, mol liter⁻¹	Rate law	Rate constant, mol^{-1} liter s^{-1}	$\Delta H^\ddagger$, kcal mol^{-1}	$\Delta S^\ddagger$, cal K^{-1} mol^{-1}	Ref.
Reductant Fe²⁺							
$[(NH_3)_5Co\mu(NH_2)Co(NH_3)_5]^{5+}$	*C. pasteurianum* $Fd_{red\,red}$ ^{d}	0.1 (pH 8.0)	k	9.6×10^5			130
$[Fe(CN)_6]^{3-}$	Horse heart cyt c(II)	0 (pH 7.0)	k	3.2×10^8			138
$[Co(phen)_3]^{3+}$	Horse heart cyt c(II)	0.5 (pH 7.0)	k	3.3×10^3			123
$[Co(ox)_3]^{3-}$	Horse heart cyt c(II)	0.5 (pH 7.0)	k	5.5			123
$[Cu(phen)_2]^{2+}$	Horse heart cyt c(II)	0.1 (pH 6.15)	k	27			142
[Cudta]	Horse heart cyt c(II)	0.1 (pH 6.15)	k	5.1×10^3			142
$[Cu(dmp)_2]^{2+}$	Horse heart cyt c(II)	0.1 (pH 6.15)	k	1.0×10^6			142
$[Fe(CN)_6]^{3+}$	Acetyl-horse heart cyt c(II)	0 (pH 7.0)	k	4.3×10^2			138
$[Co(phen)_3]^{3+}$	*P. aeruginosa* cyt c_{55}(II)	0.5 (pH 7.0)	k	6.0×10^4			123
$[Co(ox)_3]^{3-}$	*P. aeruginosa* cyt c_{55}(II)	0.5 (pH 7.0)	k	1.4			123
$[Cu(phen)_2]^{2-}$	Sperm whale myoglobin(II)	0.1 (pH 6.15)	k	4.3×10^4			142
[Cu(dta)]	Sperm whale myoglobin(II)	0.1 (pH 6.15)	k	1.8×10^5			142
$[Cu(dmp)_2]^{2+}$	Sperm whale myoglobin(II)	0.1 (pH 6.15)	k	2.8×10^6			142
$[Co(NH_3)_5N_3]^{2+}$	Fe²⁺	0.63 (DMSO)	k	1.8×10^2			95
$[Mn(QuinCN)_2]^+$	Fe²⁺	CH_3CN	k	400			97
$[Fe_4S_4(S\text{-}p\text{-}C_6H_4Me)_4]^{3-}$ ^{e}		CH_3CN	k	2.8×10^6	3.6	−17	131
$[Fe_4S_4(SCH_2Ph)_4]^{3-}$ ^{f}		CH_3CN	k	2.4×10^6			131
$[Fe_4Se_4(S\text{-}p\text{-}C_6H_4Me)_4]^{3-}$ ^{g}		CH_3CN	k	9.7×10^6			131
$[Fe(Cp\text{-}CH_3)_2]^+$		CH_3CN	k	8.3×10^6			89
$[Fe(Cp\text{-}(CH_2)_3\text{-}Cp)]^+$		CH_3CN	k	1.46×10^7			89
$[Fe(Cp\text{-}n\text{-}C_4H_9)_2]^+$		CH_3CN	k	4.9×10^6			89

$[Fe(Cp-(CH_3)_5)_2]^+$	CH_2Cl_2		k	4.4×10^7	89
$[Fe(Cp-(CH_3)_5)_2]^+$	CH_3CN		k	3.8×10^7	89
$[Fe(Cp-(CH_3)_5)_2]^+$	$(CH_3)_2CO$		k	2.4×10^7	89
$[Fe(Cp,Cp-CO_2CH_3)]^+$	CH_3CN		k	6.8×10^6	89
$[Fe(Cp,Cp-CH_2N(CH_3)_3)]^{2+}$	CH_3CN		k	1.80×10^6	89

Reductant Co⁺

$[(H_2O)_5CrCl]^{2+}$	Vit B_{12s}	0.05	k	3.80×10^3	72
$[(H_2O)_5CrNCS]^{2+}$	Vit B_{12s}	0.05	k	4.63×10^2	72
$[(H_2O)_5CrN_3]^{2+}$	Vit B_{12s}	0.05	k	1.31×10^3	72
$[(H_2O)_5CrOAc]^{2+}$	Vit B_{12s}	0.05	k	2.49×10^2	72
$[(H_2O)_5CrF]^{2+}$	Vit B_{12s}	0.05	k	8.1×10^{-1}	72
$[(H_2O)_5CrBr]^{2+}$	Vit B_{12s}	0.05	k	4.0×10^5	72
$[(H_2O)_5CrSH]^{2+}$	Vit B_{12s}	0.05	k	3.08×10^5	72
$[(H_2O)_5CrOH]^{2+}$	Vit B_{12s}	0.05	k	7.3×10^{-1}	72
$[(H_2O)_5CrOH_2]^{3+}$	Vit B_{12s}	0.05	k	1.0×10^{-1}	72

Reductant Co²⁺

$[(NH_3)_5Co\mu O_2Co(NH_3)_5]^{4+}$	0.05		k	6.0×10^3	30
$[(NH_3)_4Co\mu O_2\mu NH_2Co(NH_3)_4]^{3+}$	0.05	4.5	k	1.025×10^3	30
$[(NH_3)_5Co\mu O_2Co(NH_3)_5]^{4+}$	0.05	4.2	k	380	30
$[(NH_3)_4Co\mu O_2\mu NH_2Co(NH_3)_4]^{3+}$	0.05		k	90	30
$[(NH_3)_5Co\mu O_2Co(NH_3)_5]^{4+}$	0.05		k	188	30
$[(NH_3)_4Co\mu O_2\mu NH_2Co(NH_3)_4]^{3+}$	0.05		k	32	30
$[Ru(bipy)_3]^{3+}$	1.0		k	2.42×10^8	62
$[Ru(bipy)_3]^{3+}$	1.0		k	1.36×10^8	62
$[Fe(CN)_5(4,4'-bipy)]^{2-}$	0.1		k	6.9	57
$[Fe(CN)_5py]^{2-}$	0.1		k	4.0	57
$[Ru(NH_3)_5(4,4'-bipy)]^{3+}$	0.1		k	77	58
$[Ru(NH_3)_5py]^{3+}$	0.1		k	32	58
$[Co(ox)_3]^{3-}$	0.5		k	1.62×10^2	123

continued

Table 1.2. (continued)

Oxidant	Reductant	Medium, mol liter^{-1}	Rate law	Rate constant mol^{-1} liter^{-1} s^{-1}	$\Delta H^{\ddagger}$, kcal mol^{-1}	$\Delta S^{\ddagger}$, cal K^{-1} mol^{-1}	Ref.
Reductant Co^{2+} *(cont.)*							
Mesoporphyrin(III)	[Co(CN)$_5$]$^{3-}$	0.5	$k_1 + k_2$[CN$^-$]	40, 70 (mol^2 liter^{-2} s^{-1})			73
Deuteroporphyrin(III)	[Co(CN)$_5$]$^{3-}$	0.5	$k_1 + k_2$[CN$^-$]	120, 340 (mol^2 liter^{-2} s^{-1})			73
Protoporphyrin(III)	[Co(CN)$_5$]$^{3-}$	0.5	$k_1 + k_2$[CN$^-$]	80, 480 (mol^2 liter^{-2} s^{-1})			73
3,8-Dibromodeuteroporphyrin(III)	[Co(CN)$_5$]$^{3-}$	0.5	$k_1 + k_2$[CN$^-$]	390, 390 (mol^2 liter^{-2} s^{-1})			73
ms-Tetra(4-sulfonatophenyl) prophyrin(III)	[Co(CN)$_5$]$^{3-}$	0.5	$k_1 + k_2$[CN$^-$]	400, <70 (mol^2 liter^{-2} s^{-1})			73
3,8-Diacetyldeuteroporphyrin(III)	[Co(CN)$_5$]$^{3-}$	0.5	$k_1 + k_2$[CN$^-$]	72, 60 (mol^2 liter^{-2} s^{-1})			73
Reductant Cu$^+$							
Fe^{3+}	Cu$^+$	HClO$_4$ (pH 2.1)	k	1.3×10^7			81
[Cu(III)(H$_3$V$_4$)]$^-$	[Cu(dmp)$_2$]$^+$	0.1	k	4.2×10^6			75
[Cu(III)(H$_3$PG$_3$a)]	[Cu(dmp)$_2$]$^+$	0.1	k	1.8×10^6			75
[Cu(III)(H$_3$G$_4$)]	[Cu(dmp)$_2$]$^+$	0.1	k	3.1×10^6			75
[Cu(III)(H$_3$G$_3$a)]	[Cu(dmp)]$^+$	0.1	k	1.6×10^6			75
[Cu(III)(H$_3$G$_3$AOMe)]	[Cu(dmp)$_2$]$^+$	0.1	k	6.0×10^6			75
[Cu(III)(H$_2$L$_3$)]	[Cu(dmp)$_2$]$^+$	0.1	k	1.1×10^3			75
[Cu(III)(H$_2$GA$_2$)]	[Cu(dmp)$_2$]$^+$	0.1	k	5.2×10^7			75
[Cu(III)(H$_2$G$_2$βA)]	[Cu(dmp)$_2$]$^+$	0.1	k	2.5×10^8			75
[IrCl$_6$]$^{2-}$	[Cu(dmp)$_2$]$^+$	0.1	k	1.4×10^9			75
[Cu(III)(H$_3$V$_4$)]$^-$	[Cu(dpmp)]$^{3-}$	0.1	k	2×10^4			75
[Cu(III)(H$_3$A$_4$)]$^-$	[Cu(dpmp)]$^{3-}$	0.1	k	7×10^4			75

[Cu(III)(H$_{-3}$PG$_2$a)]	[Cu(dpmp)]$^{3-}$	0.1		k	4.8×10^5	75
[Cu(III)(H$_{-3}$AG$_3$)]$^-$	[Cu(dpmp)]$^{3-}$	0.1		k	1.7×10^5	75
[Cu(III)(H$_{-3}$G$_4$)]$^-$	[Cu(dpmp)]$^{3-}$	0.1		k	6.7×10^5	75
[Cu(III)(H$_{-3}$G$_3$a)]	[Cu(dpmp)]$^{3-}$	0.1		k	4.7×10^5	75
[Cu(III)(H$_{-3}$G$_3$AOMe)]	[Cu(dpmp)]$^{3-}$	0.1		k	8.5×10^5	75
[Cu(III)(H$_{-2}$Leu$_3$)]	[Cu(dpmp)]$^{3-}$	0.1		k	3.7×10^5	75
[Cu(III)(H$_{-2}$A$_3$)]	[Cu(dpmp)]$^{3-}$	0.1		k	3.8×10^5	75
[Cu(III)(H$_{-2}$G$_3$)]	[Cu(dpmp)]$^{3-}$	0.1		k	4.1×10^5	75
[IrCl$_6$]$^{2-}$	[Cu(dpmp)]$^{3-}$	0.1		k	3.0×10^8	75
[Fe(CN)$_6$]$^{3-}$	[CuW$_{12}$O$_{40}$]$^{7-}$	0.1		k	4.1×10^5	75
[Co(phen)$_3$]$^{3+}$	Spinach PCu(I)	0.5 (pH 6.0)		k	1.02×10^3	82
[Co(ox)$_3$]$^{3-}$	Spinach PCu(I)	0.5 (pH 7.0)		k	0.24	123
[Co(phen)$_3$]$^{3+}$	*P. aeruginosa* ACu(I)	0.5 (pH 7.0)		k	4.4×10^3	123
[Co(ox)$_3$]$^{3-}$	*P. aeruginosa* ACu(I)	0.5 (pH 7.0)		k	0.029	123
[Co(ox)$_3$]$^{3-}$	*Rhus vernicifera* StCu(I)	0.5 (pH 7.0)		k	7.3×10^2	123
Reductant Ru^{2+}						
2, R = CH$_3$	[Ru(NH$_3$)$_6$]$^{2+}$	0.15		k	0.035	23
2, R = CH$_2$F	[Ru(NH$_3$)$_6$]$^{2+}$	0.15		k	0.12	23
2, R = CHF$_2$	[Ru(NH$_3$)$_6$]$^{2+}$	0.15		k	0.21	23
2, R = CF$_3$	[Ru(NH$_3$)$_6$]$^{2+}$	0.15		k	0.36	23
trans-[Pt(NH$_3$)$_4$Cl$_2$]$^{2+}$	[Ru(NH$_3$)$_6$]$^{2+\ h}$	0.1		k	2.14×10^3	47
trans-[Pt(NH$_2$Me)$_4$Cl$_2$]$^{2+}$	[Ru(NH$_3$)$_6$]$^{2+\ h}$	0.1		k	7.5×10^3	47
[Pt(NH$_3$)$_5$Cl]$^{3+}$	[Ru(NH$_3$)$_6$]$^{2+\ h}$	0.1		k	17.7	47
[Pt(NH$_3$)$_5$Br]$^{3+}$	[Ru(NH$_3$)$_6$]$^{2+\ h}$	0.1		k	1.8×10^2	47
mer-[Pt(NH$_3$)$_3$Cl$_3$]$^+$	[Ru(NH$_3$)$_6$]$^{2+\ h}$	0.1		k	1.78×10^4	47
[PtCl$_6$]$^{2-}$	[Ru(NH$_3$)$_6$]$^{2+\ h}$	0.1		k	7.9×10^6	47
Horse heart cyt c(III)	[Ru(NH$_3$)$_5$(4,4'-bipy)]$^{2+}$	0.1		k	3.9×10^4	140
[Co(edta)]$^-$	[Ru(NH$_3$)$_5$py]$^{2+}$	0.1		k	156	58
[Co(edta)]$^-$	[Ru(NH$_3$)$_5$py]$^{2+}$	0.1		k	236	58
Fe^{3+}	[Ru(NH$_3$)$_5$py]$^{2+}$	1.0 (H$^+$)	4.7	-20	k	7.76×10^4	61

continued

Table 1.2. (continued)

Oxidant	Reductant	Medium, mol liter^{-1}	Rate law	Rate constant mol^{-1} liter s^{-1}	$\Delta H^{\ddagger}$, kcal mol^{-1}	$\Delta S^{\ddagger}$, cal K^{-1} mol^{-1}	Ref.
Reductant Ru^{2+} (cont.)							
	[Ru(NH$_3$)$_5$isn]$^{3+}$	1.0	k	6.8×10^5	1.6	-26	61
	[Ru(NH$_3$)$_5$nic]$^{2+}$	1.0 (H$^+$)	k	2.9×10^4	6.7	-15	61
	[Ru(NH$_3$)$_5$isn]$^{2+}$	1.0	k	3.3×10^5	4.1	-19	61
Fe^{3+}	[Ru(NH$_3$)$_5$isn]$^{2+}$	1.0 (H$^+$)	k	2.57×10^4	7.4	-14	61
MnO$_4^-$	[Ru(CN)$_6$]$^{2-}$	1.02	k	0.93			63
MnO$_4^-$	[HRu(CN)$_6$]$^{3-}$	1.02	k	5.03×10^3			63
MnO$_4^-$	[H$_2$Ru(CN)$_6$]$^{2-}$	1.02	k	15.4			63
[Ru$_3$O(MeCO$_2$)$_6$(py)$_3$]$^+$	[Ru$_3$O(MeCO$_2$)$_6$(py)$_3$]	CH$_2$Cl$_2$	k	1.1×10^8	4.4	-7	88
Reductant Pt^{2+}							
Fe^{3+}	[Pt(en)$_2$]$^{2+}$ ^{i}	1.0	k[Br$^-$] +	9.5×10^{-2} (mol^{-2} liter^2 s^{-1})	16.2	-2.4	48
			k[BR$^-$]^2	9.85 (mol^{-3} liters^3 s^{-1})			
Fe^{3+}	[Pt(NH$_3$)$_4$]$^{2+}$ ^{i}	1.0	k[Br$^-$] +	5.0×10^{-2} (mol^{-2} liter^2 s^{-1})	17.6	-1.9	48
			k_2[Br$^-$]^2	1.56 (mol^{-3} liter^3 s^{-1})			
Fe^{3+}	[Pt(NH$_2$Me)$_4$]$^{2+}$ ^{i}	1.0	k[Br$^-$] +	4.0×10^{-2} (mol^{-2} liter^2 s^{-1})	18.5	1.2	48
			k_2[Br$^-$]^2 +	1.09 (mol^{-3} liter^3 s^{-1})			
Fe^{3+}	[Pt(NH$_2$Et)$_4$]$^{2+}$ ^{i}	1.0	k[Br$^-$] +	9.0×10^{-3} (mol^{-2} liter^2 s^{-1})	20.5	4.5	48
			k_2[Br$^-$]^2	2.87×10^{-1} (mol^{-3} liter^3 s^{-1})			

Oxidant	Reductant	Medium	Rate law	k	$\Delta H^\ddagger$	$\Delta S^\ddagger$	Ref.
Fe^{3+}	$[Pt(NH_2Pr)_4]^{2+}$ ᶦ	1.0	$k[Br^-] + k_2[Br^-]^2$	2.0×10^{-2} (mol^{-2} liter^2 s^{-1}); 3.10×10^{-1} (mol^{-3} liter^3 s^{-1})	18.2	2.9	48
Reductant Eu^{2+}							
$[HNC_5H_4CO_2Co(NH_3)_5]^{3+}$	Eu^{2+}	1.0	k	2.4			44
$[CH_3NC_5H_4CO_2Co(NH_3)_5]^{3+}$	Eu^{2+}	1.0	k	2.4			44
$[2\text{-}CO_2Co(NH_3)_5,4\text{-}CONH_2py]^{2+}$	Eu^{2+}	2.2	k	1.22×10^3			44
$[p\text{-}H_2NCOC_5H_4N\text{-}CH_2CO_2Co(NH_3)_5]^{3+}$	Eu^{2+}	1.2	k	1.6×10^2			44
$[p\text{-}H_2NCOC_5H_4N\text{-}CH(CH_3)CO_2CO(NH_3)_5]^{5+}$							
$[\text{H-N}\langle\bigcirc\rangle=\langle\bigcirc\rangle\text{N-Co(NH}_3)_5]^{4+}$	Eu^{2+}	1.2	k	97			44
$Co(NH_3)_5]^{4+}$	Eu^{2+}	2.0	k	0.21			44
Miscellaneous reductants							
Am(V)	V(IV)	1.0 (H⁺)	k	4.88	12.7	−12.7	85
$[IrCl_6]^{2-}$	$[W_2O_4(edta)]^{2-}$	0.5	k	6.3×10^5	5.4	−13.8	39
$[Fe(phen)_3]^{3+}$	$[W_2O_4(edta)]^{2-}$	0.5	k	$>10^7$			
$[Fe(CN)_6]^{3-}$	$[W_2O_4(edta)]^{2-}$	0.1	k	0.058			
$[Co(bipy)_3]^{3+}$	$[W_2O_4(edta)]^{2-}$	0.1	k	2.0			
$[(en)_2Co\mu O_2\mu NH_2Coen_2]^{4+}$ Cd⁺		0.01 (MeOH/H₂0)	k	4×10^9			
$[Fe(bipy)_2]^{3+}$	Hg_2^{2+}	0.1 (70%MeOH)	k	0.22		31	84
Ce(IV)	U(IV)	Acetone	k		2.3	−4.35	93
Co³⁺	$[(NH_3)_5CoOClO]^{2+}$	2.1	$k[H^+]^{-1}$	9.1 (s⁻¹)	2.8	4.0	70

ᵃ 20°C.   ᶜ 18°C.   ᵉ 28°C.   ᵍ 31°C.   ᶦ 40°C.
ᵇ 50°C.   ᵈ 7°C.   ᶠ 27°C.   ʰ 5°C.

Outer-sphere association constants measured[23] for the [Fe(dipic)$_2$]$^{2-}$ reduction of binuclear cobalt(III) complexes (2), [Co(III)$_2$] with stacking ligands,

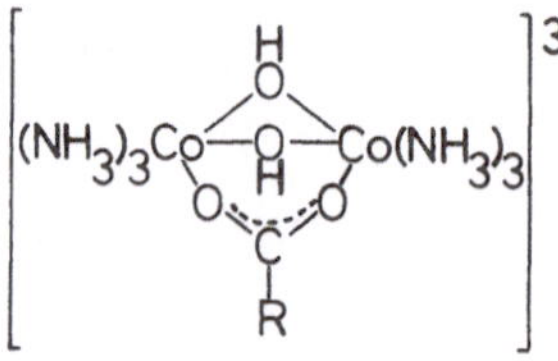

2

R = naphthalene, quinoline, or pyridine, have allowed examination of the inductive effect on electron transfer reactivity. Where the substituent R is methyl or a fluorinated methyl derivative, estimated precursor association constants vary little and thus provide a direct comparison k_{et}. A linear dependence of k_{et} on the Taft σ "inductive" parameter was found for acetato and formato complexes. Although redox potentials for these systems are not readily available, it is thought that they may be correlated with the inductive parameter. Thus —CF$_3$, the most electron-withdrawing of the groups, increases the potential most.

These systems also provide[23] a convenient source of reductant selectivity data, in this case measured by the difference in rates of the —CF$_3$ and —CH$_3$-substituted oxidants. Again, a linear correlation was discovered.

$$\log k_{CF_3} - \log k_{CH_3} = 0.95 \, E_R^\circ + 1.02 \tag{17}$$

Selectivity has a strong bearing on linear free energy plots commonly used to determine outer-sphere behavior. A dependence of the slope of these plots on the difference between E_R° values for reductants considered has been noted but does not appear to have general application.

Inner-sphere 1:1 and 2:1 picolinate-bridged complexes have been detected[24] in the reduction of [Co(III)$_2$], structure **2**, R = dipicolinate, and structure **3** by

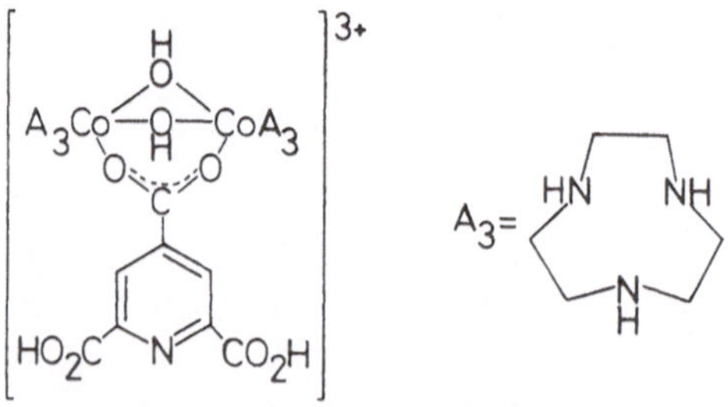

3

$[Fe(OH_2)_6]^{2+}$ and $[Ti(OH_2)_6]^{3+}$, and the rates of electron transfer, k_{et}, have been evaluated. Whereas reductions with $[Fe(OH_2)_6]^{2+}$ are susceptible to both the nature of the oxidant and the reductant, with $[Ti(OH_2)_6]^{3+}$, only the reductant has a marked effect, suggesting a chemical mechanism of electron transfer through the picolinate bridge with a transient radical intermediate.

$$[Co(III)_2dipicTi(III)] \rightleftharpoons [Co(III)_2dipic^\cdot Ti(IV)] \rightarrow products \qquad (18)$$

In an effort to examine alternative mechanisms of orbital coupling, reductions of the pyridine derivative (2) with $[Fe(dipic)_2]^{2-}$ and $[Ti(dipic)_2]^-$ were studied. In the case of $[Fe(dipic)_3]^{2-}$, outer-sphere complexation was detected and k_{et} for both inner-sphere and outer-sphere processes were similar, suggesting resonance transfer over a distance of 9–9.5 Å. No rate-limiting behavior was noted in the titanium case.

Intramolecular electron transfer from a reduced nitrobenzoate ligand to cobalt(III) in $[(NH_3)_5Co(p\text{-}O_2CC_6H_4NO_2)]^{2+}$ and 2, R = $p\text{-}C_6H_4NO_2$, has been examined.[25] The rate increases with decreasing solvent polarity as expected with increasing solvation of the nitrobenzoate radical. Activation parameters are similar to previous studies.[26] The effects of altering the bridging group in compounds of the type $[O_2NC_6H_4\text{-}X\text{-}CO_2Co(III)(NH_3)_5]^+$ can be explained[27] by good orbital overlap between the nitro radical and the carbonyl ligand or the metal centre. This is achieved where X is a large flexible chain. Conjugation in the chain can facilitate through chain electron transfer but otherwise this is markedly decreased. The lifetime[28] of the formate radical complex $[Co(NH_3)_5O_2C^\cdot]^{2+}$ toward intramolecular electron transfer is $\leqslant 10^{-6}$ s. This does not completely rule it out as an intermediate in the permanganate oxidation of $[Co(NH_3)O_2CH]^{2+}$.[29]

The reduction potentials of the complexes $[(NH_3)_4Co\mu(NH_2),$ $\mu(O_2)Co(NH_3)_4]^{4+}$ and $[(NH_3)_5Co\mu(O_2)Co(NH_3)_5]^{5+}$ have been determined[30] by cyclic voltammetry to be 0.75 and 1.0 V at 25°C, 0.05 M ionic strength. Reductions by cobalt(II)–polypyridine complexes obey Marcus theory and self-exchange rate constants of 1.0×10^{-5} mol^{-1} liter s^{-1} and 6×10^{-7} mol^{-1} liter s^{-1} have been calculated for the two complexes. These self-exchange values are somewhat unusual in that the monobridged couple is more reactive than the dibridged couple. It is suggested that charge differences in precursor complex formation might provide an explanation.

An inner-sphere mechanism is proposed[31] for the reduction of $[(en)_2Co\mu(NH_2),\mu(O_2)Co(en)_2]^{4+}$ by Cd$^+$. The reductant binds at the peroxo bridge which has a net negative charge and the successor complex decomposes with a first-order rate constant of 2.4×10^3 s^{-1} at 25°C in 0.01 M aqueous methanol. In contrast to an earlier study,[32] little or no contribution from Fe^{2+}

reduction is noted[33] in the corresponding $\mu(OH),\mu(O_2)$ complex. The rate-determining step is H^+-induced hydroxy-bridge cleavage.

Reduction by V^{2+} and $[Ru(NH_3)_6]^{2+}$ of the μ-superoxo complex $[(NH_3)_5Co(O_2)Co(NH_3)_5]^{5+}$ leads to the same μ-peroxo species $[(NH_3)_5Co(O_2)Co(NH_3)_5]^{4+}$, which decomposes[34] to give Co^{2+} and O_2 with a rate constant of 84 s^{-1} at 25°C and 0.10 M ionic strength. Protonation and complexation with Cl^- and SO_4^{2-} stabilize the complex. In contrast,[35] the complex *trans*-$[(NO_2)(en)_2Co(O_2)Co(en)_2(NO_2)]^{2+}$ hydrolyzes by two successive protonations to give H_2O_2 and *trans*-$[Co(NO_2)(OH_2)(en)_2]^{2+}$. Crystal structure,[36] electrochemical,[37] and spectroscopic[38] properties of $[\mu-O_2-Co(III)_2]$ complexes have also been considered.

The binuclear tungsten(V) complex (**4**) is a much stronger reducing agent

4

than the corresponding molybdenum species[39] which is surprising since both complexes have similar bond lengths and angles and reorganization energy is expected to be similar. Oxidation by $[IrCl_6]^{2-}$ and $[Fe(phen)_3]^{3+}$ probably involves formation of a binuclear W(V),(VI) species.

A potentiometric study of the Ti(IV)/(III) couple yields[40] an expression

$$E = 0.03 - 0.059 \ \log[Ti(III)]/[Ti(IV)] \ [H^+]^2 \tag{19}$$

for the reduction potential confirming that Ti^{3+} and TiO^{2+} are the principal species in acidic media (3.0 M ionic strength, 25°C). Reductions of osmium(III)–polypyridine complexes show reactivity by both Ti^{3+} and $TiOH^{2+}$ and the rates are in agreement with Marcus correlations indicating outer-sphere reactions. Self-exchange rates ($>3 \times 10^{-4}$ mol^{-1} liter s^{-1} for Ti^{3+} and $\geqslant 10^{-2}$ mol^{-1} liter s^{-1} for $TiOH^{2+}$) were evaluated. The rate laws for reduction of cobalt(III) complexes of the type $[(NH_3)_5CoO_2CCX_3]^{2+}$ where X is H, Cl, or F by Ti^{3+} indicate[41] that $TiOH^{2+}$ is the active species. While the acetato complex proceeds by an inner-sphere pathway in good agreement with previous studies[42] with this reagent, the chloro and fluoro species follow an outer-sphere route.

The acid-independent pathway in the reduction of cobalt(III)–amine complexes by $[Ti(edta \ H)OH_2]$ is thought[43] to represent an outer-sphere pathway, while an $[H^+]^{-1}$ dependence with $[(NH_3)_5Co(OH_2)]^{3+}$, ascribed to reaction of the conjugate base, is inner sphere. The substitution rate for H_2O coordinated on $[Ti(edta\ H)\ (OH_2)]$ is estimated to be in excess of $10^7\ s^{-1}$.

Reduction of complexes $[(NH_3)_5Co(III)L]^{3+}$, where L is a reducible ligand by Eu^{3+} and V^{2+} is subject to catalysis by the $L/L°$ couple when the ligand is released from the cobalt(II) product.[44] These catalysts have a variety of fates. Addition of excess Eu^{3+} or V^{3+} suppresses catalysis[45] by oxidation of the active species $L°$ and allows examination of the uncatalyzed reaction. For the 4-carboxamide derivative (5) a positive hydrogen dependence in the rate law is

$$\left[\begin{array}{c} CONH_2 \\ \\ N \quad CO_2Co(NH_3)_5 \end{array} \right]^{2+}$$

5

interpreted in terms of a chemical mechanism in which initial reduction of the ligand takes place and is followed by internal transfer to the metal. Complications due to complexation of released ligands, particularly picolinate derivatives, with vanadium(II) are noted at lower acidities.

Vanadous picolinate, $[V(pic)_3]^-$, is the most effective outer-sphere reductant in the heterogeneous reaction with substitution inert iron(III) oxides.[46] In removing these metal oxides from surfaces, these redox reagents may be more useful than the more conventionally used, strong acids and chelating agents.

Reductions[47] of platinum(IV) haloamine complexes by V^{2+} and $[Ru(NH_3)_6]^{2+}$ proceed by two successive one-electron changes, the first of which is rate determining. A correlation between the rate constants for the two reagents

$$\log k_V = 0.89\ \log k_{Ru} - 1.68 \tag{20}$$

is consistent with an outer-sphere reaction for vanadium(II) with no participation of halogen bridges. The reversible reaction

$$[PtL_4]^{2+} + 2Fe^{3+} + 2Br^- \rightleftharpoons trans\text{-}[PtL_4Br_2]^{2+} + 2Fe^{2+} \tag{21}$$

has been examined[48] where L is an amine ligand and also takes place in two successive one-electron steps. Trends in the platinum(IV) reduction rate constants

$$\tfrac{1}{2}en < NH_3 < NMeH_2 < NEtH_2 \sim NPrH_2 \tag{22}$$

parallel the thermodynamic change but cannot be explained by σ-donor ability alone. Steric hindrance destabilizes the higher oxidation state. Platinum(III) has been detected[49] in the reaction between $[Pt(NH_3)_4]^{2+}$ and $[PtCl_6]^{2-}$ and the $[Pt(III)(en)_2OH]$ complex prepared[50] by pulse radiolysis has been characterized.

Photolysis of $[Cr(bipy)_3]^{3+}$ in basic or alcoholic media results[51] in formation of the corresponding chromium(II) species. Selective photolysis of one polypyridyl complex in the presence of another allows determination of the Cr(III)/(II) self-exchange rate which, in methanolic solution, is close to the diffusion limit. A two-term rate law in the reduction of $[Co(NH_3)_6]^{3+}$ by $[Cr(bipy)_3]^{2+}$ is the result[52] of hydrolysis of the chromium(II) complex to give the more reactive $[Cr(bipy)_2(OH_2)_2]^{2+}$. However, addition of excess bipy has no effect on the reaction rate.

Differing rate laws in the Cr^{2+}-catalyzed substitution of Br^- in $[BrCr(OH_2)_5]^{2+}$ by Cl^- and F^- can be rationalized[53] in terms of a single mechanism in which $[XBrCr(OH_2)_4]^+$ ($X^- = Cl^-$ or F^-) is formed as an intermediate. In the fluoride case, the intermediate is formed in significant amounts.

Chromous reductions of cobalt(III) complexes with pseudoaromatic rings (**6**) show marked steric retardation by ring methyl substituents.[54] Electron trans-

6

fer takes place after Cr^{2+} complexation to the free carbonyl and is followed by rapid ring closure to give a pseudoaromatic chromium(III) species.

The extent of precursor complex formation is an important factor[55] in determining the relative rates of chromous reduction of complexes $[(NH_3)_5CoYH_n]$, where Y is O-bound ida^{3-} or $edta^{2-}$. With $[(NH_3)_5Co(ida)]$ the precursor involves chelation at Cr^{2+} and leads to the tridentate-bound ligand, whereas protonated cobalt(III) reagents yield only monodentate chromium(III) products.

Studies using the Na^+-specific chelate 18-crown-6 have established that apparent ionic strength effects in the reduction of $[Co(edta)]^-$ by $[Fe(CN)_6]^{4-}$ are best explained[56] by complexation between the latter reagent and the "inert" cation Na^+. Active reductants are $[Fe(CN)_6]^{4-}$ and $[NaFe(CN)_6]^{3-}$, while $[Na_2Fe(CN)_6]^{2-}$ was found to be virtually nonreactive. Besides reducing electrostatic interactions, the role of Na^+ may be to act as a polarizable electron transfer bridge.

An inner-sphere "dead end" complex $[edtaCo(III)NCFe(II)(CN)_4py]^{4-}$ is formed[57] in the reductions of $[Fe(CN)_5py]^{2-}$ and $[Fe(CN)_5(4,4'-bipy)]^{2-}$ by $[Co(edta)]^{2-}$. Electron transfer takes place by an outer-sphere mechanism and the second-order rate constants are in good agreement with Marcus correlations. The coresponding reactions[58] with $[Ru(NH_3)_5py]^{3+}$ and $[Ru(NH_3)_5(4,4'-bipy)]^{3+}$ are also likely to be outer sphere, and Marcus calculations, corrected for electrostatic work terms, indicate that electron transfer is primarily through the NH_3 and not the pyridine ligand. This contrasts with the reduction[17] of $[Ru(NH_3)_5py]^{3+}$ by $[Fe(CN)_6]^{4-}$ where electron transfer is through the pyridine.

Detection of ion pairing in outer-sphere electron transfer is generally possible when the reactants have large and opposite charges. Thus, in the reduction of the cobalt(III) complex **7** by $[Fe(CN)_5(4,4'-bipy)]^{3-}$, deviations from second-

$$\left[(NH_3)_5CoN \bigcirc\!\!-\!\!\bigcirc \right]^{3+}$$

7

order kinetics allow[59] calculation of K_0 and k_{et} [equations (13) and (14)]. When the π-ligand system is interrupted using 1,2-bis(4-pyridyl)-ethane instead of 4,4'-bipy on the iron(II) complex, k_{et} is higher suggesting that electron transfer does not take place through the π system but by d-orbital overlap. Indeed, k_{et} is slowest for strong field ligands at Co(III). The electron transfer rates are very similar to unimolecular processes in **8** where the inner-sphere pathway[60] has little apparent advantage in terms of rate.

$$\left[(NH_3)_5CoN \bigcirc\!\!-\!\!\bigcirc NFe(CN)_5 \right]$$

8

The reversible reaction between $[Ru(NH_3)_5(nicotinamide)]^{3+/2+}$ and $[Ru(NH_3)_5(isonicotinamide)]^{3+/2+}$ has been investigated[61] to yield a self-exchange rate for all $[Ru(NH_3)_5py]^{3+/2+}$ complexes of 4.7×10^5 mol^{-1} liter s^{-1}. Electron transfer reactions of $[Ru(NH_3)_5py]^{3+/2+}$ complexes appear for the most part to follow Marcus behavior, and reasonable agreement with the thermal electron transfer rate calculated from the IT bond of $[(NH_3)_5Ru(4,4'-bipy)-Ru(NH_3)_3]^{5+}$ was found. However, in the oxidations by Fe^{3+}, deviations from Marcus behavior which were found to be entropic in nature are considered to indicate a degree of nonadiabaticity in the electron transfer and a transmission coefficient of 5×10^{-2} was estimated.

The oxidations of $[Co(phen)_3]^{2+}$ and $[Co(bipy)_3]^{2+}$ by $[Ru(bipy)_3]^{3+}$ have a very large driving force and can be studied[62] as the dark reaction after excited state quenching of $[*Ru(bipy)_3]^{2+}$ by $[Co(phen)_3]^{3+}$ or $[Co(bipy)_3]^{3+}$. Thermal reduction of $[Ru(bipy)_3]^{3+}$ and dissociation of the cobalt(II) complexes are too slow to interfere. The reaction rates are fast but slower than predicted by calculation and the difference must lie in nonadiabaticity since there is no spin change. The mismatch between the detailed electron transfer pathway in the cross reaction compared with the self-exchange may be due to a shift from a π^* $\rightarrow \pi^*$ pathway to a mixed $d-\pi^*$ pathway.

The permanganate oxidation of $[Ru(CN)_6]^{4-}$ has a complex pH dependence[63] due to the protonation equilibria:

$$[HRu(CN)_6]^{3-} \rightleftharpoons H^+ + [Ru(CN)_6]^{4-} \qquad (pK_a = 2.53) \qquad (23)$$

$$[H_2Ru(CN)_6]^{2-} \rightleftharpoons H^+ + [HRu(CN)_6]^{3-} \qquad (pK_a = 2.46) \qquad (24)$$

The protonated ruthenium species are more reactive than $[Ru(CN)_6]^{4-}$, which indicates the importance of H^+ in the Mn(VII)/(VI) conversion with poor reducing agents. The reactivity pattern is similar to but slower than the corresponding $[Fe(CN)_6]^{4-}$ system.[64]

A higher rate[65] for the Fe^{2+} reduction of $[Co(DH)_2BN_3]$ than with its conjugate acid $[Co(DH)(DH_2)BN_3]$, where DH_2 is dimethylglyoxime and B is NH_3 or py, suggests an inner-sphere process with the oxygen as bridge site.[66] The effects of py and NH_3 on activation parameters are in line with the effects of other nonbridging ligands.

Confirmation[67,68] of the $Co_{aq}^{3+/2+}$ reduction potential in 3 M $HClO_4$ as 1.8–1.9 V was obtained in an electrochemical study of Fe^{2+} reduction. Kinetic results are in agreement with previous work on this system.[69]

The acid decomposition of chloritopentaammine cobalt(III), $[(NH_3)_5CoOClO]^{2+}$, to give ClO_2 and Co(II) is slower than the corresponding reaction of $HClO_2$.[70] The complex can be oxidized and reduced by metal ion redox agents. Oxidation by Co^{3+} has an inverse $[H^+]$ dependence, and tracer studies indicate that Co–OClO bond cleavage takes place. This contrasts with reduction by Fe^{2+} and Cr^{2+}, where CoO–ClO cleavage is important. In the latter reaction coordinated chlorite is reduced to chloride before reduction of the metal center.

Reduction of cobalt(III) to cobalt(II) in vitamin B_{12} is slow and is ascribed[71] to the lack of an electron transfer pathway through the corrin ring. Both axial positions are blocked by protein side chains preventing redox by this route. The $[Co(I)]^-$ species B_{12s} has been used[72] in the 1:1 reduction of $[(H_2O)_5CrX]^{2+}$ complexes. A marked dependence of the reaction rate on X suggests an inner-sphere mechanism to produce a $[Co(II)-X]$ product which subsequently hydrolyzes.

Parallel inner-sphere and outer-sphere pathways are noted[73] in the reduction of iron(III)(porph)(CN)$_2$ complexes, where porph is a porphyrin, by $[Co(CN)_5]^{3-}$ and $[Co(CN)_6]^{4-}$. The outer-sphere reactions, involving $[Co(CN)_6]^{4-}$, are faster than the inner-sphere route by $[Co(CN)_6]^{3-}$ consistent with a smaller reorganizational barrier.

Substitution inert copper(III)–peptide complexes have reduction potentials which are $[H^+]$ dependent such that in acidic solution, they will oxidize Mn^{2+}, Ce^{3+}, $[Fe(phen)_3]^{2+}$, and $[IrCl_6]^{3-}$.[74] Reductions of a series of complexes, $[Cu(III)(peptide)]$, with $[Cu(I)(dmp)_2]^+$, where dmp is 2,9-dimethyl-1,10-phenanthroline, are rapid, outer-sphere processes which show some adherence to Marcus behavior,[75] although the $[Cu(dmp)_2]^{2+/+}$ self-exchange rate appears to be very dependent on the cross reaction,[76,77] a lower limit of 3×10^4 mol^{-1} liter s^{-1} arising from the $[Cu(III)(peptide)]$ reactions, while a value of 9×10^8 mol^{-1} liter s^{-1} is obtained from reaction with $[IrCl_6]^{2-}$, both at 25°C and 0.10 M ionic strength.[75] Reactions of the copper(I) complex of the sterically hindered phenyl sulfonate derivative of dmp, $[Cu(dpmp)_2]^{3-}$, with $[Cu(III) (peptide)]$ are complicated by limiting rate behavior which is ascribed to activation of the copper(III) complex [reaction (25)]

$$[Cu(III)(peptide)] \rightleftharpoons [Cu(III)(peptide)^*] \tag{25}$$

prior to electron transfer. It is suggested that axial solvation[78] of the copper(III) center is required to overcome the large separation between the reacting centers. A number of these reactions are very rapid and require pulsed-flow technique.[79,80]

Pulse radiolysis has been used[81] to study the reactions between Cu^{3+}/Fe^{2+} and Cu^+/Fe^{3+} in aqueous media. Electron exchange at the tetrahedral $Cu^{2+/+}$ site in the polytungstate complex $[CuW_{12}O_{40}]^{6-/7-}$, calculated[82] from the cross reaction with $[Fe(CN)_6]^{3-/4-}$ using Marcus theory, is slow. This is surprising in view of the limited structural change possible during redox. Comparison with the corresponding $[CoW_{12}O_{40}]^{5-/6-}$ value[83] leads to the suggestion that a chemical mechanism involving the W(VI)/(V) couple is operating.

The complex $[Fe(bipy)_2]^{3+}$ oxidizes Hg_2^{2+} with a second-order rate constant of 0.22 mol^{-1} liter s^{-1} in 70% methanol at 30°C. Cleavage of the $Hg–Hg^{2+}$ bond with concomitant electron transfer is proposed as the rate-limiting step.[84]

It may be fortuitous that no $[H^+]$ dependence was found[85] in the reduction of Am(V) to Am(III) by V(IV) since opposing trends with $[H^+]$ are expected for AmO_2^+ and VO^{2+}. A possible explanation is formation of an inner-sphere complex (9) with oxygen transfer. This implies that vanadyl reduction of Am(IV) is much more rapid than disproportionation of Am(V) and Am(III).

$$O–Am(V)–O\cdots V(IV)=O$$
$$9$$

1.3.2.   Solvent Effects in Electron Transfer Reactions

Solvent effects on electron transfer rate constants have been considered[86] in a more general review. Besides extending the range of compounds which are accessible to mechanistic study, the use of nonaqueous solvents can allow investigations of effects not possible in aqueous solution.

The outer-sphere reaction (26) was monitored[87] in CH_3OH/H_2O and

$$[Co(NH_3)_5OH_2]^{3+} + [Ru(NH_3)_6]^{2+} \rightarrow [Co(NH_3)_5OH_2]^{2+} + [Ru(NH_3)_6]^{3+} \tag{26}$$

$HOCH_2CH_2OH/H_2O$ mixtures over an 80° temperature range. Plots of log k vs. T^{-1} are linear in the first case, curved in the second, and an explanation based on an increase in solvent ordering with decreasing temperature is considered most likely. This increased order induces a larger electron transfer length and reduces the transmission coefficient.

A study[88] of the self-exchange rate of the delocalized cluster $[Ru_3O(CH_3CO_2)_6(py)_3]^{+/0}$ by nmr in methylene chloride highlights problems involved in calculating electron transfer rates. The major limitations may be in estimating precursor complex formation constants rather than calculating the electron transfer rate. A statistical mechanical approach was found to give better results than the Eigen–Fuoss equation.

Ferrocene, $[Fe(Cp)_2]^{+/0}$, self-exchange has also been studied[89] by nmr in a variety of solvents. The rates do not vary with solvent dielectric as predicted by Marcus:

$$\Delta G^* = \frac{e}{4R^*} \left(\frac{1}{D_{op}} - \frac{1}{D_s} \right) \tag{27}$$

where R^* is the inter-reactant distance in the transition state and D_{op}, D_s are optical and static dielectric constants, and microscopic phenomena such as solvent orientation near the reactants require to be considered. Small ΔS^* values are consistent either with greater solvent organization in the transition state or with a degree of nonadiabaticity. Solvent effects on the oxidation of $[Fe(Cp)_2]$ by $[Co(phen)_3]^{3+}$ and Fe^{3+} have also been reported.[90]

A dielectric dependence on the rate of reaction between Fe^{2+} and $FeOH^{2+}$ in mixed alcohol–water solvents is explained[91] by the dominance of electrostatic repulsion and an Fe–Fe distance of 7–8 Å is estimated. In mixed pyridine/H_2O, reaction is between $Fe(py)_n^{2+}$ and $FeOH^{2+}$.[92] The oxidation of U(IV) to U(VI) by Ce(IV) has been studied[93] in $(CH_3)_2CO$.

Thermodynamic parameters associated[94] with formation of the bridged precursor (10) in the $[Co(NH_3)_5Cl]^{2+}$ oxidation of Fe^{2+} in DMF indicate a

$$[(NH_3)_5Co(III)ClFe(II)(DMF)_3]^{4+}$$
10

change in the Fe^{2+} geometry from octahedral to tetrahedral. A similar intermediate[95] is detected in oxidation by $[Co(NH_3)_5N_3]^{2+}$, while in $DMSO^{(96)}$ oxidation by *cis*-$[Co(en)_2Cl_2]^+$ leads to a doubly bridged intermediate in which the Fe^{2+} retains its octahedral geometry. An uncatalyzed and a $Mn(II)(QuinCN)_2$-catalyzed pathway have been observed[97] in the oxidation of Fe^{2+} by $Mn(III)(QuinCN)_2$ in acetonitrile, where QuinCN is 5-cyano-8-hydroxyquinoline.

Micellar effects on electron transfer rates have also been examined.[98,99]

1.4. Excited State Electron Transfer

Electron transfer reactions of the excited states of metal ion complexes continue to attract much attention with interest in solar energy storage[100,101] and the cleavage of H_2O to H_2 and O_2.[101–105] A useful review[106] of the properties of excited-state polypyridine complexes has appeared. The much studied $[*Ru(bipy)_3]^{2+}$ and related systems remain at the forefront of new work but other metal ion systems are attracting increasing interest (Table 1.3.) When a complex ML_3^{2+} is photochemically activated to form $*ML_3^{2+}$, excited state quenching can take place by three pathways:

$$*ML_3^{2+} + Q \xrightarrow{k_0} ML_3^{3+} + Q^- \tag{28}$$

$$*ML_3^{2+} + Q \xrightarrow{k_r} ML_3^+ + Q^+ \tag{29}$$

$$*ML_3^{2+} + Q \xrightarrow{k_e} ML_3^{2+} + Q^* \tag{30}$$

where Q is the quencher. Pathways (28) and (29) involve electron transfer, while (30) involves energy transfer. Energy transfer increases in efficiency as the spectral overlap of the donor and acceptor increases, while electron transfer rates can in general be predicted by using Marcus theory. These criteria have been used[107] to determine the quenching mechanism in reactions of a number of polypyridine complexes of Os(II), Ru(II), and Fe(II). In general if a process is thermodynamically favorable, quenching is rapid but, for example, in the quenching of $[*Os(5-Cl\ phen)_3]^{2+}$ by $[Ru(terpy)_2]^{2+}$ where both electron and energy transfers are unfavorable, the quenching rate is low ($\leq 1 \times 10^8$ mol^{-1} liter s^{-1}). Excited state lifetimes of iron(II) polypyridine complexes are much shorter than

Table 1.3. Rate Constants and Activation Parameters for Excited State Electron Transfer at 25°C

Oxidant	Reductant	Medium, mol liter^{-1}	Rate law	Rate constant, mol liter s^{-1}	$\Delta H^{\ddagger}$, kcal mol^{-1}	$\Delta S^{\ddagger}$, cal K^{-1} mol^{-1}	Ref.
[Co(bipy)$_3$]$^{3+}$	[*Ru(bipy)$_3$]$^{2+}$	1.0	k	2.27 × 10^9			62
[Co(phen)$_3$]$^{3+}$	[*Ru(bipy)$_3$]$^{2+}$	1.0	k	2.18 × 10^9			62
[(NH$_3$)$_5$CoO$_2$C-(p-NO$_2$C$_6$H$_4$)]$^{2+}$	[*Ru(bipy)$_3$]$^{2+}$	0.1	k	2.4 × 10^9			112
[(NH$_3$)$_5$CoO$_2$C-(o-NO$_2$C$_6$H$_4$)]$^{2+}$	[*Ru(bipy)$_3$]$^{2+}$	0.1	k	1.3 × 10^9			112
[(NH$_3$)$_5$CoO$_2$C-C$_6$H$_5$]$^{2+}$	[*Ru(bipy)$_3$]$^{2+}$	0.1	k	0.15 × 10^9			112
[(NH$_3$)$_5$CoO$_2$CCH$_3$]$^{2+}$	[*Ru(bipy)$_3$]$^{2+}$	0.1	k	0.21 × 10^9			112
[(NH$_3$)$_5$CoμO$_2$Co(NH$_3$)$_5$]$^{5+}$	[*Ru(bipy)$_3$]$^{2+}$	1.0	k	1.91 × 10^9			111
[(NH$_3$)$_4$CoμO$_2$,μNH$_2$Co(NH$_3$)$_4$]$^{4+}$	[*Ru(bipy)$_3$]$^{2+}$	1.0	k	4.00 × 10^9			111
[(en)$_2$CoμO$_2$,μNH$_2$Co(en)$_2$]$^{4+}$	[*Ru(bipy)$_3$]$^{2+}$	1.0	k	3.20 × 10^9			111
[(bipy)$_2$CoμO$_2$,μNH$_2$Co(bipy)$_2$]$^{4+}$	[*Ru(bipy)$_3$]$^{2+}$	0.81	k	4.47 × 10^9			111
[(phen)$_2$CoμO$_2$,μNH$_2$Co(phen)$_2$]$^{4+}$	[*Ru(bipy)$_3$]$^{2+}$	1.0	k	8.10 × 10^9			111
Ag$^+$	[*Ru(bipy)$_3$]$^{2+}$		k	3.5 × 10^6			110
Cu^{2+}	[*Ru(bipy)$_3$]$^{2+}$	0.8–1.0	k	6.2 × 10^7	2.7	-13.6	109
Eu^{3+}	[*Ru(bipy)$_3$]$^{2+}$	2.8	k	3.6 × 10^5	4.0	-19.4	109
Cu^{2+}	[*Ru(phen)$_3$]$^{2+}$	0.8–1.0	k	6.37 × 10^7	3.2	-12.2	109
Eu^{3+}	[*Ru(phen)$_3$]$^{2+}$	2.8	k	3.6 × 10^5	7.6	-7.2	109
Cu^{2+}	[*Ru(5-Br phen)$_3$]$^{2+}$	0.8–1.0	k	3.41 × 10^7	2.6	-15.3	109
Cu^{2+}	[*Ru(4,7-Me$_2$phen)$_3$]$^{2+}$	0.8–1.0	k	7.93 × 10^7	4.2	-8.1	109
Eu^{3+}	[*Ru(4,7-Me$_2$phen)$_3$]$^{2+}$	2.8	k	1.25 × 10^6	6.0	-10.5	109
Cu^{2+}	[*Ru(3,4,7,8-Me$_4$phen)$_3$]$^{2+}$	0.8–1.0	k	6.48 × 10^7	4.5	-7.6	109
[Ru(TPTZ)$_2$]$^{2+}$	[*Os(5-Cl phen)$_3$]$^{2+}$	0.5	k	2.6 × 10^9			107
[Ru(TPTZ)$_2$]$^{2+}$	[*Ru(bipy)$_3$]$^{2+}$	0.5	k	1.2 × 10^9			107
[Os(en)$_2$O$_2$]$^{2+}$	[*Ru(bipy)$_3$]$^{2+}$	0.5	k	1.67 × 10^9			108
[Os(en)$_2$H$_2$]$^{2+}$	[*Ru(bipy)$_3$]$^{2+}$	0.5	k	2.28 × 10^8			108
[Os(en)$_2$Cl$_2$]$^+$	[*Ru(bipy)$_3$]$^{2+}$	0.5	k	5.21 × 10^7			108

the corresponding osmium(II) species, which may be due to the substantial geometry change between excited and ground states. This also results in a slow $[*Fe(bipy)_3]^{2+}/[*Fe(bipy)_3]^{3+}$ self-exchange rate estimated at $<10^3$ mol^{-1} liter s^{-1}. Quenching of $[*Ru(bipy)_3]^{2+}$ emission by $[Os(en)_2O_2]^{2+}$, $[Osen_2H_2]^{2+}$, and $[Osen_2Cl_2]^+$ is qualitatively consistent with the redox properties of the osmium(III) species.[108] While the superoxide complex is reduced, the mechanistic tendency of the hydride to act as a two-electron reagent results in a low quenching rate.

Reductions of $[Co(phen)_3]^{3+}$ and $[Co(bipy)_3]^{3+}$ by $[*Ru(bipy)_3]^{2+}$ are close to diffusion controlled,[62] and calculated rate constants

$$1/k_{calc} = 1/k_D + 1/k_M \tag{31}$$

where k_D is the diffusion rate and k_M the Marcus rate, are in good agreement with observed values though some tunneling may be present. The photolysis product is the corresponding cobalt(II) complex and there is no evidence for a coordinated radical species.

Rate constants for electron transfer quenching of substituted $[*Ru(phen)_3]^{2+}$ derivatives by Eu^{3+} show[109] behavior consistent with Marcus theory with a self-exchange rate constant of 5×10^{-4} mol^{-1} liter s^{-1} for Eu$^{3+/2+}$. When Cu^{2+} is used as quencher, the rate dependence on free energy is abnormal and is explained by energy transfer in an intermediate complex. Reduction of Ag$^+$ leads[110] to the formation of the reactive Ag^0 and Ag$_2^+$ in solution. Both species participate in the fast back-reaction.

Binuclear μ-superoxo complexes of cobalt(III) react[111] with $[*Ru(bipy)_3]^{2+}$ at rates slightly higher than diffusion controlled based on their charges, possibly as a result of uneven charge distribution. The peroxo products do not decompose spontaneously to give Co^{2+} in acid media but instead participate in the rapid dark reaction. Cobalt(II) produced in the photolysis is the result of energy transfer. Quenching rates with Ti^{3+} and TiOH^{2+} are much faster than calculated[40] and an energy transfer mechanism is predicted.

Initial electron transfer[112] to orbitals on coordinated nitrobenzoate ligands is proposed for the excited state quenching of $[*Ru(bipy)_3]^{2+}$ by o- and p-nitrobenzoate complexes of Co(III)(NH$_3$)$_5$. When quenching rate constants are analyzed in terms of the mechanism

$$[*Ru(bipy)_3]^{2+} + [Co(NH_3)_5L]^{2+} \overset{k_0}{\rightleftharpoons} [*Ru(bipy)_3]^{2+} Co(NH_3)_5L^{2+}] \overset{k_{et}}{\rightarrow} \text{products} \tag{32}$$

k_{et} for the nitro-substituted reactants is an order of magnitude larger than for unsubstituted reactants for which it is proposed that acceptor orbitals are localized on the metal center. Higher quantum yields for Co(II) in these latter reactions are cited as evidence for cage recombination in the case of the nitrobenzoates. The polyelectrolyte, polybrene, is effective at expelling the $[Ru(bipy)_2(CN)_2]^+$ product of the photoelectron transfer reaction between $[*Ru(bipy)_2(CN)_2]$ and $[Fe(CN)_6]^{3-}$ from the solvent cage and thus providing a system with some potential for energy storage.[113] The back-reaction rate is close to diffusion controlled and is largely unaffected by the polybrene.

An estimate[114] of the $[Cu(dmp)_2]^{2+}|[*Cu(dmp)_2]^+$ potential of -1.4 V (vs. sce in Ch_2Cl_2 at 25°C) has been made and electron transfer quenching by the excited state has been detected with chromium(III) and cobalt(III) complexes.[114,115] The complex $[*Ru(phen)_3]^{3+}$ is a very strong oxidant ($\sim +2.00$ V). Reaction with $[Cr(CN)_6]^{3-}$ is diffusion controlled and proceeds by an energy transfer mechanism.[116]

Transition metal ions quench excited state $*Ce^{3+}$ in aqueous sulfate media[117] but only with Cu^{2+} is electron transfer detected. The products $Ce(SO_4)_n$ and Cu^+ decay by back-reaction which can be slowed dramatically by addition of CH_3CN to complex Cu^+.

1.5. Stereoselectivity in Electron Transfer

A major problem in detecting chiral discrimination in electron transfer reactions is the high rate of racemization of any chiral product by electron exchange with its reduced form.[118] In systems such as the $S_2O_8^{2-}$ oxidation of $[Co(edta)]^{2-}$ by $[Os(bipy)_3]^{3+/2+}$, $[Ru(bipy)_3]^{3+/2+}$, $[Co(bipy)_3]^{3+/2+}$, and $[Fe(phen)_3]^{3+/2+}$, where racemization is prevented, chiral discrimination has been detected. This indicates an intimate contact between the metal ion oxidant and reductant though yields are small and there is no general trend relating the absolute configurations of the reagents. Nonaqueous solvents enhance discrimination. Stereoselectivity has also been detected with a substitution inert nickel(IV) complex,[119] and the excited state quenching of Δ-$[*Ru(bipy)_3]^{2+}$ by racemic $[Co(acac)_3]$ leads to net production of Δ-$[Co(acac)_3]$ by preferential reaction with Λ-$[Co(acac)_3]$.[120]

1.6. Metalloprotein Redox Reactions

Structural studies on electron transfer metalloproteins provide an important origin for discussion of the electron transfer processes themselves.[121] The reduction potentials of a number of cytochromes c, cyt c; copper blue proteins; plastocyanin, Pc; azurin, Az; stellacyanin, St; and HiPIP, or high potential iron protein, from *Chromatium vinosum* have been determined[122] using spectroelectrochemical methods with transition metal ion complex mediators. Entropy changes associated with the redox couple ΔS°_{RC} are more negative than found with simple inorganic complexes consistent with either increased solvent structuring in the protein interior or solvent release and a more compact protein structure on reduction.

If metalloprotein redox reactions obey Marcus theory, the ratio of the rate of oxidation of reduced protein by $[Co(ox)_3]^{3-}$ and $[Co(phen)_3]^{3+}$, $R = k_{12[Co(phen)_3]^{3+}}/k_{12[Co(ox)_3]^{3-}}$, should be a constant, independent of the protein.[123] Under conditions where protein–oxidant preassociation is minimized, R values increase in the order

$$Az \sim HiPIP < cyt\ c_{551} < Pc < cyt\ c \ll St \tag{33}$$

giving a measure of the kinetic accessibility of the redox centers. Marcus self-exchange rates derived from the $[Co(ox)_3]^{3-}$ and corrected for charge effects agree well with those for $[Fe(edta)]^{2-}$ reduction and are primarily a function of the protein characteristics, notably metal-site-to-surface distance. These distances have been estimated[124] using a modification of Hopfield's approach

$$R_P = 6.2 - 0.35\ \ln k^{\infty}_{11} \tag{34}$$

where R_p is half the distance between redox centers and k^{∞}_{11} is the apparent self-exchange rate at infinite ionic strength. Values of R_p are much shorter for reactions of $[Co(phen)_3]^{3+}$ and other hydrophobic systems, which can penetrate the protein interior, compared with $[Fe(edta)]^{2-}$.

Redox inert complexes $[Cr(phen)_3]^{3+}$ and $[Cr(CN)_6]^{3-}$ have been shown[125,126] by nmr to bind at different points on the surface of the blue copper protein plastocyanin. Both sites are close to electron channels to the copper center and are the likely sites occupied by the oxidants $[Co(phen)_3]^{3+}$ and $[Fe(CN)_6]^{3-}$, which have been shown to bind to the protein.[127] The reaction of $[Co(phen)_3]^{3+}$ in inhibited by $[Cr(phen)_3]^{3+}$.[128]

The eight-iron ferredoxin, $Fd_{ox,ox'}$, from *Clostridium pasteurianum*, contains two $Fe_4S_4^*$ clusters which can be reduced[129] by pulse radiolysis to give $Fd_{ox,red}$. Oxidations of this protein by $[Co(NH_3)_5Cl]^{2+}$ proceed at a rate close to the rate of reaction of the fully reduced protein, indicating that both iron clusters react without cooperativity. Oxidations of the fully reduced protein[130] by a number of metal ion complex oxidants have also been reported and limiting kinetics consistent with metal complex binding has been observed with positively charged oxidants. It is suggested that a cysteine of an iron cluster which is close to the protein surface may be the lead in for the positively charged reaction partners.

Two studies of electron transfer reactions of Holm analog $Fe_4S_4^*$ clusters have been reported. In acetonitrile solution,[131] outer-sphere self-exchange rates of complexes $[Fe_4S^*_4(SR)_4]^{2-/3-}$ and $[Fe_4Se^*_4(SR)_4]^{2-/3-}$, where SR is a thiophenol, have been determined by nmr. The principal structural change, elongation–compression of the $Fe_4S^*_4$ cube, makes only a minor contribution to the activation energy. The rates are significantly faster than for ferredoxins or HiPIP, which may reflect the insulating properties of the proteins. A super-reduced, $8-$, form of the water-soluble cluster $[Fe_4S^*_4 (SCH_2CH_2CO_2)_4]^{6-}$ is obtained on Cr(II)edta reduction.[132] Although the $6- \rightarrow 7-$ step is too fast to measure, the $7- \rightarrow 8-$ step indicates a pH-independent reaction. The eight-iron protein from *Clostridium pasteurianum* and HiPIP could not be super-reduced.

Two pK_a values around 7 and 9 were detected by correcting for charge effects in a study[133] of the oxidation of *Chromatium vinosum* HiPIP by $[Fe(CN)_6]^{3-}$. Earlier work[134] at constant ionic strength showed no pH variation and the importance of correcting for electrostatic interaction in evaluating pH effects is stressed. Specific interactions as well as charge effects are suggested by ionic strength dependencies in a second study of $[Fe(CN)_6]^{4-/3-}$ interaction with HiPIPs from various sources.[135] The effect of the charge on the Fe_4S_4 cluster is pronounced and may be communicated to the protein surface by a hydrogen-bonded network. Structural analysis provides a likely site for electron transfer where the iron–sulfur cluster is near the protein surface. Reactions of HiPIP with cytochrome c are more complex.[136]

Ionic strength effects[137,138] in reactions of horse heart cytochrome c are poorly fit by the Debye–Hückel limiting law and are best explained using the Kirkwood–Tanford approach where the whole protein charge rather than specific site charges is involved. Binding constants between cytochrome c and iron hexacyanides have been determined.[139] Reduction by $[Ru(NH_3)_6]^{2+}$ of cytochrome c[140] and cytochrome aa$_3$[141] have been reported. The latter reaction is biphasic, consisting of a rapid initial reduction of heme a(III) followed by slow intermolecular electron transfer between two cyt a(II)a$_3$(III) molecules resulting in cyt a(II)a$_3$(II) and cyt a(III)a$_3$(III). Outer-sphere oxidations of heme proteins by copper(II) complexes have also been reported.[142]

Chapter 2

Metal-Ion–Ligand Redox Reactions

2.1. Introduction

The coverage of material in this review is for the period June 1979 to December 1980. As with previous articles of this type, it has not been possible to include every paper published. An attempt has been made, however, to cover all the major areas of investigation.

Although references to specific types of reactions and systems are provided in the various sections of this chapter, there are several reviews of a more general nature which reflect not only the range of redox reactions studied but also the newer methods available for measuring reaction rates.

Oxidation processes in solution are considered extensively in detail in a recent volume[1] in the series *Comprehensive Chemical Kinetics*. The chemistry of catalytic processes in the homogeneous phase has also been surveyed.[2] The importance of modification of the metal ion reactivity by ligand interaction has been identified for tetraza-macrocyclic systems.[3] Phenomena involved in oscillating chemical reactions continue to be of interest and a review has been published of reactions in solution.[4] The activation of small molecules by metal complexes has been considered with reference to redox processes,[5] and single-electron transfers have been reviewed in organic systems.[6] Radical pathways in reactions of transition metal organometallic complexes have been described,[7] and the role of electron and charge transfer in organometallic chemistry examined[8] in a major review. Mechanisms of redox reactions which are important in analytical chemistry have been discussed[9] as have those involving inorganic nitrogen compounds.[10]

Consideration has also been given to reaction techniques.[11] Details have been published of pulsed-flow spectroscopic methods for use in reactions with lifetimes ≥ 40 μs.[12,13] A variety of redox systems, which react too rapidly for conventional stopped-flow procedures, may now be investigated. Further details of an infrared stopped-flow system have also been provided.[14]

2.2. Metal Complexes with Inorganic Substrates

2.2.1. Reactions of Hydrazine and Hydroxylamine

Several studies have been made using both cationic and anionic oxidants. In the reaction of hydrazine with $[IrCl_6]^{2-}$, $[IrCl_5(OH_2)]^-$, $[IrCl_4(OH_2)_2]$, and $[IrBr_6]^{2-}$,[15] a first-order dependence on reactant concentrations is shown. The inverse hydrogen ion dependence observed is attributable to the protonation of the reductant. The data for the $[IrCl_6]^{2-}$ system are similar to those reported elsewhere,[16] with an increase in rate in the presence of Cl^-. N_2 is formed stoichiometrically, with no evidence for any NH_3 produced. Using *cis*-cyclohex-4-ene-1,2-dicarboxylic acid,[15] the intermediate N_2H_2 may be trapped. The mechanism proposed involves prior formation of a 1:1 $[Ir(IV)-N_2H_4]$ complex which undergoes internal electron transfer leading to $N_2H_3^+$ formation as the initial product. The outer-sphere nature of the reactions of the aquo ions is confirmed by a linear free energy relationship between the rate constants and the standard reduction potential of the Ir(IV)/Ir(III) couples. A 1:1 adduct has also been invoked in the mechanism of oxidation by $[Fe(phen)_3]^{3+}$ in sulfuric acid solutions.[17] Pu(IV) is reduced by N_2H_4, the reaction being catalyzed by Mo(VI).[18] Under the conditions used, the first (rate-determining) stage is the reaction between monomeric Mo(VI) and the reductant, with subsequent fast oxidation of Mo(V) by Pu(IV). Both N_2 and NH_3 are formed as products.

Comparative studies have been made[19] in the oxidation of N_2H_4 and NH_2OH by Co(III) complexes. The reactions of $[Co(edta)(OH_2)]^-$ with the substrates in acid media are considered to proceed via an inner-sphere mechanism with a proton bridge between the cobalt(III) species and (a filled orbital on) the nitrogen of the reductant. The rates are greater than for $[Co(ox)_3]^{3-}$ or $[Co(acac)_3]$ despite the fact that all three species have similar redox potentials. Reactions with cationic cobalt(III) species appear to be much slower. The stoichiometry of the reaction between NH_2OH and V(V) in $HClO_4$ media is dependent on the reaction conditions.[20] In the presence of excess reductant, however, the reaction is first order with respect to $[VO_2^+]$ and the dependence on hydroxylamine concentration varies between zero and one, consistent with the formation of intermediate complexes. Three reaction pathways are proposed

$$VO_2^+ \cdot NH_3OH^+ \xrightarrow{k_1} VO^{2+} + \cdot ONH_2 + H_2O \tag{1}$$

$$VO(OH)^{2+} \cdot NH_3OH^+ \xrightarrow{k_2} VO^{2+} + \cdot ONH_2 + H^+ \tag{2}$$

$$V(OH)_2^{3+} \cdot NH_3OH^+ \xrightarrow{k_3} VO^{2+} + \cdot ONH_2 + 2H^+ \tag{3}$$

followed by the rapid step in equation (4).

$$2 \cdot ONH_2 \xrightarrow{\text{fast}} N_2 + 2H_2O \tag{4}$$

Protonation of the reductant is also invoked in the reaction with $[IrCl_6]^{2-}$ in dilute (pH ~3) $HClO_4$ solutions.[21] The rate law is of the form $-d[Ir(IV)]/dt = k[Ir(IV)] [NH_3OH^+]/[H^+]$ consistent with NH_2OH as the principal reducing species.

2.2.2. Reactions of Nitrate and Nitrite

The reactions of HNO_3 and HNO_2 with iron(III) cyanide complexes have been reported.[22] Under conditions where $[Fe_2(CN)_{10}]^{4-}$ is the oxidant, the rate law is of the form $R = kH_0^{0.5}[HNO_2]^{0.5} [NO_3]^{0.5} [Fe_2(CN)_{10}^{4-}]$ where H_0 is an acidity function. The major reaction product is a mixed-valence Fe(II)/Fe(III) species incorporating NO^+ as a ligand in the inner coordination sphere of the Fe(II). The rate data are similar to those for $[Fe(CN)_6]^{3-}$ and $[Fe(CN)_4(LL)]^-$ complexes (LL = phen, bipy) and are consistent with a reaction proceeding via the formation of a NO_2 radical species

$$H^+ + HNO_2 + NO_3^- \rightleftharpoons N_2O_4 + H_2O \text{ (fast)} \tag{5}$$

$$N_2O_4 \rightleftharpoons 2NO_2 \text{ (fast)} \tag{6}$$

$$NO_2 + [Fe(CN)_6]^{3-} \xrightarrow{\text{slow}} \text{intermediate} \rightarrow \text{product} \tag{7}$$

Kinetic data for the reaction of Cr(VI) with HNO_2 have been interpreted[23] in terms of a prior formation of a nitrito complex of Cr(VI), $[CrO_3ONO]^-$, which undergoes redox recomposition either in a unimolecular process or via reaction with a second mole of HNO_2. The reaction of Pu(VI) with nitrous acid[24] exhibits a rate law somewhat similar to that for the iron(III) cyano complexes. In this case, however, the mechanism is considered to involve not only reaction of NO_2^- with PuO_2^{2+}, but also a direct process with NO_2^-.

2.2.3. Reactions of Thiocyanate

The outer-sphere reaction

$$6[IrCl_6]^{2-} + SCN^- + 4H_2O \rightarrow 6[IrCl_6]^{3-} + SO_4^{2-} + CN^- + 8H^+ \quad (8)$$

has been investigated.[25] The form of the rate law [equation (9)]

$$R = \{k_1' + k_2'[NCS^-]\}\,[Ir(IV)]\,[NCS^-] \quad (9)$$

is consistent with pathways involving ion-pair-mediated electron transfer. A scheme consistent with the findings may be written as in Scheme 1.

Scheme 1

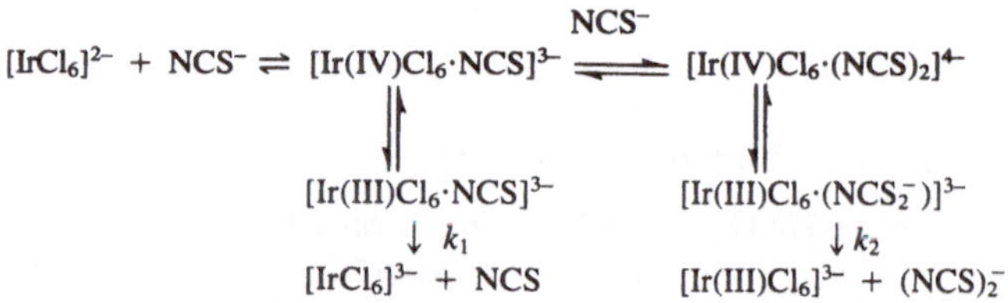

$$[IrCl_6]^{2-} + NCS^- \rightleftharpoons [Ir(IV)Cl_6 \cdot NCS]^{3-} \xrightleftharpoons{\ NCS^-\ } [Ir(IV)Cl_6 \cdot (NCS)_2]^{4-}$$

$$\downarrow \qquad\qquad\qquad\qquad\qquad \downarrow$$

$$[Ir(III)Cl_6 \cdot NCS]^{3-} \qquad\qquad [Ir(III)Cl_6 \cdot (NCS_2^-)]^{3-}$$

$$\downarrow k_1 \qquad\qquad\qquad\qquad \downarrow k_2$$

$$[IrCl_6]^{3-} + NCS \qquad\qquad [Ir(III)Cl_6]^{3-} + (NCS)_2^-$$

Using a linear free energy relationship for the k_1 pathway, the one-electron potential for the NCS/NCS^- couple has been estimated at 1.66 V. The corresponding value for $E^\circ(NCS)_2^-$ has also been evaluated at 1.29 V. The data are combined with those *(vide infra)* for iodide in an extensive examination of the self-exchange parameters for these reductants. The oxidation of NCS^- by $[Fe(CN)_6]^{3-}$ has also been studied.[26]

2.2.4. Reactions of Dithionite and Thiosulfate

The kinetics of reduction of $[Fe(CN)_6]^{3-}$ by dithionite ion, $S_2O_4^{2-}$, have been investigated.[27] Under conditions of the stoichiometric reaction ratio $[Fe(III)]:[S_2O_4^{2-}] = 2:1$ or where $S_2O_4^{2-}$ is in only slight excess (three- to ninefold), the rate-determining process is the dissociation of the reductant, with a subsequent rapid reduction of the Fe(III) complex by SO_2^-. The rate law may be written as rate $= 2k_1[S_2O_4^{2-}]$, where $k_1 = 1.7 \pm 0.3$ s^{-1} (pH $= 6.8$), $\Delta H^{\ddagger} = 11.2 \pm 0.5$ kcal mol^{-1}, and $\Delta S^{\ddagger} = -20 \pm 4$ cal K^{-1} mol^{-1}. With $S_2O_4^{2-}$ present at concentrations greater than 30:1 excess, parallel reductions of $[Fe(CN)_6]^{3-}$ by SO_2^- and $S_2O_4^{2-}$ are observed

$$R = \{k_2 K_1^{1/2} [S_2O_4^{2-}]^{1/2} + k_3[S_2O_4^{2-}]\} [Fe(CN)_6^{3-}] \tag{10}$$

$$S_2O_4^{2-} \underset{k_{-1}}{\overset{k_1}{\rightleftharpoons}} 2SO_2^{-}, \qquad K_1 = k_1/k_{-1} \tag{11}$$

with $k_2 = 2(\pm 0.6) \times 10^8 \ M^{-1} \ s^{-1}$ and $k_3 = 1(\pm 0.4) \times 10^5 \ M^{-1} \ s^{-1}$.

In a separate study[28] of the reduction of several metal complexes of Fe(III), Co(III), and Mn(III) to the corresponding bivalent species, the two-component rate law is observed [equations (12) and (13)].

$$M(III) + S_2O_4^{2-} \overset{k_\alpha}{\rightarrow} M(II) + S(IV) \tag{12}$$

$$M(III) + SO_2^{-} \overset{k_\beta}{\rightarrow} M(II) + S(IV) \tag{13}$$

Of eleven cobalt(III) complexes studied, eight are reduced solely by SO_2^{-} but for the others this pathway is swamped by the direct $S_2O_4^{2-}$ step. $[Co(terpy)_2]^{3+}$, $[Fe(edta)]^-$, and $[Mn(cydta)]^-$ are rapidly reduced by $S_2O_4^{2-}$. Differences in reactivity are observed in the reactions of cobalt(III) pentammine complexes.[29] The pyridine and ammonia species show only the k_β term which is assigned to an outer-sphere process. For azido, sulfato, trichloroacetato, and benzoate complexes, however, a bridging mechanism is proposed. Using a comparison of the rates with those for the corresponding reaction of $[Ru(NH_3)_6]^{2+}$, attempts have been made to determine the oxidant charge.

Radical intermediates have also been invoked in the mechanism of reduction of thiosulfate by Mn(VII).[30] An outer-sphere activated complex is proposed in the one-electron transfer process leading to $S_2O_3^-$ radical formation. Further rapid oxidation of the intermediate by MnO_4^- yields SO_4^{2-}. The reaction of $S_2O_3^{2-}$ with $[Mo(CN)_8]^{3-}$ is first order with respect to each reagent and independent of hydrogen ion concentration.[31] Catalysis by metal ions is observed, however, indicative of a bridging role in the activated complex.

2.2.5. Reactions of Iodide

Two studies[25,32] have been published of the reaction with $[IrCl_6]^{2-}$ which show some measure of agreement and confirm findings reported previously.[33] The reaction scheme is identical to that already described (Scheme 1) for the thiocyanate reaction. In the investigation by Adegite,[32] however, the range of iodide concentrations used was probably not sufficiently large to provide information on the $[I^-]^2$ pathway. The data have been examined in terms of the Marcus theory and a value[32] for the self-exchange rate constant (k_{11}) between $I^.$ and I_2^- has been estimated as $7 \times 10^7 \ M^{-1} \ s^{-1}$. This value although in agreement with that of $\sim 10^7\ M^{-1}\ s^{-1}$ obtained previously,[33] is probably too low, since estimates[25] of k_{11}, which lies near the diffusion-controlled limit, suggest a lower limit of $1.4 \times 10^9\ M^{-1}\ s^{-1}$ to be more accurate. An analysis of a linear free energy plot involving several charge types is consistent with a rate-limiting dissociative diffusion of the successor complex $[Ir(III)Cl_6^{3-} \cdot I^\cdot]$ rather than the electron transfer process itself. A revised value of $E^\circ(I^\cdot/I^-) = +1.33$ V has been derived from these data. In the overall reaction

$$Ir(IV) + 2I^- \underset{k_{-2}}{\overset{k_2}{\rightleftharpoons}} Ir(III) + I_2^- \tag{14}$$

k_{-2} is significantly less than diffusion controlled, a finding not unexpected in view of the significant bond lengthening attendant on reduction of I_2^- to $2I^-$. Results similar to those for the $[IrCl_6]^{2-}$ are derived for the reactions of $[IrBr_6]^{2-}$ where the lower rate constant for the k_1 pathway (28 vs. 409 $M^{-1}\ s^{-1}$) reflects the lower redox potential of the bromo complex.[25]

That the two-term expression for the rate of oxidation of NCS$^-$ is found more frequently for that substrate than for reactions involving I$^-$ is in part ascribed to the greater tendency to form X_2^- radical ions. Terms in both $[I^-]$ and $[I^-]^2$ are found, however, in the reactions with polypyridine iron(III) complexes: R = $\{k_a + k_b[I^-]\}[M(III)][I^-]$. The formation of a precursor complex $[Fe(LL)_3 \cdot I]^{2+}$ is assumed, possibly via overlap of the π orbitals on the I$^-$ with the π^* orbitals on the polypyridine ligands, although electrostatic interactions producing a simple ion pair may be preferred. Values of k_b/k_a lie in the range $(1–8) \times 10^2\ M^{-1}$ similar to those derived previously for reactions of other outer-sphere reagents.[33]

While nickel(III) cyclam, $[NiL(OH_2)_2]^{3+}$, is known to form complexes with halide ions (Cl$^-$, Br$^-$), the redox potential (0.96 V) is sufficiently high that iodide is oxidized quantitatively.[34] The rate law, however, shows only a first-order dependence in the iodide concentration. Although not conclusive, the preferred mechanism for reaction is via an outer-sphere process when rates are compared with those for halide complex formation.

2.2.6. Other Reductants

The oxidation of tellurium(IV) by cerium(IV) in perchloric acid media has been studied in the range $[H^+] = 2–5\ M$.[35] Although there is no variation with ionic strength, the reaction is retarded by one product, Ce(III), and accelerated by the other, Te(VI). Spectrophotometric evidence for a 1:1 complex between Ce(IV) and Te(IV) is confirmed by the first-order kinetic behavior of the oxidant but the order with respect to Te(IV) is fractional.

A unified view has been published of the Marcus electron transfer and Mullikan charge transfer theories as applied in organometallic chemistry.[36] In this study, electron transfer rate constants for reaction of a variety of homoleptic organometallic species such as tetraalkyl tin, tetraalkyl lead, and di-alkyl mercury with $[Fe(phen)_3]^{3+}$, $[IrCl_6]^{2-}$, and tetracyanoethylene have been compared. The iron(III) complexes cleave the organometals according to the general reaction mechanism

$$SnR_4 + [Fe(LL)_3]^{3+} \xrightarrow{k_1} SnR_4^{+\cdot} + [Fe(LL)_3]^{2+} \tag{15}$$

$$SnR_4^+ \xrightarrow{\text{fast}} SnR_3^+ + R^{\cdot} \tag{16}$$

A linear free energy relationship is exhibited in equation (17)

$$\log k_1 = 8.5E° + \text{const} \tag{17}$$

with the slope being that predicted by the Marcus theory. Also for any individual iron(III) complex, $\log k_1$ is correlated linearly with the ionization potential of the reductant. This relationship holds for a range of 10^9 in electron transfer rates and since included are those for hindered alkyl tins there is an indication that steric effects in the alkylmetal play no major role in these outer-sphere processes. Although the reaction of $[IrCl_6]^{2-}$ obeys basically the same scheme as that shown above, there is no Marcus correlation, the rate constants being dependent upon the steric properties of the ligands. It is suggested that in this case there is a contribution from an inner-sphere electron transfer to the $[IrCl_6]^{2-}$. The Mulliken theory of charge transfer in TCNE complexes has been used to evaluate the steric effects. Differences in the charge transfer transition energy in the complexes relative to that of a reference alkyl metal have been assigned to the contribution of steric hindrance to the interaction energy associated with ion pair formation in the successor complex for the inner-sphere process. In a similar study[37] the group IV metal hydrides HER_3 (E = Si,Ge,Sn; R = alkyl, phenyl) react rapidly with $[Fe(LL)_3]^{3+}$. The rates decrease in the order Sn > Ge > Si and analysis of ligand effects is consistent with electron release into the E–H σ bond as an important factor in enhancing element–hydride reactivity. Selective cleavage of only the hydrido–element bond takes place with no significant deuterium isotope effect, the ion $HER_3^{+\cdot}$ being postulated as an intermediate. The outer-sphere process is again described by Marcus cross-correlation predictions.

2.3. *Oxidation of Organic Substrates by Metal Ion Complexes*

In this section, owing to the large variety of ligands available, reactions are classified with respect to the metal ion oxidant rather than the reductant.

Several articles have been published which deal with aspects of this subject. Free radical reactions of transition metal systems have been discussed with reference to possible spin pairing with the metal center.[38] The importance of substituent effects in the oxidation–reduction properties of metal ion complexes with macrocyclic ligands has been investigated.[39] These systems are significant as models for biological process since the $E°$ value varies significantly with ring substitution while the coordination geometry of the metal ion remains essentially constant. This aspect is developed further in a review of different oxidation states of nickel[40] where the ability of the ligand to stabilize the higher states is described.

2.3.1. Chromium(VI)

Studies of alcohols and organic acids continue to be of interest. Three-electron oxidations, which are a feature of several systems, have been shown to occur in the oxidation of mandelic acid.[41] The rate law is of the form

$$R = \frac{(k_1 + k_2[\text{H}^+] + k_3[\text{MA}])[\text{Cr(VI)}]\,[\text{MA}]}{(1 + [\text{H}^+]/K_a)} \tag{18}$$

with all pathways exhibiting a three-electron exchange. In the case of the reaction which is first order in substrate, the oxidation is of the form

$$2\text{PhCH(OH)CO}_2\text{H} \xrightarrow{2\text{Cr(VI)}} \text{PhCHO} + \text{PhCO}_2\text{H} + 2\text{CO}_2 + 2\text{Cr(III)} \tag{19}$$

being the first examples of an intramolecular process where both a C–H and a C–C bond on the same carbon atom are broken simultaneously in the rate-determining step.

In the co-oxidation of 2-hydroxy-2-methylbutyric acid (HMBA) and 2-propanol (i-PrOH), which is a two-stage reaction, the participation of both Cr(VI) and Cr(V) as oxidants has been determined.[42] In the initial reaction

$$\text{HMBA} + i\text{-PrOH} + 2\text{Cr(VI)} \rightarrow \text{MeCH}_2\text{COMe}$$
$$+ \text{CO}_2 + \text{MeCOMe} + \text{Cr(V)} + \text{Cr(III)} \tag{20}$$

kinetic, product, and isotope studies are consistent with a 3e mechanism involving a ternary complex which undergoes oxidative decomposition (Scheme 2). Further

Scheme 2

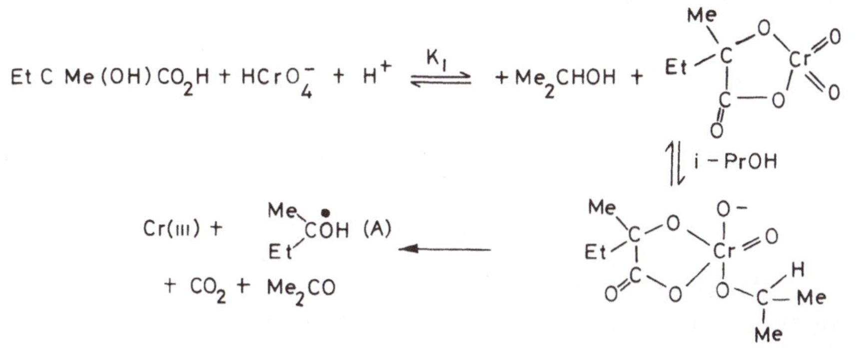

reaction of (A) or possibly CO_2^- with Cr(VI) yields Cr(V). The much slower second stage of the overall process, which may be written as

$$Cr(V) + HMBA \rightarrow EtCOMe + CO_2 + Cr(III) \tag{21}$$

appears to take place via a $(HMBA)_3$-Cr(V) species since there is exclusive oxidation of the hydroxyacid even in the presence of a 20-fold excess of i-PrOH. Cr(V) thus exhibits selectivity in reactivity toward these substrates in contrast to the Cr(IV) reactions. The unusually high deuterium isotope effects reported previously[43] for reactions of glycolic acid and the co-oxidation of glycolic acid with i-PrOH appear to be in error. Ternary complexes have been invoked in the reactions of Cr(VI) with alcohols in the presence of oxalic acid,[44] where the rate-determining decomposition of the Cr(VI) species leads directly to Cr(III). The process is inhibited by edta and other complexing agents.

The kinetics of the oxidation of malonic acid in the presence and absence of Mn(II) have been reinvestigated.[45] The rate law for the uncatalyzed reaction differs from that when Mn(II) is present and where the principal pathway is via reaction of a Mn(II)–malonic acid complex with $HCrO_4^-$.

The reaction of alkyl- and halo-substituted phenols with CrO_2Cl_2 results mainly in the formation of quinones and diphenoquinones.[45] Phenoxyl radicals are involved as intermediates. The mechanism of oxidation of α-hydroxycarboxylic acids by pyridinium chlorochromate involves a rate-limiting hydride transfer.[46] In the reaction of $HOCD_2CO_2H$, a kinetic isotope effect $(k_H/k_D) = 5.80$ has been determined. Spectroscopic evidence for Cr(IV) and Cr(V) species has been obtained in the oxidation of alkylaromatics by chromyl acetate in acetic anhydride.[47] Stopped-flow and esr studies show two stages (equations 22, 23) in the reactions of RCH_2Ph, with both rates being decreased on deuteration at the benzylic position, indicating that both processes involve oxidation of the alkylaromatic and cleavage of the C–H bond in the rate-limiting step

$$PhCH_2R + CrO_2(OAc)_2 \xrightarrow{k_1} PhCHR + Cr(V) \tag{22}$$

$$PhCH_2R + Cr(V) \xrightarrow{k_2} PhCHR + Cr(IV) \tag{23}$$

leading to the formation of a relatively stable Cr(IV) complex. The mechanism of oxidation of acetylacetone, dithizone, and other substrates in organic solvents (DMF, HMPA, DMSO) has been investigated using esr techniques.[48] There is evidence for a photochemical reduction to Cr(V) which is stabilized as a complex with the reacting ligands.

2.3.2. Vanadium(V)

The oxidation of 2-hydroxy-2-methyl propionic acid, $Me_2C(OH)CO_2H$, (hmpa) has been investigated[48] in $HClO_4$ under conditions of a large excess of reductant. Kinetic reaction orders [unity for V(V), 0–1 for ligand] are consistent with intermediate complex formation. From an analysis of the hydrogen ion dependence, the two pathways

$$VO_2^+ \cdot hmpa \xrightarrow{k_1} V(IV) + Me_2\dot{C}(OH) + CO_2 + H^+ \tag{24}$$

$$V(OH)_2^{3+} \cdot hmpa \xrightarrow{k_2} V(IV) + Me_2\dot{C}(OH) + CO_2 + 3H^+ \tag{25}$$

contribute to the rate. Thermodynamic parameters have been compared with those for reaction of other substrates by this ion. Overall entropies of activation of all substrates considered are negative and in this regard V(V) differs in its reactivity toward reductants of this type when compared with Mn(III) and Ce(IV), which show positive $\Delta S^{\ddagger}$ values. Complexes deriving from ligand interaction with proton-related vanadium(V) species have also been invoked in the reaction with glyoxal in perchlorate and nitrate media.[49] In the oxidation of aromatic azo compounds,[50] there is evidence for edta catalysis via complexing to the VO_2^+ ion. The reaction rate is diminished at increased hydrogen ion concentrations, probably owing to the dissociation of the catalytic species.

2.3.3. Manganese(III) and Manganese(VII)

A kinetic study of the oxidation of *N*-alkylphenothiazines (ptz) by manganese(III) has been made in a continuing series of investigations of the cation

$$\text{Mn (III)} + \quad\quad\quad\quad\quad\quad \longrightarrow \text{Mn(II)} + \quad\quad\quad\quad\quad\quad \tag{26}$$

radicals derived from these substrates [equation (26)]. Data derived[51] over the range $[\text{H}^+] = 0.2–1.50\ M$ are consistent with a principal reaction pathway:

$$\text{Mn}^{3+} + \text{ptz} \xrightarrow{k_1} \text{Mn(II)} + \text{ptz}^{+\cdot} \tag{27}$$

with no evidence for any contribution from the oxidation via MnOH^{2+}. Similar negligible contributions from hydrolyzed species have been reported for Fe(III), Co(III), and Np(VI), and previously only in the case of tetramethylhydrazine has the MnOH^{2+} pathway been observed to be noncontributing. It is suggested that when transferable hydrogen atoms are present, Mn(OH)^{2+} is the preferred reactant, but that in other cases a simple electron transfer to Mn^{3+} is operative. These data are of interest in that the rate constant k_1 is related to the reduction potentials of the ligands, indicating an outer-sphere process, and a lower limit of the $\text{Mn}^{3+/2+}$ self-exchange rate $(1 \times 10^{-10}\ M^{-1}\ \text{s}^{-1})$ may be derived. A similar mechanism of reaction has also been postulated in the reaction of Mn^{3+} with glycine.[52] From a product analysis and kinetic studies in 90% acetic acid, it has been suggested that the mechanism of oxidation of alkyl acetates by [Mn(OAc)₃] is via alkyl–oxygen bond rupture.[53]

The reduction of manganese(VII) by oxalate in acid media has been investigated.[54] In the presence of Mn(II), manganese(III)–oxalato complexes are formed. Formation and decomposition rates involving Mn(ox)_2^- have been derived. Several studies of the oxidation of unsaturated substrates by MnO_4^- have been published.[55,56] Unfortunately the original articles are in Hungarian and reliance has been placed on abstracts. The reactions of methylmaleic and methylfumaric acids, (H_2A) have been investigated at pH < 5. Parallel pathways involving H_2A, HA^-, and A^{2-} are observed, the rate-determining step being the *cis* attack by MnO_4^- on the double bond, resulting in an undetected cyclic Mn(V) ester. The presence of pyrophosphate in the reaction medium for oxidation of diethylmaleate or fumarate stops disproportionation of the Mn(III) product. Combination of stopped-flow and chemical quenching techniques provides data on the short-lived intermediate in the reduction of Mn(VII) by *trans*-cinnamic acid. It is suggested that a soluble Mn(IV) species is formed. Although not directly related, a study has been made of the $\text{MnO}_4^-/[\text{Ru(CN)}_6]^{4-}$ reaction[57] with an extensive determination of hydrogen ion dependence. Comparison is made with other outer-sphere MnO_4^-–reductant reactions.

2.3.4. Iron(III)

Kinetic and spectroscopic studies have been made[58] of the one equivalent metal ion oxidation of phenoxazine and phenothiazine, Ar_2NH, by iron(III) in acetonitrile solutions. The data have been analyzed kinetically using the sequence

$$Ar_2NH \xrightarrow{k_1} Ar_2NH^+ \xrightarrow{k_2} Ar_2N^+ \rightarrow product \tag{28}$$

In the presence of excess oxidant, the reaction of phenothiazine (ptz → ptz$^+$) is first order in substrate but greater than first order with respect to iron(III). Similar behavior is observed with phenoxazine, although in this system at higher [Fe(III)] (100:1 excess) the overall second-order rate constant was measured. Use of a rapid-scanning spectrometer has enabled measurements to be made of the uv–visible spectra of all the intermediates involved. In a separate study of the oxidation of phenothiazine,[59] there is evidence for reaction taking place through only the Fe^{3+} ion, with only a minor contribution if any deriving from $Fe(OH)^{2+}$ reaction. The data have been treated using a Marcus cross-correlation.

The formation of intermediates has been invoked in the oxidation benzoyl hydrazines, (L), by iron(III)–phenanthroline complexes.[60] The principal reaction is via a species of the type [Fe(III)(phen)$_2$L] with rate retardation observed when SO_4^{2-} is present.

There has been considerable recent interest in the reductions of $[Fe(CN)_6]^{3-}$. The electron exchange with *N*-propyl-1,4-dihydronicotinamide[61] is catalyzed by alkali metal ions. The increase in reaction rate is attributed to the polarizability of M^+ and the observed linear free energy relationship is discussed. An outer-sphere mechanism is postulated in the oxidation of phenothiazines.[59] A free radical mechanism involving the alcohol anion is invoked in the reaction of 1- and 2-propanol in aqueous alkaline media,[62] the kinetic order being unity for $[Fe(CN)_6]^{3-}$, OH^-, and alcohol concentrations. Catalysis by metal ions has also been observed in the presence of copper(II) and ruthenium(III) complexes. In the oxidation of α-hydroxypropionic acid in alkaline media,[63] a Cu(II)–ligand complex is formed which is oxidized slowly to a copper(III) species. Alkaline ferricyanide oxidizes butanol, the process being catalyzed by chlororuthenium complexes.[64] The rate law is consistent with oxidation of the alcohol by the Ru(III) followed by reoxidation of the catalyst by $[Fe(CN)_6]^{3-}$. The rate law is of the form:

$$-d[Fe(III)]/dt = [Ru(III)]\,(2k_1k_2[BuOH]\,[Fe(III)])/(2k_1[BuOH] + k_2[Fe(III)]) \tag{29}$$

The observed nonlinear dependence of the rate on [BuOH] is ascribed to a rate-determining alcohol-independent reoxidation of catalyst.

2.3.5. Thallium(III)

In the oxidation of acetamide, formamide, and N-methylformamide in aqueous $HClO_4$ solutions, one mole of Tl(III) reacts per mole of amide.[65] The slow reactions are complicated by a hydrolysis of the amides and mechanisms are proposed involving oxidation of ammonia and formic acid produced by the hydrolysis. There is no evidence for any dependence on T1(I) and free radical intermediates are postulated. Two consecutive one-electron steps are invoked in the reaction with acrolein (CH_2=CHCHO,Ac) in sulfate media.[66] A zero-order process is observed at high Ac:Tl(III) ratios, and is ascribed to enolization of the substrate. The rate-determining step involves the complexing with the unsaturated center rather than reaction at the carbonyl group. There is strong kinetic evidence for the formation of a nitroso intermediate in the oxidation of benzaldehyde- and acetophenone-oximes.[67] The rate of formation of the complex follows the rate law in:

$$-d[\text{oxime}]/dt = k_\alpha[\text{oxime}]\,[\text{Tl(III)}]/(1 + K[\text{Tl(III)}]) \qquad (30)$$

The decomposition of the intermediate is retarded by Tl(III) and by the presence of Cl^- ions.

2.3.6. Iridium(IV)

In an investigation of the oxidation of thiols, RSH, the kinetics of reactions of 2-thiopyrimidine and 2-thiouracil with $[IrCl_6]^{2-}$ and $[IrBr_6]^{2-}$ have been measured.[68]

$$2[IrCl_6]^{2-} + 2TP \rightarrow 2[IrCl_6]^{3-} + (TP)_2 + 2H^+ \qquad (31)$$

The rates are first order with respect to each reactant, with TU reacting about an order of magnitude more slowly than TP. The significant increase in rate at pH 2–5 is attributed to dissociation of the thiol. Under these conditions, two reactions occur:

$$RSH \rightleftharpoons RS^- + H^+ \qquad (32)$$

$$RSH + [IrX_6]^{2-} \xrightarrow{k_a} RS^{\bullet} + [IrX_6]^{3-} + H^+ \qquad (33)$$

$$RS^- + [IrX_6]^{2-} \xrightarrow{k_b} RS^{\bullet} + [IrX_6]^{3-} \qquad (34)$$

with rapid subsequent dimerization of the thiyl radicals yielding the disulfide.

In the oxidation of cyclohexanone in acidic media,[69] solvent and substrate deuterium isotope effects have been investigated. It is possible to distinguish between direct attack on the ketone and on the enol form. In these and similar systems, the apparent existence of a primary isotope effect may be misleading if the reagents are not at isotopic equilibrium. A change in the enol content of the solution by a factor of ~7 results from isotope effects on the enolisation. In D_2O, there is also an observed retardation of the redox step. The oxidations of hydroquinone and p-toluohydroquinone (H_2Q) by $[IrCl_6]^{2-}$, $[IrBr_6]^{2-}$, $[IrCl_5(OH_2)]^-$, and $[IrCl_4(OH_2)_2]$ follow a second-order rate law,[70] $R = k[H_2Q][Ir(IV)]$, the outer-sphere processes being described in terms of the Marcus cross-correlation. Iridium(IV) complexes of this type also catalyze alkane oxidation by chromium(VI) in aqueous media.[71]

2.3.7. Cerium(IV)

There is still considerable interest in the mode of oxidation of organic substrates by this reagent. Using a scanning spectrometer, the series of rapid color changes observed in the reaction with phenothiazine have been monitored.[58] With cerium(IV) in excess, a transient at 419 nm was observed which disappeared slowly, whereas with substrate in excess, several absorptions are identified with the formation of a dark green solution considered tentatively as a binuclear cation. The intermediates have been identified as the radical cation and the phenothiazinium cation [equation (35)]. The first reaction is too fast for

$$\text{(35)}$$

measurement by stopped-flow methods ($k_1 > 10^6\,M^{-1}\,s^{-1}$) but the rate of oxidation of the radical cation (k_2) has been measured. The reactions proceed with a zero-order dependence on substrate and first order in oxidant. In the case of phenoxazine, however, the rate was first order in both reagents ($k_2 = 4 \times 10^4\,M^{-1}\,s^{-1}$). In order to account for the reaction of the radical cation, it is suggested that cerium(IV) exists in two forms in acetonitrile solution, one of which is active in the reaction with the radical.

Metal-complexed ligand radicals are formed in the oxidation of iron(II) diimine complexes with cerium(IV) in sulfate media.[72,73] The reaction with $tris$-[2-pyridinal-α-methyl(methylimine)]iron(II) ([Fe(pmm)$_3$]$^{2+}$, studied at lower acid concentrations ($< 4\,M\,H_2SO_4$) is a complex process involving formation of Fe(II)– and Fe(III)–ligand oxidized species, a total of 10–11 mol of oxidant being required for reaction with each iron(II) complex. The [Fe(pmm)$_3$]$^{3+}$ species undergoes intramolecular reductions to Fe(II) and a radical which may be further oxidized by $[Fe(pmm)_3]^{3+}$ to a monoaldehyde product. Hydrogen ion dependences in the reaction of tris[glyoxal bis(methylimine)]iron(II) have been investigated.[73] The mechanism of oxidation by cerium(IV) involves not only direct formation of Fe(III) species but also intramolecular electron transfer processes of the type described above, some of which are assisted by reversible nucleophilic solvent attack on the ligand.

The nature of the mechanism of side-chain oxidation of alkyl aromatic compounds by one-electron reagents has been discussed.[74] The reaction sequence:

$$ArCH_3 + Ce(IV) \underset{k_{-1}}{\overset{k_1}{\rightleftharpoons}} ArCH_3^{+\cdot} + Ce(III) \tag{36}$$

$$ArCH_3^{+\cdot} \overset{k_2}{\rightarrow} H^+ + ArCH_2^{\cdot} \overset{Ce(IV)}{\rightarrow} \text{products} + Ce(III) \tag{37}$$

has been confirmed for the oxidation of 1,2,3-trimethyl-5-*tert*-butylbenzene by cerium(IV) in acetonitrile. Comparison with other deuterated substrates shows that the reaction proceeds without a kinetic isotope effect and there is evidence in some systems for a change-over from the rate-determining electron transfer to a rate-limiting proton transfer process.

Several intermediate complexes have been observed in the oxidation of sulfanilic acid,[75] each reacting further with cerium(IV) in a consecutive sequence. The final products are formic acid and sulfate ions. Amperometric methods have been employed in the investigation of the rates of oxidation of phenols in perchlorate media.[76] A number of papers on the reactions of alcohols and α-hydroxy-carboxylic acids have been published in the *Indian Journal of Chemistry* over the period of review, several involving transition metal [Ir(III), Ru(III)] catalyzed processes.

2.3.8. Cobalt(III)

The reaction with formic acid in equation (38)

$$2Co(III) + HCOOH \rightarrow 2Co(II) + CO_2 + 2H^+ \tag{38}$$

proceeds via the formation of intermediate complexes.[77] Although there is no stopped-flow evidence, plots of k_{obs}^{-1} against $[HCOOH]^{-1}$ are linear. The rate decreases with increasing $[H^+]$ and over the range $[H^+] = 1.0–5.0\,M$ ($HClO_4$), the oxidation proceeds via two rate-determining processes, involving the complexes Co(III)(HCOOH) and Co(OH⁻)(HCOO⁻) which undergo internal redox reactions leading to CO_2^-, which is oxidized rapidly to CO_2. An alternate formulation of the proton dissociated species is $Co(III)(OH^-)_2(HCOOH)$, which is kinetically indistinguishable.

An outer-sphere process is postulated in the oxidation of *N*-alkylphenothiazines.[78] In this reactions there is no evidence for an acid-dependent pathway. This finding is unusual for oxidations involving cobalt(III), especially those of organic substrates. In this case, the hydroxo group cannot act as a bridging group which catalyzes the electron transfer. The rate constants observed are in excess of those accepted for water exchange on cobalt(III) and a Marcus cross-correlation was attempted. Although the data show a free energy relationship, the calculated rate constants differ by about five orders of magnitude from the observed values, a situation encountered previously in these reactions. In contrast to these substrates, the oxidation of propanol and propane-diols[79] proceeds via an inner-sphere mechanism.

Oxidations by the larger outer-sphere reagent 12-tungstocobaltoate(III) ($[Co(III)W_{12}O_{40}]^{5-}$) are of interest and studies on the reaction with methyl-aromatic compounds have been reported.[80] These alkyl aromatics are acetoxylated in the α position in acetic acid media. The mechanism proposed is similar to that for the reactions of cerium(IV)[74] *(vide supra)* and a study of substituent effects shows a good relationship between the rates (relative to $R = H$) and $E°$ for the oxidation of the alkylaromatic substrate. The large deuterium isotope effect ($k_H/k_D \simeq 6$) is consistent with a rate-determining process involving concerted electron–proton transfer from the α-C–H bond to an oxygen of the heteropolyanion. The oxidation of styrene in acetic acid by $[Co(OAc)_3]$ yields different products depending on the presence or absence of any moisture, a radical pathway being postulated in the case of the latter.[81]

2.3.9. *Bromate*

The nature of chemical oscillations, especially those involving bromate ion, are currently of considerable interest. Many of these processes involve oxidation of malonic acid, and the conditions for the onset of chemical oscillation have been described,[82] following accumulation of bromomalonic acid (BrMA). The extent of formation which is critical is dependent on $[BrO_3^-]_0$ and on the

$$\text{(a)} \quad d[\text{BrMA}]/dt = k_1[\text{malonic}]^{1.3} \, [\text{BrO}_3^-]^{0.5}[\text{Ce(IV)}]^{0.7} \tag{39}$$

$$\text{(b)} \quad d[\text{BrMA}]/dt = k_1[\text{malonic}]^{0.5} \, [\text{BrO}_3^-] \, [\text{Mn(II)}]^{0.8} \tag{40}$$

metal ion present [equations (39) and (40)]. In acid media (HNO_3), the bromination of malonic acid from reaction with Br_2 (a decomposition product of bromic acid) becomes a competitive reaction with that of HOBr + malonic acid.

Further details of the mechanism have been described.[83,84] In the simplest outline, the dynamic features of the system may be defined by the kinetics of the oxidation of Ce(III) by BrO_3^-, while reduction of cerium(IV) by BrMA provides a linear feedback. Studies on the manganese(II)-catalyzed reaction in orthophosphoric acid have been published.[85] A kinetic study has also been made of the bromate–malonic-acid–ascorbic acid system.[86] A very slow reaction between ascorbic acid and BrO_3^- ($k_2 < 3 \times 10^{-2}\ M^{-1}\ s^{-1}$) is identified and a rapid process between $Br^\cdot$ and the same substrates. The resultant variations of ascorbic acid concentration show maxima and minima with time.

2.3.10. Other Oxidizing Agents

The reaction between Np(VI) and N-alkylphenothiazines is first order with respect to each reagent, and independent of $[H^+]$.[87] The radical cations produced have been identified spectroscopically. The linear relationship between $\Delta G°$ and the free energy of activation $\Delta G^{\ddagger}$ has been interpreted in terms of a Marcus theory approach. The self-exchange rate for ptz/ptz^+ was in accord with previous data,[78] but the calculated rate parameters are $\sim 10^2$ greater than the experimental values, despite an analysis of contributions to enthalpies and entropies of activation. It is not certain that these systems conform fully with the criteria for calculations of this type. Further studies on the isostructural Pu(VI) and Am(VI) as oxidants may prove helpful in delineating whether nonadiabaticity is a possible source of at least some of these discrepancies. The reduction of Np(V) $\rightarrow$ Np(IV) by ascorbic acid in the presence of Fe(II) or Fe(III) is independent of organic reductant.[88] The rate-limiting step is the reduction of Np(V) by Fe(II), the iron(III) produced reacting rapidly with the ascorbic acid. The reduction of NpO_2^+ by e^- aq, studied by pulse electrolysis,[89] leads to a partially hydrolyzed aquoneptunium(IV) species which rearranges to the fully hydrolyzed product. The reaction of PuO_2^+ by ascorbic acid leads to a Pu(III) product,[90] although parallel reactions deriving from disproportion of the Pu(V) must also be taken into account.

Studies continue into systems involving Ag(II) as an oxidant. The reactions with cyclohexanol, pentan-2-ol, and benzyl alcohol proceed via two pathways[91]:

$$Ag^{2+} + ROH \xrightarrow{k_1} Ag^+ + RO^\cdot + H^+ \tag{41}$$

$$AgOH^+ + ROH \xrightarrow{k_2} Ag^+ + RO^\cdot \tag{42}$$

followed by a rapid reaction of the radical produced. In the case of benzyl alcohol, however, $k_2 \simeq 0$, unlike the reactions for most organic substrates. In this system it is considered that the metal ion reactant attacks the aromatic ring.

Evidence has been found for complex formation between the Ag(II) and substrates in the oxidation of glycine and other amino acids.[92] This process represents the first of two steps observed, the second being the electron transfer process. Complexation rates are in the range 10^6–10^8 M^{-1} s^{-1} and are dependent on electronic factors from substituent groups on the reductants. There is a markedly slower redox process which is again dependent on structural effects within the organic substrate. The kinetics of oxidation of aliphatic aldehydes by ditelluratoargentate(III) involve a rate-determining two-electron transfer from the aldehyde to the Ag(III).[93]

Evidence has been presented for an atom transfer redox process in the reduction of Au(III) by dialkyl sulfides[94] in aqueous methanol. The two-stage reaction involves initially a substitution equilibrium which is strongly dependent on the bulkiness of the entering sulfide. A second mole of disulfide is required in the subsequent redox reaction leading to Au(I) and sulfoxide. The attack occurs via a chlorine atom of the $AuCl^-$ reagent which is transferred to the sulfur, leaving an electron pair at the metal center. The reaction is markedly solvent dependent.

The oxidation of tellurium(IV) by chloramine T is thought to proceed via a reaction with Cl_2 as the reacting species[95] with the product toluene-*p*-sulfonamide acting as an inhibitor to the process. Although $[H^+]$ does not appear to affect the rate of reduction of Te(IV) by ascorbic acid,[96] which is first order with respect to each reagent, chloride ions from the HCl medium are involved in the overall reaction.

2.4. Oxidation of Metal Ion Complexes

2.4.1. Iron(II)

Spectrophotometric and potentiometric techniques have been used to study the Fe(II)–NO_3^- reaction at high acidity [0.6–2.0 M (H^+)]. The overall reaction[97]

$$3Fe^{2+} + 4H^+ + NO_3^- \rightarrow 3Fe^{3+} + 2H_2O + NO \tag{43}$$

may be monitored by investigating absorbance changes resulting from the formation of $FeNO^{2+}$. Seven principal reactions are involved including the very slow reaction of Fe^{2+} with NO_3^-, but also those of ferrous ion with NO_2^- and HNO_2, which is assumed present in small concentrations in the reacting solutions. A computer simulation of the reaction profile has been constructed. Autocatalytic "shutdown" of the reaction occurs when $[HNO_2]$ becomes large enough that nitrous acid is consumed more rapidly by second-order disproportionation than by reaction with iron(II).

Complex side reactions have been eliminated in the corresponding reaction with iodate[98] by the presence of $CH_2:CHCH_2OH$, which reacts very rapidly with the iodine(I) formed. The overall reaction is given in equation (44),

$$4Fe(II) + IO_3^- \rightarrow 4Fe(III) + I(I) \tag{44}$$

with a kinetic rate law of the form in equation (45).

$$-d[Fe(II)]/dt = 4k_1(k_2 + k_3[Fe(II)]) [Fe(II)] [IO_3^-]/(k_{-1} + k_2 + k_3[Fe(II)]) \tag{45}$$

Rate-determining steps leading to I(IV) and I(III) are postulated, with subsequent rapid reduction of intermediate iodine species. An induction period followed by oxidation of the $[Fe(phen)_3]^{2+}$ complex is a feature of the reaction of that complex with bromate ions.[99] The rates of the corresponding reactions with Cl_2 and Br_2[100] are not affected by Cl^-, Br^-, or hydrogen ion. The oxidations occur via the one-electron transfer steps and an analysis of the data and those for other metal ion reductants has been made using a Marcus theory approach. The kinetics of the peroxodisulfate oxidation of two $[Fe(II)(\alpha\text{-diimine})_3]$ complexes have been investigated in binary aqueous-solvent mixtures.[101] The rate data have been dissected into initial and transition state energies. Comparisons between the relative contributions to these parameters for redox and substitution reactions remain a topic of interest.

In the presence of excess HNO_3, the rate of oxidation of $[Fe(CN)_6]^{4-}$ by HNO_2 is independent of the initial iron(II) complex concentration.[102] At low $[H^+]$ ($< 0.01\ M$) the rates are first order with respect to $[H^+]$, the reaction involving $[HFe(CN)_6]^{3-}$ and NO^+. Above $0.5\ M\ [H^+]$, the rate is independent of acidity. Trace metal ions (Fe^{2+}, Fe^{3+}, Cu^{2+}) catalyze the oxidation of $[Fe(CN)_6]^{4-}$ by peroxodiphosphate.[103] In the presence of edta the reduction of the copper(II) edta complex by $[HFe(CN)_6]^{3-}$ leads to a Cu(I) edta intermediate which is reoxidized by $H_2P_2O_8^{2-}$.

The kinetics of triplet methylene blue quenching by various iron(II) complexes including $[Fe(OH_2)_6]^{2+}$, $[Fe(CN)_6]^{4-}$, and $[Fe(bipy)_3]^{2+}$ have been studied at different pH's.[104] The reactions are redox controlled and the rates of the reverse electron transfer have also been determined. The properties of the outer-sphere charge transfer band in the ion pair $[Fe(CN)_6^{4-} \cdot PQ^{2+}]$ ($PQ^{2+} = 1,1'\text{-}Me_2\text{-}4,4'\text{-bipy}^{2+}$) have been used[105] to evaluate the rate constant for intermolecular reaction transfer within the ion pair. Use of a thermochemical cycle has provided a value for the rate of the overall reaction $[Fe(CN)_6]^{2-} + PQ^{2+}$ ($k_2 \sim 0.2 \times 10^{-4}\ M^{-1}\ s^{-1}$), which compares well with Marcus theory predictions. It is suggested that data on electron transfer processes may be derived from analysis of charge transfer absorption band characteristics.

2.4.2. Chromium(II)

A highly absorbing blue intermediate is formed in the reaction of chromium(II) with diprotonated 4,4'-bipyridine in aqueous perchlorate media.[106] This species has been identified as the 4,4'-bipyridine radical formed in the first stage of reaction:

$$\tag{46}$$

The first step (k_1) is an outer-sphere process, the second reaction exhibiting a two-term rate law involving a hydrogen-ion-catalyzed pathway. Intermediate green-colored complexes of 1:1 stoichiometry are also formed on mixing Cr(II) with pyrazine, complexation being accompanied by proton binding at the remote nitrogen site on the ring.[107] In the presence of a large excess of pyrazine, there is a biphasic reaction leading to redox decomposition of the 1:1 complex and to the formation of the dihydropyrazine radical cation, characterized by a 69-line epr signal:

$$\text{Cr(II)} + \text{pyz} + \text{H}^+ \rightleftharpoons [\text{Cr(pyzH)}]^{3+} \tag{47}$$

$$[\text{Cr(pyzH)}]^{3+} + \text{pyzH}^+ \underset{k_{-2} - 2}{\overset{k_2}{\rightleftharpoons}} \text{pyzH}_2^+ + [\text{Cr(III)(pyz)}]^{3+} \tag{48}$$

$k_2 \sim 14\ M^{-1}$ s.$^{-1}$ A transient species with a spectrum very closely similar to $[\text{Cr(pyzH)}]^{3+}$ is formed on reaction of Cr(II) with $[(\text{H}_3\text{N})_5\text{Co(pyr)}]^{3+}$. There is evidence for a mechanism involving Cr(II) binding followed by stepwise electron transfer to ligand and then, more slowly, from ligand to Co(III). Evidence is presented in favor of a formulation of the green complex as a Cr(III)–ligand-bound radical ion.

Chromium(III) products have also been identified in the reduction of 1,4-benzoquinone.[108] A species proposed as shown in structure **1** is formed possibly

1

via an electron transfer pathway involving concerted attack of two Cr(II) aquo ions at one of the oxygen atoms of the benzoquinone to form a doubly bridged precursor complex. This species undergoes slow hydrolysis in acid solution with proton attack at the Cr-bound oxygen of the organic substrate leading to the μ-oxo-bis(pentaaquochromium(III)) ion.

In the reaction with haloforms (CHX_3) (X = Cl, Br, or I) studied in H_2O/MeOH solvent 6 mol of Cr(II) are consumed per mole of organic substrate.[109] The reactions are first order in each reagent, independent of [H^+], and decrease in the order $k_I > k_{Br} < k_{Cl}$. An inner-sphere mechanism is postulated, the effect of C–X bond cleavage being considered more important than that of Cr–X bond formation. The stoichiometry of the corresponding reaction with CCl_4 is of reaction of 2 mol of Cr(II) per mole of CCl_4,[110] $CrCl^{2+}$ and $CrCCl_3^{2+}$ being formed in equal amounts. The reaction is again inner sphere. The rate is reduced, however, if AcOH is present in solution.

2.4.3. Vanadium(II)

In the reaction with pyruvic (pyr) and phenylglyoxylic acids (pgl),[111]

$$V^{2+} + pyr \rightleftharpoons [V(pyr)]^{2+} \underset{pyr}{\overset{}{\rightleftharpoons}} [V(pyr)_2]^{2+} \overset{k_2}{\longrightarrow} V^{3+} + pyr + pyr^{\cdot} \quad (49)$$

the first stage corresponds to a rapid equilibrium [equation (49)] involving formation of a bis(pyruvic acid) complex. A second slower reaction leads to reduction of the organic ligand. A 1:1 complex is formed in the phenylglyoxylic system. There is also in the case of $[V(pyr)]^{2+}$ evidence for a competing outer-sphere pathway [equation (50)].

$$2[V(pyr)]^{2+} \overset{slow}{\longrightarrow} 2V^{3+} + \text{dimethyltartaric acid} \quad (50)$$

Autocatalysis is exhibited in the reactions of V(II) and Eu(II) with cobalt(III) complexes containing pyridine ring systems,[112] with ligand radicals involved as the intermediates in the catalytic pathway (Scheme 3).

Scheme 3

$$V^{2+} + \text{Co(III)–ligand} \overset{k_1}{\rightarrow} \text{Co(II)} + \text{ligand} + V^{3+}$$

$$V^{2+} + \text{ligand} \underset{k_{-2}}{\overset{k_2}{\rightleftharpoons}} V^{3+} + [\text{ligand}^{\bullet}]$$

$$[\text{ligand}^{\bullet}] + \text{Co(III)–ligand} \overset{k_3}{\rightarrow} 2 \text{ ligand} + \text{Co(II)}$$

The autocatalytic behavior may be suppressed by addition of large concentrations of Eu^{3+} or V^{3+}, the latter being the more effective inhibitor. In the case of pyridine ligands with 2,3-, 2,4-, or 2,6-dicarboxylic acids, strongly absorbing complexes are observed in reaction with V(II).

The reduction of trichloracetic acid by vanadium(II)

$$2V^{2+} + Cl_3CCO_2H + H^+ \rightarrow 2V^{3+} + HCCl_2 \cdot CO_2H + Cl^- \tag{51}$$

has been investigated.[113] Variation of the rate constant with $[H^+]$ is consistent with reaction via the trichloroacetate anion in an outer-sphere process. Vanadium(II) complexes of pyrocatechol[114] and mixed $Mg(OH)_2$–$V(OH)_2$ species (pH 9–14)[115] have been invoked as reagents in the reduction of N_2 to NH_3 (via at higher pH's, N_2H_4 as an intermediate).

2.4.4. Cobalt(II)

Reactions of vitamin B_{12r} with organic halides have been studied in aqueous or methanol media.[116] The second-order reactions, ($R = 2k[Co(II)][RX]$) have been described in terms of a stepwise atom transfer process

$$B_{12r} + RX \xrightarrow{slow} X\text{--}B_{12} + R^\cdot \tag{52}$$

$$R^\cdot + B_{12r} \xrightarrow{fast} R\text{--}B_{12} \tag{53}$$

with a fast subsequent solvation of the X–B_{12} complex leading to a $B_{12}(OH_2)$ in water. With organic iodides, however, the rate law is of the form $R = 2k_1[Co(III)]^2[RI]$ with a rate-limiting step involving the reaction of B_{12r} with a $[B_{12r}{:}RI]$ precursor.

High-order reactions are also observed[117] in the reaction of pentacyanocobaltate(II) with 1,4-naphthaquinone (NQ). The initial step involves the formation of a radical anion which reacts further to yield a binuclear species (Scheme 4). The products which are unstable and were not analyzed directly are char-

Scheme 4

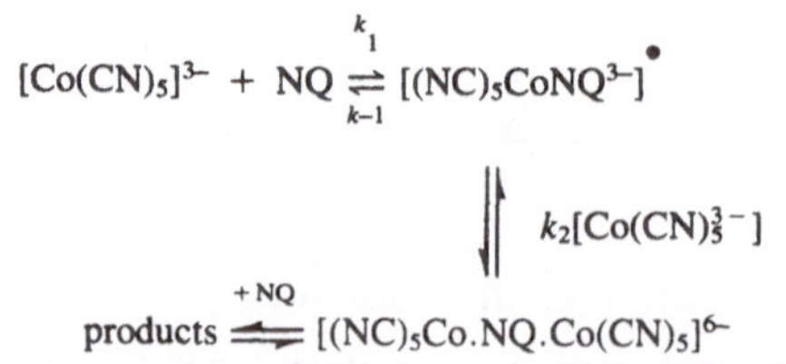

$$[Co(CN)_5]^{3-} + NQ \underset{k_{-1}}{\overset{k_1}{\rightleftharpoons}} [(NC)_5CoNQ^{3-}]^\cdot$$

$$\Big\Updownarrow \; k_2[Co(CN)_5^{3-}]$$

$$\text{products} \underset{+NQ}{\rightleftharpoons} [(NC)_5Co.NQ.Co(CN)_5]^{6-}$$

acterized by an intense band at 500 nm. A similar bridged intermediate has been postulated in the reaction with 1,4-benzoquinone.[118] A lower limit for the first (k_1) step is estimated as $\sim 2 \times 10^8 \ M^{-1} \ s^{-1}$ while the rate for formation of the binuclear species is $3.2 \times 10^4 \ M^{-1} \ s^{-1}$.

The reaction of cobalt(II) salen complexes with o-quinones[119] results in the formation of a [cobalt(III)–(salen)–(o-SQ)] complex, where (o-SQ) represents the semiquinone radical ion. (Similar reactions are observed with iron(II) and manganese(II)–salen complexes, the reactions with these metal centers being more extensive than those for cobalt(II)). In the case of 3,5-di-*tert*-butyl-*o*-benzoquinone, however, the cobalt(III) ligand radical complex has been isolated, and from ^{1}H hyperfine coupling constant studies which are assumed as diagnostic of the extent of electron transfer, the complex is described best as a low-spin cobalt(III) system with a coordinated semiquinone.

2.4.5. Other Reductants

The reaction of U^{3+} with monohaloacetic acids ($X = Br, I$) is considered to take place via an outer-sphere process.[120] Studies on these systems and with hydroxylamine[121] in binary solvent mixtures have been made. These data and those for other reactions have been used in providing a means for distinguishing been the outer- and inner-sphere processes on the basis of solvent effects. A similar analysis has been made in the characterization of an outer-sphere process in the reaction of U^{3+} with trichloroacetic acid.[122]

Enthalpies and entropies of activation have been measured in the oxidation of manganese(II)–aminopolycarboxylate complexes by Br_2 and Br_3^-.[123] An inner-sphere mechanism is postulated. In a study of the oxidation of coordinated olefins, the reaction of $[PtCl_3(C_2H_4)]^-$ with Cl_2 has been shown[124] to proceed the formation of an intermediate $[PtCl_5(CH_2CH_2Cl)]^{2-}$ which in aqueous solution undergoes aquation and dissociation, leading to the products $[PtCl_4]^{2-}$ and $ClCH_2CH_2OH$.

2.5. Intramolecular Electron Transfer

Although reference has been made above to systems where intramolecular transfer takes place, these processes have generally been on a rapid time scale following formation of the intermediate. There are some systems, however, where redox decomposition has been monitored.

The tetraglycine complex of ter-valent copper, $[Cu(III)(H_{-3}G_4)]^-$, is moderately stable in neutral solution ($t_{1/2}$ for redox decomposition ~ 5.5 h at 25°C).[125] This reaction has been investigated over a very wide pH range (0.3–13.5). In acid solution (pH 3–6), the decomposition results in the oxidation of some of the tetracoordinated glycine and Cu(II) is formed. Two reactants are involved

Scheme 5

$$[\text{Cu(III)}(\text{H}_{-3}\text{G}_4)]^- \xrightarrow{k_0}$$

$$\Big\Updownarrow \text{H}^+$$

$$[\text{Cu(III)}(\text{H}_{-3}\text{G}_4)\text{H}]^- \xrightarrow{k_1} \Bigg\} \text{ redox products}$$

(Scheme 5), which are related by an "outside" protonation on the ligand. At higher acidities an additional pathway is observed deriving from reaction of a further protonated species $[\text{Cu(H}_{-3}\text{G}_4)\text{H}_2]^+$ which rearranges to the complex ion $[\text{Cu(III)}(\text{H}_{-2}\text{G}_4)\text{H}]^+$ in which one Cu–N peptide bond is broken. Above pH 12, base amine deprotonation yields $[\text{Cu(III)}–(\text{H}_{-4}\text{G}_4)]$ prior to the redox step. The tetraglycine complex is a powerful oxidant, reacting with Br^-, Ce(III), and Mn(II) but not water. Br_2^- and $\text{HO}^\cdot$ radicals have been used to oxidize $\text{Cu(II)}\text{G}_4$ to the copper(III) species.[126] Unstable Cu(III)–carbon-bonded intermediates are formed, e.g., with $^\cdot\text{CH}_2\text{C}(\text{CH}_3)_2\text{OH}$ radicals, which decay in a first-order process which is independent of copper concentration but depends on pH.

Comparative rate studies have been made on hydridorhodoxime and its conjugate base the bis(dimethylglyoximato)rhodate ion.[127] The former species, $[\text{HRh(dmgH)}_2\text{PPh}_3]$, has a $\text{p}K_a \sim 9.5$. In more acidic media, further protonation of the oxime oxygens results in a decomposition to hydrogen and the dimeric Rh(II) derivative $[\text{Rh(dmgH)}_2\text{PPh}_3]_2$. The kinetics of the latter reaction were investigated in 1:1 H_2O–MeOH solutions in glycine buffer, the dimeric product deriving from the rapid step involving $[\text{HRh(dmgH)}_2\text{PPh}_3]$ and $[\text{H}_2\text{ORh(dmgH)}_2\text{PPh}_3]^+$ as reactants.

Dissolved SO_2 reacts virtually instantaneously with aqueous $[\text{Co(NH}_3)_5\text{OH}]^{2+}$ to form the oxygen-bonded sulfito complex $[\text{Co(NH}_3)_5\text{OSO}_2]^+$.[128] At higher pH's, however, a slow intramolecular redox reaction takes place, leading to Co(II) and SO_4^{2-} formation in a 2:1 ratio. The rate-determining step is considered

Scheme 6

$$[(\text{H}_3\text{N})_5\text{CoOSO}_2]^+ \rightarrow \text{Co}^{2+} + 5\text{NH}_3 + \text{SO}_3^-$$

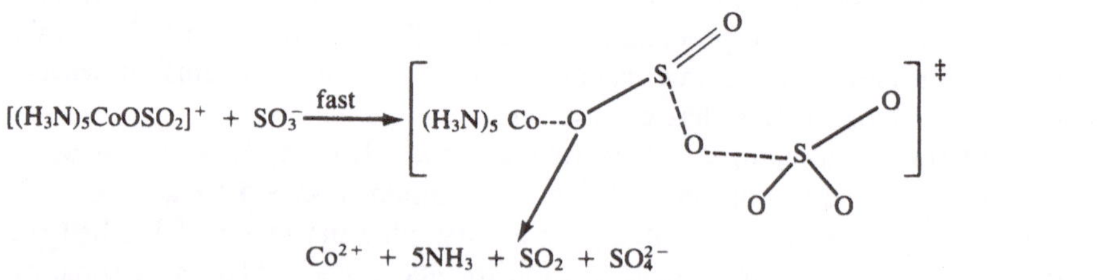

$$[(\text{H}_3\text{N})_5\text{CoOSO}_2]^+ + \text{SO}_3^- \xrightarrow{\text{fast}} \left[(\text{H}_3\text{N})_5\,\text{Co---O} \diagdown \cdots \text{S}\diagup \diagdown \text{O} \cdots \text{S} \right]^{\ddagger}$$

$$\text{Co}^{2+} + 5\text{NH}_3 + \text{SO}_2 + \text{SO}_4^{2-}$$

to be an inner-sphere one-electron transfer between the cobalt(III) and the ligand sulfite. $S_2O_6^{2-}$ has been ruled out as an appreciable product (Scheme 6).

Direct evidence has been presented[129] for both inner- and outer-sphere electron transfer in the oxidation of $[IrCl_6]^{3-}$ by OH$^\cdot$. The formation of the intermediate and its pH-dependent decay for the inner-sphere process is similar to that of the oxidation of the free ligand Cl$^-$ by HO$^\cdot$.

Part 2
Substitution and Related Reactions

Chapter 3

Reactions of Compounds of the Nonmetallic Elements

3.1. Introduction

The reactions of compounds of the nonmetals are less easy to classify straight-forwardly under the heading of "substitution reactions" than are the reactions of metal complexes. Accordingly there will be some discussion of the redox reactions of these compounds, although those involving a transition metal coreactant will usually not be included. In some cases there will be an overlap with topics usually regarded to be the province of the organic chemist.

3.2. Boron

Earlier kinetic studies[1] on the acid hydrolysis of the tetrahydroborate ion (BH_4^-) in moist acetonitrile were interpreted in terms of the accumulation of an intermediate complex of acetic acid and the tetrahydroborate anion. However, nmr measurements[2] and kinetic isotope effects[3] now show that while this intermediate is formed, and allows isotope exchange between BH_4^- and CH_3COOD, it does not accumulate. The rate-determining step involves a second intermediate $CH_3COOBH_3^-$, so that the earlier study on the hydrolysis of tetrahydroborate actually related to the hydrolysis of this species. It is suggested that this occurs

65

by spontaneous and water-catalyzed pathways, in the latter case with a rate-determining displacement of the incipient biacetate ion by water from $BH_3OC(CH_3)O.HOOCCH_3$ to give solvated BH_3.

Free radicals have been detected in the reaction between tetrahydroborate and alcohols and identified as the adduct between $^\cdot BH_2$ and the spin-trap nitrosodurene. Reactions (1)–(4) are suggested. [4]

$$BH_4^- \; + \; 2ROH \longrightarrow ROBH_2 \; + \; 2H_2 \; + \; RO^- \tag{1}$$

$$ROBH_2 \quad \overset{h\nu}{\underset{\Delta}{\Big\langle}} \quad \begin{array}{l} \longrightarrow \; ^\cdot OBH_2 \; + \; R^\cdot \;\; (R = CH_3) \\[1.5em] \longrightarrow \; ^\cdot BH_2 \;\; + \; ^\cdot OR \;\; (R > CH_3) \end{array} \tag{2,3}$$

$$ROBH_2 \; + \; 2ROH \longrightarrow B(OR)_3 \; + \; 2H_2 \tag{4}$$

Diborane, well known to undergo reaction with monodentate ligands, also reacts with chelating ligands. [5] Thus, 2,2'-bipyridyl forms the air-stable adduct $(BH_3)_2(bipy)$, while N,N,N',N'-tetramethyl-o-phenylenediamine (TMPD) reacts with equimolar amounts of diborane to give $[(TMPD)BH_2][BH_4]$. Acid hydrolysis gives the cation $(TMPD)BH_2^+$, isolated as the hexafluorophosphate. With excess diborane this ligand gives $[(TMPD)BH_2][B_2H_7]$, a stoichiometry also found for the product of reaction of 1,8-bis(dimethylaminonaphthalene).

Several other reactions involving adducts of borane with nitrogen bases have been investigated or reinvestigated. Rate constants for the acid-catalyzed hydrolysis of ammonia·borane confirm[6] the trend $k_2(NH_3.BH_3) > k_2(MeNH_2.BH_3) > k_2(Me_2NH.BH_3) > k_2(Me_3N.BH_3)$, where $k_2 = -(I/[H^+])d \ln[R_3N.BH_3]/dt$. Hydrolysis of $NH_3.BH_3$ is faster than the exchange of hydrogen bound to boron, but this is not so for $Me_3N.BH_3$. These results lead to further support for a mechanism in which there is a *cis* displacement of BH_3 via electrophilic attack of the proton at the amine nitrogen. It should be stressed that this does not preclude a mechanism involving protonation to give a five-coordinate boron species with subsequent loss of H_2, although it is less easy to reconcile with the kinetic data.

The compounds $[A_3BH][X]_2$, where A is a pyridine or a substituted pyridine and X is Br, I, or PF_6, may be synthesized[7] by direct nucleophilic displacement on $(CH_3)_3NBHBr_2$ or dihaloboron adducts of pyridine or substituted pyridines. The displacement of halide or trimethylamine from the former compound is controlled by the steric properties and basicity of the attacking amine and by the nature of the displaced halide. A rate-determining dissociation of a substituent on $(CH_3)_3NBHBr_2$ is excluded. The actual pathway may involve alternative routes depending on the sequence of replacement of halide and trimethylamine. The reactivities of compounds corresponding to these alternatives have been assessed and the following probable substitution sequence suggested[7]:

$$(CH_3)_3NBHBr_2 \rightarrow RpyBHBr_2 \rightarrow (Rpy)_2BHBr^+ \rightarrow (Rpy)_3BH^{2+}$$

The use of boron 11 nmr has shown[8] that the mixing of ammonia·borane and boron trihalides in ether gives (for BCl_3) the new compounds NH_3BH_2Cl and NH_3BHCl_2, together with NH_3BCl_3, Et_2OBH_2Cl, and Et_2OBHCl_2. Reaction with BBr_3 gives some of these products. A halogen–hydrogen exchange mechanism is postulated in both cases. Reaction with BF_3 gives as the only major products NH_3BF_3 and $(\mu\text{-}NH_2)B_2H_5$, which are suggested to result from the following scheme

$$2(NH_3BH_3) + BF_3 \xrightarrow[25°C]{Et_2O} (\mu\text{-}NH_2)B_2H_5 + NH_3BF_3 + H_2 \qquad (5)$$

$$NH_3BH_3 + BF_3 \longrightarrow NH_3BF_3 + \tfrac{1}{2}B_2H_6 \qquad (6)$$

$$NH_3BH_3 + \tfrac{1}{2}B_2H_6 \longrightarrow (\mu\text{-}NH_2)B_2H_5 + H_2 \qquad (7)$$

The application of multinuclear nmr to $B_4H_8PF_2N(CH_3)_2$[9] and $B_4H_8PF_2X$[10] (X = H,F,Cl,Br, or I) shows the presence of two geometrical isomers in both cases, except for X = H in the latter compound. At $-125°C$ rotation about the P–B bond in one isomer of the former compound (the *endo* isomer) becomes slow, with $\Delta G^{\ddagger} = 32.1$ kJ mol^{-1} for the rotation barrier, while it remains rapid for the *exo* isomer. At 80°C the two isomers interconvert rapidly on the ^{19}F nmr time scale, with $\Delta G^{\ddagger} = 78$ kJ mol^{-1}. Interconversion via ligand dissociation is excluded by the experimental value for $\Delta S^{\ddagger}$, and the mechanism is suggested to involve breaking of bonds within the boron framework, with rearrangement or rotation leading to the interconversion of the isomers.

Calculations have been carried out[11] on rearrangements of B_4H_4 and B_4F_4 as examples of the diamond–square–diamond transformation, with particular emphasis on the nature of the square mid-point structure.

B-alkyl-9-borabicyclo[3.3.1]nonanes undergo olefin–alkyl group exchange when refluxed with an olefin in THF. Kinetic and competition studies[12] support a dehydroboration–hydroboration process rather than a concerted mechanism. *Ab initio* M.O. calculations show[13] that the reaction between C_2H_4 and BH_3 proceeds through a two-step process. A loose three-center (C—B—C) π complex is formed which is then transformed into the product via a four-center transition state in a rate-determining step.

Several studies[14] have been reported on the formation of complexes between borate and catechol, including the binding of borate to anion exchange resins modified with catechol.[15] The reaction between several bidentate chelating agents (H_2L) with four boron acids, m-$NO_2PhB(OH)_2$, $PhB(OH)_2$, $B(OH)_3$, and $CH_3B(OH)_2$, has been followed by temperature-jump kinetics.[16] The overall reaction is:

$$RB(OH)_2 + \begin{array}{c} | \\ HO-C- \\ | \\ HO-C- \\ | \end{array} \rightleftharpoons \begin{array}{c} R \quad O \quad | \\ \diagdown \diagup \quad -C- \\ B^- \quad | \\ \diagup \diagdown \\ OH \quad O-C- \\ | \end{array} + H_3O^+ \tag{8}$$

Formation constants increase as the ligand and acid become more acidic. Reaction of the boric acid involves both H_2L and L^{2-}, the pattern of reactivity varying for each case. Rate constants for reaction involving H_2L vary systematically with ligand acidity and the boron substituent and have a common mechanism involving proton transfer in a rate-limiting ring closure step. Reactions involving the ligand anion and the boron acid are more diverse in character. Borate anions react more rapidly with H_2L than with L^{2-}.

3.3.   Group IV Elements

3.3.1.   Carbon

The mechanism of decomposition of phenylcarbamate anions in aqueous solution is suggested[17] to involve the scheme (9), where BH is a general acid catalyst. The rate-determining step is X, in which C–N bond fission occurs before diffusion away from the conjugate base of the catalyzing acid. Carbamate anions formed from more basic amines follow an alternative route [reaction (10)], in which step Y is rate determining.

$$RNHCO_2^- + BH \rightleftharpoons (BH.RNHCO_2^-) \rightleftharpoons (B^-.R\overset{+}{N}H_2CO_2^-)$$
$$\overset{X}{\rightleftharpoons} (B^-.RNH_2CO_2) \rightarrow B^- + RNH_2 + CO_2 \tag{9}$$

$$RNHCO_2^- + BH \rightleftharpoons (BH.RNHCO_2^-) \rightleftharpoons (B^-.R\overset{+}{N}H_2CO_2^-)$$
$$\rightleftharpoons B^- + R\overset{+}{N}H_2CO_2^- \overset{Y}{\rightarrow} B^- + RNH_2 + CO_2 \tag{10}$$

3.3.2. Silicon

The stereochemistry and mechanistic implication for nucleophilic substitution at silicon continue to receive attention, and have been the subject of a comprehensive review.[18] The nucleophiles HMPA, DMSO, DMF, and Ph_3PO catalyze the hydrolysis of triorganochlorosilanes and also their racemization. Stable 1:1 ionic adducts of HMPA with Me_3SiBr have been isolated in the past and suggested to be possible intermediates in the racemization process. However, it has now been shown[19] that such ionic 1:1 adducts are not usually formed and that the reported example was a special case which should not be generalized into a mechanism involving displacement of halides by the nucleophile. The previously accepted mechanism involving the formation of five- and six-coordinate species still seems more feasible.

Tert-butyldimethylsilyl vinyl ethers, $t\text{-}BuMe_2SiOC(R)\!\!=\!\!CH_2$, undergo hydrolysis in acetonitrile–water mixtures with general and specific acid catalysis and a linear dependence of $\log k_{H+}$ on $\sigma_P^+(R)$, showing that the reaction proceeds *via* the Ad_E2 mechanism with rate-determining proton transfer to carbon. In contrast, the rate constants for the hydrolysis of the trimethylsilyl analogs are not correlated by $\sigma_P^+(R)$, but still depend on proton and general acid concentrations. The rate-determining step in this case is nucleophilic attack on silicon with either synchronous proton transfer and water attack or an equilibrium protonation and general base-assisted attack of water.[20] Kinetic studies have also been reported on the cleavage of picolyl- and quinolylmethyl-trimethylsilanes by sodium methoxide in methanol[21] and the base-catalyzed solvolysis of acylsilanes in THF–aqueous-NaOH mixtures.[22]

The γ-radiation-induced chain reactions in solution of R_3SiH (R = Me or Et) in CCl_4–cyclohexane mixtures gives R_3SiCl, $CHCl_3$, and RCl as the main products. These result from the abstraction of hydrogen from the two silanes by $^{\cdot}CCl_3$ radicals. The alkylsilanes are much more reactive than Cl_3SiH in hydrogen atom transfer reactions with $^{\cdot}CCl_3$ radicals.[23] The formation of silyl radicals in the photolysis of aryldisilanes has been detected by spin-trapping techniques.[24]

The reaction of silanes with an alcohol is catalyzed by

$$(EtO)_3SiH + EtOH \rightarrow (EtO)_3SiOEt + H_2 \qquad (11)$$

the complexes $[Fe(C_2H_4)(dppe)_2]$, $[FeH_2(PMePh_2)_4]$, and $[FeH_2(N_2)(PEtPH_2)_3]$, where dppe = 1,2-bis(diphenylphosphino)ethane. The latter two complexes are particularly effective and both involve induction periods and the formation of iron silyl complexes,[25] as shown below for $[Fe(H_2)(PMePh_2)_4]$ and R = EtO.

$$FeH_2L_4 \rightarrow [FeH_2L_3] + L$$

$$[FeH_2L_3] + R_3SiH \rightarrow [FeH_3(SiR_3)L_3]$$

$$[FeH_3(R_3Si)L_3] + EtOH \rightarrow [FeH_3(R_3Si \cdot EtOH)L_3]$$

$$[FeH_3(R_3Si \cdot EtOH)L_3] \rightarrow [FeH_4L_3] + [R_3Si(OEt)]$$

$$[FeH_4L_3] \rightarrow [FeH_2L_3] + H_2$$

The relative rate constants for the addition of triethylsilane to alkynes, catalysed by *trans*-di-μ-hydridobis(tricyclohexylphosphine)bis(silyl or germyl)diplatinum complexes, increase as the π-acceptor capacity of the alkyne increases.[26] Some interesting reactions have been described [27] between tris(trimethylsilyl)methyl silicon halides, $(Me_3Si)_3CSiR_2X$, and electrophilic reagents such as $AgNO_3$, AgOAc, AgO_2CCF_3, $Hg(NO_3)_2$, $HgCl_2$ and $HgBr_2$ in alcohols, acetic acid, trifluoroacetic acid or in mixtures of these. The results are accounted for by the formation of a cationic intermediate through abstraction of halide from $(Me_3Si)CSiR_2X$ by the electrophile. The intermediate may involve a methyl group bridging the 1- and 3-silicon atoms, with the nucleophile subsequently attacking at either of these atoms.

The kinetics of reversible dimerisation of *o*-phenylenedioxydimethylsilane has been followed by ^1H nmr spectroscopy.[28] This involves the formation of a ten-membered ring (equation (12)), with an activation energy of only 57 kJ mol^{-1}.

$$(12)$$

The rate of polymerization of silicic acid produced by the hydrolysis of tetramethoxysilane (followed by silicon-29 nmr) has turned out to be slow.[29] Indeed at pH 3.5–4.0 monosilicic acid may still be detected after several weeks. There is no direct relationship between disappearance of monomeric silicic acid and production of silica gel.[30]

3.3.3. Germanium

Conductivity studies on triorganohalogermanes with a range of nucleophiles fail to show the existence of ionic 1:1 adducts in which the halide is displaced.[19]

Here too it appears that the racemization of these compounds, induced by nucleophiles, results from the formation of higher coordination number species. Rate constants have been measured for the loss of tritiated organogermanes, $(XC_6H_4)_3GeT$ ($X = m$-Cl, p-Cl, m-Me, p-Me, o-Me, p-OMe, or o-OMe) and Ph_3GeT, Ph_2GeHT, and $PhGeH_2T$.[31] For $(XC_6H_4)_3GeT$ there is a good linear correlation between log k_{rel} (k_{rel} = rate constant relative to Ph_3GeH) and σ. For the compounds $(XC_6H_4)Ph_2GeT$ (X = p-NO$_2$, p-CN, m-Cl, or p-F) a plot of log k_{rel} against σ^- constants is better than that against σ constants. These results imply that there is substantial delocalization of charge from the anionic germanium center into the aromatic rings and hence the existence of p_π–p_π bonding between Ge and C.

The hydrolysis of the trisoxalatogermanate(IV) anion is acid catalyzed, and may involve the protonation and subsequent ring opening of an oxalate species.[32] Loss of the oxalate group leads to the formation of $[GeO(C_2O_4)_2]^{2-}$ *via* $Ge(OH_2)_2(C_2O_4)_2$. The species $[GeO(C_2O_4)_2]^{2-}$ undergoes photochemical reaction to give Ge(II).

3.4. Group V Elements

3.4.1. Nitrogen

Nitric acid oxidizes hydroxylamine to dinitrogen monoxide and nitrous acid in proportions that depend upon the reaction conditions.[33] The production of HNO_2 is a maximum at ca. 4–5 mol dm^{-3} [HNO_3] and is a function of [NH_3OH^+]. The reaction, which only takes place at [HNO_3] >2.5 mol dm^{-3}, is autocatalytic, with an induction period that is eliminated by added nitrous acid. The mechanism involves the oxidation of hydroxylamine by dinitrogen tetroxide to nitrous acid via HNO as an intermediate. The nitrous acid reacts with hydroxylamine to give dinitrogen monooxide.

Rate law (13),

$$\text{rate} = kh^{1/2}[HNO_2]^{1/2}[\text{nitrate}]^{1/2}[Fe_2(CN)_{10}^{4-}] \tag{13}$$

established[34] for the reaction of $[Fe_2(CN)_{10}^{4-}]$ with nitrous and nitric acids, may be interpreted most readily by the rate-determining reaction of the Fe(III) species with nitrogen dioxide formed by the homolysis of dinitrogen tetroxide. The product appears to be a binuclear Fe(III)/Fe(II) cyanide complex with a NO^+ group in the coordination shell of the Fe(II).

The rate constant (k_1) for the formation of the nitronium ion from nitric acid in aqueous sulfuric, perchloric, and methanesulfonic acid [equation (14)] has been obtained[35] by studying the kinetics of nitration of the reactive amine anisole, present in high concentration. This means that the reaction of the nitronium ion with amine is then fast [equation (15)]. Studies on the reverse reaction (16) show that two molecules of water are needed in the hydration

$$HNO_3 \underset{k_{-1}}{\overset{k_1}{\rightleftharpoons}} NO_2^+ \tag{14}$$

$$NO_2^+ + A \xrightarrow{\text{fast}} \text{products} \tag{15}$$

$$H_2O + H_2O + NO_2^+ \longrightarrow H_3O^+ + HNO_3 \tag{16}$$

of the nitronium ion, the second one being involved in general base catalysis. Ionization of nitric acid will involve general acid catalysis.

There has been much interest in the reactivity of nitronium salts, for example, toward amides in acetonitrile,[36] the oxidation of alkanes in acetonitrile (via a carbenium ion intermediate),[37] and toward organic sulfides, phosphines, arsines, and stibines.[38] Reaction with diaryl, aryl alkyl, and dialkyl sulfides give sulfoxides, but a few percent of the ring C-nitro products were found for diphenyl sulfide. This is a result of the equilibrium between nitrito onium ions $>\overset{+}{X}$–ONO and nitronium ions $>\overset{+}{X}$–NO$_2$. This is the first demonstration of the ambident reactivity of the nitronium ion in solution chemistry, and reflects the relative oxidizing and nitrating activity of the NO_2^+ ion toward substrates. The equilibrium was confirmed by carrying out *trans*-nitrosation of *N,N*-dimethylaniline in the presence of onium salts.

The oxidation of nitrous acid by chlorite to give nitrate and chloride involves uncatalyzed and chloride-catalyzed pathways.[39] The mechanism involves oxygen atom transfer from chlorite to nitrous acid, generating peroxonitrite and hypochlorous acid. The former species isomerizes to nitrate and the latter species oxidizes nitrous acid to nitrate. The Cl⁻-catalyzed reaction involves oxidation by $HOCl_2^-$. The effect of solvent composition (water–dioxan) on the nitrite–chlorate reaction[40] is in accord with previous views on the mechanism of this well-studied reaction.

The reaction of nitrous acid with several amminehaloosmium(III) complexes involves the oxidation of Os(III) to Os(IV) with reduction of nitrous acid to nitrogen monooxide [equation (17)], and the subsequent reductive diazotization of the Os(IV) species by the nitrogen monooxide produced in (17), the deprotonated species being active [equation (19)]. The conversion of a

$$[Os(NH_3)_4I_2]^+ + HNO_2 + H^+ \rightarrow [Os(NH_3)_4I_2]^{2+} + NO + H_2O \tag{17}$$

$$[Os(NH_3)_4I_2]^{2+} \rightleftharpoons [Os(NH_3)_3I_2(NH_2)]^+ + H^+ \tag{18}$$

$$[Os(NH_3)_3I_2(NH_2)]^+ + NO \rightarrow [Os(NH_3)_3I_2(N_2)]^+ + H_2O \tag{19}$$

second ammine group to a dinitrogen ligand occurs more easily than the first reaction, as a result of the remarkable enhancement of the acidity of the Os(III) ammine by the presence of the dinitrogen ligand. Thus the pK_a values of $[Os(NH_3)_6]^{3+}$ and $[Os(NH_3)_5N_2]^{3+}$ are ~16 and 6.6, respectively, a truly remarkable observation.[41]

The oxidation of hexacyanoferrate(II) by nitrous acid involves an encounter-controlled reaction between $[HFe(CN)_6]^{3-}$ and the nitrosonium ion.[42] More conventional nitrosations have also received much attention, as have the use of specific nitrosating agents and the transfer of NO^+ groups between molecules. The rate constant for the reaction of nitrous acid and methanol (*O*-nitrosation) is significantly lower than that expected for a diffusion-controlled reaction.[43] Solvent-jump relaxation techniques have been used in the study of the NOCl—1-butanol reaction in CCl_4—HOAC mixtures.[44]

The nitrosation of several amines in aqueous solution by dissolved gaseous nitrosyl chloride has been studied.[45] Provided the amine is in excess, the nitrosation is quantitative, and there is no complication from hydrolysis of the nitrosyl chloride. The reactivity of amines more basic than *N*-methyl-4-nitroaniline (pK_a = 1.49) toward nitrosyl chloride is largely independent of their basicity, and therefore these amines are presumed to react on encounter. Studies on reactions of NOCl with secondary amines[46] and olefins[47] have also been reported. The iodine-catalyzed *N*-nitrosation of piperidine, morpholine, and *N*-methylpiperazine by NO is independent of the amine and involves a rate-determining formation of nitrosyl iodide, followed by rapid reaction with the amine.[48]

Hydrogen isotope exchange at the nitrogen atom of *N,N*-dimethylanilinium ions is catalyzed by acid[49] and nitrosonium ion.[50] The latter reaction has been studied at acidities for which the subsequent chemical reaction between NO^+ and amines is much slower. The mechanism postulates the formation of a loose complex between NO^+ and the anilinium ion, followed by proton loss from the nitrogen. This process is related to the direct diazotization of some anilinium ions, and to the nitrous acid-catalyzed nitration and oxidation of the *N,N*-dimethylanilinium ion.

A number of other nitrosations have been studied, often in an environmental or biological context. These include as substrates: *N*-acetylamino acid methyl esters (cysteine, tyrosine, histidine, tryptophan),[51] 4-nitrophenol,[52] cimetidine,[53] tertiary amines,[54] and methylurea.[55,56] In the last case an explanation is put forward for the lack of catalytic activity of added bromide or thiocyanate.[55] The denitrosation reaction was also examined, using hydrazine as a trap for free nitrous acid. A range of denitrosation reactions have been studied[57] together with the reactivity of thiourea, alkylthioureas, cysteine, glutathione, *S*-methylcysteine and methionine toward *N*-methyl-*N*-nitrosoaniline.[58] The results are consistent with an initial rate-determining *S*-nitrosation of the thiourea by the protonated nitrosamine. Alicyclic nitrosamines and nitrosoamino acids have also been considered as transnitrosating agents.[59]

The suggestion that added nitrite ion stabilizes the trioxodinitrate ion at higher pH values through recombination with the nitrosyl ion formed by cleavage of the trioxodinitrate anion [equation (20)]:

$$N_2O_3^{2-} \rightleftharpoons NO^- + NO_2^- \tag{20}$$

has been confirmed by the use of ^{15}N labels.[60] Studies with nitrogen-15-labeled nitrite and trioxodinitrate have shown the remarkable complexity of the chemistry of $N_2O_3^{2-}$. *Ab initio* calculations have been reported on the rearrangements HNO $\rightarrow$ NOH and FNO $\rightarrow$ NOF,[61] and XNO $\rightarrow$ NOX.[62]

The base-catalyzed decomposition of nitroamine (nitramide) is of classical significance in the history of acid–base catalysis, and is generally believed to involve the base-catalyzed elimination of H^+ and OH^- from the *aci* form of nitroamine [equation (21)].

$$NH_2NO_2 \rightleftharpoons NH\!\!=\!\!NO_2H \xrightarrow{B} N_2O + BH^+ + OH^- \tag{21}$$

However, it seems that this reaction is more complicated than originally thought. Thus kinetic studies[63] have shown that there is a second pathway for the decomposition, involving the decomposition of the nitroamine anion [equation (22)]:

$$NH_2NO_2 \rightleftharpoons H^+ + NHNO_2^- \xrightarrow{B} N_2O + H_2O \tag{22}$$

Furthermore, studies in the more strongly basic regions, together with a consideration of isotope effects,[64] have revealed new mechanistic details for pathway (21). Neutral primary amine catalysts and negatively charged oxygen bases give linear Brønsted relationships for increasing catalyst strength, but eventually at higher catalyst strength show curvature. These results suggest a mechanism for the traditional decomposition pathway in which proton transfer precedes heavy atom reorganization. A scheme is shown in equations (23)–(25).

$$B + HN\!\!=\!\!NO_2H \rightleftharpoons B.HN\!\!=\!\!NO_2H \tag{23}$$

$$B.HN\!\!=\!\!NO_2H \rightleftharpoons \overset{+}{B}H.\overset{-}{N}\!\!=\!\!NO_2H \tag{24}$$

$$\overset{+}{B}H.\overset{-}{N}\!\!=\!\!NO_2H \rightarrow BH^+ + N_2O + OH^- \tag{25}$$

For weak catalysts, the reverse of step (24) is fast and therefore heavy atom reorganization is rate determining. For stronger catalysts, proton transfer to the base is favored, the reverse of (24) is slow, and proton transfer begins to be rate determining. Kinetic hydrogen isotope effects on the decomposition of nitroamine catalyzed by negatively charged oxygen bases were much smaller than previously reported (k^H/k^D ca.~ 2–3 rather than 10).

The oxidation of hydroxylamine to N_2 and N_2O by aquovanadium(V) ions in the acidity range 1–5 mol dm^{-3} occurs[65] by two pathways involving $VO_2^+.NH_3OH^+$ and $V(OH)_2^{3+}.NH_3OH^+$.

Ab initio SCF calculations on disproportionation reactions of diimide (N_2H_2) to form hydrazine and dinitrogen have been concerned with concerted hydrogen transfer pathways between *cis* and *cis* forms and *cis* and *trans* forms, but a two-step hydrogen atom transfer process involving N_2H and N_2H_3 radicals was shown to be energetically feasible.[66]

Esr studies wth organic spin traps have shown that azide radicals ($N_3^·$) are formed in the reaction[67,68] of hydroxyl radicals with azide ions, and that the radicals $N_3^·$ and OCN$^·$ are formed by persulfate oxidations of azide and cyanide.[68] However, azide radicals could not be detected[69] in the photochemical reductive *cis* elimination of *cis*-diazidobis(triphenylphosphine)platinum(II) to give Pt(PPh$_3$)$_2$, even though they have been observed in the esr spectra of the photolysis products of other azido complexes. It is suggested, therefore, that the azide ligands are cleaved off as N_6 (hexaazabenzene). Calculations have shown that N_6 is slightly stabilized and could therefore be stable at the low temperatures of these experiments.

The reactions between hydrazine and hexachloroiridate(IV), monoaquopentachloro-, diaquotetrachloro-, and hexabromoiridate(IV) are first order in both reactants and inversely dependent on [H$^+$]. The reaction of [IrCl$_6$]$^{2-}$ is catalyzed by chloride ion. The hydrazine is converted to N_2 in stoichiometric amounts, and diimide is suggested to be an intermediate. Evidence was obtained by trapping N_2H_2 with an unsaturated dicarboxylic acid.[70] The following reactions are suggested:

$$N_2H_4 + Ir(IV) \rightleftharpoons \text{intermediate}$$

$$\text{intermediate} \rightarrow Ir(III) + N_2H_3^· + H^+$$

$$Ir(III) + N_2H_3^· + H^+ \rightarrow \text{intermediate}$$

$$Ir(IV) + N_2H_3^· \rightarrow Ir(III) + N_2H_2 + H^+$$

$$2Ir(IV) + N_2H_2 \rightarrow 2Ir(III) + N_2 + 2H^+$$

$$2N_2H_3^· \rightarrow N_4H_6$$

$$N_4H_6 \rightarrow 2NH_3 + N_2$$

$$2N_2H_3^· \rightarrow N_2H_4 + N_2H_2$$

The formation of diimide is suggested[71] to be the slow step in the oxidation of hydrazine by molybdenum(VI). The kinetics of the formation of chelating carbene complexes from hydrazine and hexakismethylisonitrile–ferrate(II) have been reported.[72]

Several reports on reactions in liquid ammonia have been published, including the proceedings of the Fifth International Conference on excess electrons and metal–ammonia solutions,[73] proton exchange at cobalt(III) ammine complexes,[74] and the reaction of the ammoniated electron with thiosulfate.[75]

3.4.2. Phosphorus

The subject of pentacoordinated phosphorus has been considered in depth in two volumes[76] entitled *Structure and Spectroscopy* and *Reaction Mechanisms*. The latter volume covers two major topics, namely, a comparison of the Berry pseudorotation process and the turnstile mechanism for explaining the interconversion of rotamers, and the consideration of reactions (from simple to enzymatic) which involve five-coordinate phosphorus intermediates.

Halogen exchange reactions of phosphorus(V) halides and oxohalides only take place in the presence of a suitable alkali metal halide, as demonstrated[77] by the use of ^{31}P nmr techniques. This technique has also been much used in the study of the hydrolysis of various phosphate species, including some where dramatic catalytic effects have been observed. Thus complex formation[78] between $N_4Co(III)(aq)$ species $[N_4 = (en)_2, tren, or (NH_2(CH_2)_3NH_2)_2]$ and pyrophosphate results in an enhancement of the hydrolysis of pyrophosphate to orthophosphate by a factor of about 10^5 at 25°C and pH 7. The reactive species in solution is postulated to be a 3:1 $N_4Co(III)$:pyrophosphate complex, an observation that opens up interesting possibilities for the study of polynuclear complexes in the hydrolysis of polyphosphates in general. The hydrolysis[79] of bidentate triphosphate in the well-defined inert complex $[Co(NH_3)_4H_nP_3O_{10}]$ only proceeds at two thirds the rate of that for the free ligand. An interesting feature of this work was the observation that when the hydrolysis took place in frozen solution, the ^{31}P nmr spectra of the thawed solution showed the presence of $[Co(NH_3)_4(H_nPO_4)(H_nP_2O_7)]$ and $[Co(NH_3)_4(H_nP_2O_7)]$ as intermediate products of hydrolysis. The former species corresponds to hydrolysis of the phosphate linkage in the six-membered chelate ring [equation 26)], and the latter to hydrolysis between the terminal, uncoordinated phosphate group and the chelate ring. The reaction in frozen solution allowed the hydrolysis of the phosphate chain to proceed at a modest rate but greatly retarded the rate of dissociation of the hydrolysis products from the Co(III), possibly as a result of the differing volumes of activation of the dissociation and hydrolysis reactions.

The (p-nitrophenylphosphato)pentaamminecobalt(III) cation undergoes[80]

$$(26)$$

base catalysis to generate *p*-nitrophenolate ion, *p*-nitrophenylphosphate ion, and hydroxo(phosphoramido)tetraamminecobalt(III). The ester hydrolysis is accelerated about 10^8-fold compared to the uncoordinated *p*-nitrophenylphosphate. Product distribution, ^{18}O tracer, and ^{31}P nmr studies imply the participation of a five-coordinate aminophosphorane, generated by intramolecular attack of a deprotonated, coordinated ammonia at the phosphorus center [equation (27)]. This gives *p*-nitrophenolate and hydroxo(phosphoramido)tetraamminecobalt(III), while the *p*-nitrophenylphosphate results from a competing, conventional Co–O bond rupture (S_N1cB). This same base hydrolysis mechanism also slowly hydrolyzes the product complex to give phosphoramidate anion. The accelerating effect in the base hydrolysis probably arises from the proximity of the NH_2^- group to the phosphorus center and the rapid decay of the aminophosphorane so formed, despite the strain introduced by ring formation in the aminophosphorane. The reaction

$$(27)$$

also presents an interesting example of the transfer of a phosphoryl group from oxygen to nitrogen and as such may be relevant to amino transferase chemistry.

The catalytic effect of an intramolecular amino function on phosphate ester hydrolysis has also been examined for phosphorylethanolamine diesters.[81] A nucleophilic role for the amine has been implicated with a rate enhancement of 10^6–10^7.

Methyltetraphenoxyphosphorane and the *p*-nitrophenoxy analog rapidly disproportionate on mixing to give all five possible phosphoranes, $CH_3P(OPh)_n(OAr)_{4-n}$. All disproportionation equilibria have been characterized[82] along with rate constants for their dissociation to phosphonium salts and for the corresponding associations.[83]

Hydroxyphosphoranes are thought to be intermediates in the hydrolysis of esters of oxo acids of phosphorus. It is of considerable interest that a stable, acidic five-coordinate phosphorus hydroxy acid has been prepared (structure **1**).

Another hydroxy acid has also been prepared (structure 2) but this is less stable and equilibrates in solution with its open-chain tautomer (structure 3).[84]

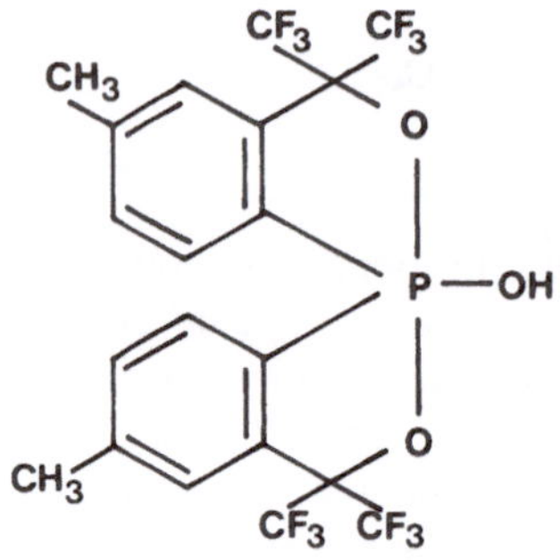

The formation of six-coordinate intermediates in displacement reactions at phosphorus has been demonstrated conclusively for reactions between chlorophosphoranes and trimethylsilyl- or benzyltri-n-butylammonium azides.[85] The formation of six-coordinate anions from some spirophosphoranes and nucleophiles is under kinetic control, so that the less stable *trans* isomers are formed initially.[86]

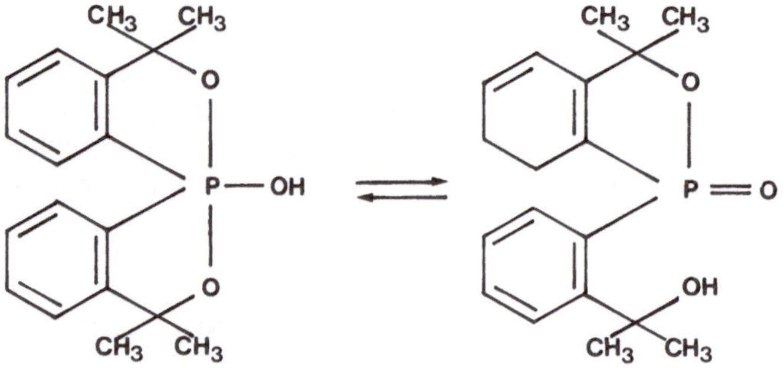

An associative A2 mechanism is operative for the hydrolysis of diphenylphosphinic amide $Ph_2P(O)NH_2$ and some analogs in dilute perchloric acid–dioxan, as shown by the retardation of rate produced by the presence of *ortho*-methyl substituents in the aryl groups.[87]

A useful survey of the chemistry of phosphite radicals, prepared by pulse radiolysis, has been presented.[88] These can take part in oxidation, reduction, substitution, and addition reactions. Reports have been published on the oxidation by peroxodiphosphate of alkyl aryl sulfides,[89] methyl phenyl sulfoxide,[90] and hexacyanoferrate(II).[91] The oxidation of dialkyl sulfides by peroxomonophosphoric acid[92] involves nucleophilic attack of the sulfide on the peroxo oxygen, followed by oxygen–oxygen bond fission to give the sulfoxide and phosphate.

Reactions between primary or secondary polyfluoroalkyl phosphines with nucleophiles have been discussed in terms of either a phosphaalkene intermediate or a nucleophilically initiated hydride ion shift from phosphorus to carbon. Evidence for the correctness of the former possibility has been presented[93] for the alkaline hydrolysis of $(CF_3)_2PH$ by the trapping of the intermediate $CF_3P{=}CF_2$ [equation (28)]:

$$(CF_3)_2PH \xrightarrow{\text{MeO}^-} (CF_3)_2P^- \xrightarrow[-F^-]{} CF_3P{=}CF_2 \xrightarrow{\text{MeOH}} CF_3P(OMe)CHF_2 \quad (28)$$

The phosphoalkene $CF_2{=}PH$ and the phosphoalkyne $FC{\equiv}P$ have been identified as intermediates in the reaction between CF_3PH_2 and solid KOH.[94]

Hydridocyclotriphosphazenes are formed[95] during the reactions of organocopper reagents with halocyclotriphosphazenes [equation (29)]. These reactions

$$(29)$$

were designed for the controlled replacement of halogens by alkyl groups, and the formation of the hydrido species was unexpected. The hydrogen is derived from the alcohol, and a metallophosphazene (structure **4**) is postulated to be an intermediate.

4

The compound $(NPCl_2)_4$ reacts with t-butylamine to give nongeminal structures, while $(NPCl_2)_3$ gives geminal products exclusively. This difference in stereochemistry has always been attributed to the greater reactivity of the former compound, a postulate that has now been confirmed by kinetic studies.[96] The factors leading to the formation of "bicyclic phosphazenes" in the reactions of $N_4P_4Cl_6(NHR_2)_2$ and the mixed amino derivative have been assessed.[97]

Fluorocyclophosphazenes $(NPF_2)_{3,4}$ differ from the chloro compounds in that only one fluoro atom per phosphorus is replaced in reactions with amines such as methylamine, n-butylamine, and dimethylamine. However, nucleophiles such as C_6H_5ONa, CF_3CH_2ONa, and $(CH_3)_2NLi$ readily replace all the fluorine atoms, and so it appears that the observations on the fluoro compounds and amines merely reflects the poor-leaving group ability of fluorine and the low nucleophilicity of the amines.[98]

3.4.3. Arsenic

Periodate catalyzes oxygen exchange between arsenate ions and water due to the formation of a condensed arsenato-periodate species, for which kinetic data have been presented.[99] Alcohol exchange with trialkylarsenates appears to be[100] an associative ligand exchange. A wide range of rates were obtained with different types of alcohol, but in general the reactions were much faster than those for phosphate esters.

The mould *Scopulariopsis brevicaulis* is able to convert arsenate into organoarsenate. A model has been put forward[101] for this biological methylation based on the alkylation of arsenate by sodium methyl sulfate [equation (30)].

$$OAs(V)(OH)_3 \xrightarrow{2e} As(III)(OH)_3 \xrightarrow{CH_3^+} CH_3As(V)(O)(OH)_2$$

$$\xrightarrow{CH_3^+} CH_3As(OH)_2 \xrightarrow[etc.]{CH_3^+} (CH_3)_3As \qquad (30)$$

3.4.4. Antimony

The irradiation with visible light of an oxygen-containing solution with hydrochloric acid containing $[Ru(bipy)_3]^{2+}$ and antimony(III) gives hydrogen peroxide and Sb(V). It is suggested that the Sb(III) is converted to Sb(IV), which then disproportionates to Sb(III) and Sb(V).[102]

3.5. Group VI Elements

3.5.1. Oxygen

There is much interest at present in the reactivity of excited states of dioxygen and various oxygen species such as superoxide ion and the perhydroxyl radical, and hydroxyl radicals. The photooxidation of acetone azine proceeds via a free radical pathway initiated by electron transfer taking place from the azine to singlet oxygen ($^1O_2$). This generates superoxide and azine cation radical, which reacts further to give tetramethyl-1,2-dioxa-4,5-diazine.[103] When aqueous solutions in which $[H_2O_2]/[reductant] > 10^3$ are subjected to flash photolysis, then all primary ˙OH radicals generated by photolysis of hydrogen peroxide are scavenged by hydrogen peroxide giving $HO_2^{\cdot}$ radicals. Oxidation of several substrates under these conditions has been reported[104] and pathways identified for ascorbic acid, involving reaction with $HO_2^{\cdot}$ and $O_2^{\cdot-}$. Cautionary notes have been expressed over assumptions, commonly made, that superoxide ion acts as an oxidant. For acidic reducing substrates such as ascorbic acid and 3,5-di-*tert*-butylcathechol, superoxide actually acts as a Brønsted base abstracting a proton to give substrate anion and dismutation species HO_2^- and $^1O_2$, the latter then oxidizing the substrate anion.[105] However, hydrophenazines, reduced flavins, and hydroxylamines are oxidized by $O_2^{\cdot-}$ in DMF solutions. These are all substrates that are susceptible to oxidation via a hydrogen atom transfer mechanism. Superoxide ion thus oxidizes hydroxylamine to NO, and hydrazine to dinitrogen, although the latter reaction appears to involve a complex pathway with a $3N_2H_4:4O_2^-$ stoichiometry.[106] Nitrotoluenes in DMF have been oxidized by electrogenerated superoxide ion to give the corresponding nitrobenzoic acid via the nitrobenzaldehyde.[107]

The reaction between hydrogen peroxide and hypochlorous acid has been studied in the pH range 4–13 in the presence and absence of acceptors for ˙OH and $O_2^{\cdot-}$ radicals.[108] The results are complex and differ from those reported earlier.[109] Several studies on the kinetics of the heterogeneous catalysis of the decomposition of hydrogen peroxide have been reported, including Fe(III)/Al(III)hydroxide-oxides,[110] platinum and charcoal,[111] and lead dioxide.[112] Some electrochemical investigations on H_2O_2 in nonaqueous and alkaline aqueous media have been described.[113] Reactions between hydrogen peroxide and transition metal complexes are not included in this section.

3.5.2. Sulfur

Reactions of peroxodisulfate are often catalyzed by Ag(I), and the kinetics of the Ag(I)-catalyzed decomposition of peroxodisulfate ion to dioxygen have been investigated.[114] The kinetics are not influenced by added cerium(III) but the formation of O_2 is inhibited by Ce(III). The reaction is greatly accelerated by acrylamide and/or Cu(II) and this results in the absorption of oxygen. The Ag(I)-catalyzed reaction is thought to involve the formation of Ag(II) and $SO_4^{\cdot-}$, and the following scheme has been proposed:

$$Ag(I) + S_2O_8^{2-} \rightarrow Ag(II) + SO_4^{\cdot-} + SO_4^{2-}$$

$$SO_4^{\cdot-} + H_2O \rightarrow HO^{\cdot} + HSO_4^-$$

$$Ag(I) + HO^{\cdot} \rightarrow Ag(II) + OH^-$$

$$Ag(I) + SO_4^{\cdot -} \rightarrow Ag(II) + SO_4^{2-}$$

$$2Ag(II) \rightarrow Ag(I) + Ag(III)$$

$$H_2O + Ag(III) \rightarrow Ag(I) + 1/2O_2$$

However, to accommodate the catalysis by Cu(II) [probably owing to the reduction of Cu(II) to Cu(I) by $HO_2^{\cdot}$] an alternative scheme seems feasible:

$$2^{\cdot}OH \rightleftharpoons H_2O_2$$

$$H_2O_2 + Ag(II) \rightarrow Ag(I) + HO_2^{\cdot} + H^+$$

$$HO_2^{\cdot} + Ag(II) \rightarrow Ag(I) + O_2 + H^+$$

$$Cu(II) + HO_2^{\cdot} \rightarrow Cu(I) + O_2 + H^+$$

$$Cu(I) + S_2O_8^{2-} \rightarrow Cu(II) + SO_4^{\cdot -} + SO_4^{2-}$$

The kinetics of oxidation of indigocarmine by $K_2S_2O_8$ are zero order in the dye, reflecting a rate-determining cleavage of $S_2O_8^{2-}$ to give $SO_4^{\cdot -}$ radical ions.[115] Pulse radiolysis of a solution of styrene and $S_2O_8^{2-}$ gives a transient species formed from styrene and the $SO_4^{\cdot -}$ radical.[116] There have been reports on the kinetics of oxidation by peroxodisulfate of Tl(I),[117] low-spin iron(II) complexes in binary aqueous mixtures,[118] and *cis*-[Coen$_2$(NO$_2$)NCS]$^+$ cations.[119]

Azide is oxidized by peroxomonosulfate to give dinitrogen and dinitrogen oxide, while azidopentaamminechromium(III) is oxidized to [Cr(NH$_3$)$_5$NO]$^{2+}$, generating a NO$^-$ nitrosyl complex. Oxygen-18 tracer experiments show that the reaction proceeds by transfer of a terminal peroxide oxygen of the peroxymonosulfate to the reductant.[120] The oxidation by HSO_5^- of the tris(bipyridyl)iron(II) complex has been reported.[121]

The octacyanomolybdate(V) anion is oxidized by thiosulfate in the pH range 4.0–5.1 with the formation of $S_4O_6^{2-}$.[122] The reduction of tetranitromethane by thiosulfate is catalyzed by Cu(II), and the following mechanism is postulated[123]:

$$C(NO_2)_4 + S_2O_3^{2-} \rightarrow C(NO_2)_3^- + NO_2 + S_2O_3^-$$

$$Cu(II) + S_2O_3^{2-} \rightarrow Cu(I) + S_2O_3^-$$

$$C(NO_2)_4 + Cu(I) \rightarrow C(NO_2)_3^- + NO_2 + Cu(II)$$

$$2S_2O_3^- \rightarrow S_4O_6^{2-}$$

The possible role of soot and water in the atmospheric oxidation of SO_2 has been explored by studying the catalytic oxidation of SO_2 by O_2 on carbon in aqueous suspensions.[124] A complex rate law was obtained, which was interpreted in terms of several equilibria involving the adsorption of O_2 and S(IV) and the formation of $C_x \cdot O_2 \cdot 2S(IV)$, which reacts to give C_x + 2S(VI). The oxidation of thiocyanate to sulfate and cyanate by alkaline hexacyanoferrate(III) with OsO_4 catalyst has been discussed briefly.[125]

The oxidation of several disulfides by gold(III) in aqueous solution proceeds[126] by S–S bond scission to form the sulfenic acid, the gold(III) being reduced to the metal. Oxidation of 1-β-D-thioglucose disulfide may involve a different pathway, namely, the formation of a sulfonic acid which then undergoes hydrolysis by nucleophilic attack to give glucose and sulfite.

A sixth century Chinese alchemical recipe for the solublization of cinnabar for medicinal use suggests a mixture of potassium nitrate, copper sulfate, and vinegar. The success of this procedure has been attributed[127] to a chloride impurity in the nitre and the eventual formation of chlorine!

Details on the following nucleophilic substitutions at sulfur have been published: the carboxylate-ion-catalyzed hydrolysis of a sulfite ester, bis(p-nitrophenyl)sulfite, where the mixed anhydride intermediate has been detected,[128] and the hydrolysis of arenesulfinamides $(R^1C_6H_4SONHC_6H_4R^2)$.[129] In the latter case the hydrolysis proceeds at 25°C by both an acid-catalyzed (A-2) and an H^+-dependent, nucleophile-catalyzed reaction, but at higher temperatures an H^+-independent nucleophilic catalysis could also be identified.

The phenomenon of rupture of a sulfur–carbon bond by ˙OH radical attack on sulfoxides is of long standing, but the reaction has been investigated further by pulse radiolysis and by the use of a conductivity detection method with a time resolution of about 50 ns. Electrophilic addition of the hydroxyl radical to the sulfoxide gives a transient adduct $R_2SO(˙OH)$ which decays into a sulfinic acid RSO_2H and a radical R˙. In mixed sulfoxides the probability of radical split-off is governed by the stability of the leaving radical. Other reactions of hydroxyl radicals may compete, depending on the nature of the R groups.[130]

Sulfur organic radical cations are often stabilized by accepting an electron pair from a second sulfur atom:[131]

$$\overset{+}{S}˙ + :S \rightleftharpoons S^+ \because S \tag{31}$$

A similar stabilization has been observed for halide ions, and is greatest when the electronegativity difference between S and X is minimum. Thus the stabilization is greatest for $R_2S \because I$. Rate constants for the formation and decay of $R_2S \because S$ are given.[132]

3.5.3. Selenium

The species formed in solutions of selenium dioxide in hydrochloric acid (4–12 M), often called dichloroselenious acid, is actually a hydrate of seleninyl chloride ($SeOCl_2$), as shown by Raman studies.[133] At higher [HCl] the pentachloroselenate(V) anion is formed:

$$SeOCl_2 + 3HCl \rightleftharpoons H_3O^+ + SeCl_5^- \tag{32}$$

3.5.4. Tellurium

The Os(VIII)-catalyzed oxidation of Te(IV) to Te(VI) by $[M(CN)_8]^{4-}$ (M = Mo or W) in alkaline solutions occurs by intermediate formation of Te(V).[134]

3.6. Group VII Elements

3.6.1. Fluorine

Further steps have been taken toward the characterization of the fluoroxysulfate anion $[SO_4F]^-$. In addition to the report of the crystal structure of the rubidium salt[135] there has been a useful survey of its decomposition kinetics and general reactivity toward a range of oxidizable substrates. Aqueous fluoroxysulfate decomposes in a complex process to give dioxygen, hydrogen peroxide, and peroxomonosulfate. The product H_2O_2 and O_2 both contain one atom from the solvent water and one atom from the fluoroxysulfate. The $[HSO_5]^-$ contains an atom of solvent oxygen in its terminal peroxide position. It is suggested[136] that $[SO_4F]^-$ reacts with water to give $[HSO_5]^-$ and H_2O_2, where the H_2O_2 probably reacts further with the $[SO_4F]^-$ by a free radical chain reaction to produce O_2. The redox reactions studied were the oxidation of chlorite to ClO_2; $[Co(NH_3)_5ClO_2]^{2+}$ to ClO_2 and a mixture of $[Co(NH_3)_5F]^{2+}$ and $[Co(NH_3)_5H_2O]^{3+}$; Cr^{2+} to Cr^{3+} and $[CrF]^{2+}$ plus a small amount of polynuclear Cr(III) species; and I^- to I_2. So far it appears that oxygen atom transfer is not important in the reactions of $[SO_4F]^-$ but that fluorine transfer does take place. Indeed it has been suggested that any reductant that reacts rapidly with fluoroxysulfate must be a good acceptor for an F atom or F^+ ion.

Another fluorine compound to receive attention is "fluorine perchlorate,"[137] $FClO_4$, now prepared in pure and much more stable form by the thermal decomposition of NF_4ClO_4. Its reaction with fluorocarbons to give alkyl perchlorates has been investigated. Reaction with the unsymmetrical olefin $CF_3CF{=}CF_2$ gives two isomers, indicating that the O–F bond in $FOClO_3$ is of low polarity.

3.6.2. Chlorine

Irradiation of aqueous solutions of perchlorate at 184.9 nm or γ irradiation leads to the formation of $O(^3P)$ atom, the concentration of which may be estimated by measuring the ethylene produced in its reaction with cyclopentene. The $O(^3P)$ atom does not react with perchlorate. It is formed by a direct action mechanism[138] [equations (33) and (34)] and dimerizes to give dioxygen.

$$ClO_4^- \xrightarrow{h\nu} [ClO_4^-]^* \tag{33}$$

$$[ClO_4^-]^* \longrightarrow ClO_3^- + O(^3P) \tag{34}$$

The oxidation of ammonia by hypochlorite,[139] ultimately to dinitrogen, is a well-known reaction. When carried out under uv irradiation the rate of reaction is increased a hundred-fold but the general stoichiometry and course of the reaction are unchanged except that there are rather larger amounts of nitrite formed. Of particular interest is the stage in the reaction when bond formation between nitrogen atoms occurs. It is thought that this occurs after chloramine formation and reactions (35) and (36) are suggested as candidates.

$$Cl_2N^- + NH_2Cl \rightarrow Cl_2N{-}NH_2 + Cl^- \tag{35}$$

$$Cl_2N^- + NCl_3 \rightarrow Cl_2N{-}NCl_2 + Cl^- \tag{36}$$

The kinetics of oxidation of isobutyraldehyde by aqueous chlorine has been studied.[140] The iodine–chlorine reaction in carbon tetrachloride requires catalytic amounts of water, and the rate law is first order in chlorine, iodine and water.[141] The following reaction scheme is suggested[142]:

$$I_2 + Cl_2 + H_2O \rightleftharpoons \text{``}I_2O\text{''} + 2HCl \tag{37}$$

$$\text{``}I_2O\text{''} + H_2O \rightarrow 2IOH \tag{38}$$

$$2IOH + 2HCl \rightarrow 2ICl + 2H_2O \tag{39}$$

3.6.3. Bromine

Without doubt the main interest here lies in the oscillating reactions involving bromate (the Belousov–Zhabotinskii reaction), i.e., the catalytic oxidative bromination with acidic bromate of (usually) aliphatic polycarboxylic acids or polyketones. The best-studied system involves malonic acid and a catalyst, usually cerium. The oxidation of Ce(III) by acidic bromate is inhibited periodically by bromide whenever the concentration of bromide exceeds a critical value.

There is no doubt that these fascinating reactions are now pretty well understood, and indeed computer simulations are now being attempted, although there is still a great deal of uncertainty over the values of certain rate constants. The state of the art is reflected in a series of short reviews by the leading workers in this area,[143] while five different classes of reaction showing oscillation have been discussed in the light of a general mechanism.[144] Other aspects of these reactions have been reviewed,[145–149] while specific topics covered include the effect of O_2[150] and the mechanism of uncatalyzed reactions with bromate.[151–154] The kinetics and mechanism of the Ce(IV)—bromate reaction has been well studied[155] and has implications for the Belousov–Zhabotinskii oscillating reaction. The values of the rate constant are sensitive to the distribution of the Ce(IV) among its various sulfato complexes. This work shows that at least two rate constants used in the scheme for the oscillating reaction are given values that are substantially different from their true value. Some data on the Ce(IV)–chlorous acid reaction are also reported.

3.6.4. Iodine

Sulfanilic acid is oxidized by periodate to azobenzene-4,4'-disulfonic acid in a reaction that involves nucleophilic attack of the aromatic amine group on the iodine, followed by loss of iodate and the formation of a nitrene, which then dimerizes.[156]

The hydrolysis of dilute aqueous solutions of iodine between pH 7 and 10 involves[157] a rapid hydrolysis to give hypoiodous acid, which disproportionates slowly to give iodate and iodide, with overall rate law:

$$-d(\Sigma(I_2))/dt = k[\text{HOI}]^2 + k' [\text{HOI}][\text{OI}^-] \tag{40}$$

Other reactions of iodine studied include the oxidation of isopropanol[158] (for which, at pH 9.18, the rate is independent of $[I_2]$ and involves a rate-determining loss of a proton), and the oxidation of substituted p-phenylenediamines to quinone diimines in a mechanism involving a two-electron step, the iodine attacking the —NR_2 group. This latter work was carried out with a stopped-flow apparatus having five mixing chambers to allow premixing of selected reactants and pH jumps.[159]

Chapter 4

Substitution Reactions of Inert Metal Complexes— Coordination Numbers 4 and 5

4.1. Introduction

Among topics reviewed are solvents and their role in determining inorganic mechanisms,[1] reaction kinetics and solvation in nonaqueous solvents,[2] and initial state and transition state solvation in inorganic reactions.[3,4] The principles of high-pressure kinetics have been briefly outlined[5] and the general question "When is an intermediate not an intermediate?" has been discussed[6] with brief reference to ligand exchange reactions. Reviews on more specific reaction types include substitution reactions of square-planar complexes involving polydentate ligands[7] and reactivity patterns of planar bis(dithiolene)complexes,[8] the latter with reference to Rh, Co, Ni, and Cu in various oxidation states.

During the period under review the general pattern of activity has mainly followed earlier trends and is still dominated by work on the reactions of square-planar complexes of platinum(II) and palladium(II). Although many of the basic features of these reactions are now well established, there are several anomalies and some significant gaps in our knowledge. For example, the comfortable view that associative mechanisms are generally operative is disturbed by the well-known results of investigations into reactions of sterically-hindered Pd(II) complexes and of some anation reactions of $[Pt(dien)H_2O]^{2+}$. In addition, although there has been a great deal of discussion on the *trans* effect in the reactions of Pt(II) complexes there have been comparatively few systematic kinetic studies on the effect of varying the ligand *trans* to the leaving group. Some current contributions to resolving these problems are discussed below. Two approaches, the investigation of solvent effects and the determination of activation volumes, have been particularly prominent in the recent literature. The first, when it aims to differentiate the initial state and transition state solvent effects, involves lengthy measurements often hampered by solubility problems and difficulties in assigning single-ion transfer parameters. The second requires the use of sophisticated apparatus and careful measurements if reliable conclusions are to be reached. Progress has been made in both these areas, and particularly significant contributions have been made by combining measurements of pressure effects with variations in solvent so that the total activation volume can be resolved into components arising from changes in solvation and changes in bond lengths and angles. Particular examples of these aspects of mechanistic studies are discussed more fully in the following sections.

4.2. Square-Planar Complexes

4.2.1. Platinum(II)

The reactions under consideration in this section are summarized in Table 4.1. Some years ago Gray and Olcott[9] in an often-quoted paper reported the surprising result that the anation reactions of $[Pt(dien)(H_2O)]^{2+}$ by Cl^- and by NO_2^- follow a two-term rate law, and they suggested that this could be explained by the occurrence of a dissociative process. These reactions have now been reinvestigated[10] and the authors conclude that, within experimental error, rates of these anations can be represented by a single-term equation, and therefore that Cl^- and NO_2^- do not show anomalous behavior. Another reinvestigation,[11] this time into the pressure dependence of aquation reactions of $[Pt(dien)X]^+$ complexes (X = Cl or Br), reveals that the earlier report[12] of a marked dependence of activation volume on pressure is incorrect. This new result casts doubt upon the value of such pressure dependence ($\Delta\beta_1^{\ddagger}$) as a sound mechanistic parameter. The new experiments show that $\Delta V_1^{\ddagger}$ varies only slightly with ionic strength—the values of $\Delta V_1^{\ddagger}$ are between -9 and -10 cm^3 mol^{-1} for variations in ionic strength from zero to 1.0 M.

Palmer and Kelm and coworkers have carried out comprehensive investigations into several substitution reactions of Pt(II) complexes by measuring the activation parameters ($\Delta H^{\ddagger}$, $\Delta S^{\ddagger}$, and $\Delta V^{\ddagger}$) and solvent effects. In the first group

TABLE 4.1. *Summary of Reactions of Pt(II) Complexes*

	Reagents[a]	Conditions[b]	Parameters measured[c]	Ref.
(a)	$[Pt(dien)(H_2O)]^{2+} + Y$ $(Y = Cl^-, NO_2^-$, and other nucleophiles)	15°C, $\mu = 0.1\ M$	k	10
(b)	$[Pt(dien)X]^+ + H_2O$ $(X = Cl$ or $Br)$	25°C, $\mu = 0{-}1\ M$	$\Delta V^{\ddagger}$	11
(c)	$trans\text{-}[Pt(py)_2Cl(NO_2)] + py$	Acetone, nitromethane, MeOH, EtOH, CH_2Cl_2		
	$cis\text{-}[Pt(py)_2Cl(NO_2)] + py$	MeOH	$\Delta H^{\ddagger}, \Delta S^{\ddagger}, \Delta V^{\ddagger}$	13
	$trans\text{-}[Pt(PEt_3)_2Cl_2] + Y$ $(Y = Br^-$ or $py)$	MeOH		
(d)	$cis\text{-}$ and $trans\text{-}[Pt(2,4,6\text{-}$ $Me_3C_6H_2)(PEt_3)_2Br] + tu$	EtOH, DMSO, acetone	$\Delta H^{\ddagger}, \Delta S^{\ddagger}, \Delta V^{\ddagger}$	17
(e)	$[PtCl_4]^{2-} + H_2O$	Aq. MeOH, aq.t-BuOH	k_{aq}	19
	$[PtCl_3(H_2O)]^- + Cl^-$		k_{Cl}	
(f)	$[PtCl_4]^{2-} + H_2O$	25°C	Time for establishing equilibrium	20
(g)	$[PtCl_4]^{2-} + CN^-$	25–35°C, borate buffer	$\Delta H^{\ddagger}, \Delta S^{\ddagger}$	21
(h)	$[Pt(NO_2)_4]^{2-} + CN^-$	25–35°C	k	22
(i)	$[PtCl_n(H_2O)_{4-n}]^{2-n} + C_2H_4$	25°C, 1.0 M	k	23
(j)	$trans\text{-}[PtCl_2(S\text{-}mbn)L] + ol$ $(L =$ derivative of py or aniline; $ol = cis\text{-}1,2\text{-}$ dichloroethylene, 2,3-dimethyl-2-butene or mbn)	8–25°C Benzene, CH_2Cl_2	$\Delta H^{\ddagger}, \Delta S^{\ddagger}$	24
(k)	$cis\text{-}[Pt(4CN\text{-}py)_2Cl_2] + Y$ $(Y = tu, SCN^-$, or $I^-)$	25°C Aq. MeOH, THF, CH_3CN, DMSO	k	25
(l)	$[Pt(SEt_2)Cl_3]^- + am$ $(am =$ one of ten amines, mainly py derivatives)	30°C, MeOH	k	26
(m)	$[PtCl_3(basH_2)]^+ \rightarrow$ ring closure $[PtCl_2(basH)]^+ \rightarrow$ ring closure	19–40°C, 0.4 M	$\Delta H^{\ddagger}, \Delta S^{\ddagger}$	28
(n)	$[Pt(en)(glycine)Br]^+ \rightarrow$ ring closure	19–50°C, pH 5.0–7.3	k	29

[a] dien, $NH_2CH_2CH_2NHCH_2CH_2NH_2$; py, pyridine; mbn, 2-Me-2-butene; tu, thiourea; en, $NH_2CH_2CH_2NH_2$; bas, $S(CH_2CH_2NH_2)_2$.
[b] Temperature, ionic strength, solvent (water unless otherwise specified).
[c] k, rate constants for the reactions shown.

of reactions,[13] involving the substrates *cis*- and *trans*-[Pt(py)$_2$(Cl)(NO$_2$)], and *trans*-[Pt(PEt$_3$)$_2$Cl$_2$] [Table 4.1(c)], most of which were studied by using conductivity measurements, the k_1 term in the normal two-term rate equation is virtually negligible so that activation parameters refer to the k_2 path (except that the reactions in acetone do not go to completion and the rate constants given are therefore composite values for the forward and reverse reactions). The activation parameters $\Delta H^{\ddagger}$ (~ 50 kJ mol^{-1}) and $\Delta S^{\ddagger}$ (large negative values) show no marked variation with solvent or with the character of the substrate. However, the activation volumes (known with greater precision than the $\Delta S^{\ddagger}$ values) show clear trends, and a plot of $\Delta V^{\ddagger}_{exp}$ against the solvent electrostriction parameter (a function of the dependence of dielectric constant on pressure) is linear. This allows an interpretation of $\Delta V^{\ddagger}_{exp}$ as the sum of $\Delta V^{\ddagger}_{intr}$ and $\Delta V^{\ddagger}_{solv}$ where these two components refer, respectively, to the part of $\Delta V^{\ddagger}_{exp}$ involving changes in bond lengths and angles, and the part involving solvation changes. The authors conclude that potentially the quantity $\Delta V^{\ddagger}_{exp}$ contains more explicit information as to the intimate nature of the reaction mechanism than does $\Delta S^{\ddagger}$. Also, the linear relationship between $\Delta V^{\ddagger}$ and $\Delta S^{\ddagger}$ postulated for some reactions of octahedral complexes is not found in these reactions, a fact that may be worth bearing in mind in view of the recent discussions on the interpretation of activation volumes in octahedral substitutions.[14-16]

Another group of reactions [17] [Table 4.1(d)] involves the sterically hindered bromomesityl-bis(triethylphosphine) complexes previously investigated with methanol as solvent.[18] Activation parameters ($\Delta H^{\ddagger}$, $\Delta S^{\ddagger}$, and $\Delta V^{\ddagger}$) were measured for the k_1 path and for the k_2 path in these reactions, except that no reliable estimates could be made of k_1 for the *trans* isomer in ethanol. As expected, steric hindrance is more important in the reactions of the *cis* isomer [$k_2(cis)/k_2(trans) < 0.1$] which gives larger $\Delta H^{\ddagger}_2$ values and more negative values of $\Delta V^{\ddagger}_2$. Values of activation parameters derived from the k_1 term in the rate equation support the conclusion that an associative mechanism is operative. In the reactions with DMSO (k_1 path) the *trans* isomer reacts more slowly than the *cis* isomer. This surprising result is interpreted in terms of formation of Pt–O bonds (*cis* isomer) and Pt–S bonds (*trans* isomer).

There have been several reports on reactions of the tetrachloroplatinate(II) ion and one on tetranitroplatinate(II) [Table 4.1(e)–(h)]. Measurement of the rate of aquation of [PtCl$_4$]$^{2-}$ and of the rate of anation of the product, [PtCl$_3$(H$_2$O)]$^-$, in aqueous methanol and aqueous *t*-butanol leads to a comparison with other transition metal chloro-complexes and a somewhat elaborate rationalization of the results in terms of the reactant charge product difference and solvent permittivity effects.[19] The catalytic effect of [Pt(DMSO)Cl$_3$]$^-$ on the aquation of [PtCl$_4$]$^{2-}$ is very briefly reported.[20]

Elding and Gröning[23] have investigated a complicated system, the substitution reactions of ethene with substrates of general formula [PtCl$_n$(H$_2$O)$_{4-n}$]$^{2-n}$

(see Table 4.2). Comparison of the second-order rate constants for substitution reactions of several complexes shows that ethene as an entering group is similar to DMSO, possibly because of steric requirements for the formation of the transition state. Spectra (200–300 nm) of several ethene–chloro–aqua complexes are given. The kinetic results are discussed in terms of the general stoichiometric mechanism in which aqua-complexes are reactive enough to be steady-state intermediates in associative solvent paths. The authors give alternative explanations for earlier reports of unreactive solvento–Pt(II) intermediates and conclude that there is as yet no conclusive evidence for such species. Miya *et al.*[24] have also studied substitution reactions of Pt(II)–olefin complexes, [PtCl$_2$L(ol)] (L = derivative of pyridine or aniline), in this case by the unusual method of using an optically active substrate and measuring the decrease in CD strength on replacement of the asymmetrically coordinated prochiral olefin with a non-prochiral olefin. Evidence is given for a single-path mechanism (when L = pyridine derivative) and for a two-path mechanism (when L = substituted aniline). One interesting and unexplained result is that for the reactions of *trans*-[PtCl$_2$(S-mbn)(4-Xpy)] with mbn (X = CN or H, mbn = 2-Me-2-butene) the $\Delta S^{\ddagger}$ values are positive, whereas the reactions of the same complexes with *cis*-1,2-dichloroethylene give the kind of negative $\Delta S^{\ddagger}$ values more usually found for substitution reactions of Pt(II) complexes.

Owing to its greater solubility in water the complex *cis*-[Pt(4CN-py)$_2$Cl$_2$] is more suitable than [Pt(bipy)Cl$_2$] as a substrate for substitution reactions designed to investigate solvent effects. Reactions of this *cis* complex with thiourea and other entering groups have been studied [25] and the results analyzed into initial state and transition state effects without, in the case of thiourea, encountering the difficulties of assigning single-ion transfer parameters for charged reactants. The transition state effects are smaller than the initial state effects but they are not insignificant. These issues are discussed more fully elsewhere.[3,4] Tobe *et al.*[26] have continued their study of kinetic *trans* effects and now compare

Table 4.2. *Rate Constants at (k) at 25°C and* μ = *1.00 M for Reactions of Ethene with Various Substrates*[23]

Substrate	Leaving group	$10^3 k/M^{-1}$ s^{-1}
[PtCl$_4$]$^{2-}$	Cl$^-$	1.1 ± 0.1
[PtCl$_3$(H$_2$O)]$^-$	H$_2$O	9.1 ± 0.8
trans-[PtCl$_2$(H$_2$O)$_2$]	H$_2$O	4.0 ± 0.4
trans-[PtCl$_2$(H$_2$O)$_2$]	Cl$^-$	5.9 ± 0.4
cis-[PtCl$_2$(H$_2$O)$_2$]	H$_2$O	36 ± 5
[PtCl(H$_2$O)$_3$]$^+$	*trans*-H$_2$O	35 ± 3
[Pt(H$_2$O)$_4$]$^{2+}$	H$_2$O	10 ± 1

the *trans* effect of SEt_2 with that of SMe_2, reported earlier.[27] In general the SEt_2 complexes are more reactive but the authors, in a detailed discussion, point out that the relative reactivities are dependent on the nature of the complex and of the entering group and that several factors must be taken into account when attempting to interpret these results. Other reported aspects of the *trans* effect and *trans* influence include the ^{195}Pt nmr of ^{15}N-labeled amine complexes,[28] the use of ^1H and ^{13}C nmr (and X-ray techniques) to study Pt(II) complexes with Schiff-base ligands[29] and X-ray structures of arsine complexes.[30]

Kinetics of ring closure of the ligand $S(CH_2CH_2NH_2)_2$ (bas) have been studied[31] by using the complexes $[PtCl_3(basH_2)]Cl$ and $[PtCl_2(basHCl)]$. As expected, the observed rate constants show an inverse acid dependence and, by a full analysis of the data, it is shown that in basic media the first ring closure is about 20 times faster than the second. The values of activation parameters for the first and second closures are, respectively, $\Delta H^{\ddagger} = 14.2 \pm 0.3$ and 15.7 ± 0.1 kcal mol^{-1}, and $\Delta S^{\ddagger} = -8.2 \pm 0.8$ and -9.0 ± 0.5 cal deg^{-1} mol^{-1}. These values of $\Delta S^{\ddagger}$ are rather more positive than those normally found for substitution reactions of Pt(II) complexes, consistent with an intramolecular mechanism. Comparison of the activation parameters for the two steps suggests that more strain is involved in reaching the transition state for the second step, a conclusion that is consistent with results from other ring-closure reactions involving amine ligands [e.g., $MeN(CH_2CH_2NH_2)_2$]. The kinetics of glycine ring closure has also been reported.[32] In the more general area of Pt(II) substitutions an attempt has been made to develop equations (based on published data) for calculating rate constants.[33] The method used is to represent the total free energy of activation as the sum of contributions from the three nonreacting ligands (T, C_1, and C_2) and the leaving and entering groups (R and N, respectively) in the general reaction [equation (1)].

$$\begin{array}{c} C_1 \diagdown \qquad \diagup R \\ Pt \\ T \diagup \qquad \diagdown C_2 \end{array} \quad + N \longrightarrow \quad \begin{array}{c} C_1 \diagdown \qquad \diagup N \\ Pt \\ T \diagup \qquad \diagdown C_2 \end{array} \quad + R \qquad (1)$$

One set of empirical constants is obtained for various groupings of T, C_1, and C_2 and two other sets (parameters R and N, respectively) for various leaving and entering groups. These empirical constants are then used quite successfully to calculate rate constants for several reactions (other than those from which the constants were obtained), the parameter R for reactions in which the transition state energy is determined by the bond-breaking process, and the parameter N for reactions in which the energy of bond formation is more important.

4.2.2. Palladium(II)

Palmer and Kelm[34,35] have carried out extensive investigations into the reactions of the sterically hindered complexes of Pd(II) containing the ligand Et₄dien [Table 4.3(a) and (b)]. In aqueous solutions[34] the activation volumes ($\Delta V_{exp}^{\ddagger}$) corresponding to the k_1 path (independent of entering group concentration) were found to be an approximately linear function of the overall volume changes for the reactions:

$$[Pd(Et_4dien)X]^+ + H_2O \rightarrow [Pd(Et_4dien)(H_2O)]^{2+} + X^- \qquad (2)$$

From the gradient of this plot it is concluded that bond breaking is about 50% complete in the transition state and, making use of this result, the total activation volume can be apportioned between the bond-breaking and bond-making processes. For the latter the calculated value, $(\Delta V^{\ddagger})_1$, is essentially constant for a series of leaving groups and its mean value ($-15.5 \pm 0.9 \text{ cm}^3 \text{ mol}^{-1}$) is very close to the estimated value of the volume of a water molecule in the second coordination sphere of a singly charged complex cation. These results give support to the I_a mechanism. A corresponding detailed analysis was not possible for the nonaqueous solvents[35] (because supporting volume data—partial molar volumes of the reacting species or the total volume change—were not obtainable), but the trend in $\Delta V^{\ddagger}$ values (MeOH ~ EtOH > H₂O > DMSO > DMF) parallels the order of decreasing cross-sectional area at the oxygen atom in these solvent molecules. Once again an associative interchange mechanism is suggested as being most likely. A different approach to the problem of the mechanism of substitution in these sterically hindered Pd(II) complexes has been made by Blandamer *et al.*[36] They have examined salt effects on the reactions of [Pd(Et₄dien)Cl]⁺ with alkali metal and alkylammonium bromides, and they report the striking sensitivity of the first-order rate constants to the concentration of added salt and the nature of the added cation. These effects are not explicable on the basis of a Brønsted–Bjerrum analysis using the Debye–Hückel limiting law. Unaware of the result of Palmer and Kelm[34] they have proceeded on the assumption that the mechanism is dissociative and that the transition state is larger than the initial state. By combining the results of solubility measurements of the substrate in solutions of KCl, [NEt₄]Cl, and [NBu₄ⁿ]Cl with the observed salt effects they have rationalized their results on this basis. Although Palmer and Kelm[35] do not report values of $\Delta V^{\ddagger}$ for the specific reaction studied by Blandamer *et al.*[36] their $\Delta V^{\ddagger}$ values for a series of similar reactions are all within a small range and are all negative. It therefore seems probable that, under the conditions used by Palmer and Kelm the activation volume for the reaction of [Pd(Et₄dien)Cl]⁺ with Br⁻ would be negative, contrary to the assumption made by Blandamer *et al.* However, in the presence of high concentrations of

Table 4.3. Summary of Reactions of Pd(II) Complexes

Reagents[a]	Conditions[b]	Parameters measured[c]	Ref.
(a) $[Pd(Et_4dien)X]^+$ + Y^-		$\Delta H^{\ddagger}$, $\Delta S^{\ddagger}$, $\Delta V^{\ddagger}$	34
(X = Cl, Br, N_3, I, NCS, or NH_3^+; Y = N_3, I, or Br)			
(b) $[Pd(Et_4dien)I]^+$ + Br^-	MeOH, EtOH, DMSO, DMF, CH_3CN	$\Delta H^{\ddagger}$, $\Delta S^{\ddagger}$, $\Delta V^{\ddagger}$	35
(c) $[Pd(Et_4dien)Cl]^+$ + Br^-	25°C, various salt concentrations	k	36
(d) $[Pd(L–L)Cl_2]$ + Y	25°C, DMF	k	37
($L–L$, derivative of 1,10-phenanthroline; Y = Br^-, I^-, SCN^-, or thiourea)			
(e) $[Pd(acac)_2]$ + am	25°C, MeOH, benzene, THF	k	38
(acac = acetylacetone anion, am = alkylamine)			
(f) $[Pd(Fo)X]$ + Y	25°C, MeOH, CH_3CN, DMF, acetone, 1,4-dioxane	k	39
(Fo = 1-(2-hydroxyphenyl)-3,5-diphenylformazan; X = NH_3 or pyridine; Y = PPh_3, Me_2CS, or SCN^-)			

[a] Et_4dien, $Et_2NCH_2CH_2NHCH_2CH_2NEt_2$.
[b] Temperature, solvent (water unless otherwise specified).
[c] k, rate constants for the reactions shown.

alkylammonium salts, which appear to have remarkable specific effects, a different result may be found.

Substitution reactions of $[Pd(L-L)Cl_2]$ [Table 4.3(d)] obey the usual two-term rate law for replacement of the first chloride[37] (the second chloride is replaced rapidly). The discriminating abilities of these complexes (calculated from n_{Pd}^0 values obtained using the standard substrate *trans*-$[Pd(PR_3)_2(NO_2)_2]$) increase with decreasing pK_a of the ligand (L–L) and, as in the case of some Pt(II) complexes, the behavior of thiourea as entering group is anomalous.

An interesting solvent effect is revealed in a study[38] of the reactions of $[Pd(acac)_2]$ with alkylamines [Table 4.3(e)]. One acetylacetonate ligand is replaced in each case but whereas the rate equation is of the normal two-term form in methanol ($k_{obs} = k_0 + k_1$ [am]), in benzene and in THF the solvent path is not observed but $k_{obs} = k_1[am] + k_2[am]^2$. These results are accounted for by suggesting that the substrate may be involved in hydrogen-bonding interactions with methanol, but in benzene and THF where such interactions with the solvent are not possible a corresponding interaction with the amine occurs, thus leading to a term in the rate equation that involves the square of the amine concentration. Evidence is given that Pd–C bonded species are not intermediates in these reactions. Solvent effects have also been investigated[39] for the reactions summarized in Table 4.3(f). The normal two-term rate law was found and the initial state and transition state solvent effects were determined. Nonspecific solvation effects appear to be dominant and an examination of the interdependence of the variable factors (entering group, leaving group, and solvent) leads to the conclusion that the bond-making and bond-breaking processes are synchronous.

4.2.3. Nickel(II)

Most of the work reported on reactions of Ni(II) complexes is considered in other sections of this book (e.g., substitution reactions of octahedral complexes and reactions of labile complexes). However, a few papers are of special interest here, including one in which there is a rare comparison of square-planar complexes of Pt(II), Pd(II), and Ni(II).[40] The reactions studied are the replacement of various coordinated bidentate ligands containing sulfur donor atoms (ethyl xanthate, dialkyldithiophosphates, diphenyldithiocarbamate, monothioacetylacetonate) by dithiocarbamate in acetone solution. In all, 24 reactions were studied and several different stoichiometries and types of kinetic behavior were found. In many cases the substrate, $M(S–S)_2$, reacted with the nucleophile, $(S'–S')^-$, to give the product $M(S'–S')_2$ in a second-order process and no intermediates were detected. In other cases a mixed-ligand intermediate, $M(S–S)(S'–S')$, was observed and the total reaction to form $M(S'–S')_2$ was followed in two stages. In a few cases an intermediate containing three ligand molecules was detected but no definite conclusion could be reached as to whether such intermediates were 4- or 5-coordinate, i.e., whether they were of types **1, 2,** or **3.**

It was pointed out that the same kind of difficulty arises in interpreting the earlier results of Pearson and Sweigart.[41] The results in Table 4.4 were obtained for the reactions of dithiophosphates of Pt, Pd, and Ni with the *N,N*-diethylthio-carbamate ion. The trend in activation parameters is consistent with a greater degree of bond formation for Ni(II) and a more significant contribution of bond breaking for Pt(II). The effects of attacking nucleophile and leaving group, and the relationship between rates of substitution and Lewis acidity, is discussed for the Ni(II) complexes. Bidentate ligands containing sulfur (and selenium) donor atoms have also been used in a kinetic study[42] of the formation of mixed-ligand complexes in reactions of the type

$$[Cu(S{-}S)_2]^{2-} + [M(S'{-}S')_2]^{2-} \rightleftharpoons [Cu(S{-}S)(S'{-}S')]^{2-}$$
$$+ [M(S'{-}S')(S{-}S)]^{2-} \quad (3)$$

The reactions (M = Ni, Pd, or Pt) were followed by determining the changes in epr spectra of the Cu(II) species. The fact that the exchange is inhibited by free ligand and accelerated by solvated Cu(II) ions is not in accord with the mechanism proposed earlier[43] for this type of reaction. It is suggested that a chain mechanism operates but that the chain carriers are the monoligand complexes [Cu(S—S) and M(S'—S')] rather than the free ligands. Two papers by

Table 4.4. Kinetic Data for the Reactions of Dithiophosphate Complexes of Pt, Pd, and Ni with N,N-Diethylthiocarbamate[40]

	Pt(II)	Pd(II)	Ni(II)
Relative second-order rate constants (25°C, μ = 0.1 *M* in acetone)	1	1900	9×10^5
$\Delta H^{\ddagger}$/kJ mol^{-1}	76 $\pm$ 2	42 $\pm$ 1	20 $\pm$ 1
$\Delta S^{\ddagger}$/J mol^{-1} K^{-1}	-12 ± 7	-62 ± 3	-86 ± 5

Cusumano report reactions of square-planar Ni(II) chelates with bipy and other bidentate heterocyclic amines. In the first[44] rate constants were measured for the forward and reverse reactions:

$$[Ni(X\text{-}xan)_2] + bipy \underset{k_r}{\overset{k_f}{\rightleftharpoons}} [Ni(X\text{-}xan)_2(bipy)] \tag{4}$$

(X-xan = O-substituted xanthate) in benzene at 25°C. This reaction represents the addition of a bidentate ligand to give an octahedral complex. The latter presumably has the *cis* configuration so a substitution process is involved in the reaction. Values of k_f and k_r are dependent upon the nature of the substituent X, as are the equilibrium constants (calculated from k_f/k_r as they are too large to be measured directly with accuracy). For the reaction in which X = cyclohexyl the activation parameters are $\Delta H_f^{\ddagger} = 5.4 \pm 0.6$ kcal mol^{-1}, $\Delta H_r^{\ddagger} = 24 \pm 1$ kcal mol^{-1}, $\Delta S_f^{\ddagger} = -21 \pm 4$ cal K^{-1} mol^{-1} and $\Delta S_r^{\ddagger} = 26 \pm 3$ cal K^{-1} mol^{-1}. These values reflect the nature of the processes involved and are similar to values found for analogous reactions of Ni(II) substrates when steric factors were absent. The substituent effects are discussed in terms of Taft's σ^* parameters. Reactions

4

of the square-planar complex **4** with bidentate heterocyclic amines in wet chlorobenzene also yields octahedral products[45] (see Scheme 1).

Values of $\Delta H^{\ddagger}$, $\Delta S^{\ddagger}$, ΔH° and ΔS° are given and a value of less than unity for the gradient of the free energy relationship [log(rate constant) as ordinate versus log(equilibrium constant)] implies an I_a mechanism. The planar complexes [NiX$_2$L$_2$] (X = Cl, Br, I, or NCS; L = pyridine derivative with methyl groups in the 2- and 6-position) are uniquely inert to substitution and are not even attacked by strong mineral acids.[46] This lack of reactivity is attributed to steric hindrance.

Scheme 1

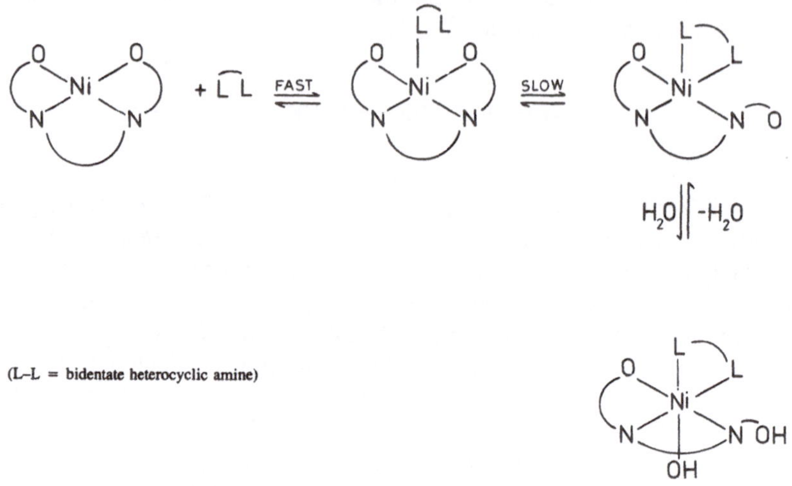

(L–L = bidentate heterocyclic amine)

4.2.4. Gold(III)

Cis and trans effects in square-planar gold(III) complexes are the subject of a review.[47] Skibsted[48] has carried out a thorough investigation of the reactions in which the four ammonia ligands in $[Au(NH_3)_4]^{3+}$ are replaced successively with Br^-. Two distinct stages are observed experimentally, the first rapid and the second slow. The data from the rapid stage do not give rise to simple pseudo-first-order rate constants but they fit a scheme of two consecutive reactions (having comparable rates) which are identified as the successive replacement of the first two NH_3 ligands. The author draws attention to (and resolves) the problem in systems of this kind of assigning the calculated rate constants to the two consecutive steps. (Incidentally, in this connection he refers to an early treatment[49] of this problem which was overlooked by others[50,51]). The replacement of one NH_3 ligand from $[Au(NH_3)_3Br]^{2+}$ gives trans-$[Au(NH_3)_2Br_2]^+$, and the reaction of the latter by a pseudo-first-order process to give the tribromoammine is identified as the experimentally observed slow stage of the overall process. The fourth NH_3 ligand is displaced rapidly. From the dependence of the pseudo-first-order rate constants on $[Br^-]$, and from the temperature dependence, the second-order rate constants and activation parameters shown in Table 4.5 were obtained for the successive replacement of NH_3 from the tetra-ammine. Alexander and Holper[52] report the effect of solvent composition (aqueous MeOH) on the kinetics of replacement of Cl^- by Br^- in $[Au(Et_4dien-H)Cl]^+$ (cf. Ref. 53). They find that the results obey a two-term

Table 4.5. Kinetic Data for the Replacement of Ammonia Ligands from Au(III) Complexes[48]

Substrate	k/M^{-1} s^{-1} (25°C)	$\Delta H^{\ddagger}/kJ$ mol^{-1}	$\Delta S^{\ddagger}/J$ mol^{-1} K^{-1}
$[Au(NH_3)_4]^{3+}$	3.4 ± 0.08	73 ± 3	8 ± 3
$[Au(NH_3)_3Br]^{2+}$	6.5 ± 0.4	69 ± 3	2 ± 10
trans-$[Au(NH_3)_2Br_2]^+$	$9.3 \pm 0.3 \times 10^{-5}$	88 ± 3	-26 ± 8

rate law but the contribution from the bromide-dependent path is fairly small— so the present results are not really in conflict with the earlier report[54] that the complex reacts at rates that are almost independent of the concentration of the entering group. The correlations between the rate constants and the Grunwald–Winstein solvent Y values are consistent with the view that both reaction paths are associative. The very complicated reactions

$$[AuN_nCl_x] + Cl^- \longrightarrow [AuN_{n-1}Cl_{x+1}]^- + N \tag{5}$$

(where N denotes a nucleotide or nucleoside) have been investigated kinetically in 1:9 v/v DMSO–MeOH.[55]

4.2.5. Miscellaneous

Reactions of square-planar (and tetragonal) complexes other than those considered above are listed here as a cross-reference to the sections on labile and organometallic complexes. These include the following reactions of Rh(I)[56,57]:

$$[Rh(Cl)(CH_3CN)(PPh_3)_2] + PPh_3 \longrightarrow [Rh(Cl)(PPh_3)_3] + CH_3CN \tag{6}$$

$$[Rh(CO)_2Cl_2]^- + py \longrightarrow [Rh(CO)_2Cl_2(py)]^+ \longrightarrow [RhCl_3py_3] \tag{7}$$

[in the report[57] of reaction (7) it is not clear what species is reduced when the Rh(I) is converted to Rh(III)], and Ir(I)[58]:

$$[(cod)IrCl(pic)] + am \longrightarrow [(cod)Ir(am)] + Cl^- + pic \tag{8}$$

(cod = 1,5-cyclo-octadiene, pic = 2-picoline, am = 2,2'-bipyridyl or 1,10-phenanthroline).

Fast reactions of Cu(II) include studies on ligand exchange in bis(salicylaldimato) complexes,[59,60] the formation of macrocyclic complexes,[61] the displacement of macrocyclic ligands,[62] and the replacement of water by ammonia in the Cu(II) aqua-ion, followed by the T-jump method.[63] The displacement by Cu(II) of macrocyclic ligands from Zn(II) complexes has also been reported.[64]

4.2.6. Isomerization

The mechanisms of *cis–trans* isomerization of Pt(II) and Pd(II) complexes have been the subject of a great deal of study and controversy for many years. The arguments continue, but some clarification of the issues has been brought about by the publication of a welcome review by Anderson and Cross.* The types of mechanism hitherto proposed have also been reviewed briefly by Favez et al.[66] as an introduction to their study of the *cis–trans* isomerization of $[PtX_2L_2]$ in the presence of PR_3 (L = PR_3 with R = Me, Et, *n*-Bu, or *p*-tolyl; X = Cl, Br, or I) in dichloromethane. The authors point out that, as much of the kinetic data available are open to more than one interpretation, experimental investigation should concentrate on the characterization of intermediates in solution by the most direct method and, for that reason they have used ^{31}P nmr spectroscopy. With this and other spectroscopic techniques they have identified in solution the four-coordinate species $[PtXL_3]^+$ and the five-coordinate $[PtI_2(PMe_3)_3]$ and $[PtXL_4]^+$. In the case of $[PtCl(PMe_3)_3]Cl$ an X-ray crystal structure analysis shows this to contain a four-coordinate cation. The various dynamic processes involving these species are also investigated and the final conclusion is that the isomerization occurs by a double-displacement mechanism of the type

$$cis\text{-}[PtX_2L_2] + L \overset{fast}{\rightleftharpoons} [PtXL_3]^+ + X^- \qquad (9)$$

$$[PtXL_3]^+ + X^- \overset{slow}{\rightleftharpoons} trans\text{-}[PtX_2L_2] + L \qquad (10)$$

A brief report[67] of the *cis–trans* isomerization of $[PtCl_2(PhCN)_2]$ in chloroform shows this to be very slow (k_c = 3.8 × 10^{-6} s^{-1}, k_t = 2.9 × 10^{-7} s^{-1} at ~ 25°C.) but no mechanism is proposed. Several other papers deal with the subject of *cis–trans* isomerization of square-planar complexes but these all involve organometallic compounds and are reviewed in a later section of this book.

---

*See Ref. 65; cf also the Corrigenda (following p. 411 of *Chem. Soc. Rev.* 9) which gives corrections to this article.

4.3. Tetrahedral Complexes

The mechanism of oxygen (^{18}O) exchange between chromate ions and water appears to be very complicated.[68] Chloride ion is not a catalyst but at least five processes are involved and the individual rate constants (s^{-1} or M^{-1} s^{-1} at 25°C, $\mu = 1.0$ M) have been evaluated for these (Scheme 2). This scheme does not

Scheme 2

$CrO_4^{2-} + H_2*O$	$\rightleftharpoons$	$CrO_3*O^{2-} + H_2O$	$k_1 = 3.2 \pm 0.2 \times 10^{-7}$
$HCrO_4^- + H_2*O$	$\rightleftharpoons$	$HCrO_3*O^- + H_2O$	$k_2 = 2.3 \pm 0.2 \times 10^{-3}$
$HCrO_4^- + H_2*O + H^+$	$\rightleftharpoons$	$HCrO_3*O^- + H_2O + H^+$	$k_3 = 7.3 \pm 0.9 \times 10^6$
$HCrO_4^- + CrO_4^{2-}$	$\rightleftharpoons$	$Cr_2O_7^{2-} + OH^-$	$k_4 \approx 10^{-3}$
$2HCrO_4^-$	$\rightleftharpoons$	$Cr_2O_7^{2-} + H_2O$	$k_5 = 9.0 \pm 1.1$

account for a discrepancy between values of k_5 determined by this and other methods, and it is considered that there may be further reactions involved with third-order rate constants. The rapid ligand exchange reactions of tetrahedral bis(5-thio- and selenopyrazolonato)Ni(II) complexes of the type in equation (11) have been investigated by nmr measurements at low temperature[69]:

$$[NiL_2] + [NiL_2'] \longrightarrow 2[NiLL'] \tag{11}$$

4.4. Five-Coordinate Complexes

This field of study is represented by a variety of metals but most of the complexes are labile or organometallic and they are reviewed elsewhere in this book. This section is therefore to be regarded merely as a convenient cross-reference. Coates *et al.*[70] have continued studies on five-coordinate Cu(II) complexes of the "umbrella" ligand (L), $N(CH_2CH_2NMe_2)_3$. These complexes cannot be described as "inert" but their reactions are considerably slower (rate constants in the range 1–60 s^{-1} at 283–308 K) than normally encountered for Cu(II), and are considered to occur by a dissociative interchange mechanism:

$$[CuL(H_2O)]^{2+} + Y^- \rightleftharpoons \{CuL(H_2O \ldots Y\}^+ \rightleftharpoons [CuLY]^+ + H_2O \tag{12}$$

where $Y^- = NCO^-$, Cl^-, or Br^-. Kinetic studies on the following fast reactions have also been reported[71]:

$$[NiX(QAS)]^+ + Y^- \xrightarrow{\text{MeOH}} [NiY(QAS)]^+ + X^- \tag{13}$$

(where QAS = tris-(o-diphenylarsino)phenyl; X = Br for Y = NO_2, I, NCS, N_3, or CN; X = I, NO_2, Cl, N_3, NCS, or Br for Y = CN) and for the reactions of the Co(I) complex $[Co(dmgBF_2)_2P(C_4H_9)_3]^-$ with acrylonitrile, methyl mercaptide, and cyanide.[72] The exchange of monoolefins with $[Fe(CO)_4(PhCH{=}CH_2)]$ is slow and dissociative, whereas the exchange of diolefins with $[Fe(CO)_3(diolefin)]$ is predominantly associative.[73]

Chapter 5

Substitution Reactions of Inert Metal Complexes— Coordination Numbers 6 and Above

5.1. Introduction

This review follows on from the previous report by the present authors on this area.[1] It covers the literature up to approximately December 1980, with a few references beyond that date. There has been the customary varied activity in the area of substitution kinetics covered in this chapter. In this first section we indicate briefly regions in which we feel that the most useful and significant progress has been made in the time covered by this volume.

The use of activation volumes in the diagnosis of mechanism has continued to provide much valuable information. Activation volumes for substitution at octahedral complexes have formed the subject of a well-referenced review,[2] in which the importance both of intrinsic and of solvation contributions is recognized. The topics of most relevance to this chapter include isomerization and racemization reactions of cobalt(III) complexes, aquation of cobalt(III) and of iron(II) complexes, and base hydrolysis of cobalt(III) complexes. Merbach's continuing investigations[3] into the effects of pressure on rates of solvent exchange at $2+$ and $3+$ transition metal cations, while not being always strictly relevant to this chapter, have provided the very important generalization that substitution mechanisms appear to change from associative to dissociative as one traverses the first row of transition metals, both for $2+$ and $3+$ ions. Langford[4] has expressed some reservations about the derivation of this generalization from the determined activation volumes. Many specific instances of the usefulness of $\Delta V^{\ddagger}$ values in mechanistic diagnosis will be found in the appropriate sections of this chapter.

Another fundamental aspect which continues to exercise the ingenuity of kineticists is the search for possible five-coordinate intermediates in substitution at cobalt(III). Recently a variety of approaches, involving especially kinetic and stereochemical experiments on base hydrolysis and on induced aquation, have been employed (see Section 5.7.3).

As the majority of substitution reactions of low-spin d^6 complexes of cobalt(III) and iron(II) appear to proceed by dissociative mechanisms, the well-established existence of second-order pathways for some such reactions, especially of diimine—iron(II) species, has prompted considerable research and discussion. The operation of the much-debated mechanism of initial attack by the nucleophile at the coordinated diimine has proved remarkably difficult to prove or disprove with any certainty. Progress in this field has been limited of late, with various key experiments proving impossible to execute or merely providing further equivocal evidence (Section 5.4.2).

Much more progress has been made in the field of photochemistry. Further work on chromium(III) complexes increases our understanding of this much studied area of photoreactivity, while the developing field of photochemical substitution at rhodium(III) has begun to exhibit coherent patterns (Section 5.8.10). Here there is usually a difference of reaction stereochemistry between photochemical and thermal activation of *cis* complexes of the ammine– or amine–halide type, with photochemical substitution generally giving a *trans* product, but thermal substitution proceeding without isomerization. A general picture of photochemical substitution at rhodium(III) is emerging, in which the reaction sequence involves ligand loss, giving a five-coordinate triplet intermediate of square-pyramidal geometry. Such intermediates have only a very transitory existence, of course, but persist long enough for the ligands to take up their preferred axial or equatorial positions.

Another area in which considerable recent activity is beginning to give some sort of consistent pattern is that of solvent effects on reactivity of transition metal complexes, both for substitution and for redox reactions. The general aspects of initial state and transition state contributions to such solvent effects have been well reviewed by Buncel and Wilson,[5] though their examples are drawn almost exclusively from organic chemistry. A similar but somewhat briefer treatment of inorganic systems, perforce concentrating on transition metal complexes, has also appeared.[6] This review amplifies and extends an earlier outline of this same topic.$^{(7)}$ Later in this chapter we discuss in some detail the most recent treatments of this sort, in dealing with Wells's analysis of solvent effects on reactivity for complex formation reactions (Section 5.7.4.2) and for solvolysis of the *trans*-[Co(py)$_4$Cl$_2$]$^+$ cation (Section 5.7.1.3). At the moment one can be fairly confident of the qualitative conclusions of such analyses, but accurately quantitative dissection of trends into initial state and transition state components lies some way into the future. This statement is particularly true for reactions involving ions, for in such cases the kinetic interpretation is based on single-ion thermodynamic parameters whose derivation is still a matter of controversy.

5.2. Chromium

5.2.1. Introduction

Mechanistic studies of chromium chemistry tend to concentrate on complexes of the inert $+3$ oxidation state which are, therefore, dealt with first. A few references to work involving other oxidation states are included at the end. There have been no major reviews of mechanistic chromium chemistry during this period, although a useful article$^{(8)}$ deals with the preparations and reactivities of fluoro-containing complexes of chromium(III). Further reviews deal with the determination of volumes of activation, and their use for elucidating inorganic reaction mechanisms, and these are particularly useful for an understanding of chromium(III) substitutions.$^{(2,9)}$

5.2.2. Aquation and Solvolysis of Chromium(III) Complexes

5.2.2.1. Unidentate Leaving Groups

5.2.2.1.1. Aquo Complexes. Aquation of complexes of the type [Cr(H$_2$O)$_5$CH$_2$X]$^{2+}$ and [Cr(H$_2$O)$_5$CHX$_2$]$^{2+}$ (X = Cl, Br, or I) have been studied in 0.02–1.0 mol dm^{-3} [H$^+$], and $I = 1.0$ mol dm^{-3}, with light and oxygen absent. For [Cr(H$_2$O)$_5$CH$_2$X]$^{2+}$ ions, when X = Cl or Br, the products of the reaction are [Cr(H$_2$O)$_6$]$^{3+}$, X$^-$, and methanol, whereas when X = I, iodomethane is the major product formed. The [Cr(H$_2$O)$_5$CHX$_2$]$^{2+}$ ions react to give [Cr(H$_2$O)$_6$]$^{3+}$, X$^-$, CO, HCO$_2$H, and H$_2$.$^{(11)}$ The rate data from these studies are collected in Table 5.1. In the case of [Cr(H$_2$O)$_5$CH$_2$X]$^{2+}$ ions, a simple acid-independent first-order rate law is observed. The chloro and bromo complexes form [Cr(H$_2$O)$_5$CH$_2$OH]$^{2+}$ ion by S_N2 attack of water at the CH$_2$X group:

$$[(H_2O)_5CrCH_2X]^{2+} + H_2O \longrightarrow [(H_2O)_5CrCH_2OH]^{2+} + HX \quad (1)$$

$$[(H_2O)_5CrCH_2OH]^{2+} + H_3O^+ \longrightarrow [Cr(H_2O)_6]^{3+} + CH_3OH \quad (2)$$

Table 5.1. Rate Constants and Activation Parameters at 298.2 K ($I = 1.0$, $NaClO_4/HClO_4$) for the Aquation of $[Cr(H_2O)_5CH_2X]^{2+}$ and $[Cr(H_2O)_5CHX_2]^{2+}$ Ions ($X = Cl$, Br, or I)[(10,11)]

Complex	$10^7k/s^{-1}$	$\Delta H^{\ddagger}/kcal\ mol^{-1}$	$\Delta S^{\ddagger}/cal\ K^{-1}\ mol^{-1}$
$CrCH_2Cl^{2+}$	5.6	26.1 ± 0.2	8.6 ± 0.5
$CrCH_2Br^{2+}$	24	24.3 ± 1.3	-2.7 ± 4.1
$CrCH_2I^{2+}$	4.5	25.4 ± 0.2	-2.4 ± 0.7
$CrCHCl_2^{2+}$	41	24.4 ± 1.5	1.2 ± 1.1
$CrCHBr_2^{2+}$	14	27.3 ± 1.5	6.3 ± 4.6
$CrCHI_2^{2+}$	0.85^a	27.3 ± 1.6	0.8 ± 1.7

^{a} These data refer to the acid-independent pathway; an inverse [H$^+$] dependence was also measured.

The $[Cr(H_2O)_5CH_2I]^{2+}$ ion is postulated to react via a carbanion-type transition state in which a proton is transferred from the incoming water molecule to the leaving CH_2I^- group to give MeI and $[Cr(H_2O)_5OH]^{2+}$, which rapidly protonates.

In the case of $[Cr(H_2O)_5CHX_2]^{2+}$ ions, a simple acid-independent first-order rate law is observed again, except when $X = I$ for which the rate law is: rate $= (k_0 + k_{-1}[H^+]^{-1})\ [Cr(H_2O)_5CHI_2^{2+}]$; at 298 K, $10^9k_0 = 2.0\ s^{-1}$, $\Delta H^{\ddagger} = 28.9 \pm 1.2\ kcal\ mol^{-1}$ and $\Delta S^{\ddagger} = -3.2 \pm 3.4\ cal\ K^{-1}\ mol^{-1}$. The following mechanism is proposed for the acid-independent reactions. Consistent with Scheme 1 is the observation that the reactions of $[Cr(H_2O)_5CDX_2]^{2+}$ ions give HD as

Scheme 1

$[CrCHX_2]^{2+} + H_2O$	$\xrightarrow{slow}$	$[CrCH(OH)X]^{2+} + HX$	
$[CrCH(OH)X]^{2+}$	$\rightarrow$	$[CrCHO]^{2+} + HX$	
$[CrCHO]^{2+} + H_2O$	$\rightarrow$	$[CrCO]^{2+} + H_2 + OH^-$	
$[CrCO]^{2+}$	$\rightarrow$	$Cr^{3+} + CO$	
$[CrCO]^{2+} + H_2O$	$\rightarrow$	$Cr^{3+} + HCO_2H$	

the major hydrogen product, and there is a negligible kinetic isotope effect. The inverse acid dependence in the case of $[Cr(H_2O)_5CHI_2]^{2+}$ ion could arise either from the participation of the $[Cr(H_2O)_4(OH)CHI_2]^{2+}$ conjugate base, or from initial attack by OH$^-$ rather than H_2O. Since it is known that the ratio k_{OH}/k_{H_2O} decreases for nucleophilic attack at alkyl halides along the series RI > RBr > RCl, it seems likely that this is the reason for the absence of an inverse acid dependence for the chloro and bromo species.

Two studies involving mixed DMSO–H_2O solvates of Cr(III) have been reported.[(12,13)] Continuing an investigation of the labilization of Cr(III) complexes by coordinated oxyanions, the reaction of the nitratopenta-aquochromium(III) ion in acidic water–DMSO mixtures has been reported. Up to five DMSO molecules become coordinated to Cr(III), depending on the DMSO concentration, but $[Cr(DMSO)_6]^{3+}$ ion is not observed to form. The most likely explanation of the results is that nitrogen–oxygen bond fission is involved, and that the coordinated nitrate ion catalyzes H_2O–DMSO exchange prior to N–O fission. At 298 K, the half-life for solvent exchange between $[Cr(H_2O)_5ONO_2]^{2+}$ and DMSO is ~10 s in acidic solution, a remarkably rapid reaction for the normally inert Cr(III). Loss of the nitrate ion from the mixed solvento complexes is much slower ($t_{1/2}$ ~3 h). Complexes with HSO_3^- and HCO_3^- also form rapidly with aquo-chromium(III) complexes, without Cr–O bond breaking, and these oxyanion species also undergo solvent exchange much more rapidly with DMSO.$^{(12,13)}$ The effect of nitrite ion has also been studied, and the formation constant for the reaction between CO_2 and $[Cr(H_2O)_6]^{3+}$ ion to give $[Cr(H_2O)_5OCO_2H]^{2+}$ deduced from kinetic studies.$^{(13)}$

In two other studies the catalyzed loss of ligands from $[Cr(H_2O)_5X]^{2+}$ ions has been reported.$^{(14,15)}$ The chromium(II)-catalyzed substitution of Br^- in $[Cr(H_2O)_5Br]^{2+}$ by F^- and Cl^- has been studied.$^{(14)}$ In the presence of Cl^- ions, a third-order rate law, rate $= k_{Cl}[Cr^{2+}][Cl^-][CrBr^{2+}]$ is reported, whereas substitution by F^- ions involves the more complex rate law, rate $= (r[HF]/[H^+])[Cr^{2+}][CrBr^{2+}]/(1 + s[HF]/[H^+])$, where r and s are constants; at 298 K, $\Sigma[\text{anions}] = 1.0$ mol dm^{-3}, $k_{Cl} = 16.1 \pm 0.7$ dm^6 mol^{-2} s^{-1} ($\Delta H^{\ddagger} = 10.8$ kcal mol^{-1}, $\Delta S^{\ddagger} = -17$ cal K^{-1} mol^{-1}), $r = 3.06 \pm 0.18$ dm^3 mol^{-1} s^{-1}, and $s = 2.3 \pm 0.5$. An inner-sphere redox mechanism is proposed as:

$$Cr^{2+} + X^-/HX \rightleftharpoons CrX^+ (+H^+)\ (K_1) \qquad \text{rapid} \qquad (3)$$

$$CrBr^{2+} + CrX^+ \rightleftharpoons \{CrBrCrX^{3+}\}^{\ddagger} \rightleftharpoons Cr^{2+} + CrXBr^+ \qquad (k_2,k_{-2}) \quad (4)$$

$$CrXBr^+ + Cr^{2+} \rightleftharpoons \{BrCrXCr^{3+}\}^{\ddagger} \rightleftharpoons Br^- + Cr^{2+} + CrX^{2+} \qquad (k_3) \quad (5)$$

With this scheme, $k_{Cl} = K_1k_2/(1 + k_{-2}/k_3)$, and $r = K_1K_2k_3$, $s = K_1K_2$. Indirect support for the proposed mechanism comes from the synthesis, in low yields, of $[CrClBr]^+$ ion from the reaction of Cr^{2+} with aqueous bromine in 0.5 mol dm^{-3} [HCl].

The attack by H_2O_2 of coordinated thiolato ligands (L) in $[Cr(H_2O)_5L]^{n+}$ (L $= SCH_2CH_2NH_2$ or $SC_6H_4NH_3$), $[Cr(H_2O)_4(SCH_2CO_2)]^+$, and $[M(en)_2L]^{n+}$ (M $=$ Co(III) or Cr(III), L $= [SCH_2CH_2NH_2]^-$, $[SCH_2CO_2]^{2-}$, or $[SC(Me)_2CO_2]^{2-}$; M $=$ Cr(III), L $= [SCH_2CH_2CO_2]^{2-}$) has been investigated. All of the Cr(III) complexes undergo Cr–S bond fission via a presumed unstable sulfenato intermediate, with nucleophilic attack by the coordinated thiolate group at the O–O bond of H_2O_2. Coordinated thiolates are good nucleophiles especially when coordinated to Co(III) rather than Cr(III).$^{(15)}$ Selected rate data at 298 K for the aquation (k_1) and attack by H_2O_2 (k_2) are collected in Table 5.2.

Table 5.2. Rate Constants and Activation Parameters for the Aquation (k_1) and
Reaction with H_2O_2 (k_2) of Thiolato–Chromium(III) Complexes[15]

Complex	10^4k, s^{-1}	k_2, dm^3 mol^{-1} s^{-1}	$\Delta H^{\ddagger}$ (k_2), kcal mol^{-1}	$\Delta S^{\ddagger}$ (k_2), cal K^{-1} mol^{-1}
[Cr(en)$_2$(SCH$_2$CO$_2$)]$^+$	11.1	1.13	9.30	-27.0
[Cr(en)$_2$(SCH$_2$CH$_2$CO$_2$)]$^+$	10.4	1.19	9.99	-24.7
[Cr(en)$_2$(SCMe$_2$CO$_2$)]$^+$	7.94	0.16	10.65	-26.6
[Cr(en)$_2$(SCH$_2$CH$_2$NH$_2$)]$^{2+}$		1.13	9.71	-25.8
[Cr(H$_2$O)$_5$(SCH$_2$CH$_2$NH$_3$)]$^{3+}$	2.30^a	0.176^a	10.6	-26.5
[Cr(H$_2$O)$_5$(SC$_6$H$_4$NH$_3$)]$^{3+}$		0.0451^b	10.8	-28.5
[Cr(H$_2$O)$_4$(SCH$_2$CO$_2$)]$^+$	2.10^c	0.445^c	10.4	-25.6

^{a} 25.5°C. ^{b}26.0°C. ^{c}26.6°C.

5.2.2.1.2. Ammine and Amine Complexes. Following the synthesis of [Cr(NH$_3$)$_5$CN]$^{2+}$ ion by the reaction between CN$^-$ and [Cr(NH$_3$)$_5$(DMSO)]$^{2+}$ ions in DMSO, the kinetics of the acid-catalyzed hydrolysis were examined from 5×10^{-4} to 2.0 mol dm^{-3} [H$^+$], and ionic strengths *(I)* in the range 0.5–2.0 mol dm^{-3}. The usual acid-independent and acid-dependent pathways were ob-

Scheme 2

$$[\text{Cr(NH}_3)_5\text{CN}]^{2+} \quad + \quad 2\text{H}_2\text{O} \quad \xrightarrow{k_o} \quad [\text{Cr(NH}_3)_5\text{OH}_2]^{3+} \quad + \quad \text{HCN} \quad + \quad \text{OH}^-$$

$$[\text{Cr(NH}_3)_5\text{CN}]^{2+} \quad + \quad \text{H}^+ \quad \overset{K}{\rightleftharpoons} \quad [\text{Cr(NH}_3)_5\text{CNH}]^{3+}$$

$$[\text{Cr(NH}_3)_5\text{CNH}]^{3+} \quad + \quad \text{H}_2\text{O} \quad \xrightarrow{k_H} \quad [\text{Cr(NH}_3)_5\text{OH}_2]^{3+} \quad + \quad \text{HCN}$$

served (Scheme 2): $k_{obs} = k_0 + k_H K[\text{H}^+]/(1 + K[\text{H}^+])$, and at 298 K, $I = 1.0$ mol dm^{-3}, $10^3 k_H K = 2.62 \pm 0.03$ dm^3 mol^{-1} s^{-1} ($\Delta H^{\ddagger} = 19.2 \pm 0.8$ kcal mol^{-1}), $10^2 k_H = 3.3 \pm 0.7$ s^{-1} ($\Delta H^{\ddagger} = 15.3 \pm 1.3$ kcal mol^{-1}, $\Delta S^{\ddagger} = -14.0 \pm 4.3$ cal K^{-1} mol^{-1}), and $K = 0.079 \pm 0.017$ dm^3 mol^{-1}. Values of k_H were found to decrease with ionic strength by a factor of ~2.5, and K to increase by a factor of 6–7. At $I = 2.0$ mol dm^{-3}, $k_0 < 10^{-7}$ s^{-1}, $10^2 k_H = 1.47 \pm 0.04$ s^{-1}, $K = 0.279 \pm 0.009$ dm^3 mol^{-1}, and the acid-dependent pathway dominates over the acid-independent pathway. [Cr(NH$_3$)$_5$CN]$^{2+}$ ion was observed to be more robust at high pH to base-catalyzed decomposition than many other acidopentamminechromium(III) ions.[16]

Other kinetic studies of [Cr(NH$_3$)$_5$X]$^{2+}$ ions reported recently include a computer-based spectrophotometric investigation of the halide complexes,[17] and a study of the effect of alcohol (ROH; R = Me, Et, Pr, or Bu)–water mixtures on the rate of aquation of the bromo complex and its Co(III) analog.[18] The effect of ion pairing by dicarboxylates (succinate, malonate, malate, tartrate, maleate, and phthalate), and sulfate ion on the rate of aquation of $[M(NH_3)_5X]^{2+}$ ions (X = Cl or Br) has been investigated in 10% EtOH–H_2O (M = Co) and in aqueous solution (M = Cr).[19]

The rate of aquation of the $[Cr(NH_3)_5F]^{2+}$ ion increases linearly with $[Al^{3+}]$, with $10^4 k_{Al} = 7.5$ dm^3 mol^{-1} s^{-1} at 308 K, pH 3.0 and $I = 2.0$ mol dm^{-3}. Decreasing the pH causes a reduction in the observed rate due to protonation of the coordinated fluoride ion (Scheme 3), k_2 was found to be negligibly small

Scheme 3

$$[Cr(NH_3)_5F]^{2+} \quad + \quad Al^{3+} \quad \xrightarrow[(H_2O)]{k_1} \quad [Cr(NH_3)_5OH_2]^{3+} \quad + \quad AlF^{2+}$$

$$\Updownarrow K_H$$

$$[Cr(NH_3)_5FH]^{3+} \quad + \quad Al^{3+} \quad \xrightarrow[(H_2O)]{k_2} \quad [Cr(NH_3)_5OH_2]^{3+} \quad + \quad AlF^{2+} \quad + \quad H^+$$

such that $k_{obs} = k_1[Al^{3+}]/(1 + K_H[H^+])$, where at 308 K the protonation constant, $pK_4 = 3.0 \pm 0.1$, and $\log(k_1[Al^{3+}]) = -4.00 \pm 0.05$. The value of pK_H is rather surprising since for free fluoride ion at 308 K, $pK_a = 3.2 \pm 0.2$. The effect of ion pairing by ClO_4^- and $[Co(CN)_6]^{3-}$ ions was also investigated briefly.[20]

For the oxidation of coordinated azide ion in $[Cr(NH_3)_5N_3]^{2+}$ by peroxymonosulfate, the rate law is $k[Cr(NH_3)_5N_3^{2+}][HSO_5^-]$, and at 298 K, $10^2 k = 1.3$ dm^3 mol^{-1} s^{-1}. Oxidation of free azide ion by peroxymonosulphate was also studied over a wide pH range.[21]

Results obtained at 298 K for the rates of water exchange with *cis*- and *trans*-$[Cr(NH_3)_4(OH_2)_2]^{3+}$, *fac*-$[Cr(NH_3)_3(OH)_3]^{3+}$, and *trans*-$[Cr(NH_3)_2(OH_2)_4]^{3+}$ ions are collected in Table 5.3, together with data for $[Cr(H_2O)_6]^{3+}$ and

Table 5.3. Rate Data at 298.2 K, I = 1.0 mol dm⁻³ (Perchlorate), for Water Exchange with Aquaamminechromium(III) Complexes[22]

Complex	$10^5 k$, s^{-1} ^{a}	$\Delta H^\ddagger$, kJ mol^{-1}	$\Delta S^\ddagger$, JK^{-1} mol^{-1}
$[Cr(NH_3)_5OH_2]^{3+}$	5.75 ± 0.06	99.1 ± 1.4	$+ 7 \pm 5$
cis-$[Cr(NH_3)_4(OH_2)_2]^{3+}$	5.92 ± 0.13	95.1 ± 1.9	$- 7 \pm 6$
trans-$[Cr(NH_3)_4(OH_2)_2]^{3+}$	1.17 ± 0.03	98.5 ± 2.1	$- 9 \pm 7$
fac-$[Cr(NH_3)_3(OH_2)_3]^{3+}$	5.78 ± 0.09	95.1 ± 1.3	$- 7 \pm 5$
trans-$[Cr(NH_3)_2(OH_2)_4]^{3+}$	0.997 ± 0.013	97.0 ± 1.1	$- 15 \pm 4$
$[Cr(OH_2)_6]^{3+}$	0.246 ± 0.012	109.6 ± 1.3	$+16 \pm 5$

^{a} These rate constants have not and should not be statistically corrected for reasons explained by the authors.

$[Cr(NH_3)_5OH_2]^{3+}$ ions.[22] The reactions were studied at high acidity where they are acid independent. For a series of reactions of the type:

$$[CrL_5X] + H_2O \longrightarrow [CrL_5OH_2] + X \text{ (charges omitted)} \qquad (6)$$

where CrL_5 is $[Cr(NH_3)_5]$ *cis-* or *trans-*$[Cr(NH_3)_4(OH_2)]$, *fac-*$[Cr(NH_3)_3(OH_2)_2]$, or $[Cr(H_2O)_5]$, and $X = I^-$, Br^-, Cl^-, NCS^-, OH_2, and NH_3, it was found possible to divide values of $\Delta G^{\ddagger}$ into two terms, one dependent on CrL_5, and the other on X:

$$\Delta G^{\ddagger} = \Delta G^{\ddagger}(ML_5) + \Delta G^{\ddagger}_X \qquad (7)$$

Values of $\Delta G^{\ddagger}_X - \Delta G^{\ddagger}_{OH_2}$, in kJ mol^{-1}, are -7, -1, 0, $+6$, $+14$, $+18$, and the relative reactivity ratios $k_X/k_{H_2O} = 17$, 1.5, 1. 1/11, 1/280, and 1/1400 for $X = I^-$, Br^-, OH_2, Cl^-, NCS^-, and NH_3 respectively. Plots of $\Delta G^{\ddagger}$ for CrL_5X against values for $[Cr(H_2O)_5X]$ are linear, with the kinetic *trans* effect of co-ordinated NH_3 evident.[22]

Solvent exchange rates are reported for $[Cr(NH_3)_5OH_2]^{3+}$ ion in DMSO and $[Cr(NH_3)_5(DMSO)]^{3+}$ ion in H_2O.[23] For the former, solvolysis, reaction $\Delta H^{\ddagger} = 22.74$ kcal mol^{-1} and $\Delta S^{\ddagger} = -2.98$ cal K^{-1} mol^{-1}, whereas for the aquation reaction $\Delta H^{\ddagger} = 20.30$ kcal mol^{-1} and $\Delta S^{\ddagger} = -11.92$ cal K^{-1} mol^{-1}. The rates did not vary significantly with ionic strength and acidity (typically 0.1 mol dm^{-3} [HClO$_4$]), and there is no loss of ammonia. The rate constants are $10^4 k = 3.2$ s^{-1} (49.5°C) and $10^4 k = 6.1$ (50.0°C) for the aquation and solvolysis reactions, respectively. The following ions were also prepared or characterized spectrally in solution: *trans-*$[Cr(NH_3)_4S_1S_2]^{3+}$ ($S_1 = S_2 = $ DMSO; $S_1 = H_2O$, $S_2 = $ DMSO) and *cis-*$[Cr(NH_3)_4S_1S_2]^{3+}$ ($S_1 = S_2 = $ DMF; $S_1 = $ DMSO, $S_2 = $ DMF; $S_1 = H_2O$, $S_2 = $ DMSO or DMF), and the spectrochemical series for Cr(III) is reported to be $H_2O < $ DMF $< $ DMSO $< $ Cl.[23]

Arrhenius parameters for the acid hydrolysis of $[Cr(RNH_2)_6]^{3+}$ ions are reported to be $\log_{10}(A/s^{-1}) = 8.7$ and 14.7, $E_a = 21.8$ and 29.7 kcal mol^{-1} when R $= $ H and Me, respectively.[24] The reactions are acid independent up to pH 6, and in perchloric acid $[Cr(RNH_2)_5OH_2]^{3+}$ ion is the only chromium product. For $[Cr(NH_3)_6]^{3+}$ ion in nitric acid, however, $[Cr(NH_3)_5ONO_2]^{2+}$ is reported to be the immediate product (i.e., no aquo complex is formed) and similar behavior is observed in H_2SO_4, HCl, and HBr solutions.

References to recent studies of *cis-* and *trans-*$[Cr(en)_2(X)Y]^+$ ions are collected in Table 5.4.[25-28] For loss of the first cyanide ion from *cis-*$[Cr(en)_2(CN)_2]^+$ ion, the acid-independent pathway is negligible and $k_{obs} = k_H K[H^+]/(1 + K[H^+])$, where $1 \gg K[H^+]$. Plots of k_{obs} versus $[H^+]$ were linear up to 0.3 mol dm^{-3} [H$^+$], and the value of $k_H K$ found to be 25 times larger than for the

Table 5.4. Recent Studies of cis- and trans-[Cr(en)₂(X)Y]⁺ Ions (Rate Data at 298 K; Y⁻ = Leaving Ion)

Ref.	X	Y	I, mol dm^{-3}	$10^5 k_0$, s^{-1}	$10^4 k_H K$, dm^3 mol^{-1} s^{-1}	$\Delta H^\ddagger$, kJ mol^{-1}	$\Delta S^\ddagger$, JK^{-1} mol^{-1}
25	*cis*	CN⁻	CN⁻	1.0	0	199	78 ± 2	−16 ± 4
26	*cis*	Cl⁻	Br⁻	0.1	117		61.2 ± 3.5	−95 ± 11
27	*trans*	Br	ONO	Aquates with simultaneous loss of Br⁻ and ONO⁻ when [H⁺] = 10⁻³ mol dm⁻³. 10⁴k = 1.7 s⁻¹ at 283 K.				
28	*cis*	Cl	Cl	S_N2 mechanism for ³⁶Cl⁻ exchange study.				

corresponding *cis*-[Cr(H₂O)₄(CN)₂]⁺ ion (for the aquo complex $\Delta H^\ddagger$ = 86 kJ mol⁻¹, $\Delta S^\ddagger$ = −23 J K⁻¹ mol⁻¹). The acid hydrolysis reaction proceeds with retention of configuration, and Ag⁺ ion was observed to promote cyano–isocyano linkage isomerism[25]:

$$\text{Cr—CN} + \text{Ag}^+ \rightarrow \text{Cr—NC—Ag}^+ \tag{8}$$

Thus *cis*-[Cr(en)₂(CN)₂]⁺ ion in the presence of Ag⁺ results in isomerization of both CN⁻ ions to give a visible spectrum resembling that of other Cr–N₆ chromophores.[25]

The aquation rate for *cis*-[Cr(en)₂(Cl)Br]⁺ ion (Br⁻ loss) is intermediate between that of the corresponding dichloro and dibromo complexes, and the rate data fit the linear free energy relationship which correlates $\Delta G^\ddagger$ values for corresponding cobalt and chromium complexes.[26]

The *trans*-[Cr(en)₂(ONO)Br]⁺ ion is reported to aquate in 0.1 mol dm⁻³ [NaClO₄]/10⁻³ mol dm⁻³ [H⁺] with simultaneous loss of NO₂⁻ (major pathway) and Br⁻ ions. [27] At 283 K, 10⁴k = 1.7 s⁻¹, and it is calculated that for loss of Br⁻, 10⁵k = 4.5 s⁻¹, and for loss of NO₂⁻, 10⁵k = 13 s⁻¹.

An S_N2 mechanism is postulated for radio-chloride exchange with *cis*-[Cr(en)₂Cl₂]⁺ ion.[28]

The acid and base hydrolysis, and ligand field photolysis of [Cr(tren)F₂]⁺ ion, and the solvolysis of [M(tren)Cl₂]⁺ ions (M = Cr or Co) in DMSO, DMF, *N*-methylformamide, and formamide solutions[25] have been reported (tren = β,β′,β″-triamino triethylamine). For the thermal loss of the first F⁻ ion from [Cr(tren)F₂]⁺, rate = $k_H K$[H⁺][complex], where at 298 K, 10²$k_H K$ = 2.40 dm³ mol⁻ s⁻¹, $\Delta H^\ddagger$ = 14.5 ± 0.5 kcal mol⁻¹ and $\Delta S^\ddagger$ = −15.2 ± 1.4 cal K⁻¹. The rate increases in D₂O because the protonation constant, K, is larger. The extrapolated acid-independent rate constant for aquation at 298 K is estimated to be $10^5k = 1.5 \pm 0.5$ s^{-1}. The product, α-[Cr(tren)(H$_2$O)F]$^{2+}$ ion, aquates further with rupture of one of the Cr–N terminal bonds:

$$[Cr(tren)(H_2O)F]^{2+} + H_3O^+ \rightarrow [Cr(trenH)(H_2O)_2F]^{3+} \tag{9}$$

The rate data at 298 K for this reaction are $10^4k = 6.79 \pm 0.30$ s^{-1}, $\Delta H^{\ddagger} = 18.90 \pm 0.50$ kcal mol^{-1}, and $\Delta S^{\ddagger} = -9.7 \pm 1.9$ cal K^{-1} mol^{-1} ($I = 0.5$ mol dm^{-3}). Ligand field photolysis of [Cr(tren)F$_2$]$^+$ ions in acidic solution gives [Cr(tren)(H$_2$O)F]$^{2+}$ ion with a quantum yield of 0.21.[29]

A dissociative mechanism is proposed for the solvolysis of [M(tren)Cl$_2$]$^+$ ions (M = Co or Cr) in dipolar aprotic solvents.[30] The rate data are collected in Table 5.5. Despite the differences in $\Delta S^{\ddagger}$, cal K^{-1} mol^{-1}, for Cr(III) compared with Co(III) which are -11.5 ± 3.7, -10.2 ± 3.5, -17.9 ± 4.0, and -14.0 ± 4.0 in DMSO, DMF, N-methyl formamide, and formamide respectively, a dissociative mechanism is favored. The postulated dissociative mechanism for this Cr(III) complex is believed to arise from steric hindrance by the tren to the more usual nucleophilic attack. Since the Cl$^-$ ion is lost from a cis position with respect to the tertiary N atom of tren, this is believed to be in line with a dissociative mechanism. Values of $\Delta V^{\ddagger}$ are being obtained to confirm this behavior. The rate constants increase with changes in solvent in the order DMF > DMSO > formamide > H$_2$O > N-methylformamide, in line with the changes in the dielectric constants, and as expected for a dissociative mechanism.[30]

Following the synthesis of [Cr(L)(R)(OH$_2$)]$^{2+}$ ions (L = 1,4,8,12-tetra-azacyclopentadecane, [15]aneN$_4$; R = alkyl), from the reactions of RX with [Cr(L)]$^{2+}$, the kinetics of the Hg^{2+}- and MeHg$^+$-catalyzed S_E2 solvolysis reactions, to give [Cr(L)(OH$_2$)]$^{3+}$ ions, have been investigated.[31] The rate data are collected in Table 5.6. A family of bridged bis(benzylchromium) cations

Table 5.5. Comparison of Activation Parameters for the Solvolysis of [Co(tren)Cl$_2$]$^+$ and [Cr(tren)Cl$_2$]$^+$ Ions in Nonaqueous Solvents at 298 K[30]

	[Co(tren)Cl$_2$]$^+$		[Cr(tren)Cl$_2$]$^+$	
Solvent	$\Delta H^{\ddagger}$, kcal mol^{-1}	$\Delta S^{\ddagger}$, cal K^{-1} mol^{-1}	$\Delta H^{\ddagger}$, kcal mol^{-1}	$\Delta S^{\ddagger}$, cal K^{-1} mol^{-1}
Formamide	20.1	-4.9	14.9	-18.9
DMSO	21.2	-7.0	17.7	-18.5
DMF	24.6	4.0	22.0	-6.2
NMF	17.1	-11.2	10.4	-29.1
H$_2$O	17.9	-10.4	12.3	-23.0

Table 5.6. Kinetic Data at 298.2 K for the Hg^{2+} and $MeHg^+$ Catalyzed Loss of R Groups from $[Cr(L)(R)(OH_2)]^{2+}$ Ions (L = 1,4,8,12-tetraazacyclopentadecane, $[15]aneN_4)^{(31)}$

R	$k(MeHg^+)$, dm³ mol⁻¹ s⁻¹	$k(Hg^{2+})$, dm³ mol⁻¹ s⁻¹	$\Delta H\ddagger$ (Hg^{2+}), kJ mol⁻¹	$\Delta S\ddagger$ (Hg^{2+}), J mol⁻¹ K⁻¹
Me	1.63×10^3	3.1×10^6		
Et	9.9^a	2.5×19^3	30.2 ± 1.2	-77 ± 4
PhCH₂	5.2	1.1×10^3	33.4 ± 1.6	-75 ± 5
Prⁿ		82.1	48.8 ± 1.5	-45 ± 5
Buⁿ		48.8	33.5 ± 2.1	-100 ± 7
C₅H₁₁		43.3	29.7 ± 1.2	-114 ± 4
Prⁱ		4.3×10^{-3}		
C₆H₁₁		1.6×10^{-3}		
Adamantyl		3.1×10^{-3}		

^{a} For the reaction with $EtHg^+$, $k = 8.2$ dm³ mol⁻¹ s⁻¹.

analagous to structure **1** has also been prepared,$^{(32)}$ and homolysis of a Cr—C

$$\left[(H_2O)_5 CrCH_2 - \bigcirc - CH_2 Cr(OH_2)_5\right]^{4+}$$

1

bond shown to be the rate-determining step in the redox reactions with Fe^{3+} and Cr^{2+}.

5.2.2.2. Multidentate Leaving Groups

The aquation of α-$[Cr(tren)(H_2O)F]^{2+}$ to give $[Cr(trenH)(H_2O)_2F]^{3+}$ ion$^{(24)}$ was described in Section 5.2.2.1.2. Cr–N bond cleavage also occurs during the acid hydrolysis of $[Cr(tren)(ox)]^+$ ion.

$$[Cr(tren)(ox)]^+ + 4H_2O \rightarrow [Cr(H_2O)_4(ox)]^+ + trenH_4^{4+} \qquad (10)$$

Intermediate, partially coordinated tren complexes of the type $[Cr(trenH)(ox)OH_2]^{2+}$, $[Cr(trenH_2)(ox)(OH_2)_2]^{3+}$, and $[Cr(trenH_3)(ox)(OH_2)_3]^{4+}$ were isolated from the reaction mixture by ion exchange chromatography.$^{(33)}$ The acid-independent rate data for the stepwise unwinding of the tren molecule are collected in Table 5.7. Fission of the first Cr–N bond in the strained tren complex occurs 20 times faster than in the less strained α-$[Cr(trien)(ox)]^+$ ion, and subsequent reactions are much slower. From studies in concentrated hydro-

Table 5.7. Rate Data for the Stepwise Loss of β,β',β''-Triaminotriethylamine (tren) from $[Cr(tren)(ox)]^+$ Ion[33]

Complex	$\Delta H^{\ddagger}$, kcal mol^{-1}	$\Delta S^{\ddagger}$, cal K^{-1} mol^{-1}
$[Cr(tren)(ox)]^+$	19.0 ± 1.3	-4.8 ± 1.9
$[Cr(trenH)(ox)(OH_2)]^{2+}$	19.5 ± 1.1	-14.7 ± 1.8
$[Cr(trenH_2)(ox)(OH_2)_2]^{3+}$	25.0 ± 1.5	-4.3 ± 2.0
$[Cr(trenH_3)(ox)(OH_2)_3]^{4+}$	26.0 ± 1.5	-4.7 ± 2.0

chloric acid solutions, the following mechanism is proposed for the formation of $[Cr(tren)Cl_2]^+$ ion (Scheme 4). Cr–N bond dissociation is the rate-determining

Scheme 4

$$[Cr(tren)(ox)]^+ \quad + \quad H^+ \quad \overset{rapid}{\rightleftharpoons} \quad [Cr(tren)(oxH)]^{2+}$$

$$[Cr(tren)(oxH)]^{2+} \quad + \quad Cl^- \quad \rightarrow \quad [Cr(tren)(oxH)Cl]^+$$

$$[Cr(tren)(oxH)Cl]^+ \quad + \quad H^+ \quad \rightleftharpoons \quad [Cr(tren)(oxH_2)Cl]^{2+}$$

$$[Cr(tren)(oxH_2)Cl]^{2+} \quad + \quad Cl^- \quad \rightarrow \quad [Cr(tren)Cl_2]^+ \quad + \quad oxH_2$$

step in the acid hydrolysis of the *trans*-$[Cr(nta)_2]^{3-}$ ion[34] (Scheme 5). From

Scheme 5

$$[Cr(nta)_2]^{3-} \quad + H^+ \quad \overset{K_1}{\rightleftharpoons} \quad [Cr(nta)(ntaH)]^{2-} \qquad rapid$$

$$[Cr(nta)(ntaH)]^{2-} \quad + H^+ \quad \overset{K_2}{\rightleftharpoons} \quad [Cr(ntaH)_2]^- \qquad rapid$$

$$[Cr(nta)_2]^{3-} \quad \overset{k_1}{\underset{k_2}{\rightleftharpoons}} \quad [Cr(nta)(nta\text{-}N)OH_2]^{3-}$$

$$[Cr(nta)(nta\text{-}N)OH_2]^{3-} \quad \overset{k_3}{\underset{(H^+)}{\rightarrow}} \quad [Cr(nta)(H_2O)_2]^{3-} \quad + ntaH_3$$

$$[Cr(nta)(ntaH)]^{2-} \quad \overset{k_4}{\underset{H^+}{\rightarrow}} \quad [Cr(nta)(H_2O)_2]^{3-} \quad + ntaH_3$$

$$[Cr(ntaH)_2]^- \quad \overset{k_5}{\underset{(H^+)}{\rightarrow}} \quad [Cr(nta)(H_2O)_2]^{3-} \quad + ntaH_3$$

$$[Cr(ntaH)_2]^- \quad \overset{k_6}{\underset{k_7}{\rightleftharpoons}} \quad [Cr(ntaH)(ntaH_2\text{-}O)]$$

$$[Cr(ntaH)(ntaH_2\text{-}O)] \quad \overset{k_8}{\rightarrow} \quad [Cr(nta)(H_2O)_2]^{3-} \quad + ntaH_3$$

studies between pH 0.50 and 5.00, the pseudo-first-order rate constant (k_{obs}) was found to vary with $[H^+]$ as follows: $k_{obs} = (k_1 + k_4 K_1[H^+] + k_5 K_1 K_2[H^+]^2 + k_H K_1 K_2[H^+]^3) / (1 + K_1[H^+] + K_1 K_2[H^+]^2)$, where at 298.2 K $10^4 k_1 = 1.28$ s^{-1} ($\Delta H^{\ddagger} = 14.3$ kcal mol^{-1}, $\Delta S^{\ddagger} = -28.4$ cal K^{-1} mol^{-1}), $10^3 k_4 = 6.26$

s^{-1} ($\Delta H^{\ddagger} = 1.47$ kcal mol^{-1}, $\Delta S^{\ddagger} = -19.4$ cal K^{-1} mol^{-1}), $10^3 k_5 = 3.06$ s^{-1} ($\Delta H^{\ddagger} = 14.8$ kcal mol^{-1}, $\Delta S^{\ddagger} = -20.3$ cal K^{-1} mol^{-1}) and $10^2 k_H = 1.66$ s^{-1} ($\Delta H^{\ddagger} = 16.5$ kcal mol^{-1}, $\Delta S^{\ddagger} = -11.4$ cal K^{-1} mol^{-1}). K_1 and K_2 are interpreted as equilibrium constants for the protonation of free carboxylate groups [$\Delta H^{\circ}(K_1) = 1.6$ and $\Delta H^{\circ}(K_2) = -1.6$ kcal mol^{-1}; $\Delta S^{\circ}(K_1) = 20.0$, $\Delta S^{\circ}(K_2) = 3.3$ cal K^{-1} mol^{-1}]. nta–N and ntaH$_2$–O represent nta ligands which have uncoordinated N and carboxylate groups, respectively. $k_H = k_6 k_8/(k_7 + k_8)$, and log $K_1 = 3.18$, log $K_2 = 1.89$ at 293.2 K.

5.2.2.2.1. Tris(chelates). Acid hydrolysis of tris(salicaldehydato)-chromium(III), [Cr(SA)$_3$],[35] and the 5-substituted derivatives, [Cr(SAX)$_3$],[36] have been investigated. Consecutive loss of the three ligands is observed with $k^{\mathrm{I}} > k^{\mathrm{II}} > k^{\mathrm{III}}$. The observed rate constant for each stage varies with acidity according to the equation $k_{\mathrm{obs}} = k_0 + k[\mathrm{H}^+]$:

$$[\mathrm{Cr(SA)_3}] \xrightarrow{k^{\mathrm{I}}} [\mathrm{Cr(SA)_2(H_2O)_2}]^+ \xrightarrow{k^{\mathrm{II}}} [\mathrm{Cr(SA)(H_2O)_4}]^{2+} \xrightarrow{k^{\mathrm{III}}} [\mathrm{Cr(H_2O)_6}]^{3+} \tag{11}$$

For each step the mechanism shown in Scheme 6 is proposed, with $k = k_1 K_1$.

Scheme 6

[Cr(SA)$_{3-n}$]$^{n+}$	$\xrightarrow{k_0}$	[Cr(SA)$_{2-n}$]$^{(n+1)+}$ + SAH
$\updownarrow K_1$		
[Cr(SAH)(SA)$_{2-n}$]$^{(n+1)+}$	$\xrightarrow{k_1}$	[Cr(SA)$_{2-n}$]$^{(n-1)+}$ + SAH

Protonation is believed to occur at the phenolic oxygen (q.v.). Sulfate and nitrate ions were found to strongly catalyze the second- and third-step reactions, whereas Cl$^-$ and Br$^-$ have little effect. The first stage (k^{I}) is not affected by oxo anions, and it was found that the acid-dependent pathway is the one subject to anion catalysis. The following mechanism is proposed for [Cr(SA)$_2$(H$_2$O)$_2$]$^+$ ion (X$^-$ = anion) (Scheme 7). With this scheme, $k_{\mathrm{obs}} = k_0 + k_1 K[\mathrm{H}^+] + k_2 K_1 K_2[\mathrm{X}^-]$

Scheme 7

[Cr(SA)$_2$(H$_2$O)$_2$]$^+$	+	H$^+$	$\overset{K_1}{\rightleftharpoons}$	[Cr(SAH)(SA)(H$_2$O)$_2$]$^{2+}$
[Cr(SAH)(SA)(H$_2$O)$_2$]$^{2+}$	+	X$^-$	$\overset{K_2}{\rightleftharpoons}$ ion pair $\overset{k_2}{\rightarrow}$	[Cr(SA)(H$_2$O)$_3$X]$^+$ + SAH
[Cr(SA)(H$_2$O)$_3$X]$^+$	+	H$_2$O	$\rightarrow$	[Cr(SA)(H$_2$O)$_4$]$^{2+}$ + X$^-$

/ $(1 + K_1[\mathrm{H}^+] + K_1 K_2[\mathrm{H}^+][\mathrm{X}^-])$. The rate-determining step for the sulfate catalysis is shown to be sulfate anation (k_2), by comparison with results for the anation of [Cr(H$_2$O)$_6$]$^{3+}$ ion by sulfate.[35]

Table 5.8. Effect of 5-Substituents (X) on the Rate of the Acid-Dependent Pathway of the Second and Third Stages of the Hydrolysis of tris-(5-X-Salicaldehydato)–Chromium(III)[36]

X	$[Cr(SAX)_2(H_2O)_2]^+$ 10^4k, K_1, s^{-1} at 41°C[a]	$[Cr(SAX)(H_2O)_4]^{2+}$ 10^5k, K_1, s^{-1} at 80°C[b]
MeO	6.08 ± 0.75	10.30 ± 0.08
Me	4.75 ± 0.90	8.10 ± 0.07
H	5.52 ± 0.39	6.83 ± 0.14
Br	4.69 ± 0.46	5.97 ± 0.18
NO₂	1.53 ± 0.08	1.78 ± 0.02

[a] $I = 2.0$ mol dm⁻³.
[b] $I = 3.0$ mol dm⁻³.

The effect of 5-substituents (X) on the rate of the acid-dependent pathway for $[Cr(SAX)_2(H_2O)_2]^+$ and $[Cr(SAX)(H_2O)_4]^{2+}$ ions is shown in Table 5.8. The rate data fit with the Hammett substituent constants σ_p rather than σ_m as expected if protonation occurs at the phenolic oxygen atom.[36]

A very complex behavior has been reported for the aquation of $[Cr(ox)_3]^{3-}$ ion in the reversed micellar systems benzene—dodecylammonium propionate or octylammonium tetradeconoate containing solubilized water.[37] The results differ markedly from an earlier investigation,[38] and the reasons for the discrepancy and the complex rate behavior is not clear.

Two groups have studied the rate of exchange of acac with $[M(acac)_3]$ (M = Co or Cr[39]; M = Co,Cr,Ru, or Rh[40]). The former group studied the reaction in DMSO and propanol, and the effect of acidity (CH_2ClCO_2H) in dioxane and DMSO.[39] The latter group studied the reactions in acac solution and found the pseudo-first-order rate constants to decrease in the order Co > Cr > Ru > Rh. The rate constants were found to be 10^5k (s^{-1}) = 2.4(93°C), 5.6(117°C), 9.5(158°C), and 2.4 (185°C) respectively. Activation parameters (Table 5.9) indicate an S_N1 mechanism for Co(III) and an S_N2 mechanism for the other metal ions.

Table 5.9. Activation Parameters at 298 K for the Exchange of Pentane-2,4-dione (acac) with $[M(acac)_3]$[40]

M	$\Delta H^{\ddagger}$, kcal mol⁻¹	$\Delta S^{\ddagger}$, cal K⁻¹ mol⁻¹
Cr	28.8 ± 0.7	−5.0 ± 2.2
Co	36.4 ± 2.2	+19.2 ± 6.1
Ru	27.4 ± 1.8	−14.6 ± 4.5
Rh	28.3 ± 0.5	−19.0 ± 1.4

5.2.2.3. Bridged Dichromium(III) Complexes

The interesting μ-carbonato-complexes of Rh(III) and Cr(III) having the structure [LMμ(OH)$_2$(OCO$_2$)ML] (M = metal, L = 1,4,7-triazacylcononane, [9]aneN$_3$), have been reported, and the geometry of the Cr(III) complex established by X-ray cystallography.[41] The acid-catalyzed decarboxylation reactions proceed with a rate law, rate = (k_0 + k_1[H$^+$])[complex]; at 303.2 K, I = 1.0 mol dm^{-3}, 10^4k_0 = 0.31 ± 0.05 s^{-1}, and 10^4k_1 = 0.60 ± 0.08 dm^3 mol^{-1} s^{-1} ($\Delta H^{\ddagger}$ = 15.9 ± 0.5 kcal mol^{-1}; $\Delta S^{\ddagger}$ = −25.4 ± 1.8 cal K^{-1} mol^{-1}) for the Cr(III) complex. For the Rh(III) complex very similar data were observed for k_1 ($k_0 \simeq 0$).

Oxidation of [Cr(H$_2$O)$_6$]$^{2+}$ ion by benzoquinone in acidic solution gives two bridged species of the type [(H$_2$O)$_4$Cr(OH)(HQ)Cr(OH$_2$)$_4$]$^{4+}$ (major species HQ = hydroquinone) and [(H$_2$O)$_5$Cr-O-Cr(OH$_2$)$_5$]$^{4+}$ (minor species). The hydroquinone dimer is postulated to have structure 2. This dimer aquates too slowly

2

to be the precursor of the μ-oxo-bridged dimer. A maximum yield of the μ-oxo-dimer is obtained at high acidities, and a mechanism for its formation involving oxidation by protonated benzoquinone followed by rapid loss of hydroquinone is invoked. Some trimeric species were also isolated.[42]

A family of bridged bis(benzylchromium) cations analogous to structure 1 have also been prepared and homolysis of the Cr–C bond investigated.[32]

5.2.3. Formation of Chromium(III) Complexes

5.2.3.1. Reactions of Hexakis(solvento)chromium(III) Ions

5.2.3.1.1. [Cr(H$_2$O)$_6$]$^{3+}$. Job's method of continuous variations was used to show that [Cr(OAc)$_2$(H$_2$O)$_4$]$^+$ ion is the product of the reaction between [Cr(H$_2$O)$_6$]$^{3+}$ and acetate ions. Formation of the monocomplex is rate determining and an I_a mechanism is proposed for the reaction between [Cr(H$_2$O)$_6$]$^{3+}$ and AcO$^-$ ions to give [Cr(OAc)(H$_2$O)$_5$]$^{2+}$, and an I_d mechanism for the reaction between $[Cr(H_2O)_5OH]^{2+}$ and OAc^- to give $[Cr(OAc(OH)(H_2O)_4]^+$ ion.[43] This conclusion was reached from a comparison of the rate data for the interchange steps (k_{an} for Cr^{3+}, $OAc^- \rightarrow CrOAc^{2+}$ and k'_{an} for $CrOH^{2+}$, $OAc^- \rightarrow Cr(OAc)OH^+$) with that for solvent exchange (Table 5.10). Thus k_{an} was found to be significantly larger than the solvent exchange rate (k_{ex}), in line with an I_a mechanism, whereas k'_{an} is very similar to the rate of exchange of H_2O with $[Cr(H_2O)_5OH]^{2+}$ ion, as expected for an I_d mechanism. Interestingly, analogous conclusions were reached from studies of the rate of exchange of H_2O with $[Fe(H_2O)_6]^{3+}$ and $[Fe(H_2O)_5OH]^{2+}$ ion which have been found[44] to proceed with I_a and I_d mechanisms, respectively, by measurements of $\Delta V^{\ddagger}$.

From studies of the rate of formation of $[Cr(H_2O)_5L]^+$ and $[Cr(H_2O)_5LH]^{2+}$ ions (L^{2-} = phthalate ion) over a narrow range of pH ($10^3[H^+] = 0.25$–3.16) and temperature (30–40°C), the activation parameters for the interchange reactions were found to be $\Delta H^{\ddagger} = 93$ and 94 kJ mol^{-1} and $\Delta S^{\ddagger} = -5$ and -2 J K^{-1} mol^{-1} for the reactions Cr^{3+}, $L^{2-} \rightarrow CrL^+$, and Cr^{3+}, $LH^- \rightarrow Cr(HL)^{2+}$, respectively.[45] As in the study of the reactions of acetate ion, significant ion pairing was observed with $\Delta H^{\circ} = -8$ and -8 kJ mol^{-1} and $\Delta S^{\circ} = -11$ and -19 K^{-1} mol^{-1}, respectively. An I_a mechanism is proposed.[45]

Unlike the studies of $[Cr(H_2O)_6]^{3+}$ reacting with carboxylate ions, the reaction with 1,10-phenonthroline in aqueous ethanol is reported to be independent of pH in the range 3.5–5.0. A simple second-order rate law was observed.[46] However, for the reaction with nitrilotri-acetate (nta) ions in the same solvent mixture, $k_{obs} = kK_E[ntaH^{2-}]/(1 + K_E[ntaH^{2-}])$, where K_E is the ion pair formation constant.[47] An I_a mechanism is favored.

Oxo anions are known to catalyze loss of ligands from aquo–chromium(III) complexes as discussed in Section 5.2.2.1.2. Two groups have studied the reaction of Cr(III) with xylenol orange, in the presence[48] and absence[49] of

Table 5.10. Activation Parameters for the Following Interchange Reactions:

$$[Cr(H_2O)_6]^{3+}, OAc^- \xrightarrow{k_{an}} [Cr(H_2O_5OAc]^{2+} + H_2O$$

$$and\ [Cr(H_2O)_5OH]^{2+}, OAc^- \xrightarrow{k'_{an}}$$
$$[Cr(H_2O)_4(OAc)OH]^+ + H_2O$$

	$\Delta H^{\ddagger}$, kJ mol^{-1}	$\Delta S^{\ddagger}$, J K^{-1} mol^{-1}
k_{an}	103	31[b]
k'_{an}	104	41[c]
k_{ex} [a]	109	+1

[a] Solvent exchange with $[Cr(H_2O)_6]^{3+}$.
[b] $\Delta S^{\ddagger}(k_{an}K_{ip}) = -66$ J K^{-1} mol^{-1}.
[c] $\Delta S^{\ddagger}(k'_{an}K_{ip}) = -78$ J K^{-1} mol^{-1}.

oxoanions, and found that whereas CO_3^{2-} ion increases the rate of reaction with chromium(III), Cl^-, NO_3^-, and SO_4^{2-} ions have only slight catalytic activity.

Reaction of $[Cr(H_2O)_6]^{2+}$ with thio-bis(ethylene nitrilo)tetraacetic acid (TEDTA) gives a violet complex **3** in which TEDTA is quinquedentate. At pH

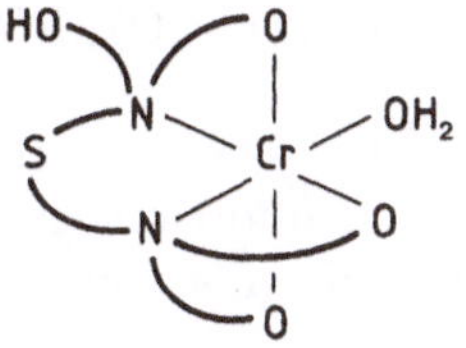

3

4.0–4.5, **3** was observed to rearrange to a pink complex **4** in which TEDTA is sexidentate. **3** was also found to dimerize in the presence of base to give **5**, which then reacted further to give the green product **6**.[50]

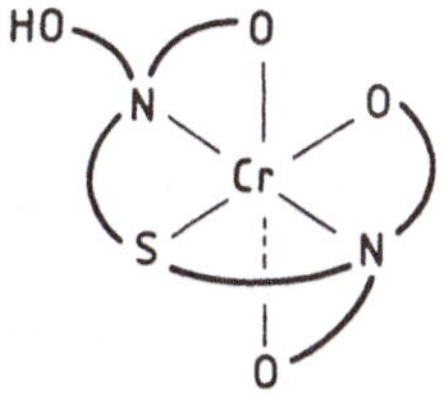

4

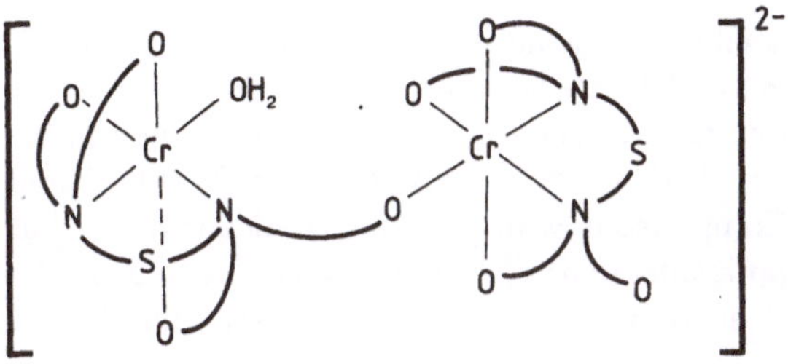

5

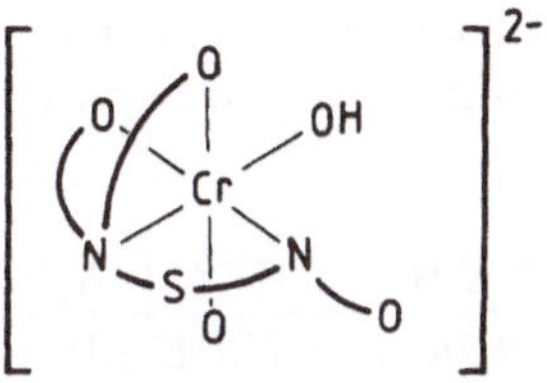

6

The reactions of Cr^{3+} with $[Fe(CN)_6]^{4-}$ ion is reported to be first order in [Cr(III)] and zero order in $[Fe(CN)_6^{4-}]$. The product is formulated as $[CrFe(CN)_5OH]^-$, although polarography indicates the formation of $[Cr(CN)_2]^+$ ion as well.[51]

References to the anation of $[Cr(H_2O)_6]^{3+}$ by sulfate ion[35] and the formation of the μ-oxo-bridged dimer[42] $[(H_2O)_5CrOCr(OH_2)_5]^{4+}$ were given previously.

5.2.3.1.2. [Cr(DMSO)₆]³⁺. The rate-determining step in the reaction of $[Cr(DMSO)_6]^{3+}$ ion with excess o-tolylbiguanidine (L) to give $[CrL_3]^{3+}$ has been shown to be the formation of the monocomplex $[Cr(L)(DMSO)_4]^{3+}$ ion[52].

5.2.3.2. Formation of Mixed-Ligand Complexes

5.2.3.2.1. Ammine and Amine Complexes. Formation of the phosphate complex, $[Cr(NH_3)_5(H_2PO_4)]^{2+}$, from the reaction of $[Cr(NH_3)_5OH]^{3+}$ ion with phosphate (pH 1–2) proceeds via parallel ion pair and ion dipole interchange pathways (Scheme 8). From kinetic studies at $I = 1.0$ mol dm^{-3} [LiClO$_4$], the

Scheme 8

$$[Cr(NH_3)_5OH_2]^{3+} \quad + \quad H_3PO_4 \quad \overset{K_1}{\rightleftharpoons} \quad [Cr(NH_3)_5OH_2]^{3+}, H_3PO_4$$

$$\downarrow \; k_1(-H_3O^+)$$

$$[Cr(NH_3)_5H_2O_4]^{2+}$$

$$\uparrow \; k_2(-H_2O)$$

$$[Cr(NH_3)_5OH_2]^{3+} \quad + \quad H_2PO_4^- \quad \overset{K_2}{\rightleftharpoons} \quad [Cr(NH_3)_5OH_2]^{3+}, H_2PO_4^-$$

outer-sphere association constants were found to be $K_1 = 0.32$ and $K_2 = 1.8$ dm^3 mol^{-1}, with K_1 and K_2 showing very little temperature dependence ($\Delta H° \sim 0$). The interchange rate constants, k_1 and k_2, were found to be approximately equal, with $10^4 k = 1.45$ s^{-1} at 323.2 K ($\Delta H^{\ddagger} = 25.0$ kcal mol^{-1}, $\Delta S^{\ddagger} = 1.1$ cal K^{-1} mol^{-1}). Comparisons with previous studies indicate that whereas an I_a mechanism is applicable to $[Cr(H_2O)_6]^{3+}$ anations and an I_d mechanism to $[Co(NH_3)_5OH_2]^{3+}$ anations, the behavior of $[Cr(NH_3)_5OH_2]^{3+}$ ion is somewhere between these two.[53]

Reaction of cis-$[Cr(en)(ox)(OH_2)_2]_2^+$ ion with oxalate (ox^{2-}/Hox$^-$) gives predominantly $[Cr(ox)_2(en)]^-$ ion, but some $[Cr(ox)(enH)(OH_2)_2]_2$ is also formed. These products are unstable under these conditions and $[Cr(ox)_3]^{3-}$ ion is eventually formed.[54] The pseudo-first-order rate constant was observed to vary as follows:

$$k_{obs} = \frac{k_1 K_c[H^+]([H^+] + K_2')[ox^{2-}]}{K_2'([H^+] + K_{a1}) + K_c[H^+]([H^+] + K_2')[ox^{2-}]} \tag{12}$$

The mechanism proposed (Scheme 9) involves two ion pairs (IP-1 and IP-2).

Scheme 9

$$cis\text{-}[Cr(ox)(en)(OH_2)_2]^+ \quad \overset{K_{a1}}{\rightleftharpoons} \quad [Cr(ox)(en)(OH)(OH_2)] \quad + \quad H^+$$

$$cis\text{-}[Cr(ox)(en)(OH_2)]^+ \quad + \quad Hox^- \quad \overset{K_B}{\rightleftharpoons} \quad [IP-1]$$

$$cis\text{-}[(Cr(ox)(en)(OH_2)]^+ \quad + \quad ox^{2-} \quad \overset{K_C}{\rightleftharpoons} \quad [IP-2]^-$$

$$[IP-1] \quad \overset{K_2'}{\rightleftharpoons} \quad [IP-2]^- \quad + \quad H^+$$

$$[IP-1] \text{ or } [IP-2]^- \quad \overset{k_1}{\rightarrow} \quad [Cr(ox)_2(en)]^-$$

At 298.2 K and $I = 0.8$ mol dm^{-3}, $pK_{a1} = 5.92$, $10^4K_2' = 7.25$, $K_B = 1.35$ dm^3 mol^{-1}, $K_C = 4.70$ dm^3 mol^{-1} and $10^3k_1 = 1.44$ s^{-1} ($\Delta H^{\ddagger} = 18.1$ kcal mol^{-1}, $\Delta S^{\ddagger} = -10.7$ cal K^{-1} mol^{-1}) and an I_a mechanism is involved. Comparison with previous studies shows that oxalate anation reactions of oxalato–chromium(III) complexes have relatively low $\Delta H^{\ddagger}$ and negative $\Delta S^{\ddagger}$, whereas oxalate anation of complexes which do not have an oxalate group present initially have high $\Delta H^{\ddagger}$ and positive $\Delta S^{\ddagger}$ values. It is postulated that in the reactions of oxalato complexes, dissociation of one end of an oxalate ligand is involved.

5.2.3.2.2. Porphyrins. The mechanism in (Scheme 10) has been found for the reaction of $[Cr(TPPS)(OH_2)_2]^{3-}$ ion (TPPS^{6-} = *meso*-tetrakis(p-sulfan-

Scheme 10

$$[Cr(TPPS)(OH_2)_2]^{3-} \quad \overset{k_1}{\underset{k_{-1}}{\rightleftharpoons}} \quad [Cr(TPPS)(OH_2)(NCS)]^{4-}$$

$$K_{a1} + H^+ \downarrow\uparrow -H^+ \qquad\qquad +H^+ \downarrow\uparrow -H^+ K_{a1}^{NCS}$$

$$[Cr(TPPS)(OH_2)OH]^{4-} \quad \overset{k_4}{\underset{k_{-4}}{\rightleftharpoons}} \quad [Cr(TPPS)(OH)(NCS)]^{5-}$$

$$K_{a2} + H^+ \downarrow\uparrow -H^+$$
$$[Cr(TPPS)(OH)_2]^{5-}$$
$$K_1 = k_1/k_{-1}; \, K_4 = k_4/k_{-4}$$

sulfanatophenyl)porphinate ion) with NCS$^-$ between pH 1–13 ($I = 1.0$ mol dm^{-3}[NaClO$_4$]). From equilibrium studies at 298.2 K, $pK_{a1} = 7.63$, $pK_{a2} = 11.45$, and $K_1 = 2.52 \pm 0.16$ dm^3 mol^{-1}. Kinetic studies gave $pK_{a1} = 8.13$, $K_4 = 0.69$ dm^3 mol^{-1}, $10^3k_1 = 4.67 \pm 0.04$ dm^3 mol^{-1} s^{-1} ($\Delta H^{\ddagger} = 16.8$ kcal mol^{-1}, $\Delta S^{\ddagger} = -12.8$ cal K^{-1} mol^{-1}), $10^3k_{-1} = 1.91$ s^{-1} ($\Delta H^{\ddagger} = 15.7$ kcal mol^{-1}, $\Delta S^{\ddagger} = -18.5$ cal K^{-1} mol^{-1}), $k_4 = 28.5$ dm^3 mol^{-1} s^{-1} ($\Delta H^{\ddagger} = 16.7$ kcal mol^{-1}, $\Delta S^{\ddagger} = +4.1$ cal K^{-1} mol^{-1}), and $k_{-4} = 41.3$ s^{-1}. TPPS^{6-} ion labilizes Cr(III) by a factor of $\sim10^2$ compared with a factor of 10^9 for Co(III). The hydroxy complexes are even more labile, $[Cr(TPPS)(OH)(OH_2)]^{4-}$ being 6.1×10^3 times as labile as $[Cr(TPPS)(OH_2)_2]^{3-}$ ion, and 9.5×10^5 times as labile as $[Cr(NH_5)_5OH_2]^{3+}$ ion. A dissociative mechanism is favored.[55]

In *trans*-[Cr(salen)(OH$_2$)$_2$]$^+$ ion the Cr atom sits above the plane of the four N atoms with Cr–OH$_2$ bond lengths of 192.3 and 208.5 pm. Anation by N$_3^-$ or NCS$^-$ involves the formation of a mono complex only, on the stopped-flow timescale (Scheme 11). At 303.2 K, k_1 = 0.10 and 0.20 dm^3 mol^{-1} s^{-1},

Scheme 11

$$[\text{Cr(salen)(OH}_2)_2]^+ \quad + \quad \text{X}^- \underset{k_{-1}}{\overset{k_1}{\rightleftharpoons}} [\text{Cr(salen)(OH}_2)\text{X}]$$

$$\updownarrow K_a \qquad\qquad\qquad\qquad\qquad\qquad \updownarrow K_a'$$

$$[\text{Cr(salen)(OH)}_2(\text{OH})] \quad + \quad \text{X}^- \underset{k_{-2}}{\overset{k_2}{\rightleftharpoons}} [\text{Cr(salen)(OH)X}]^-$$

$k_2\text{K}_a$ = 4.2 × 10^{-7}, and 3.8 × 10^{-8} s^{-1}, k_{-1} = 0.024 and 0.001 s^{-1}, and $k_{-2}K_a'$ = 5.2 × 10^{-7} and 1.1 × 10^{-8} s^{-1} for X$^-$ = NCS$^-$ and N$_3^-$, respectively.[56]

5.2.3.2.3. Amino Acid and Carboxylato Complexes. Rate data for the formation (k_f) and dissociation (k_d) of [Cr(X)Y]$^{(n-2)-}$ ions from the reactions of [Cr(Y)(OH$_2$)]$^{(n-2)-}$ ions from the reactions of [Cr(Y)(OH$_2$)]$^{(n-3)-}$ ions {H$_4$Y = edta, H$_3$Y = N-(hydroxyethyl)ethylenediamine-N,N',N'-triacetic acid (hedtra), ethylenediamine-N,N',N'-triacetic acid (edtra), and N-methylethylenediamine-N,N',N'-triacetic acid (medtra)} with anions X$^-$ (NO$_2^-$,N$_3^-$,NCS$^-$) are collected in Table 5.11. The accelerated rates of anation and aquation are ascribed to transient coordination of the pendant groups of ligands Y^{n-}. Nitrite ion probably reacts without Cr–O bond fission in the case of [Cr(hedtra)(H$_2$O)]. However, for [Cr(medtra)(H$_2$O)] nitrosation proceeds via parallel reactions, one with and one without Cr–O bond cleavage.[57]

The reaction of [Cr(ox)$_2$(gly)]$^{2-}$ ion (gly = glycinate ion) with oxalate ion (ox^{2-}) gives a mixture of [Cr(ox)$_3$]$^{3-}$ and [Cr(ox)$_2$(OH$_2$)$_2$]$^-$ ions, with pronounced catalysis by NO$_3^-$ ions. The mechanism in Scheme 12 was proposed.[58]

Table 5.11. Rate Data at 298.2 K (I = 1.0 mol dm^{-3}) for the Formation (k_f) and Dissociation (k_d) of [Cr(III)(X)Y]$^{(n-2)-}$ Ions[57]

Y	X	k_f, dm^3 mol^{-1} s^{-1}	k_d, s^{-1}
edta^{4-}	NCS$^-$	13.7 ± 0.6	26.8 ± 1.9
edtaH^{3-}	NCS$^-$	0.773 ± 0.042	3.17 ± 0.33 × 10^{-2}
hedtra^{3-}	NCS$^-$	3.32 ± 0.11	0.244 ± 0.019
hedtra^{3-}	NO$_2^-$	12 ± 2	6.24 ± 0.32
hedtra^{3-}	N$_3^-$	19.8 ± 1.2^a	0.189 ± 0.012^b
edtra^{3-}	NCS$^-$	2.95 ± 0.10 × 10^{-2}	2.40 ± 0.17 × 10^{-3}
medtra^{3-}	NO$_2^-$	1.7 ± 0.9 × 10^{-3}	8 ± 3 × 10^{-4}

^{a} $\Delta H_f^{\ddagger}$ = 48.1 ± 1.2 kJ mol^{-1}; $\Delta S_f^{\ddagger}$ = −57 ± 5 J K^{-1} mol^{-1}.
^{b} $\Delta H_d^{\ddagger}$ = 60.4 ± 1.2 kJ mol^{-1}; $\Delta S_d^{\ddagger}$ = −55 ± 4 J K^{-1} mol^{-1}.

Scheme 12

$$[Cr(ox)_2(gly)]^{2-} + H_3O^+ \xrightarrow{k'} [Cr(ox)_2(O_2CCH_2NH_3)(OH_2)]^-$$
$$\updownarrow k_a$$
$$[Cr(ox)_2(gly)]^{2-} + H_2O \xrightarrow{k''} [Cr(ox)_2(O_2CCH_2NH_3)(OH)]^{2-}$$
$$[Cr(ox)_2(O_2CCH_2NH_3)(OH_2)]^- + H_3O^+ \xrightarrow{k'_c} cis\text{-}[Cr(ox)_2(OH_2)_2]^- + glyH_2^+$$
$$[Cr(ox)_2(O_2CCH_2NH_3)(OH_2)]^- + oxH_2 \xrightarrow{k'_d} [Cr(ox)_3]^{3-} + glyH_2^+ + H_3O^+$$
$$[Cr(ox)_2(O_2CCH_2NH_3)(OH_2)]^- + NO_3^- + H^+ \xrightarrow{k'_e} [Cr(ox)_2(ONO_2)(OH_2)]^{2-} + glyH_2^+$$
$$cis\text{-}[Cr(ox)_2(OH_2)_2]^- + oxH_2 \underset{k_2}{\overset{k_1}{\rightleftharpoons}} [Cr(ox)_2(O_2CCO_2H)(H_2O)]^{2-} + H_3O^+$$
$$k_4 \updownarrow k_3$$
$$[Cr(ox)_3]^{3-} + H_3O^+$$
$$[Cr(ox)_2(OH_2)(ONO_2)]^{2-} + oxH_2 \underset{k_2'}{\overset{k_1'}{\rightleftharpoons}} [Cr(ox)_2(O_2CCO_2H)(ONO_2)]^{3-} + H_3O^+$$
$$[Cr(ox)_2(O_2CCO_2H)(ONO_2)]^{3-} \underset{k_4'}{\overset{k_3'}{\rightleftharpoons}} [Cr(ox)_3]^{3-} + H^- + NO_3^-$$

5.2.3.3. *Formation of Cr(III) Complexes from Cr(II) or Cr(IV)*

Most of these reactions are considered in more detail in the sections on redox reactions.

$[Cr(pyz)(OH_2)_5]^{3+}$ ion (pyz = pyrazine) is formed from the second-order reaction between aqueous Cr(II) and pyrazine. A transient pyrazine–radical complex of Cr(III) was also observed.[59] Reaction of $[Cr(H_2O)_6]^{2+}$ with 4,4'-bipyridine gives a transient blue species assigned to the 4,4'-bipyridine radical.[60] The chromium product is $[Cr(H_2O)_6]^{3+}$.

Reduction of $[Co(NH_3)_5Y]^{M+}$ ions by $[Cr(H_2O)_6]^{2+}$ gives the aminopolycarboxylato complexes $[Cr(OH_2)_5Y]^{n+}$ (Y = polyaminocarboxylates).[61]

A radical mechanism was found for the formation of alkyl–chromium(III) complex $[Cr(L)R]^{2+}$ from the reactions of $[Cr(L)]^{2+}$ with alkyl halides, RX (L = tetraazamacrocyclic ligand [15]aneN$_4$).[62]

The diperoxochromium(IV) complex, $[Cr(O_2)_2(en)(OH_2)]$, reacts with iodide ion in weakly acidic solutions on the stopped-flow timescale to give $[Cr(OH_2)_4(en)]^{3+}$ ion:

$$2[Cr(O_2)_2(en)(OH_2)] + 10I^-$$
$$+ 16H^+ \rightarrow 2\,[Cr(OH_2)_4(en)]^{3+} + 5I_2 + 2H_2O \quad (13)$$

A two-step mechanism was observed, fast formation of $[Cr(IV)(O_2)(en)I]^-$ ion being followed by the slower redox reaction.[63]

5.2.4. Chromium(III) Photochemistry

5.2.4.1. Ammine Complexes

Ligand field photoaquation of $[Cr(NH_3)_6]^{3+}$ and $[Cr(NH_3)_5X]^{2+}$ (X = Cl, Br, or NCS) ions have been investigated at high pressures. Values of $\Delta V^{\ddagger}$ cm^3 mol^{-1} were found to be -6.4 ± 0.3 for replacement of NH_3 by H_2O (average value), and -13.0 ± 0.5, -12.2 ± 0.3, and -9.8 ± 0.2 for loss of Cl$^-$, Br$^-$, and NCS$^-$, respectively. The data support an I_a mechanism.$^{(64)}$

Whereas thermal aquation of $[Cr(NH_3)_5CN]^{2+}$ ion proceeds with loss of CN$^-$ ion, ligand field photolysis results in the loss of equatorial *(cis)* NH$_3$ only, to give a 2:1 mixture of *cis-* and *trans-*$[Cr(NH_3)_4(OH_2)CN]^{2+}$ ions. The quantum yield is 0.33, and unlike other $[Cr(NH_3)_5X]^{2+}$ ions, the $^4B_2$ state is the lowest quartet excited state when X = CN. Charge transfer excitation was also investigated, and the same products and quantum yield found as for ligand field excitation, indicating that efficient crossing from charge transfer to ligand field states occurs.$^{(65)}$

Quantum yields are reported$^{(66)}$ for the 254-nm photoinduced bridge cleavage of the rhodochromium(III) ion, $[(H_3N)_5Cr(OH)Cr(NH_3)_5]^{5+}$. A detailed ligand field analysis of the photosubstitution reactions of $[Cr(NH_3)_5F]^{2+}$ ion has appeared.$^{(67)}$

5.2.4.2. Amine Complexes

Three products were detected from the photolysis of *trans-*$[Cr(en)_2F_2]^+$ ion. The dominant product is $[Cr(en)(enH)(OH_2)F_2]^{2+}$ in which the fluoride ions remain in *trans* positions, although a small amount of an isomer of this ion was separated, together with some *cis-*$[Cr(en)_2(OH_2F]^{2+}$ ion.$^{(68)}$

The photoaquation of $[Cr(en)_n(NH_3)_{6-2n}]^{3+}$ ions (n = 1, 2, or 3) has been investigated. When n = 1 or 2, the quantum yield for loss of en is larger than expected on statistical grounds. For example, when n = 2 the relative rate of loss of en and NH$_3$ by photoaquation is 2.83, and when n = 1 the ratio is 1.64. The reason for the easier loss of en is ascribed to steric strain in the chelate rings rather than factors associated with different excited states.$^{(69)}$

The 366–578-nm photoaquation of *trans-*$[Cr(en)_2(NCS)F]^+$ ion at 283.2 and 293.2 K, results mainly in the loss of NCS$^-$ ion (ϕ = 0.22–0.28), with some loss of en (ϕ = 0.05–0.11). The results agree well with the theory of Vanquickenborne and Ceulemans, the main factor favoring loss of NCS$^-$ rather than F$^-$ ion being the π bonding of F$^-$ in the ^4E excited state. The $^4B_2$ state is also involved.$^{(70)}$

Photoaquation of *trans-*$[Cr(en)_2(CN)_2]^+$ ion in acidic solution gives only

$[Cr(en)(enH)(OH_2)(CN)_2]^+$ ($\phi = 0.6$), whereas the *cis* isomer reacts with chelate ring opening ($\phi = 0.4$) and loss of CN^- ion ($\phi = 0.1$) as expected from the theory of Vanquickenborne and Ceulemans.[71]

The photophysics of the quenching of the excited states of *trans*-$[Cr(en)_2(NCS)F]^+$ ion, by $[Cr(CN)_6]^{3-}$ or $[Cr(ox)_3]^{3-}$ ions, has been investigated using Stern–Volmer plots. Two quartet states are shown to be involved in addition to the doublet state.[72] The concentration dependence of the quenching of ^2E excited states of $[CrL_3]^{3+}$ ions (L = bipy or phen) has also been investigated by laser flash photolysis.[73]

The quantum yield for the photoracemization of Δ- and Λ-$[Cr(en)_3]^{3+}$ ion in acid solution is 0.015. Although the data do not rule out a bond rupture mechanism, the twist mechanism is favored.[74]

Previous studies of the photoaquation of $[Cr(bipy)_3]^{3+}$ ion indicated that a seven-coordinate aquo complex is involved. In contrast, photosolvolysis in DMF solution proceeds by reduction to the +2 oxidation state (L = bipy)[75]:

$$*[CrL_3]^{3+} + e^- \xrightleftharpoons{DMF} [CrL_3]^{2+} \tag{14}$$

$$[CrL_3]^{2+} + 2DMF \rightleftharpoons [CrL_2(DMF)_2]^{2+} + L \tag{15}$$

$$[CrL_2(DMF)_2]^{2+} + [CrL_3]^{3+} \rightleftharpoons [CrL_2(DMF)_2]^{3+} + [CrL_3]^{2+} \tag{16}$$

Ligand field photolysis of $[Cr(tren)F_2]^+$ ion leads to loss of F^- ion with $\phi = 0.21$.[24]

5.2.4.3. Other Chromium(III) Complexes

The photoinduced loss of pentane-2,4-dione (acac) from $[Cr(acac)_3]$ in 50% aqueous ethanol is opposed by the thermal anation reaction. The reaction was studied at 16 wavelengths in the range 230–730 nm. Between 350 and 730 nm the low quantum yields ($\phi \sim 0.01$) are both wavelength and pH independent (pH 0–13). At 250 nm the quantum yield increases by a factor of 4, but there is no evidence for a redox process.[76]

5.2.5. Isomerization and Racemization Reactions

The kinetics of racemization of Λ-$[Cr(ox)_2(phen)]^-$ ion has been investigated in the presence of Cu^{2+} ions at $I = 2.0$ mol dm^{-3}. A rate law of the type rate $= (k_1 + k_2 [Cu(II)]_f)$ complex was observed where $[Cu(II)]_f$ is the uncomplexed copper(II) concentration. Perchlorate, nitrate, chloride, and sulfate copper(II) salts were used, and the rates found to decrease in the order $ClO_4^- > NO_3^- > Cl^- \gg SO_4^{2-}$ due to ion pairing. Thus Cu^{2+} ion is a much more effective catalyst than CuX^+ ions (X = anion).[77]

The *trans–cis* isomerization of $[Cr(mal)_2(H_2O)_2]^-$ ion (mal = malonate^{2-}) has been investigated in mixed aqueous–organic solvents (MeOH, EtOH, Me_2CO, and dioxane). The rates increase with a decrease in the water concentration, but are independent of the organic diluent used. This contrasts with the behavior of the corresponding oxalato complexes, and the rate-determining step is traced to the loss of H_2O.[78]

The Ag^+-catalyzed linkage isomerism of CN^- ion in *cis*-$[Cr(en)_2(CN)_2]^+$ ion,[20] and the photoracemization of Δ- and Λ-$[Cr(en)_3]^{3+}$ ions[74] were considered previously.

5.2.6. Base Hydrolysis

The base hydrolysis of the anion of Reinecke's salt, *trans*-$[Cr(NCS)_4(NH_3)_2]^-$ ion has been studied in aqueous[79] and methanolic[80] solution. A conjugate base mechanism is proposed, with an I_a or A mechanism favored in methanol. From studies between 0.05 and 0.47 mol dm^{-3} $[OMe^-]$ in MeOH, a rate law of the type, rate = $(k_1 + k_2[MeO])$ [complex] was found. At 323.2 K, $k_1 = 2.18 \pm 0.16$ s^{-1} and $k_2 = 5.71 \pm 0.55$ dm^3 mol^{-1} s^{-1} $[\Delta H^{\ddagger}(k_2) = 24.8 \pm 2$ kcal mol^{-1}, $\Delta S^{\ddagger}(k_2) = 0.7 \pm 3.5$ cal K^{-1} mol^{-1}]. Loss of both NCS^- ion and NH_3 was detected.

Base hydrolysis of $[Cr(tren)F_2]^+$ ion results in loss of F^-, with $k_{OH} = 158.4$ dm^3 mol^{-1} s^{-1}, $\Delta H^{\ddagger} = 22.75 \pm 1.60$ kcal mol^{-1} and $\Delta S^{\ddagger} = 27.8 \pm 3.0$ cal K^{-1} mol^{-1}. The product is α-$[Cr(tren)(OH)F]^+$ ion, and an I_a mechanism is favored for the conjugate base. The F^- ion which is removed comes from a hydrophilic region in the ion, and the second F^- ion, which is located in a hydrophobic region, is not readily lost.[24]

5.2.7. Solids

Synthesis of amine–chromium(III) complexes often involves controlled solid-phase heating, with ammonium halide catalysts employed to break up the crystal lattice. Recently the effect of ammonium chloride on the deamination of $[Cr(en)_3]X_3$ has been investigated. When X = Cl^-, *cis*-$[Cr(en)_2X_2]X$ is known to form,[81] and when X = NCS^- the *trans* product is produced.[82] However, in the presence of a large amount of NH_4NCS some *cis*-dithiocyanato complex is also formed. Bridged complexes are believed to be involved [e.g., $Cr_3(en)_7Cl_9$].

Cis–trans isomerization reactions of $[CrX_2(aa)(bb)]X$ and $[CrBr_2(aa)_2]Br$ (aa, bb = en, tn) have been studied in the solid-phase and activation parameters obtained.[83,84]

5.2.8. Other Chromium Oxidation States

5.2.8.1. Chromium(II)

High-spin Cr(II) compounds are Jahn–Teller distorted and invariably very labile. The low-spin $[CrL_3]^{2+}$ ions (e.g., L = bipy, phen) are more inert and can be investigated by stopped-flow methods. Recent examples include L = bipy with substituents in the 3,3′ positions (3,3′-CHCH and 3-Me).[85]

5.2.8.2. Chromium(VI)

Oxygen exchange with $[CrO_4]^{2-}$ ion has been investigated between pH 7 and 12 at 298.2 K, I = 0.2 and 1.0 mol dm^{-3}. The following rate law is reported[86]: rate = $k_1[CrO_4^{2-}]$ + $k_2[HCrO_4^-]$ + $k_3[H^+][HCrO_4^-]$ + $k_4[HCrO_4^-][CrO_4^{2-}]$ + $k_5[HCrO_4^-]^2$.

5.3. Group VII Elements

There is relatively little kinetic activity to report, at least in the field of slow substitution reactions. In oxidation state IV these elements have a d^3 electron configuration, and thus are likely candidates for slow substitution. However manganese(IV) is strongly oxidizing and the number of ligands which can coexist with it is very small. The only simple complex which is fairly stable in aqueous solution is $[MnF_6]^{2-}$. This hydrolyzes quickly in alkaline solution, but has a reasonable lifetime in neutral aqueous solution. Technetium(IV) and rhenium(IV) are stable in a redox sense, but of course the radioactive character of technetium severely restricts the number of laboratories which can undertake kinetic studies on compounds of this element. The growing importance of technetium complexes as radiopharmaceuticals has resulted in the appearance of a few papers on substitution kinetics, some of them difficult to classify in the present book owing to a lack of knowledge of coordination number (and even composition at times!). Substitution at technetium(V) is of preparative importance and mechanistic interest *(vide infra)*. Technetium(VI) and rhenium(VI) compounds are prone to disproportionation; the very stable TcO_4^- and ReO_4^- anions are tetrahedral, and have very little substitution chemistry.

5.3.1. Technetium

There seems to be some disagreement as to the rapidity with which hexahalogenotechnetates(IV) undergo hydrolysis. The aquation of $[TcCl_6]^{2-}$ is very slow in strong acid,[87] less slow[88] or fairly fast[89, 90] in dilute acid, and rapid in neutral solution,[90] to take a majority or consensus view. The conversion of $[TcCl_6]^{2-}$ into $[TcBr_6]^{2-}$ is slow,[87] as is the reaction of $[TcBr_6]^{2-}$ with nitrilotriacetate.[91] The latter reaction proceeds through intermediates containing bromide, hydroxide, and nta variously coordinated to the technetium en route to the oxo-bridged di- and trinuclear products. Some kinetic indications can be gleaned from a qualitative study of the hydrolysis of $[Tc_2Cl_8]^{3-}$ in aqueous hydrochloric acid.[92] This system is complicated by disproportionation and oxidation reactions; the kinetic complexity is demonstrated by the published absorbance vs. time plots, which suggest that five processes may be taking place.

The chloride ligands in dichloro[hydro-tris(1-pyrazolyl)borato]oxotechnetium(V) are inert to substitution.[93] This inertness is presumably attributable to electrostatics, as technetium(V) is a d^2 center with only modest crystal field stabilization. Rates of substitution of the chlorides in the $[TcOCl_4]^-$ anion are not slow enough to cause difficulties in the preparation of pharmaceuticals of the type $[TcO(LL)_2]^{n-}$ by ligand exchange. These oxotechnetium(V) complexes, also obtainable by tin(II) reduction of TcO_4^- in the presence of the ligand LL (a reaction undoubtedly of complicated mechanism), may be five- or six-coordinate in solution. The complexes with LL = citrate or gluconate are themselves precursors for analogous complexes with more exotic ligands, again obtained by ligand exchange. A recent study of the conversion of the bis(gluconate) complex into the bis(*meso*-dimercaptosuccinate) complex showed this to be kinetically a one-step process.[94] However, the conversion of the bis(gluconate) complex into the bis(2,3-dimercaptopropanesulfonate) complex takes place in two kinetically distinct stages.[95] An earlier investigation of the reaction of the bis(citrate) complex with cysteine also found evidence for a TcO(citrate)(cysteine) intermediate.[96] The observed variation of rate with pH has been attributed to protonation of the oxoligand.[94]

5.3.2. Rhenium

Rate constants have been determined for mercury(II)-catalyzed aquation of the hexachlororhenate(IV) anion in a range of binary aqueous solvent mixtures. With the aid of ancillary information on Gibbs free energies of transfer for the reactant ions from water into the solvent mixtures it proved possible to analyze the observed reactivity trends into initial state and transition state components.[97] The results are discussed below, in conjunction with a parallel study of mercury(II) catalysis of aquation of chloro–cobalt(III) complexes.

5.4. Iron

The low-spin complexes of iron(II) form a small but important group of substrates for kinetic studies of slow substitution processes. Recent work in this area will be treated in order of ligand type, starting with pentacyano complexes, followed by diimine complexes, in turn followed by complexes with such ligands as porphyrins and phthalocyanines. In the section of diimine ligands the current state of the controversy concerning nucleophilic attack at coordinated diimine ligands will be summarized, with relevant material relating to diimine complexes of other metals also mentioned. Some substitution reactions at iron(III) are relatively slow, some are related to iron(II) analogs. This section therefore ends with a selection of references to kinetic studies on iron(III) complexes.

5.4.1. Pentacyanoferrate(II) Complexes

Substitution reactions of these complexes generally proceed by a limiting dissociative (D) mechanism (Scheme 13). Kinetic investigations of such reactions

Scheme 13

$$[Fe(CN)_5L]^{3-} \underset{k_{-1}}{\overset{k_1}{\rightleftharpoons}} [Fe(CN)_5]^{3-} + L$$
$$k_2 \downarrow + L'$$
$$[Fe(CN)_5L']^{3-}$$

usually establish limiting rate constants for dissociation of the starting complex, k_1, and the discrimination ratio for the $[Fe(CN)_5]^{3-}$ intermediate, k_2/k_{-1}. Values of k_1, with values for the activation parameters $\Delta H^{\ddagger}$ and $\Delta S^{\ddagger}$ where determined, are listed in Table 5.12.[98-104] This table includes recent determinations and,

Table 5.12. *Kinetic Parameters for Substitution at Pentacyanoferrates(II); Rate Constants and Activation Parameters Referring to Limiting Rates for Loss of L from* $[Fe(CN)_5L]^{n-}$, *and in All Cases Referring to 298.2 K and Aqueous Solution*

L	$10^3 k$, s^{-1}	$\Delta H^{\ddagger}$, kJ mol^{-1}	$\Delta S^{\ddagger}$, J K^{-1} mol^{-1}	Ref.
Pyridine	1.1	104	+46	98
Pyrrolidine, **7**	3.72			99
	1.27^a			
	18.02^b			
Histamine, **8** (pH 9)	0.74	120	+96	100
L-Histidine	0.53	105	+46	101
Imidazole, **9**	1.33	102	+42	101
	2.2	90	+4	102
Thiourea	39.0	69	−38	103
dpm, **10**	1.3			104
$[Co(NH_3)_5(dpm)]^{3+}$	6.63			104

^{a} In 2 M KI.
^{b} In 2 M Et$_4$NBr.

for comparison, earlier values for selected complexes. Table 5.12 shows a relatively small range of k_1 values, especially if the rather rapid loss of the S-bonded ligand thiourea[103] is discounted. Even coordination of the $[Co(NH_3)_5]^{3+}$ moiety to the other end of dipyridylmethane, dpm (10), has only a small effect on the rate constant for iron–nitrogen bond breaking.[104]

In the particular case of substitution reactions of $[Fe(CN)_5(OH_2)]^{3-}$, ligand replacement is formally a complex formation reaction, and indeed the standard Eigen–Wilkins kinetic pattern and mechanism applies, albeit to what appears to be a negatively charged aquo ion! Values of observed second-order rate constants, with $\Delta H^{\ddagger}$ and $\Delta S^{\ddagger}$ values where available, are listed in Table 5.13.[103–107] As is well known, replacement of water is much faster than replacement of most other ligands in pentacyanoferrates(II). Also the rate of replacement varies much more with nature of the incoming group. In particular, reaction with positively charged groups is faster than reaction with uncharged ligands, owing to the larger

Table 5.13. Second-Order Rate Constants, and Activation Parameters, for Reactions of Pentacyanoaquoferrate(II) and Its Dinuclear Analog with Nucleophiles, in Aqueous Solution at 298.2 K

Nucleophile	k_2, dm^3 mol^{-1} s^{-1}	$\Delta H^{\ddagger}$, kJ mol^{-1}	$\Delta S^{\ddagger}$, J K^{-1} mol^{-1}	Ref.
$[Fe(CN)_5(OH_2)]^{3-}$				
Thiourea	202	45	-50	103
Dimethylthiourea	280	67	-29	103
N-Methylpyrazinium, 11	2350			105
$[Co(NH_3)_5(dpm)]^{3+}$ ^{a}	3900			104
4-Cyanopyridine	620			106
	760			107
Pyridine	365			106
	458			107
$[(NC)_5Fe\text{-}NC\text{-}Fe(CN)_4(OH_2)]^{6-}$				
Pyridine				
Nitrosobenzene	~ 1000	70^b	$+30^b$	107
3-,4-Cyanopyridines				

^{a} dpm = 10.
^{b} Measured for 3-cyanopyridine as entering group.

11

value of the outer-sphere association constants in the former case. This trend is illustrated by results in Table 5.13, for the rate of reaction with $[Co(NH_3)_5(dpm)]^{3+}$ is greater than with the N-methylpyrazinium cation, in turn greater than rates of reaction with uncharged pyridine ligands. It is also interesting to note that the rate constant for reaction with 4-cyanopyridine is approximately twice that for reaction with pyridine; the former ligand is known to be able to react with $[Fe(CN)_5(OH_2)]^{3-}$ at either nitrogen atom.[106] In addition to the quantitative kinetic information available on the systems detailed in Table 5.13, some semi-quantitative kinetic information can be deduced from the formation time scales quoted for color development in formation reactions of $[Fe(CN)_5(OH_2)]^{3-}$ with the pharmaceutically relevant thioligands thiamazole (methimazole, **12**), 6-mercaptopurine (**13**), and thiopental (**14**).[108] However, it is not clear whether these

12   **13**   **14**

ligands bond to the iron through sulfur or through nitrogen, and the chemistry is further complicated by protonation equilibria for each ligand.

The difference in mechanism, D for ligand replacement except when the leaving group is water (I_d mechanism), adumbrated above, is also indicated by the different isokinetic plots obtained for the two types of reaction and illustrated in connection with the investigation of the kinetics of dissociation and of formation of thiourea complexes.[103] The slope of the linear free energy plot connecting rates and equilibrium constants for reaction of pentacyanoaquoferrate(II) with a series of cobalt(III) complexes also indicates a dissociative mechanism for these reactions.[104]

The dependence of rate constants on pH for reaction of pentacyanoaquoferrate(II) with thiourea and with dimethyl and diallyl thiourea can be analyzed in terms of rate constants for reaction of aquo, hydroxo, and protonated forms of the iron(II) complex. Relative reactivities toward thiourea are[103]

$$[Fe(CN)_5(OH_2)]^{3-} > [Fe(CN)_5(OH)]^{4-} \gg [Fe(CN)_4(CNH)(OH_2)]^{2-}$$

Much greater reactivity of the aquo and hydroxo forms than of the protonated form has previously been established for reaction with the N-methylpyrazinium cation.[105] From the kinetic study of the reaction with thiourea estimates were made for protonation constants,[103] but these estimates have subsequently been challenged.[109]

Effects of added potassium or tetraalkylammonium halides (up to 2 mol dm^{-3}) on substitution at the $[Fe(CN)_5(pyrrolidine)]^{3-}$ anion (pyrrolidine = 7) in aqueous solution have been described. The potassium halides retard the reaction somewhat, the tetraalkylammonium halides lead to a slightly more marked acceleration. The pattern here[99] is similar to that established earlier for aquation of the $[Fe(5NO_2phen)_3]^{2+}$ [110] and $[Fe(bipy)_3]^{2+}$ [111] cations. The kinetic results for the pyrrolidine complex are discussed in terms of the effects of the added salts on the structure of the water, and the reflection of these effects in initial state and transition state hydration. The reordering of solvent water needed during transition state formation is described as "less demanding" when there is more solvent structure, as there is then more void space between solvent water molecules.[99] Unfortunately it is not possible to separate the established trends into initial state and transition state components here, and thus put these authors' hypothesis on a quasiquantitative basis. Such dissections are possible when dealing with binary aqueous, and nonaqueous, solvent media [cf., e.g., the discussion of solvent effects on aquation of *trans*-$[Co(py)_4Cl_2]^+$ in the cobalt(III) section below] as more or less reasonable estimates for the thermodynamic parameters for transfer of ions are available for such solvent media. The estimation of analogous quantities for transfer of ions for water into aqueous salt solutions is a subject still in its infancy. There is one aspect of the kinetic results relating to salt effects on substitution rates for $[Fe(CN)_5(pyrrolidine)]^{3-}$ that is puzzling, and that is that the relative effects of the three tetraalkylammonium cations employed were

$$NEt_4^+ > NBu_4^{n+} > NMe_4^+$$

Though this order is unexpected, it should be remarked that there is at least one other kinetic study which has reported just this order of effectiveness, viz. depolymerization of diacetone.[112] The role of solvent structure and consequences for initial state and transition state hydration have also been discussed in relation to the replacement of histamine and related ligands from their pentacyanoferrate(II) complexes by pyridine.[100] Again the discussion centers on the interaction between the solvent and the leaving group.

There has been some controversy over the structure of the binuclear "dimer" of $[Fe(CN)_5(OH_2)]^{3-}$. Structures involving either one or two cyanide bridges, **15** or **16**, have been proposed. In an extensive study of the kinetics of reaction of this "dimer" with several nucleophiles (Table 5.13), competition ratios were

$[(NC)_5Fe—NC—Fe(CN)_5(OH_2)]^{6-}$

$$\left[(NC)_4Fe \underset{CN}{\overset{CN}{\diagup\diagdown}} Fe(CN)_4 \right]^{6-}$$

15 **16**

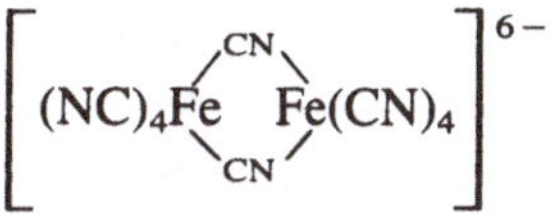

determined. The close similarity between competition ratios for the "dimer" and for $[Fe(CN)_5(OH_2)]^{3-}$ suggested the presence of an aquo ligand in the former, and thus favored structure **15**.[107]

Reaction of $[Fe(CN)_5(NO)]^{2-}$ with ammonia and with ethylenediamine has been studied over a range of pH.[113] At low concentrations of ammonia or of ethylenediamine, and pH < 12.5, the initial step is attack at the coordinated nitrosyl by hydroxide. Only at high concentrations of ammonia (ethylenediamine) and high pH is there attack at the nitrosyl by the ammonia (ethylenediamine). Hydroxide is thus a much better nucleophile than ammonia (ethylenediamine) toward nitrosyl coordinated to iron(II). Hydroxylamine is also a considerably more potent nucleophile than ammonia in such situations; on the other hand, trimethylamine does not attack at nitrosyl bonded to iron(II). Whereas the attack of ammonia at the coordinated nitrosyl results in its replacement by a fairly simple mechanism, when the initial attack is by hydroxide it seems that a reaction sequence of four steps is involved (Scheme 14).

Scheme 14

$$-NO \xrightarrow{OH^-} -NO_2H \xrightarrow[fast]{OH^-} -NO_2 \rightarrow -OH_2 \xrightarrow{NH_3} -NH_3.$$

5.4.2. Diimine Complexes

Iron(II) complexes of the diimine (**17**) ligands 2,2'-bipyridyl, bipy (**18**), and 1,10-phenanthroline, phen (**19**), and their substituted derivatives, have been

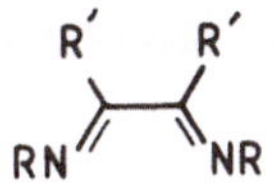

17 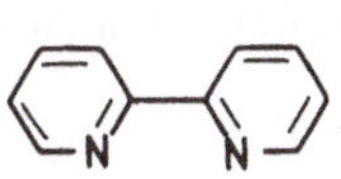 **18** 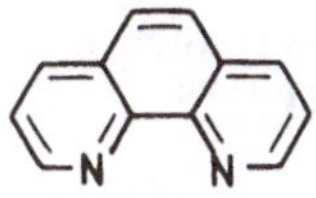 **19**

the subject of innumerable kinetic studies during the past decades, and are still actively being studied. The question of the detailed mechanism of their reaction with strong nucleophiles such as hydroxide and cyanide remains unsettled; the present situation is described in a following section. Although this mechanistic question is yet unresolved, the kinetic behavior of these complexes is clearly established and is increasingly being used to probe various aspects of medium effects on reactivity. This too is discussed below, but before dealing with these two aspects we shall report on a few more straightforward studies.

Several years ago it was shown that the hexadentate Schiff base derived from phenyl 2-pyridyl ketone and triethylenetetramine, hxsbp (**20a**), formed a

$$\text{N} \overset{\displaystyle R}{\underset{\displaystyle }{\diagdown}} \text{NCH}_2\text{CH}_2\text{NHCH}_2\text{CH}_2\text{NHCH}_2\text{CH}_2\text{N} \overset{\displaystyle R}{\underset{\displaystyle }{\diagup}} \text{N}$$

(20a) R = Ph

(20b) R = H

20

stable low-spin iron(II) complex; the kinetics of several substitution reactions of this complex were reported.[114] Now, eight years after submission, a companion study of the iron(II) complex of the hexadentate Schiff base derived from pyridine 2-aldehyde, hxsbh (**20b**), has at last appeared.[115] One point of interest about these complexes is that they are remarkably inert to substitution, despite the presence of two aliphatic nitrogens between the two diimine moieties; complexes of the type $[\text{Fe(bipy)}_2(\text{NH}_3)_2]$ are high spin and very labile. In view of Goedken's subsequent demonstration of facile aerial oxidation of coordinated ethylenediamine in $[\text{Fe(CN)}_4(\text{en})]^{2-}$ to coordinated diimine,[116] we have recently carried out an X-ray crystal structure determination on $[\text{Fe(hxsbh)}]^{2+}$, as its thiocyanate.[117] This has shown that the hxsbh ligand does indeed exist in this complex in the form shown in (**20b**); aerial oxidation has not taken place in this case. This is perhaps surprising—a hexadentate Schiff base containing three diimine moieties would surely form an even more stable and substitution-inert iron(II) complex. The chelate effects of these ligands are thus clearly of great importance in determining the high degree of inertness of their iron(II) complexes. Substitution takes place a little less slowly at the hxsbh complex than at the hxsbp complex, as one would expect, but both complexes undergo substitution much more slowly than analogous complexes containing three bidentate Schiff bases of type **21**.[118]

$$\text{N} \overset{\displaystyle R}{\underset{\displaystyle N}{\diagdown}} \longrightarrow \text{X}$$

21

Many years ago it was demonstrated that the rate law for the reaction of $[Fe(5NO_2phen)_3]^{2+}$, and of analogous sulfonato-substituted complexes, with peroxodisulfate took the form shown in equation (17):

$$-d[Fe(5NO_2phen)_3{}^{2+}]/dt = \{k_1 + k_2[S_2O_8{}^{2-}]\}[Fe(5NO_2phen)_3{}^{2+}] \quad (17)$$

The k_1 term in this rate law was ascribed to rate-determining dissociation of the complex, with subsequent redox steps fast.[119] Now the rate law for reaction of the 2,2'-bipyridyl complex with peroxomonosulfate has been found to be:

$$-d[Fe(bipy)_3{}^{2+}]/dt = k_1[Fe(bipy)_3{}^{2+}] \quad (18)$$

Hence for this oxidant and substrate, the sole reaction path is by initial rate-determining dissociation.[120] This situation parallels that reported for peroxodisulfate oxidation of the maleonitriledithiolato–cobalt(II) complex,[121] and for peroxodiphosphate oxidation of low-spin iron(II) complexes.[122] Substitution processes also appear to play a small part in the complicated reaction mechanism for cerium(IV) oxidation of tris[2-pyridinal-α-methyl(methylimine)]iron(II), the low-spin iron(II) complex of the Schiff base **22**, an aliphatic analog of **21**.[123]

22

In a study of racemisation of tris(bipy) and tris(phen) complexes of iron(II) and nickel(II) in the solid state, further evidence was obtained to support the assignment of an associative-type mechanism, with a transition state or intermediate of coordination number 7, for racemization at nickel(II), but of racemization via a trigonal prism intermediate for the iron(II) complexes.[124] A useful freeze-drying technique has been described for the isolation of optically active solids in preparations of halides of cations of this type.[125] Different kinetic patterns, and therefore the likelihood of different mechanisms, have been established for the racemization of $[Fe(phen)_3]^{2+}$ and $[Ni(phen)_3]^{2+}$ in water containing small amounts of nitrobenzene. The extent of association between the cations and the nitrobenzene was estimated from determination of the solubilities of nitrobenzene in aqueous solutions of $[M(phen)_3]Br_2$.[126] Complementary determinations of the solubilities of $[M(phen)_3]Br_2$ in aqueous nitrobenzene might have permitted an initial-state–transition-state dissection to have been carried out.

Some years ago it was established that activation volumes for reaction of

$[Fe(bipy)_3]^{2+}$ and $[Fe(phen)_3]^{2+}$ with hydroxide and with cyanide were large and positive, all being around $+20 \text{ cm}^3 \text{ mol}^{-1}$.[127] This was surprising as, despite arguments over the actual site of attack of the nucleophile (cf. below), it was generally accepted that the second-order terms in the rate laws could be assigned with confidence to bimolecular processes—for which one would expect $\Delta V^{\neq}$ to be about $-10 \text{ cm}^3 \text{ mol}^{-1}$ on organic precedent.[128] A mechanism analogous to the Eigen–Wilkins mechanism for complex formation was suggested to account for the observed activation volumes, with dissociative activation within an ion pair being invoked as dominant.[129] However, the role of solvation was neglected in this interpretation and, as subsequent experiments have shown, may make an important contribution to the overall observed $\Delta V^{\neq}$. In order to vary the solvent medium greatly it is also necessary to vary the substrate, switching from iron(II)–diimine cations to the (still d^6) nonelectrolyte $[Mo(CO)_4(bipy)]$ for non-aqueous kinetics. So in the study under discussion, $\Delta V^{\neq}$ values were determined for the reaction of $[Fe(4Me \text{ phen})_3]^{2+}$ with cyanide in water and in $33^1/_3\%$ methanol, but for $[Mo(CO)_4(bipy)]$ with cyanide in methanol and in dimethyl sulfoxide. Values of $+10$ and $+13 \text{ cm}^3 \text{ mol}^{-1}$ for reaction in water and in $33^1/_3\%$ methanol are not, considering the apparatus used, significantly different. This is not surprising if solvation of the cyanide is a key factor, for cyanide is strongly selectively hydrated in binary aqueous mixtures. However, cyanide is much less solvated in methanol, and hardly solvated at all in dimethyl sulfoxide. Thus in going from water to methanol and to dimethyl sulfoxide, the release of electrostricted solvent as the cyanide enters the transition state will make a progressively smaller contribution to the observed activation volume, allowing this to approach more closely the "gas phase" value. Indeed $\Delta V^{\neq}$ is $+4 \text{ cm}^3 \text{ mol}^{-1}$ in methanol, and $-9 \text{ cm}^3 \text{ mol}^{-1}$ in dimethyl sulfoxide.[130] Despite this neat trend according to solvation changes and apparent success in rationalizing the $\Delta V^{\neq}$ values, it should be pointed out that, apart from the unfortunate need to change compounds halfway, there are other somewhat unsatisfactory features. The main one is the sheer size of the difference between the activation volumes in water and in dimethyl sulfoxide, which corresponds to an incredibly large number of water molecules being transferred from the cyanide hydration shell to bulk solvent in the process of transition state formation. Another is the un-expectedly large difference between the activation volumes for cyanide attack at this 4-methyl compound and at the parent cation. This difference is in the same direction as that found for aquation of the 4,7-dimethyl and unsubstituted cations,[131] but is several times larger for only half the substituent effect. It may be that this is signaling the danger of assuming negligible solvation differences for the tris[diimine]iron(II), or diiminemolybdenum(0), moieties in going from the initial to the transition states in the various solvent media.

Bis(diimine) complexes of the type $[Fe(phen)_2Cl_2]$ solvolyze very quickly in aqueous solution, but are stable in the solid state and in at least some aprotic solvents. Therefore although aquation of, say, $[Fe(phen)_3]^{2+}$ in water or in water-rich solvent mixtures leads to Fe_{aq}^{2+} via rate-limiting loss of the first phen ligand, there is a real possibility that in nonaqueous media species of the type $[Fe(phen)_2X_2]$ may be persistent intermediates or even final products. Indeed dissolution of $[Fe(phen)_3]Cl_2$ in 1,2-dichloroethane leads rapidly to the formation of $[Fe(phen)_2Cl_2]$.[132] The rapidity of reaction reported here may be ascribed to the very high chemical potential of Cl^- in 1,2-dichloroethane.[133] A related but more complicated system is provided by the solvolysis of $[Fe(phen)_3]^{2+}$ in the presence of chloride and/or bisulfate in acidic dimethyl sulfoxide. The authors of the original paper[134] interpreted their results in terms of solvolysis of ion pairs, with their relative proportions and reactivities determining the observed rate of dissociation. However, it has now been claimed that the kinetics observed have to be interpreted in terms of a yet more complicated mechanism, in which the compound $[Fe(phen)_2Cl_2]$ also appears as a major product.[135]

A much smaller difference in solvent properties underlies the recent investigation of dissociation of $[Fe(bipy)_3]^{2+}$ and $[Fe(phen)_3]^{2+}$ in H_2O and in D_2O. Rate constants decreased by between 10% and 20% on going from H_2O to D_2O, but differences between activation parameters $\Delta H^{\ddagger}$ and $\Delta S^{\ddagger}$ in the two solvents were too small to be considered significant. The author discusses the structural properties of H_2O and D_2O, and recognizes that their reflection in initial state and transition state solvation determines reactivities.[136] However, no data are presented to enable an analysis of reactivities in terms of initial state and transition state contributions to be possible. Such analyses are possible for a number of systems featuring $[Fe(phen)_3]^{2+}$ and closely related complexes; recent examples will be discussed in the remainder of this section.

Several years ago reactivity trends for the reaction of $[Fe(bipy)_3]^{2+}$ with cyanide in binary aqueous solvent mixtures were discussed in terms of initial state and transition state contributions.[137] The lack of any knowledge of the effect of solvent composition on the chemical potential of the cyanide ion meant that chloride ion had to be used as model. Subsequent estimation of transfer chemical potentials for cyanide ion from solubility measurements on potassium cyanide[138] has meant that a more satisfactory analysis can be undertaken. Whereas it originally appeared that the changing chemical potential of the cyanide, like hydroxide, has a major effect on rate constant trends, it now appears that the chemical potential of the cyanide ion changes very little with solvent composition in water-rich mixtures, and that the modest increase in rate observed on going from water into such solvent mixtures is due to a modest change in the difference between the chemical potentials of the $[Fe(bipy)_3]^{2+}$ and of the transition state. However, both complex and transition state are stabilized greatly by transfer from water into, for instance, aqueous methanol. It is of interest to compare this pattern with that established for peroxodisulfate oxidation of $[Fe(phen)_3]^{2+}$ in binary aqueous mixtures containing oxidation-resistant cosolvents. Here the marked stabilization of the complex on transfer from water into binary solvent mixture is opposed by a marked destabilization of the $2-$ peroxodisulfate anion. The total solvent effect on the initial state is thus fairly small, but is still larger than the effect on the transition state.[139]

Ever since Gillard's suggestion[140] that several puzzling features of the kinetics and reactivity patterns of substitution reactions of diimine complexes could be explained by invoking attack by the nucleophile initially at the coordinated ligand, there has been continuing controversy on this matter.[141] At the time of writing there is no sign of a tidy solution acceptable to all parties. Indeed it is beginning to look as though direct attack at the metal may be the normal route for diimine complexes of platinum(II) and palladium(II), but that initial attack at the ligand may well be the mechanism for the majority of iron(II) and ruthenium(II) complexes of this type. A paper on the kinetics of hydroxide attack at $[\text{Fe}(5\text{NO}_2\text{phen})_3]^{2+}$ and at $[\text{Ru}(5\text{NO}_2\text{phen})_3]^{2+}$ (see Section 5.5.1 for reaction of $[\text{Ru}(5\text{NO}_2\text{phen})_3]^{2+}$ with the CH_2NO_2^- anion) compares rate constants and equilibrium constants for these reactions with those for hydroxide attack at analogous quaternized organic molecules and ions. This paper contains a valuable table summarizing kinetic and equilibrium data for a range of inorganic and organic reactions of this type.[142] The importance of organic analogs in this area is emphasized by the study of hydroxide attack at 5-nitro-1,10-phenanthroline itself,[143] and by the inclusion of a section on inorganic diimine complexes in a review of organic heterocyclic pseudobases.[144]

In reactions of the bipyrimidine, bipym (**23**), complex $[\text{Fe}(\text{bipym})_3]^{2+}$ with cyanide and with azide in aqueous solution it is possible to obtain forward and reverse rate constants for the ligand-substituted intermediate. Kinetic data are reported, as are ΔH° and ΔS° values for the equilibrium between complex and nucleophile.[145] Whereas azide does not displace phen from $[\text{Fe}(\text{phen})_3]^{2+}$, it does react with the weaker bipym complex. The complex $[\text{Fe}(\text{bipz})_3]^{2+}$, where bipz = 2,2'-bipyrazine (**24**), reacts with hydroxide again via attack at the ligand.

In this case such attack is followed by ligand cleavage. Reaction of $[\text{Fe}(\text{box})_3]^{2+}$, box = **25,** with hydroxide also results in ligand cleavage. Kinetic results for reactions of both these complexes are given.[146]

Ligand attack is not restricted to these diimine ligands, nor indeed to substitution reactions. Reversible attack at coordinated dimethylglyoxime has been suggested in the reaction of hydroxide with the cobalt(II) complex $[\text{Co}(\text{dmgH})_2]$.[147] Oxidation of $[\text{Fe}(4,7\text{-dihydroxyphen})_3]^{2+}$ by dioxygen is not thought to involve direct interaction of oxygen with the central iron atom, but rather by attack of the dioxygen on one of the hydroxy substituents.[148] This is, of course, slightly different from the reactions mentioned so far in this section in that attack is taking place at a ring substituent rather than at an actual ring atom.

5.4.3. Other Low-Spin Iron(II) Complexes

Another important group of low-spin iron(II) complexes comprises those with tetradentate cyclic ligands, which generally have four nitrogen donor atoms, for example porphyrins (porph) and phthalocyanines (pc). Here kinetic studies of substitution generally are concerned with the axial ligands in the fifth and sixth coordination positions. The mechanism of substitution in reactions of iron(II) phthalocyanine complexes,

$$Fe(pc)LL' + L'' \rightarrow Fe(pc)LL'' + L' \tag{19}$$

where L, L', and L'' are all nitrogen or phosphorus bases, is limiting dissociative (D), but the transient Fe(pc)L intermediate is very reactive and not particularly discriminating. It is, for example, considerably less discriminatory than Fe(porph)L intermediates. The substitution kinetics of the reactions of equation (19) were studied in acetone solution, with a view to examining axial ligand labilities, leaving group effects, and *trans* effects. The results were compared with those from analogous porphyrin complexes, and discussed in terms of σ- and π-bonding effects and consequences. In fact the phthalocyanine and porphyrin complexes show considerable similarities, but these two groups may be contrasted with complexes of other macrocycles, and of analogous oxime complexes. Differences are thought to reside in π-bonding differences.[149] The effects of tetramethyl versus tetraphenyl substitution have been assessed for the reactions of the respective bis[acetonitrile]tetraazamacrocycleiron(II) complexes with imidazole and with N-methylimidazole. The nature of the substituents has a significant effect on activation enthalpies. These reactions proceed by limiting dissociative mechanisms, as for the phthalocyanine and porphyrin complexes mentioned above. However it is difficult to carry out direct comparisons between these tetra-substituted derivatives and the phthalocyanine and porphyrin systems, since there are important difference, of solvent medium and of charge on complex, between the two groups.[150]

The stepwise replacement of acetonitrile by trimethyl phosphite in $[Fe(MeCN)_6]^{2+}$ has been monitored by nmr and by ultraviolet–visible spectroscopy. It is possible to derive some approximate kinetic data from these observations. The coordinated acetonitrile is replaced fairly slowly—indeed it takes several days to replace the fifth acetonitrile to obtain $[Fe(MeCN)\{P(OMe)_3\}_5]^{2+}$.[151] The kinetics of formation of chelating dicarbene products from the reaction of the hexakisisonitrile complex $[Fe(CNMe)_6]^{2+}$ with hydrazine and with methyl-hydrazine have been monitored. These reactions are first order both in iron(II) complex and in (methyl)hydrazine; rate constants and $\Delta H^{\ddagger}$ and $\Delta S^{\ddagger}$ values are reported. The reaction of $[Fe(CNMe)_6]^{2+}$ with methylamine exhibits more complicated kinetics, and it proved impossible to suggest an unequivocal mechanism.[152] *Cis–trans* isomerization is extremely slow for the complexes $[Fe(BH_3CN)_2\{P(OR)_3\}_4]$, R = alkyl, in such solvents as acetonitrile, benzene, methylene chloride, or chloroform—indeed in some cases is undetectable after a week. Admission of even a trace of air to the system reduces isomerization times to a matter of hours, clearly implicating the intermediacy of iron(III).[153]

5.4.4. Iron(III) Complexes

Several kinetic studies have dealt with iron(III) complexes and their reactions which complement iron(II) studies mentioned above. Such studies will be reported in this section; such reactions as complex formation at Fe_{aq}^{3+} are dealt with at appropriate places elsewhere in this volume.

The rate law for dissociation of imidazole from the pentacyanoferrate(III) complex $[Fe(CN)_5(im)]^{3-}$ is:

$$-d[Fe(CN)_5(im)^{3-}]/dt = \{k_1 + k_2[H^+]\}[Fe(CN)_5(im)^{3-}] \qquad (20)$$

The k_1 path corresponds to solvent-assisted dissociation. Formation of this pentacyanoimidazolatoferrate(III) complex follows a simple second-order rate law:

$$+d[Fe(CN)_5(im)^{3-}]/dt = k_2'[Fe(CN)_5(OH_2)^{2-}][HIm] \qquad (21)$$

Activation parameters $\Delta H^{\ddagger}$ and $\Delta S^{\ddagger}$ were determined for the k_1 and k_2' terms in the above rate laws. The mechanisms both of dissociation and formation are said to be I_d [154]; the mechanism of dissociation of this pentacyanoferrate(III) complex therefore differs from that, D, usual for the pentacyanoferrate(II) series *(vide supra)*. Reaction of $[Fe_2(CN)_{10}]^{4-}$, $[Fe(bipy)(CN)_4]^-$, or $[Fe(phen)(CN)_4]^-$ with dinitrogen tetroxide (in the form of a nitrous-acid–nitric-acid mixture) is a complicated mixture of substitution and redox processes. The N_2O_4 provides the nitrosyl group coordinated to the iron in the product; the kinetics suggest rate-determining attack of $\cdot NO_2$ on the cyanoferrates(III).[155]

The kinetics of displacement of chloride by substituted imidazoles from iron(III)–porphyrin complexes have been examined. Intermediates have been characterized, and the role of hydrogen bonding between the imidazole and the leaving chloride probed.[156]

Exchange of acetylacetone with $[Fe(acac)_3]$ is a relatively slow process. In the presence of small amounts of water, the observed first-order rate constant is given by:

$$k_{obs} = k_1 + k_2[OH_2] \tag{22}$$

Both k_1 and k_2 paths are assigned S_N2 mechanisms, with nucleophilic attack by water and acacH, the former being considerably more nucleophilic here. The assignment of mechanism is made primarily on the basis of the activation entropies found for each path.[157] Exchange of acac with complexes [M(acac)$_3$] has also been studied for M = Cr, Ru, Co, and Rh; these systems are covered in Section 5.2.2.2.1. (p. 116) and Section 5.7.5.2. (p. 165).

5.5. Ruthenium

Although the majority of substitution studies in the period covered by this report concern ruthenium(II), there have been a few kinetic studies of substitution at ruthenium(III), and two at ruthenium(IV).

5.5.1. Ruthenium(II)

Rate constants for the reactions of [Ru(NH$_3$)$_5$(OH$_2$)]$^{2+}$ with ethyl glycinate, methyl sarcosinate, and methylamine have been determined at 298.2 K in aqueous solution. These have been compared with the rate constant for reaction of [Ru(NH$_3$)$_5$(OH$_2$)]$^{2+}$ with ammonia, to show that alkyl substitution in ammonia reduces the affinity of ruthenium(II) for such ligands.[158] A link between "classical" formation reactions of this type and organometallic systems is provided by a study of reactions of [Ru(NH$_3$)$_5$(OH$_2$)]$^{2+}$ with a selection of alkenes and alkynes.[159]

Kinetics of aquation of [Ru(LLLL)X$_2$], with LLLL = cyclam, 2,3,2-tet, en$_2$, or (NH$_3$)$_4$, and X = Cl or Br, have been followed by cyclic voltammetry. Rate constants, and activation parameters ($\Delta H^{\ddagger}$ and $\Delta S^{\ddagger}$) have been evaluated, and compared with kinetic parameters for reactions of analogous compounds of ruthenium(III) and cobalt(III). Similar trends obtain for all three sets of complexes. There is retention of stereochemistry, rates decrease as the extent of chelation in LLLL increases, and *trans* complexes are less labile than *cis* analogs. Reactivities are determined by solvation of the initial and transition states, by nephelauxetic effects, and by σ-*trans* effects. A limiting dissociative (*D*) mechanism is proposed for the ruthenium complexes, with square-pyramidal geometry for the transient intermediate [cf. rhodium(III) photochemistry below, Section 5.8.10].[160] Differences in isomer lability have also been described for

[Ru(terpy)(PPh$_3$)$_2$Cl$_2$], where chloride substitution occurs very much more rapidly in the *cis* than in the *trans* isomer. This difference is ascribed to the large *trans* effect of the triphenylphosphine, which is perforce *trans* to a chloride ligand in the *cis*-dichloro isomer.[161] [Ru(bipy)$_2$Cl$_2$] appears, not unexpectedly, to exist only in the *cis* form in dimethyl sulfoxide solution, from nmr studies. The activation energy for solvolysis of this compound, giving *cis*-[Ru(bipy)$_2$(dmso)Cl]$^+$, is 19 kcal mol^{-1}; regrettably no rate constants are given in this brief communication.[162] A preliminary report has appeared on base labilization of S-bonded dimethyl sulfoxide in [Ru(dmso)$_6$]$^{2+}$ and in [Ru(dmso)$_4$Cl$_2$], studied by ^{1}H nmr spectroscopy of aqueous solutions containing these species.[163]

Three isomers have been isolated of the compound [Ru(LL)$_2$Cl$_2$], where the bidentate ligand LL is 2-(phenylazo)pyridine (**26**). The least stable *(trans)* isomer isomerizes at relatively high temperatures to the C$_2$-*cis* isomer, **27**, by a

26　　　　　　　　　　　　27

simple edge twist. The rate constant at 414 K in *m*-dichlorobenzene is 7.9 × 10^{-5} s^{-1}; activation parameters are $\Delta H^{\ddagger} = 33$ kcal mol^{-1} and $\Delta S^{\ddagger} = +1$ cal deg^{-1} mol^{-1}.[164] Linkage isomerization of coordinated sulfur dioxide in [Ru(NH$_3$)$_4$(SO$_2$)Cl]Cl in the solid state has been described.[165] Another solid state reaction of some interest to solution kineticists is substitution in [Ru(NH$_3$)$_5$(N$_2$)]X$_2$, where X = Cl, Br, or I. In water the dinitrogen ligand is easily replaced by water, but in the solid state ammonia is displaced by halide when these salts are heated. Another contrast with solution studies is that the mechanism of substitution appears to change character completely in going from the chloride (dissociative) to the bromide and iodide (associative); mechanism assignment here follows from the determined activation parameters.[166]

Mechanisms of substitution at coordinated diimines have been discussed, with some mention of ruthenium(II), in the section on iron(II) above (Section 5.4.2). One system involving only ruthenium(II) is the attack of the CH$_2$NO$_2^-$ anion at 5-nitro-1,10-phenanthroline coordinated to ruthenium(II) in [Ru(5NO$_2$phen)$_3$]$^{2+}$ and in [Ru(5NO$_2$phen)(bipy)$_2$]$^{2+}$. Nmr studies show that the nucleophile attacks at the 6-position of the 5NO$_2$phen in the latter compound, and that three CH$_2$NO$_2^-$ groups attach themselves to the former cation, one at the 6-position of each ligand.[167]

The remainder of the references to ruthenium(II) complexes concern photochemical substitution. Quantum yields of [Ru(bipy)$_2$(dmf)Br]$^+$ and [Ru(bipy)$_2$Br$_2$]

from photoanation of $[Ru(bipy)_3]^{2+}$ on irradiation in dimethylformamide at 458 nm are linearly dependent on the concentration of bromide. It is claimed that ion pair formation is an essential prerequisite for photosubstitution, and that the dibromo product mentioned above is derived from ion triplets. The results here[168] parallel closely those obtained in an earlier study of the analogous thiocyanate system.[169] Though the ion pair (ion triplet) interchange model is fully consistent with the experimental results, it is impossible to rule out completely a mechanism involving monodentate bipyridyl intermediates. It is a pity that, as so often, parallel experiments with 1,10-phenanthroline complexes were not conducted; the rigid 1,10-phenanthroline ligand is extremely unlikely to act in a monodentate fashion. Complexes of the type $[Ru(bipy)_2X_2]$ and $[Ru(bipy)_2LX]^+$ can readily be synthesized by photochemical substitution from $[Ru(bipy)_2(py)_2]^{2+}$ using a low-polarity solvent such as acetone or dichloromethane. Under these conditions, substitution at the bis(pyridine) complex takes place by a simple dissociative mechanism, with a quantum yield (0.18) which is independent of the nature and concentration of the incoming group.[170] Information on acid–base reactions of the excited state of $[Ru(bipy)_2(CN)_2]$, and of its phen analog, is relevant to kinetic studies of these species.[171]

Photolysis of a range of complexes *trans*-$[Ru(NH_3)_4LL']^{2+}$, with L and L' taken variously from a range of nitrogen donor ligands such as pyridine, pyrazine, and ammonia, in the visible range in aqueous solution results in predominant loss of ammonia. The relative quantum yields for complexes containing various ligands L and L' are consistent with the patterns established and rationalized for photosubstitution in similar complexes in the $[Ru(NH_3)_5L]^{2+}$ series.[172]

Isomerization of the ruthenium(II)–isonitrile complex $[Ru(PPh_3)_2(CNBu_2^t)_2Cl_2]$ takes place on irradiation in the range 254–436 nm. The mechanism is thought to include photoelimination of one triphenylphosphine ligand, followed by dimerization of the square-pyramidal intermediate thus produced.[173] The intermediacy of a transient square-pyramidal species here recalls the postulate of such a species in the thermal aquation of dihalogenoruthenium(II) complexes discussed near the beginning of this section.

The first step in the reduction of bis[histidinato]cobalt(III) by hexacyanoruthenium(II) is in fact a substitution process, attack of the cobalt(III) at cyanide coordinated to the ruthenium.[174]

5.5.2. Ruthenium(III)

The activation volume for aquation of the $[Ru(NH_3)_5Cl]^{2+}$ cation is -30 cm^3 mol^{-1} (at 333 K), for the reverse reaction, chloride anation of $[Ru(NH_3)_5(OH_2)]^{3+}$, it is -20 cm^3 mol^{-1}. In order to explain these very large values, a limiting associative (A) mechanism is proposed both for aquation and for anation. In a way the anation value is the more striking, since release of electrostricted water on forming the $2+$ transition state from the $3+$ and $1-$ reactants will make a significant positive contribution to the initial state to transition state volume change. The intrinsic (i.e., ignoring solvent) activation volume may therefore be even larger than $-30 \text{ cm}^3 \text{ mol}^{-1}$ for the anation reaction! The assignment of A mechanisms for these substitutions is in line with proposals for other substitutions at ruthenium(III); only base hydrolysis gives evidence which might indicate a different mechanism.[175]

To complement the photosubstitution studies on ruthenium(II) complexes *(vide supra)*, there has been one report of photochemical studies on ruthenium(III) complexes. This deals with the *cis* and *trans* isomers of the $[\text{Ru(en)}_2\text{Cl}_2]^+$ and $[\text{Ru(en)}_2(\text{OH}_2)\text{Cl}]^{2+}$ cations. Under irradiation at ligand field frequencies, aqueous solutions of these cations undergo isomerization and aquation. Both dichloro complexes lose one chloride; both give a *cis–trans* mixture for the chloroaquo product. Irradiation of either *cis*- or *trans*-chloroaquo complex gives a mixture of both isomers. Chloride loss from the dichloro complexes is characterized by low quantum yields, and there is an even less photoefficient reaction of the *cis*-chloroaquo complex which seems to give a μ-hydroxo–ruthenium(III) dimer. The main activity in current photochemical studies in this area seems to be the observation and rationalization of the stereochemical courses of photosubstitutions. In the present case, the results are discussed in terms of an idealized model with three distinct steps, chloride loss, then any rearrangement of a five-coordinate transient intermediate, then water attack at this five-coordinate intermediate. It is the postulate of a distinct five-coordinate intermediate that causes most unease in this simple sequence. The discussion in the present paper, both of results and of theories, of course includes much comparison with photosubstitution at rhodium(III), cobalt(III), and chromium(III).[176]

An example of a kinetic study of a coordinated ligand reaction at ruthenium(III) is provided by that of hydrolysis of *p*-nitrophenylacetate catalyzed by $[\text{Ru(NH}_3)_5(\text{im})]^{2+}$. This complex is a very effective catalyst, being 10,000 times more effective than its cobalt(III) equivalent.[177] Acetylacetone exchange at $[\text{Ru(acac)}_3]$ will be discussed under cobalt(III) (see Section 5.7.5.2). The rate law for ruthenium trichloride catalysis of oxidation of (substituted) phenols by periodate in alkaline aqueous solution is claimed to indicate preequilibrium formation of a ruthenium(III)–phenol complex. This is assumed to arise from reaction between the phenol and RuOH_{aq}^{2+} present in aqueous solutions of ruthenium trichloride.[178] The substitution reaction

$$\text{Ru(III)} + \text{H}_2 \rightarrow \text{Ru(III)(H}^-) + \text{H}^+ \tag{23}$$

is a key component in the autocatalytic reduction of hexachlororuthenium(IV) by dihydrogen in chloride media.[179]

5.5.3. Ruthenium(IV)

The nitrile ligands in $[Ru(PhCN)Cl_5]^-$ and its acetonitrile analog are labile with respect to substitution, for example, replacement by dimethylformamide.[180] The complexes $[Ru(NO)(NO_3)_x(OH_2)_{5-x}]^{(3-x)+}$ contain ruthenium(IV) if the nitrosyl group is acting as NO^-. They dissociate readily in water. Rate constants have been determined for some stages of these stepwise hydrolyses.[181]

5.6. Osmium

Preparative results reported for replacement of dimethyl sulfoxide coordinated to osmium(II) by phosphorus and arsenic bases may provide a basis for a kinetic study some time,[182] but all genuine kinetic studies of substitution at osmium during this period covered by this report refer to complexes of osmium(IV). The halogenoammine complexes $[Os(NH_3)_4X_2]^{2+}$ and $[Os(NH_3)_3X_3]^+$ aquate slowly in acid solution. At high pH's these complexes undergo disproportionation. Isomerization of *cis*-$[Os(NH_3)_4Cl_2]^{2+}$ and of *cis*-$[Os(NH_3)_4Br_2]^{2+}$ occurs surprisingly rapidly. The pH dependence of isomerization rates implicates the intermediacy of an amido species.[183] The kinetics of hydrolysis of $[Os(NO)X_5]^{2-}$ anions, with X = Cl, Br, or I, have been studied in aqueous solution over the range 298–348 K. Rate constants have been obtained for the first step of each hydrolysis sequence, which is the loss of the halide ligand *trans* to the nitrosyl ligand. Hydrolysis rates decrease in the order I > Br > Cl; these osmium complexes aquate less readily than their ruthenium analogs. Some results are also reported on the effects of added salts, such as sodium perchlorate, sodium nitrate, or ammonium chloride, on rates and extent of hydrolysis.[184] These complexes are classified as osmium(IV) here by taking the nitrosyl ligand as NO^- (cf. ruthenium above); the authors of the paper under discussion neatly duck the oxidation state question by referring to Os(NO)(III) throughout!

Hexahalogenoosmium anions $[OsX_6]^{2-}$ undoubtedly contain osmium in oxidation state $4+$; it is with this currently popular class of complexes that the remaining references in this section deal. Rate constants and activation parameters ($\Delta H^{\ddagger}$ and $\Delta S^{\ddagger}$) have been determined in acidic aqueous solution for 13 aquation and anation reactions of the type represented by the forward and reverse processes of the equilibrium of equation (24):

$$[OsL_4IX]^{2-} + H_2O \rightleftharpoons [OsL_4I(OH_2)]^- + X^- \qquad (24)$$

with X = F, Cl, Br, or I, and $L_4 = Cl_4$ or I_4 (equatorial). An associative mechanism is assumed, *cis* and *trans* effects are assessed, and isokinetic plots of these and earlier kinetic results are presented.$^{(185)}$ The $[OsI_6]^{2-}$ anion does not react with pyridine either under thermal or photochemical conditions, unless the reaction is carried out in a nonpolar solvent such as benzene. A solution of trilaurylethylammonium hexaiodoosmate(IV) in benzene reacts with pyridine to give the $[OsI_5(py)]^-$ anion.$^{(186)}$ Mercury(II) is, as one would expect, a good catalyst for removal of iodide from the mixed halogenoosmates(IV) mentioned above. The mercury(II) attacks preferentially at iodide; both Hg^{2+} and HgI^+ can act as catalysts. The anions cis-$[OsCl_4I_2]^{2-}$ and fac-$[OsCl_3I_3]^{2-}$ isomerize during mercury(II)-induced aquation to give products containing the greatly favored linear I–Os–OH$_2$ unit. Whereas mercury(II) is a very good catalyst for halide removal from osmium(IV), alkaline earth metal(II) cations are very feeble. However they do have a measurable effect, which increases in the order Mg^{2+} < Ca^{2+} < Sr^{2+} < Ba^{2+}. The Zn^{2+} and Al^{3+} cations also have small but detectable catalytic effects, lying between those of Sr^{2+} and Ba^{2+}. Quantitative kinetic information is sparse in this paper—most of the kinetic data are presented in one graph.$^{(187)}$ The final reference of relevance to these hexahalogenosmate(IV) substitution processes is an extensive paper on the characterization of chloro-aquoosmium(IV) species, in which ultraviolet–visible absorption data are given.$^{(188)}$ Such information is of obvious relevance to kinetic studies, both in the characterization of intermediates and products, and in the selection of suitable wavelengths for kinetic monitoring.

5.7.　Cobalt(III)

The level of kinetic activity in this already well-studied area remains remarkably high, with a lot of interesting and important research as well as a lot of routine data collection.

5.7.1.　Aquation and Solvolysis

These reactions still provide many new references, which in the following pages are arranged partly by ligand type and partly by phenomenon probed. Arrangement is by relation of ideas rather than fully systematic.

5.7.1.1　Polydentate Amine–Halide Complexes

In a paper devoted to three aspects of substitution kinetics of $[Co(amine)_2(dien)Cl]^{2+}$ cations, aquation is represented by the temperature dependence of rate constants for the *unsym-fac-cis* (**28**) and *sym-fac-cis* (**29**) isomers of the cations with amine = NH$_3$.$^{(189)}$ The *cis*-α isomer of $[Co(trien)(imidH)Cl]^{2+}$ aquates five times more slowly than the *cis*-β_2 isomer. Rate constants and ac

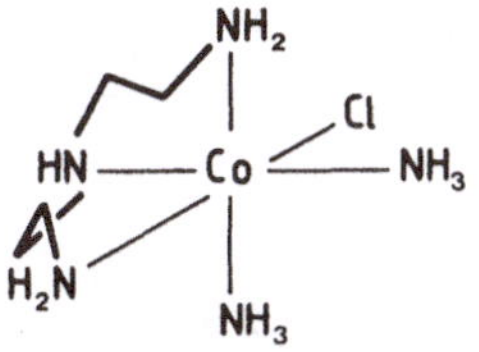

28

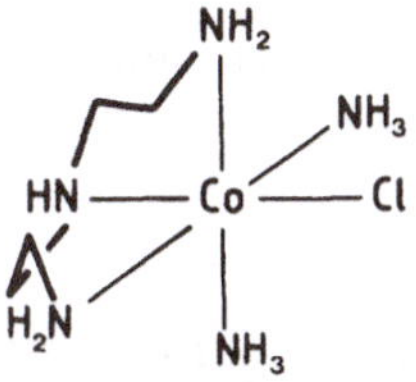

29

tivation parameters ($\Delta H^{\ddagger}$ and $\Delta S^{\ddagger}$) for aquation of these two complexes are compared with those for closely related complexes of tetradentate and bidentate amines.[190] It is normal practice to monitor kinetics of aquation of cobalt(III) complexes using ultraviolet–visible spectroscopy. There are sometimes difficulties when the spectra of reactants and products are very similar. It has now been shown that such difficulties can in principle be avoided by determining rate constants by the use of polarographic techniques. Reasonably good agreement between polarographically determined rate constants and earlier ultraviolet–visible results has been shown for such complexes as $[Co(NH_3)_5X]^{2+}$, *trans*-$[Co(en)_2X_2]^+$, and *trans*-$[Co(NH_3)_4X_2]^+$, in each case with X = Cl⁻ or Br⁻. This paper includes the first determinations of the rate constant for aquation of the cyclohexanediamine complex *trans*-$[Co(chxn)_2Br_2]^+$.[191]

An X-ray crystal structure of the complex $[Co(\eta^2\text{-tameH})(\eta^3\text{-tame})Cl]^{3+}$, where tame = 1,1,1-tris(aminomethyl)ethane, has proved the presence of a bidentate tame ligand with one terminal nitrogen protonated (30). Equilibrium

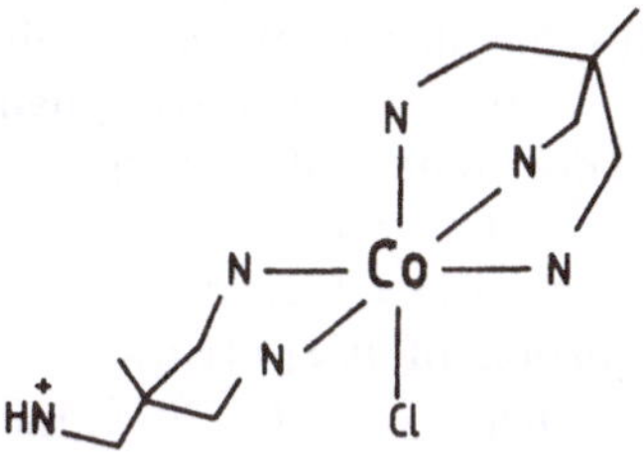

30

deprotonation of this —NH₃⁺ group is invoked to explain the acid dependence of rates of aquation of this $[Co(\eta^2\text{-tameH})(\eta^3\text{-tame})Cl]^{3+}$ cation in the pH range 7.6–8.8.[192] A crystal structure determination was again necessary to establish the stereochemistry of the $\delta\lambda\text{-}(-)_{589}\text{-}\Delta(S)\text{-}cis\text{-}$(1-aminopropan-2-ol-N)chlorobis(1,2-diaminoethane)cobalt(III) cation, shown schematically as **31**. Aquation of this complex presumably proceeds *via* the aquo complex **32** *en route* to the product **33**, but there is no kinetic evidence for the intermediacy of **32**. A parallel study of the kinetics of mercury(II)-catalyzed aquation of **31** *(vide infra)* provided a value for the rate constant for the ring closure step **32** to **33**.

As this is some 20 times faster than the aquation of **31**, the nonobservation of **32** as an intermediate in the **31** to **33** reaction is understandable.$^{(193)}$

5.7.1.2. Activation Volumes

These have been determined for spontaneous and for silver(I)-catalyzed aquation of the $[Co(NH_3)_5Br]^{2+}$ cation, in the absence and presence of polyelectrolytes. The results are set out in Section 5.7.3 below.$^{(194)}$ Activation volumes have also been determined for aquation, and for formation (Section 5.7.4, *q.v.*), of the pentacyanocobaltate(III) complexes $[Co(CN)_5X]^{3-}$. Values of $\Delta V^{\ddagger}$ are $+7.8$, $+7.6$, and $+14.0$ cm^3 mol^{-1} for X = Cl, Br, and I, respectively, at an ionic strength of 1.0 mol dm^{-3} and a temperature of 313.2 K. For the chloride, $\Delta V^{\ddagger}$ drops slightly, to $+9.1$ cm^3 mol^{-1}, when the temperature is decreased to 298.2 K. The observed values can be rationalized satisfactorily in terms of intrinsic and solvation volume changes operating in a *D* mechanism. The acid-catalyzed aquation of the azido-complex $[Co(CN)_5(N_3)]^{3-}$ has an activation volume of $+16.8$ cm^3 mol^{-1} in dilute perchloric acid (0.1 mol dm^{-3}).$^{(195)}$

Activation volumes for aquation of $[Co(tren)Cl_2]^+$ and of *cis*-β-$[Co(trien)Cl_2]^+$ are $+7.3$ and $+3.0$ cm^3 mol^{-1}, respectively.$^{(196)}$ In this paper various factors affecting activation volumes for aquation of amine–halide–cobalt(III) complexes are discussed, including hydration of the leaving halide, contortions of polydentate ligands, and the likely volume of a transient five-coordinate cobalt(III) species such as $[Co(NH_3)_5]^{3+}$. One way of effecting a simplification of interpretation is to work with uncharged leaving groups, such as urea. Whereas activation volumes for aquation of $[Co(NH_3)_5X]^{n+}$ complexes with charged leaving groups are markedly negative, e.g., -18.5 and -10.6 cm^3 mol^{-1} for X = SO_4^{2+} and Cl$^-$, respectively, when the leaving group is uncharged $\Delta V^{\ddagger}$ is much closer to zero, e.g., -1.7 cm^3 mol^{-1} for X = dmso. Now $\Delta V^{\ddagger}$ values of $+1.3$ and $+3.5$ cm^3 mol^{-1} have been determined for aquation of the complexes $[Co(NH_3)_5(urea)]^{3+}$ and *trans*-$[Co(dmgH)_2(urea)Cl]$.$^{(197)}$ Further kinetic information on the uncharged leaving group dmso has been reported,$^{(198)}$ but only at ambient pressure.

5.7.1.3. Solvent Effects

Rate constants for aquation of $[Co(NH_3)_5(dmf)]^{3+}$ have been determined over a temperature range of 40 K for eight solvent mixtures with x_{DMF} up to

0.7.[199] Rates of aquation of $[Co(NH_3)_5Br]^{2+}$, and of its chromium(III) analog, have been determined in methanol–, ethanol–, isopropanol–, and t-butyl-alcohol–water mixtures at various temperatures. The well-known compensation effect is apparent for all these cosolvents. In the aqueous methanol and aqueous ethanol series of solvent mixtures there is correlation of rate constants with "solvent ionising power."[200] A correlation of rate constants with solvent polarity was established in a much more limited investigation of aquation of α-*cis*-[Co(edda)(OH$_2$)Cl], studied in water and in 20% methanol, ethanol, and acetone. The variation of rate constants, and of the activation parameters $\Delta H^{\ddagger}$ and $\Delta S^{\ddagger}$, with solvent composition was claimed to indicate an I_d mechanism.[201]

Treatments of solvent effects on rates and activation parameters of the type referred to in the preceding paragraph suffer from the limitation that kinetic parameters are composite quantities. In principle it is much more satisfactory to try to analyze reactivity trends in terms of initial state and transition state components, though in practice this is not often straightforward, especially when charged reactants are involved. For many years Wells has been engaged in establishing transfer chemical potentials for ions for transfer from water into binary aqueous solvent mixtures. Sets of values are available for such cosolvents as methanol,[202, 203] t-butyl alcohol,[204] acetone,[203] dioxan,[205] ethylene glycol,[206] and dimethyl sulfoxide,[207] in all cases based on equivalent calculations for the transfer chemical potential of the solvated proton. In recent years Wells has started to use his single-ion values to analyze reactivity trends into initial state and transition state components, for example, for solvolyses of a range of organic halides and of (dissociative) aquations of a variety of chloro–transition-metal complexes.[208] Recently Wells has turned his attention to anation reactions of metal ions and complexes (Section 5.7.4), and to solvolysis of the *trans*-[Co(py)$_4$Cl$_2$]$^+$ cation,[209] discussed here. In all cases it is assumed that the leaving group is completely removed from the metal ion in the transition state, in other words that a limiting D mechanism operates. This simplification allows the use of transfer chemical potentials for chloride ion as one moiety of the transition state in solvolysis of chloro complexes. This simple model exaggerates somewhat the role of solvation of the leaving group, but may not have a marked effect on general conclusions. The cycle shown in Scheme 15 is used to connect

Scheme 15

Water: $[Co(py)_4Cl_2]^+ \xrightarrow{\delta_m \Delta G^{\ddagger}} [Co(py)_4Cl]^+ + Cl^-$

$\delta_m\mu^{\circ}\{Co(py)_4Cl_2^+\} \downarrow \quad \delta_m\mu^{\circ}\{Co(py)_4Cl^+\} \downarrow \qquad \downarrow \delta_m\mu^{\circ}\{Cl^-\}$

Mixed solvent: $[Co(py)_4Cl_2]^+ \xrightarrow[\delta_m \Delta G^{\ddagger}]{} [Co(py)_4Cl]^+ + Cl^-$

the various kinetic and thermodynamic components of the solvent effects. In order to determine the relative importance of initial state and transition state solvation effects it is necessary either to know transfer chemical potentials for the initial state cation, or to make some reasonable assumption about these and/ or transition state chemical potentials. In the paper under discussion, Wells uses trends established[202-206] for M^+ and M^{2+} cations to deduce that transition state solvent effects dominate. Now while simple aquo ions M^+ and M^{2+} seem not unreasonable models for, say, $[Co(NH_3)_4Cl_2]^+$ and $[Co(NH_3)_4Cl]^{2+}$, they seem less attractive for the initial and transition states in the present system, where the four pyridine ligands with their strongly hydrophobic and organophilic peripheries are likely to favor transfer into more organic media. In fact it is easy to assess Wells's model, by measuring solubilities of *trans*-$[Co(py)_4Cl_2]Cl$ in water and in mixed solvents, and thence, with the aid of Wells's estimates for $\delta_m\mu°(Cl^-)$, to obtain values for transfer of the cation. We have done this for methanol–water mixtures,[210] and, from this information and Wells's kinetic results, have also estimated transfer chemical potentials for the transition state (charge separation unspecified), $\delta_m\mu^\ddagger$. The results of our estimations are compared with Wells's model ions in Figure 5.1. The marked increase in solubility of *trans*-$[Co(py)_4Cl_2]Cl$ with increasing methanol content suggests that an aquo ion might not prove a good model for this particular complex cation; Figure 5.1 confirms this suspicion. However, as so often in this area, the conclusion from

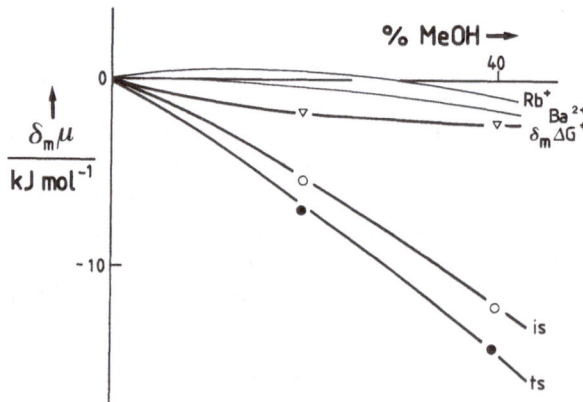

Figure 5.1 Initial state (is) and transition state (ts) transfer chemical potential variations for solvolysis of *trans*-$[Co(py)_4Cl_2]^+$ in aqueous methanol. Kinetic data are from Ref. 209, solubility data from Ref. 210; $\delta_m\mu°$ (is) values were obtained by using $\delta_m\mu°(Cl^-)$ values interpolated from Ref. 202. Transfer chemical potentials for Rb^+ and for Ba^{2+} (also from Ref. 202) are shown for comparison. All values are on the molar scale, at 298.2 K.

the two different treatments is the same! It should be added that this comforting agreement is lucky. If one changes one's single-ion assumptions only very slightly, then the transfer chemical potentials for the M^+ and M^{2+} aquo ions can very soon decrease in magnitude to zero and become positive, changing one's conclusions about the relative importance of initial state and transition state solvation accordingly, as indicated in Figure 5.2. We do not wish to denigrate Wells's achievements, but only wish to emphasize how important it is in this area, where the key species, the transition state, is unobservable, to keep a watch on assumptions and to assess the consequences of making even small alterations to them.

There is another important feature of Wells's investigation of the aquation of the *trans*-[Co(py)$_4$Cl$_2$]$^+$ cation that is relevant to the present section. Plots of $\Delta H^\ddagger$ and of $\Delta S^\ddagger$ against solvent composition for methanol–water mixtures show extrema in the water-rich region whose positions can be rationalized in terms of solvent structural features.[209]

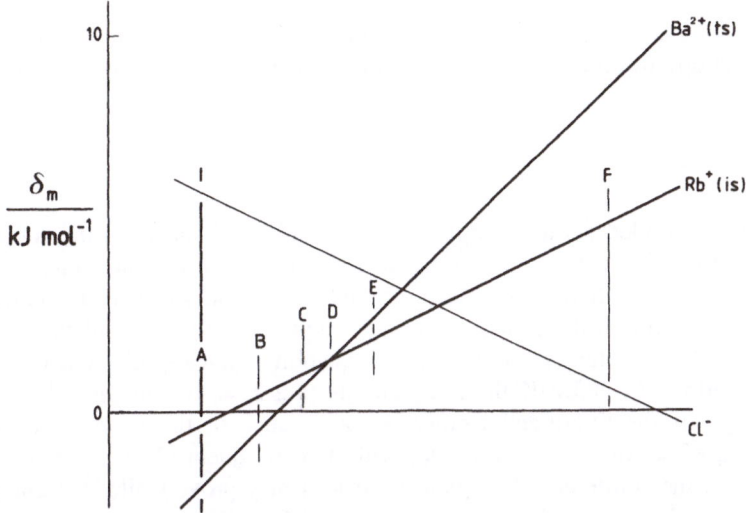

Figure 5.2 The effect of varying single-ion splitting assumptions on conclusions regarding the relative importance of initial state and transition state contributions to reactivity trends for solvolysis of *trans*-[Co(py)$_4$Cl$_2$]$^+$ in aqueous methanol. The effect on $\delta_m\mu°(Ba^{2+}$ = initial state model) and on $\delta_m\mu°(Rb^+$ = transition state model) of decreasing $\delta_m\mu°(Cl^-)$ from Wells's value (Ref. 202) is shown left to right. The conclusions change thus: at A (Wells's treatment, Ref. 209) ts > is; at B |ts| = |is|, but signs opposite; at C ts < is; at D ts = is; at E ts > is; at F ts ≫ is. All values are on the molar scale, at 298.2 K.

5.7.1.4. Salt Effects

It has been known for many years that rate constants for aquation of cationic cobalt(III)–amine or ammine–halide complexes, for example $[Co(NH_3)_5Cl]^{2+}$, are significantly affected by the addition of salts of $2-$ anions. These effects are ascribed to the different reactivities of the ion pairs present in significant quantities in such solutions, even in aqueous media. The extent of ion pairing is expected to increase on addition of, for instance, an alcohol to aqueous media. It has recently been shown that the addition of even 10% of ethanol is sufficient to have a marked effect on observed aquation rate constants. This has been demonstrated for added sulfates and for a range of added dicarboxylates. Attempts were made to correlate effects on rate constants with ion-pairing constants.[211] The addition of oxalate results in rate enhancement; the determination of rate constants as a function of concentration of oxalate and of temperature has been carried out. From these results it proved possible to calculate not only equilibrium constants for ion pair formation and ion pair reactivities but also the enthalpies and entropies corresponding to these equilibrium and kinetic parameters.[212] In both of these investigations attention has been focused on solute–solute interactions, with scant consideration given to the possible effects of the added ions on the solvent [contrast the study of salt effects on substitution rates of the pentacyanopyrrolidineferrate(II) anion mentioned above, Section 5.4.1.].

5.7.1.5. Oxo Anion Complexes

The S-bonded sulfito complex $[Co(NH_3)_5(SO_3)]^+$ has been the subject of several studies, but there has hitherto been no kinetic examination of its O-bonded isomer. At high pH's intramolecular redox processes become dominant, but in acid solution the kinetic pattern for aquation can be established. In the pH range 3.3–4.6 the rate constant for aquation is linearly dependent on acid concentration. At 283.2 K the rate constant is 2.2×10^6 dm^3 mol^{-1} s^{-1}. One advantage of the O-bonded isomer is the absence of the very strong *trans*-labilizing effect of S-bonded sulfite, with its consequent kinetic complications when both the sulfite and the ligand *trans* to it may prove labile.[213] The preparation and characterization of *trans*-$[Co(en)(S_2O_3)(OH_2)]^+$ in aqueous solution has permitted assessment of the *trans* effect of S-bonded thiosulfate. The sequence of decreasing kinetic *trans* effect at cobalt(III), $SO_4^{2+} > RSO_2^- > S_2O_3^{2-}$ parallels the *trans* influence of these ligands as established from crystal structure determinations of appropriate cobalt(III) complexes.[214] The marked *trans*-labilizing effect of S-bonded thiosulfate is reflected in the remarkable ease of the *cis* to *trans* isomerization involved in the first stage of aquation of the newly characterized O,S-bonded thiosulfato-bis[ethylenediaminecobalt(III)] cation, which gives the above-mentioned aquo–thiosulfato complex:

$$[Co(en)_2(S_2O_3)]^+ \rightarrow trans\text{-}[Co(en)_2(OH_2)(S_2O_3)]^+ \qquad (25)$$

Rate constants and activation parameters $\Delta H^{\ddagger}$ and $\Delta S^{\ddagger}$ are reported for this reaction (22.3 kcal mol^{-1} and zero, respectively, in 0.1 mol dm^{-3} aqueous lithium perchlorate).[215]

Aquation of the chlorito complex $[Co(NH_3)_5(ClO_2)]^{2+}$ forms part of the overall substitution and redox reactivity pattern established for this complex. Indeed the mechanism of aquation involves an internal redox process. Rate constants and activation parameters for aquation of this chlorito complex in acidic aqueous solution are $k_{298} = 8.0 \times 10^{-6}$ s^{-1}, $\Delta H^{\ddagger} = 105$ kJ mol^{-1} and $\Delta S^{\ddagger} = +13$ J K^{-1} mol^{-1}. Although this complex could hardly be described as robust, the chlorite anion is considerably less fragile when coordinated to cobalt(III) than when bonded to H$^+$.[216] Cobalt(III)–amine complexes of organic carboxylato ligands are usually stable and inert; the complex cis-$[Co(en)_2(imidH)(salicylato)]^{2+}$ is no exception. In moderately acid solution the imidazole ligand is protonated, but not the salicylato ligand. The rate law for aquation is

$$-d[\text{complex}]/dt = \{k_1 + k_2[\text{H}^+]\}[\text{complex}] \qquad (26)$$

A dissociative interchange mechanism is proposed for both aquation paths indicated by this rate law.[217]

The elusive species $[Co(NO_3)_6]^{3-}$ has been isolated. Its half-life in acetonitrile solution is about $\frac{1}{2}$ hour; repeat scan spectra of its solvolysis in acetonitrile exhibit clean isosbestic points. The one-step process thereby indicated is thought to involve electron transfer as an essential component of dissociation (cf. chlorite above.[218]

The kinetic behavior of carbonato complexes is discussed in a separate section below (Section 5.7.7).

5.7.1.6. *Miscellaneous*

The activation energy for solvolysis of the phenoxy radical complex of cobalt(III) in methanol is surprisingly high, 110 kJ mol^{-1}. Solvent effects on reactivity have been monitored for this complex, in particular for its reaction with a partially hindered bis phenol.[219] However, there is no evidence for alkylperoxy radicals coordinated to cobalt(III) as transient intermediates in the reaction of $[Co(acac)_2]$ with *tert*-alkylperoxy radicals.[220]

The results of CNDO/2 calculations on relative bond strengths in mono-1,10-phenanthroline, mono-2,2'-bipyridyl, and monoaliphatic diimine complexes of cobalt(III) should prove useful in discussion of rates of substitution and aquation of cobalt(III)–diimine complexes.[221] Presumably these authors will shortly turn their attention to the much more studied iron(II) analogs; they do mention the molybdenum(0) analog [Mo(CO)$_4$(bipy)].

5.7.2. Base Hydrolysis and Base Solvolysis

Rate constants and activation parameters E_a and $\Delta S^{\ddagger}$ are reported for base hydrolysis of the *sym-fac-cis*-[Co(amine)$_2$(dien)X]$^{2+}$ cations, structure **29,** for (amine)$_2$ = (NH$_3$)$_2$, X = Cl or Br, and for (amine)$_2$ = en, X = Cl.[189] Kinetic parameters for base hydrolysis of *cis*-α and of *cis*-β$_2$-[Co(trien)(imidH)Cl]$^{2+}$ have been obtained, and compared with those for closely related trien and bis(ethylenediamine) complexes of cobalt(III).[190]

Base hydrolysis of *trans*-[Co(en)$_2$(CO$_3$)Cl] results in loss of chloride, in contrast to loss of carbonate in acid solution (cf. Section 5.7.7). This base hydrolysis follows first-order kinetics; repeat scan spectra show clean isosbestic points at 444 and 556 nm. Activation parameters $\Delta H^{\ddagger}$ and $\Delta S^{\ddagger}$ are 31.1 and +27 for the acid-independent path, and 25.2 and +25 for the acid-dependent path (units of kcal mol^{-1} and cal deg^{-1} mol^{-1}).[222]

Both kinetics and stereochemistry of base hydrolysis of the cobalt(III)–zwitterionic ligand complexes *trans*-[Co(en)$_2$X($^-$O$_2$CCHRNH$_3$$^+$)]$^{2+}$ with X = Cl or Br and R = H, Me, or Et have been investigated. The amino acid ligands, glycine, alanine, and aminobutyric acid, respectively, are bonded through carboxylate–oxygen. In each case there is about 33% isomerization to *cis* product, while in the case of the glycine complex there is also some isomerization to give N-bonded glycine in the product. Activation enthalpies lie in the range 93–97 kJ mol^{-1}, and activation entropies in the ranges 76–87 J K^{-1} mol^{-1} for chloro complexes, 92–101 J K^{-1} mol^{-1} for bromo-complexes.[223] These results extend earlier work on complexes *trans*-[Co(en)$_2$Cl(RCO$_2$)]$^+$ where increasing steric hindrance by substituents R was found to increase the *trans:cis* product ratio.[224] Related sulfonato complexes *cis*-[Co(en)$_2$Cl(H$_2$N(CH$_2$)$_n$SO$_3$)]$^+$, with n = 1 or 2, are much more reactive than the carboxylato analogs, or indeed than with analogous alkylamine complexes, since it is much easier to generate conjugate bases from the sulfonato complexes. However the conjugate base of the NH$_2$CH$_2$SO$_3$$^-$ complex cannot react by intramolecular attack of the sulfonato group at the cobalt, whereas the conjugate base of *cis*-[Co(en)$_2$Cl(gly)]$^+$ does react by attack of the carboxylate moiety at the cobalt. Rate constants and activation parameters $\Delta H^{\ddagger}$ and $\Delta S^{\ddagger}$ were determined for the aminosulfonato complex with n = 1.[225]

The mechanism of base hydrolysis of the [Co(NH$_3$)$_5$(*p*-nitrophenylphosphato)]$^{2+}$ cation is complicated, involving reactions of the coordinated ligand (see Section 5.7.11) as well as base hydrolysis of the complex itself. This base hydrolysis takes place by the usual S_N1CB mechanism.$^{(226)}$

Base hydrolysis of optically active *mer*-[Co(dien)(1,3-diaminopropan-2-ol)Cl]$^{2+}$ results in complete racemization. Such a stereochemical result has not been reported before for base hydrolysis of cobalt(III) complexes. However the dien ligand remains 100% in the meridional geometry. The optically active hydroxo analog racemizes at least 10^4 times more slowly. These three observations can be accommodated readily in the classical π stabilization hypothesis of Basolo and Pearson.$^{(227)}$

Activation volumes for base hydrolysis of the complexes α- and β-[Co(edda)(NH$_3$)$_2$]$^+$ and α-[Co(edda)(NO$_2$)$_2$]$^-$, $+16.6$, $+22.3$, and $+11.9$ cm^3 mol^{-1}, respectively, are claimed all to be consistent with the operation of the S_N1CB mechanism for base hydrolysis.$^{(228)}$ Activation volumes for base hydrolysis of the [Co(NH$_3$)$_5$Br]$^{2+}$ cation are smaller in aqueous polyelectrolytes than in water, but are still markedly positive. Activation volumes and entropies suggest that the effect of the macroions is to dehydrate the $2+$ reactant. There is a correlation of activation volumes and activation entropies for a variety of substitutions of this type in polyelectrolytes;$^{(229)}$ compare the earlier correlation of activation volumes and entropies for a range of substitutions in aqueous solution.$^{(230)}$

Although it seems certain that the majority of base hydrolysis reactions of cobalt(III)–ammine or amine–halide and related cations proceed by the S_N1CB route, including the systems mentioned so far in this section, there is some evidence that base hydrolysis of certain cobalt(III)–macrocyclic-amine–dihalide cations may proceed by an E_2 mechanism. This mechanism is concerted, with synchronous cleavage of the cobalt–chloride and nitrogen–hydrogen giving a five-coordinate intermediate directly rather than through a six-coordinate conjugate base$^{(231)}$ (Scheme 16).

Scheme 16

Several pieces of evidence suggest the operation of this E_2 mechanism for base hydrolysis of the *cis*-[Co(cyclen)Cl$_2$]$^+$ cation (cyclen = 1,4,7,10-tetraazacyclododecane, **34**. In the first place, the large difference in rate constant between this complex and its bromo analog is too large to be explained convincingly on the basis of an S_N1CB mechanism, but perfectly acceptable if the E_2 mechanism operates. Secondly, a comparison of the kinetic parameters k, $\Delta H^{\ddagger}$, and $\Delta S^{\ddagger}$

34

for acid aquation of this cyclen complex and its α-trien, en$_2$, and cyclam analogs reveals only small differences, but a parallel comparison for base hydrolysis shows that the rate constant for the cyclen complex is much larger than for the others, while $\Delta H^{\ddagger}$ for the cyclen complex is unusually low. Further evidence pertaining to the question of the operation of the S_N1CB or E_2 mechanisms comes from Brønsted coefficients and from hydrogen–deuterium isotope effects.[232] Base hydrolysis of the *cis*-[Co(en)$_2$(py)Cl]$^{2+}$ cation could take place by the S_N1CB or E_2 mechanisms just discussed, or by attack of hydroxide at the co-ordinated pyridine ligand.[140] The fast rate of base hydrolysis of this complex might be taken as suggesting this last route,[233] but a recent investigation of the secondary isotope effect (C$_5$D$_5$N *vice* C$_5$H$_5$N) in this reaction failed to provide positive evidence in favor of the ligand-attack route. It should perhaps be added that the absence of a large secondary isotope effect does not rule out the ligand-attack route; the negative result can be interpreted by any of the three mechanisms.[234] In the base hydrolysis of this pyridine complex interpretation is complicated by the choice between abstracting a proton and attaching hydroxide to the coordinated ligand. In the case of base hydrolysis of the *cis-α*- and *cis-β$_2$*-[Co(trien)(imidH)Cl]$^{2+}$ cations, interpretation of base hydrolysis kinetics is complicated by a choice of positions from which a proton may be extracted. Nonetheless the kinetic results have been compared with those from similar trien and en$_2$ complexes of cobalt(III).[217]

There has been disagreement[235] over the interpretation of kinetic results[236] for base hydrolysis of the [Co(en)$_2$(ox)]$^+$ cation in aqueous solution. This disagreement seems to have arisen from the marked difference between the tendency of Co(OH)$_3$ to precipitate from solutions of this cation and from solutions of [Co(NH$_3$)$_4$(ox)]$^+$. Whereas Co(OH)$_3$ precipitates readily from solutions containing this latter tetraammine complex, it does not precipitate from solutions containing [Co(en)$_2$(ox)]$^+$, nor indeed from solutions containing the [Co(en)$_3$]$^{3+}$ cation.[237]

The remainder of the references in this section relate to base solvolysis in nonaqueous media, where the principles are the same but details different. Rate constants and activation parameters ($\Delta H^{\ddagger}$ and $\Delta S^{\ddagger}$) have been reported for the reaction of *cis*- and *trans*-[Co(en)$_2$Cl$_2$]$^+$ and of *trans*-[Co(en)$_2$Br$_2$]$^+$ with the bases piperidine, diethylamine, butylamine, cyclohexylamine, and benzylamine in methanol. An S_N1CB mechanism is proposed in each case, principally on the basis of the large positive activation entropies.[238] These studies were extended, for the *trans*-[Co(en)$_2$Cl$_2$]$^+$ cation, to methanol + dimethylformamide solvent mixtures, and to pure dimethylformamide. The observed trend of rates—reaction is very much slower in pure dimethylformamide—was discussed in terms of an S_N1CB mechanism and catalysis by methoxide produced by addition of base.[239]

Perhaps the most rewarding nonaqueous solvent for the study of base solvolysis kinetics is liquid ammonia. Here it is possible to split observed rate constants into equilibrium constants for conjugate base formation and rate constants for the subsequent onward reaction of the conjugate base, a split achieved only extremely rarely in aqueous systems. The most recent contribution in this area deals with rates of proton exchange with the complexes *trans*-[Co(2,3,2-tet)Cl$_2$]$^+$ (2,3,2-tet = **35**), [Co(NH$_3$)$_5$Cl]$^{2+}$, [Co(NH$_3$)$_5$F]$^{2+}$, and [Co(NH$_3$)$_6$]$^{3+}$.

$$
\begin{array}{ll}
\mathrm{HN-CH_2-CH_2-CH_2-NH} & \\
\;\;\;| & \;\;\;| \\
\mathrm{H_2C} & \mathrm{CH_2} \\
\;\;\;| & \;\;\;| \\
\mathrm{H_2C} & \mathrm{CH_2} \\
\;\;\;| & \;\;\;| \\
\mathrm{H_2N} & \mathrm{NH_2}
\end{array}
$$

35

In all cases the observed first-order rate constant is composite, as shown in:

$$k_{\mathrm{obs}} = k_1 + k_2[\mathrm{H}^+] \tag{27}$$

indicating proton exchange concurrently with the amine and amido forms of the complexes. Although the acidities of these complexes are all similar, proton exchange rates differ markedly. In the case of the pentaammine complexes it is possible to monitor the rates of exchange of protons with ammonia ligands *cis* and *trans* to the halide individually and observe significantly different rate constants for exchange at the two different types of ammonia ligand.[240]

5.7.3. *Catalyzed Aquation*

Acid-catalyzed aquation has already been covered at appropriate points in Section 5.7.1. Most of the present section deals with catalysis of removal of halide from cobalt(III) with the aid of mercury(II). The final paragraphs of the section deal with the question of catalytic removal of a variety of groups and the relation of the results to the question of the possible intermediacy of transient five-coordinated intermediates.

In general the scheme for mercury(II)-catalyzed aquation is as in

Scheme 17

$$[M\text{——}Cl]^{n+} + Hg^{2+} \rightleftharpoons [M\text{——}Cl\text{——}Hg]^{(n+2)+} \rightarrow M^{(n+1)+} + HgCl^+$$

in which the binuclear species may be anything from a stable intermediate to a transition state. The discussion of mercury(II)-catalyzed aquation of complexes *trans*-$[Co(en)_2X(^-O_2CCHRNH_3^+)]^{2+}$ suggests that the presence of the $—NH_3^+$ group in the ligand should make the reaction approximate fairly closely to a one-step bimolecular process. Second-order rate constants and activation parameters ($\Delta H^{\ddagger}$ and $\Delta S^{\ddagger}$) are reported. There is considerable stereochemical change during the reaction, for the products are about 85% *cis*.[223] Mercury(II)-catalyzed aquation of *cis*-α-$[Co(trien)(imidH)Cl]^{2+}$ is about five times faster than of the *cis*-β_2 isomer; this ratio of reactivities is similar to that for uncatalyzed aquation.[190] A large amount of kinetic data, rate constants and activation parameters (E_a, $\log A$, and $\Delta S^{\ddagger}$), has been reported for mercury(II)-catalyzed aquation of *unsym-fac-cis*-$[Co(dien)(amine)_2Cl]^{2+}$ cations, **28** above, with amine = ammonia, alkylamine, benzylamine, or (substituted) pyridine. Although the kinetic pattern appears to be simple second order for all these reactions, the authors recognize that their kinetic parameters may be composite, consisting of the equilibrium constant for formation of a transient binuclear intermediate and the actual rate constant for dissociation of this intermediate. There is a second reason why these kinetic parameters may be composite quantities, and that is they may refer to minor catalytic pathways involving $HgCl^+$ and perhaps also $HgCl_2$ in parallel with the major Hg^{2+} catalysis.[189] The rate law for mercury(II) halide catalysis of bromide loss from $[Co(CN)_5Br]^{3-}$, in which

$$k_{\text{obs}} = k_1 + k_2[HgX^+] + k_3[HgX_2] \tag{28}$$

is interpreted in terms of the equilibrium

$$2HgX_2 \rightleftharpoons HgX^+ + HgX_3^- \tag{29}$$

Catalysis by HgX_3^- is negligible, while k_2 is about 10,000 times larger than k_3, which seems entirely plausible.[241] Rapid removal of chloride from the bis(ethylenediamine)-1-aminopropanol complex **31** by mercury(II) is followed by slow cyclization of the aquo species **32** to give **33**. As pointed out in Section 5.7.1, uncatalyzed aquation of **31** gives **33** in a one-step process, since under these conditions release of chloride is considerably slower than the rate of cyclization of **32**.[193]

Mercury(II)-catalyzed aquation of $[Co(NH_3)_5Cl]^{2+}$ and of *trans*-$[Co(en)_2Cl_2]^+$, and also of $[ReCl_6]^{2-}$ (Section 5.3.2), follows second-order kinetics, so it may be assumed that any binuclear intermediate is either transient or a very close approximation to a transition state. Second-order rate constants for catalyzed aquation of the cobalt(III) complexes change very little as the proportion of organic cosolvent is varied for reaction in binary aqueous solvent mixtures. This behavior contrasts with the marked increase in rates for the $[ReCl_6]^{2-}$ anion as the proportion of organic cosolvent is increased. This difference can be ascribed to charge reinforcement on transition state formation for the cobalt(III) complexes, charge cancellation for the rhenium(IV) complex. There will be considerable release of electrostricted solvent on going from the initial state to the transition state for the $[ReCl_6]^{2-}$ reaction. When these reactivity trends are analyzed into initial state and transition state contributions, it becomes apparent that the very small variation of rate with solvent composition for mercury(II)-catalyzed aquation of $[Co(NH_3)_5Cl]^{2+}$ in ethanol–water mixtures represents the difference between a large destabilization of the initial state and a marginally larger destabilization of the transition state. Of course the greater sensitivity of rate to solvent composition for $[ReCl_6]^{2-}$ indicates a bigger difference between initial state and transition state changes. Again both are destabilized on adding ethanol, but the transition state is destabilized only to about half the extent of the initial state.[97] In these systems, where all the components are hydrophilic, reagents and transition states alike are destabilized, to various degrees, on adding a cosolvent such as an alcohol to water. To obtain different patterns of reactivity it is necessary to turn to complexes with hydrophobic exteriors, such as $[Fe(phen)_3]^{2+}$ or $[Fe(bipy)_3]^{2+}$,[137,138] or to organic molecules [cf., for example, mercury(II) catalysis of solvolysis of the pharmaceutical fenclorac[242]].

Sodium methanesulfonate or sodium benzenesulfonate retard mercury(II) catalysis of aquation of the *cis*-$[Co(phen)_2(OH_2)Cl]^{2+}$ cation. This unusual result—one would have expected the opposite, as indeed one finds for mercury(II) catalysis of $[Co(NH_3)_5Cl]^{2+}$—was ascribed to hydrophobic interactions between the organic sulfonate and the 1,10-phenanthroline ligands.[243] The effects of added polyelectrolytes have also been examined for silver(I) catalysis. Anionic polyelectrolytes enhance rates of silver(I) catalysis of aquation of $[Co(NH_3)_5Br]^{2+}$, as expected. However if these systems are studied under high pressures, then retardation is observed. This suggests that desolvation of the transition state is an important factor in the polyelectrolyte media.[194]

Another type of catalysis that has been studied for many years is that of nitrous acid catalysis of solvolysis of azido complexes. Nitrate ion competition and nitrosation rates have been monitored for the $[Co(NH_3)_5N_3]^{2+}$ cation, in aqueous solutions containing up to 1 mol dm^{-3} nitrate. The rate law established is:

$$-d[Co(NH_3)_5N_3^{2+}]/dt = \{k_1 + k_2[NO_3^-]\}[H^+][HNO_2][Co(NH_3)_5N_3^{2+}] \quad (30)$$

It can be accommodated by several mechanisms, including one which involves the intermediacy of a five-coordinate cobalt species for the nitrate-independent path. The problems associated with interpreting results of competition experiments are reviewed, in relation to mercury(II) catalysis as well as to nitrous-acid–azide reactions, and ideas for resolving some of the ambiguities are presented. A combination of rate law and competition evidence still supports the intermediacy of five-coordinate cobalt(III) complexes in some at least of these catalyzed aquations.$^{(244)}$ Another mode of catalysis involves permanganate oxidation of coordinated dimethyl sulfoxide. Product ratios from permanganate-assisted aquation of $[Co(NH_3)_5(dmso)]^{3+}$ and nitrous acid-assisted aquation of $[Co(NH_3)_5N_3]^{2+}$ in water and in aqueous ethanol suggest that these two reactions proceed via different intermediates, and further that $[Co(NH_3)_5]^{3+}$ is not the intermediate in either case! It is proposed that both reactions proceed by dissociative interchange reactions of intermediates containing dimethyl sulfone and N$_4$O ligands, respectively.$^{(245)}$ Similarly it has been shown that mercury(II)-catalyzed aquation of $[Co(tren)(NH_3)Cl]^{2+}$ and nitrous acid-assisted aquation of $[Co(tren)(NH_3)N_3]^{2+}$, in both cases investigated for the two isomers with primary and tertiary nitrogens respectively *trans* to chloride or azide, do not proceed through the common intermediate $[Co(tren)(NH_3)]^{3+}$. Again nitrate ion was used in the competition experiments in this investigation.$^{(246)}$ In contrast to these references, there is a brief conference report which mentions evidence for a simple D mechanism for induced aquations at cobalt(III).$^{(247)}$

5.7.4. Formation

5.7.4.1. In Aqueous Solution

Activation parameters for reaction of $[Co(NH_3)_5(OH_2)]^{3+}$ with salicylate are $\Delta H^{\ddagger} = 129$ kJ mol^{-1} and $\Delta S^{\ddagger} = +69$ J K^{-1} mol^{-1}. These values refer to reaction of the aquo complex with the salicylate monoanion (salH$^-$).$^{(248)}$ The $[Co(NH_3)_5(OH_2)]^{3+}$ cation reacts with all three forms of succinate, succ^{2-}, succH$^-$, and succH$_2$. Rate constants have been determined in the temperature range 323.2–338.6 K. The mechanism involves outer-sphere association followed by rate-determining interchange and water loss, in other words the classical

Eigen–Tamm–Wilkins mechanism for complex formation.[249] The same mechanism is thought to operate in the formation of the oxalate complex from cis-$[Co(en)_2(NH_3)(OH_2)]^{3+}$. Again all three forms of the ligand, ox^{2-}, oxH^-, and oxH_2, react with the aquo complex. The outer-sphere association constant for the oxH^- anion is 1.5, for the ox^{2-} anion 5.8 dm^3 mol^{-1}.[250] The kinetic results here are compared with those established several years ago for the reaction of $[Co(NH_3)_5(OH_2)]^{3+}$ with oxalate.[251] The formation of the sulfito complex $[Co(NH_3)_5(SO_3)]^+$ is mentioned in Section 5.7.7. below.[213]

There have been a few reports over the years on nitrate catalysis of substitution at cobalt(III) and at chromium(III). Now it has been shown that nitrate has a catalytic effect on oxalate anation of α-cis-$[Co(edda)(OH_2)_2]^+$. The outer-sphere association constant appears to be larger when nitrate is present. The possibility of the formation of a transient five-coordinate intermediate under the influence of the nitrate (cf. the end of the previous section, 5.7.3.) should not be ruled out.[252]

Reaction of the newly prepared and characterized cation $trans$-$[Co(en)_2(S_2O_3)(OH_2)]^+$, containing S-bonded thiosulfate, with nitrite or thiocyanate follows a second-order rate law. It is much more likely that this is to be explained by a dissociative interchange than by an associative mechanism. Rate constants and activation parameters ($\Delta H^{\ddagger}$ and $\Delta S^{\ddagger}$) are reported. The S-thiosulfate ligand has a considerably smaller $trans$-labilizing effect than S-sulfite. The hydroxo complex $trans$-$[Co(en)_2(S_2O_3)(OH)]$ is less reactive than the aquo complex.[214]

Substitution at pentacyanocobaltate(III), as at pentacyanoferrate(II) (Section 5.4.1), is generally thought to proceed by a limiting dissociative, D, mechanism. Activation volumes of $+8.4$, $+9.4$, and $+8.2$ cm^3 mol^{-1} for reaction of $[Co(CN)_5(OH_2)]^{2-}$ with bromide, iodide, and thiosulfate, respectively, are consistent with a D mechanism, in that they are equal within experimental uncertainty and are markedly positive. Such estimates for the partial molar volume of $[Co(CN)_5]^{2-}$ as can be derived from these anation results and from the other substitution results detailed elsewhere (Section 5.7.1 above) are all equal, as they should be for the operation of a D mechanism in all cases, with the same intermediate.[195]

Oxidation of formic acid by Co^{3+} aq is believed[253] to involve substitution at Co^{3+} aq as first step. However it did not prove possible to estimate formation rate constants in this study; they were in fact assumed to be much larger than rate constants for the subsequent redox steps. Reaction of cis-$[Co(en)_2(OH_2)_2]^{3+}$ with tungstate takes place much more rapidly than other formation reactions of aquo–cobalt(III) complexes, and indeed falls within the stopped-flow range. These reactions cannot therefore involve cobalt–oxygen bond breaking, but must rather be substitution at tungsten(VI).[254]

5.7.4.2. In Binary Aqueous Solvents

Rate constants have been determined as a function of pH, temperature, and incoming ligand concentration for the reaction of cis-$[Co(en)_2(OH_2)_2]^{3+}$ with ethylenediamine in ethanol + water solvent mixtures. Parallel S_N1 and S_N2 paths are proposed, corresponding to the two terms of equation (31):

$$k_{obs} = k_1 + k_2[en] \tag{31}$$

The role of various protonated and deprotonated forms of the reactants is discussed.[255]

Wells has analyzed solvent effects on rates of anation into initial state and transition state contributions for a variety of systems in a detailed and extensive (85 references) paper. The reactions of direct relevance to this section involve anation at $[Co(NH_3)_5(OH_2)]^{3+}$, but anation at $[Co(dmgH)_2(OH_2)L]^+$ (Section 5.7.8) and rhodium(III) analogs (Section 5.8.9), and indeed at Ni_{aq}^{2+}, is clearly closely connected. The organic cosolvents are methanol, ethanol, t-butyl alcohol, ethylene glycol, dioxan, acetone, and, in one case, dimethyl sulfoxide. The same free energy cycle is used here as in the chloro–cobalt(III) systems discussed in Section 5.7.1.3; the mechanistic assumption is complete metal–water bond breaking on transition state formation, in other words a D mechanism rather than the universally accepted I_d mechanism. The exaggeratedly dissociative assumption may well not have a significant effect on the general conclusions. Whereas the transition state was generally found to dominate in aquation of cobalt(III)–amine–halide[208] and $trans$-$[Co(py)_4Cl_2]^+$ (Section 5.7.1.3)[209] aquation, the conclusions are less uniform for anation. At low mole fractions of organic cosolvents, for example, in the composition region where addition of small quantities of alcohols promote water structure, either transition state or initial state solvation may dominate. At higher mole fractions of cosolvent, initial state solvation changes generally dominate.[256] Wells's treatment does not require knowledge of transfer chemical potentials for such species as $[Co(NH_3)_5]^{3+}$ or Ni_{aq}^{2+}, but his deductions do rely on his values of $\delta_m\mu°(M^{2+})$ being negative for transfer from water into the various mixed solvents. In fact Wells's values for $\delta_m\mu°(M^{2+})$ are only available for M = main group metal; there are no values published for M = Co or Ni as yet. However, it is possible to calculate values of $\delta_m\mu°(Fe^{2+})$ for transfer from water into, e.g., aqueous methanol from published values for stability constants for formation of $[Fe(phen)_3^{2+}]$,[257] solubilities of 1,10-phenanthroline,[258] solubilities of $[Fe(phen)_3](ClO_4)_2$,[259] and published $\delta_m\mu°(ClO_4^-)$ values[203] [equation (32)].

$$\delta_m\mu°(Fe^{2+}) = \delta_m\mu°[Fe(phen)_3^{2+}] - 3\delta_m\mu°(phen) - RT\delta_m(\ln \beta_3) \tag{32}$$

Estimates of $\delta_m\mu^\circ(Fe^{2+})$ can also be made in a very similar manner from $Fe(bipy)_3^{2+}$ stability constants,[260] 2,2'-bipyridyl solubilities,[261] and published estimates for $\delta_m\mu^\circ[Fe(bipy)_3^{2+}]$.[138] If we now make the very reasonable assumption that $\delta_m\mu^\circ(Ni^{2+}) = \delta_m\mu^\circ(Fe^{2+})$ we can analyze solvent effects on reactivity into initial state and transition state contributions for formation reactions of Ni^{2+}. The result of such an analysis for the reaction of Ni^{2+} with 2,2'-bipyridyl in methanol + water mixtures is shown in Figure 5.3. The conclusion reached here, that initial state effects are rather larger than transition state effects, is the same as that reached by Wells for this reaction. The differences between these are somewhat larger in Wells's treatment, owing to his assumption of a D rather than an I_d mechanism. One feature of Figure 5.3 which does not emerge from Wells's approach is the dominant role played by the 2,2'-bipyridyl. It is

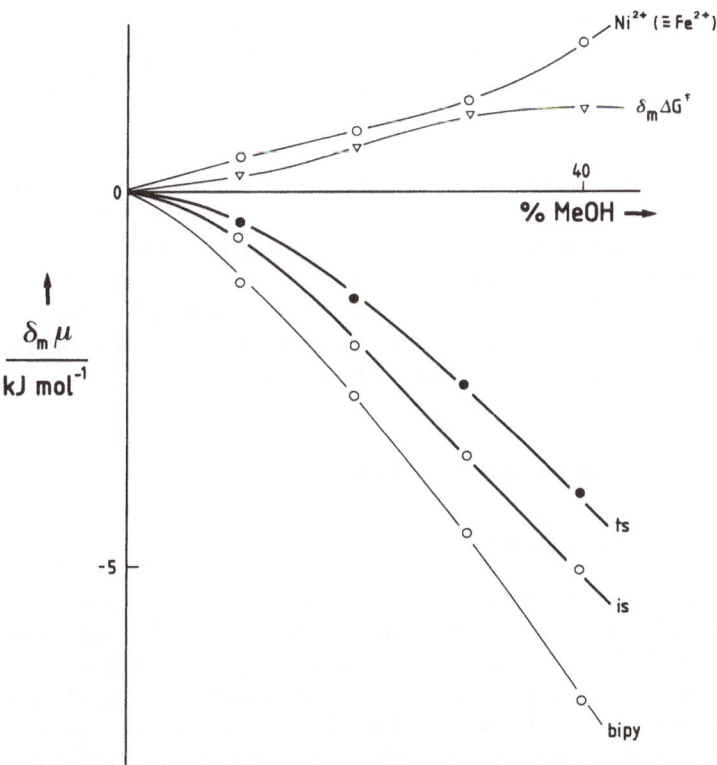

Figure 5.3   Initial state (is) and transition state (ts) transfer chemical potential variation for complex formation between Ni^{2+} and 2,2'-bipyridyl (bipy) in aqueous methanol (sources of data and estimates as given in the text). All values are on the molar scale, at 298.2 K.

this ligand, not the Ni^{2+}, which leads to stabilization of the initial state on transfer from water into aqueous methanol! The importance of largely hydrophobic ligands in determining the sign and magnitudes of transfer chemical potentials for complexes, and for initial and transition states, is clear both here and in the case of solvolysis of the *trans*-[Co(py)$_4$Cl$_2$]$^+$ cation discussed in Section 5.7.1.3. It is interesting to contrast the dominance of the 2,2'-bipyridyl (Figure 5.3) in this free energy analysis with the apparent dominance of the Ni^{2+} in a parallel enthalpy analysis of this same complex formation reaction.[262] It is not possible to extend the analysis to entropies, since the single-ion assumptions involved in the G and H analyses are not the same.

5.7.4.3.   In Nonaqueous Solvents

Both references here deal with anation of [Co(NH$_3$)$_5$(dmso)]$^{3+}$ in dimethyl sulfoxide. Rate constants have been determined at 318.4 K for reaction with chloride, azide, and thiocyanate. These rate constants consist, as usual, of an outer-sphere association constant and a rate constant for ligand interchange. Values of the former range from 5 to 22 dm^3 mol^{-1}. There is a range of about ten times in the interchange rate constants, which themselves are not quite the same as the rate constant for dmso exchange at the free [Co(NH$_3$)$_5$(dmso)]$^{3+}$ cation. However the operation of the usual I_d mechanism is not in question, as it is demonstrated that dmso exchange rates in [Co(NH$_3$)$_5$(dmso)]$^{3+}$, X$^-$ ion pairs vary significantly with the nature of the anion X$^-$.[263] A similar picture has been built up for bromide anation at this same [Co(NH$_3$)$_5$(dmso)]$^{3+}$ complex. Here a comparison of anation and solvent exchange rates indicates that internal return of an almost-left dmso molecule is relatively slow.[264]

5.7.5.   Solvent and Ligand Exchange

5.7.5.1.   Solvent Exchange

As indicated in Section 5.7.4.3, the effects of ion pairing on rates of dimethyl sulfoxide exchange with the [Co(NH$_3$)$_5$(dmso)]$^{3+}$ cation have been established.[263,264] The small accelerating effects of ion pairing are listed in Table 5.14. Activation enthalpies and activation entropies for dimethylformamide exchange with [Co(NH$_3$)$_5$(dmf)]$^{3+}$ both decrease as the proportion of dimethylformamide increases in water + dimethylformamide solvent mixtures. These trends are discussed in terms of an interchange mechanism and the effects of interactions between free and coordinated dimethylformamide molecules on the outer-sphere association constant.[199]

CNDO/2 calculations on S_N1 and S_N2 exchange of ammonia with [Co(NH$_3$)$_6$]$^{3+}$

Table 5.14. Specific Rates of Bond Fission, k_{sp}, within $[Co(NH_3)_5(dmso)]^{3+}$ Ion Pairs[263]

Gegenion	$10^5 k_{sp}$, s^{-1}
Chloride	> Azide
Azide	21
Bromide	11–12
Cf. free cation	12.6

indicate that the bimolecular mechanism is less favorable, at least for a pentagonal bipyramidal transition state.[265]

5.7.5.2. Ligand Exchange

The kinetics of chloride exchange with the *trans*-[Co(en)$_2$(CN)Cl]$^+$ cation have been investigated in methanol, dimethyl sulfoxide, dimethylformamide, and ethylene glycol. Rate laws, rate constants, and activation parameters (E_a and $\Delta S^{\ddagger}$ were determined for each solvent. The first-order dependence on chloride in dimethylformamide was ascribed to ion pair formation in an I_d mechanism. However an I_a mechanism was thought to operate in the other solvents. This assignment was based on intuitive feelings about relative solvating abilities, the dependence of rate constants on dielectric constants, parallels with chloride exchange at *trans*-[Co(en)$_2$Cl$_2$]$^+$, and activation entropies.[266] ^{81}Br nmr spectra of D$_2$O solutions of [Co(tecyclen)(OH$_2$)Br]Br$_2$ are claimed to show that exchange of bromide in this complex takes place at a rate corresponding to the nmr frequency at room temperature. The ligand tecyclen is 2R,5R,8R,11R-tetraethyl-1,4,7,10-tetraazacyclododecane (**36**); a crystal structure determination of the

$$
\begin{array}{ccc}
\text{HN} - \text{CHEt} - \text{CH}_2 - \text{NH} \\
| \qquad\qquad\qquad | \\
\text{H}_2\text{C} \qquad\qquad\qquad \text{CHEt} \\
| \qquad\qquad\qquad | \\
\text{EtHC} \qquad\qquad\qquad \text{CH}_2 \\
| \qquad\qquad\qquad | \\
\text{HN} - \text{CH}_2 - \text{CHEt} - \text{NH}
\end{array}
$$

36

bromide salt was used to assign stereochemistry as there are 16 possible diastereoisomers.[267]

The rate of exchange of nitrite with the 2-amino-2-methyl-3-butanoneoxime (LLH, **37**) complex [Co(LL)(LLH)(NO$_2$)$_2$] is independent of the concentration

37

of nitrite or of hydrogen ions; exchange is first order in complex. In water the half-life is about 16 h at 313 K, while the activation parameters are $\Delta H^{\ddagger} = 29.4$ kcal mol^{-1} and $\Delta S^{\ddagger} = +13$ cal deg^{-1} mol^{-1}. Rates are almost identical in H_2O, D_2O, and methanol, which renders a mechanism involving nitrite loss and recombination unlikely. Addition of cobalt(II) has no effect on exchange rates, which rules out a redox mechanism. Thus the most likely mechanism is intramolecular, perhaps via a transient nitrito intermediate.[268]

Kinetics of exchange of acetylacetone with the complexes [M(acac)$_3$], M = Co, Rh, Ru, and Cr, have been examined in solvent acetylacetone, over the temperature range 358–463 K. The order of decreasing rates is

$$M = Co(III) > Cr(III) > Ru(III) > Rh(III)$$

Activation parameters and deuterium isotope effects indicate that there is a difference of mechanism between [Co(acac)$_3$] and the other three complexes. In all cases it is thought that the mechanism involves the generation of a monodentate acac intermediate, but this rate-determining step is thought to be dissociative for cobalt ($\Delta H^{\ddagger} = 36.4$ kcal mol^{-1}; $\Delta S^{\ddagger}$ positive) but associative for the other metals ($27.4 \leqslant \Delta H^{\ddagger} \leqslant 28.8$ kcal mol^{-1}; $\Delta S^{\ddagger}$ negative).[269] The exchange of acetylacetone with [Co(acac)$_3$] has also been studied in acetonitrile[270] and in dimethyl sulfoxide and dioxan.[271] In the latter investigation effects of added protons were studied by addition of chloroacetic acid.

Ligand exchange kinetics for [Co(Et$_2$dtc)$_3$] have been monitored in dioxan and in dimethylformamide. Ligand exchange is very slow, with rates first order in complex, zero order in added ligand, indicating an S_N1 mechanism. In dimethylformamide the activation parameters were found to be $E_a = 88.4$ (± 8.9) kJ mol^{-1} and $\Delta S^{\ddagger} = +48.8$ (± 25.2) J K^{-1} mol^{-1}.[272] The positive activation entropy supports the assignment of a dissociative mechanism.

5.7.5.3. Ligand Replacement

Whereas tris[O,O'-dimethylphosphorodithioatocobalt(III)] reacts with 1,2-bis(diphenylphosphino)ethane by a two-stage process, the reaction of the analogous diisobutyl complex to the bis(dithio-mono-diphosphine) complex has the kinetic characteristics of a simple one-stage process. This reaction follows second-order kinetics, but the authors nonetheless propose a dissociative mechanism.[273] There is a brief mention of the replacement of iodide by imidazole in a cobalt(III)–porphyrin complex in a paper primarily concerned with the oxidation of cobalt(II)–porphyrin species by oxygen and iodine.[274] The development of a deep red-orange color in solutions containing bis[histidinatocobalt(III)] and $[Ru(CN)_6]^{4-}$ exposed to visible light is attributed to the generation of an Ru(II)–C–N–Co(III) species with a strong intramolecular charge transfer absorption by attack of cyanide ligand at the cobalt(III) in the bis(histidinato) complex.[174]

5.7.6. Racemization and Isomerization

The scattered references to these closely related classes of reaction are gathered into three groups of connected studies in this section.

5.7.6.1. Cis ⇌ Trans and Related Interconversions

The rate for *trans* to *cis* isomerization of the $[Co(en)_2(OAc)(OH_2)]^{2+}$ cation decreases on applying pressure. The activation volume for this isomerization is $+7.9$ cm^3 mol^{-1} in 0.05 mol dm^{-3} perchloric acid, decreasing to $+5.6$ as the concentration of perchloric acid increases to 1.0 mol dm^{-3}. Volumes, enthalpies, and entropies of activation, and added salt effects, all indicate an I_d mechanism (leaving group water) for this isomerization as for isomerizations of similar cobalt(III) complexes.[275] Volumes and entropies of activation are positive both in H$_2$O and in D$_2$O for *trans* to *cis* isomerization of *trans*-$[Co(en)_2(OH_2)]^{3+}$ and of *trans*-$[Co(en)_2(OSeO_2H)(OH_2)]^{2+}$ as well as of the acetato complex named above. The recent report of rate constants as a function of temperature and pressure in D$_2$O shows that activation entropies are slightly more positive, activation volumes slightly less positive, in D$_2$O than in H$_2$O. Differential solvation changes on going from the initial state to the transition state are invoked, and discussed in relation to solvent structure, but the requisite thermodynamic data are not yet available to enable a dissection of solvent effects into initial state and transition state components to be carried out.[276]

The strong labilizing effect of bonded sulfur appears to lead to very easy *cis* to *trans* interconversion during aquation of the O,S-bonded thiosulfate complex $[Co(en)_2(S_2O_3)]^+$, for the product is *trans*-$[Co(en)_2(OH_2)(S_2O_3)]^+$.[215] Isomerization of the bis(thiosulfato) complex $[Co(en)_2(S_2O_3)_2]^-$, which of course contains monodentate thiosulfate ligands, has been studied in an attempt to avoid ion-pairing problems which have plagued some earlier investigations of *cis* ⇌ *trans* isomerizations of $[Co(en)_2X_2]^+$ complexes in basic solution.[277]

The slow isomerization $\Lambda \rightarrow \Delta$-*cis*-$\beta_2$-[Co(ebms)(aa)], subsequent to rapid oxidation by dioxygen of the cobalt(II) complex of the dianion of *N,N'*-ethylenebis(α-methylsalicylideneamine) (ebms) in the presence of an *l*-amino-acid (aaH), is a key step in the kinetics and mechanism of formation of this class of cobalt(III) complexes. This conclusion emerged from a study of stereoselectivity and reactivity in these systems.[278]

On heating solid cis-α-[Co(trien)Cl$_2$]Cl to about 430 K, slow isomerization to the cis-β isomer takes place. The presence of the tetradentate ligand rules out a tetragonal twist, but permits a trigonal twist. Thus this observation is important in providing a rare unequivocal demonstration of the operation of the trigonal twist in an isomerization.[279]

As ever, *mer* $\rightleftharpoons$ *fac* interconversions are very much more rarely described than *cis* $\rightleftharpoons$ trans. However during the period under review one extensive report on *mer* $\rightleftharpoons$ *fac* isomerizations did appear. In this, rate constants and activation parameters (E_a and $\Delta S^{\ddagger}$) were established for isomerization of a series of cations *mer*-[Co(dien)(amine)(OH$_2$)$_2$]$^{3+}$, with amine = NH$_3$, RNH$_2$, BzNH$_2$, py, or 3,5-Me$_2$py. It is interesting that the NH$_3$, BzNH$_2$, and four RNH$_2$ complexes had isomerization rate constants within the narrow range $1.3-2.4 \times 10^{-3}$ s^{-1}, although the complexes of NH$_3$ and BzNH$_2$ have considerably higher activation energies (100 and 96 kJ mol^{-1}, respectively) than the other four amine complexes (81–84 kJ mol^{-1}). The pyridine complexes isomerize somewhat more slowly than the NH$_3$, RNH$_2$, and BzNH$_2$ complexes.[189]

^{13}C nmr studies on solutions containing [Co($\pm$-bn)$_3$]$^{3+}$ and [Co(*meso*-bn)$_3$]$^{3+}$ show that interchange processes are much more rapid in the latter. The $\pm$-bn complex is relatively rigid, whereas the *meso*-bn complex is more flexible, with a degree of conformational lability bestowed by relatively easy $\delta \rightleftharpoons \lambda$ conformation interconversion for the coordinated ligand. Added tetrahedral oxoanions exert a significant effect, demonstrated for the *fac*-[Co(meso-bn)$_3$]$^{3+}$ cation, owing to specific interactions with equatorial methyl groups.[280]

5.7.6.2. Linkage Isomerization

The most important feature of yet another investigation of nitrito to nitro isomerization of complexes [M(NH$_3$)$_5$(ONO)]$^{2+}$, with M = Rh and Ir (*vide infra*) as well as Co, is the discovery of a hitherto undetected base catalysed pathway [equation (33)].

$$k_{obs} = k_1 + k_2[OH^-] \qquad (33)$$

Values of k_1, k_2, $\Delta H_1^{\ddagger}, \Delta H_2^{\ddagger}, \Delta S_1^{\ddagger}$, and $\Delta S_2^{\ddagger}$ are given for all three nitrito complexes. Tracer studies, using oxygen-18 labeling, showed both k_1 and k_2 paths to be intramolecular. For the cobalt complex, a value of $+27$ cm^3 mol^{-1} for $\Delta V_2^{\ddagger}$ suggests a conjugate base preequilibrium mechanism for the k_2 path. This assignment of mechanism is consistent with the observed reactivity sequence for k_2:

$$Co : Rh : Ir = 95 : 9 : 1.$$

A similar rate law operates for isomerization of the *cis* and *trans* forms of $[Co(en)_2(ONO)_2]^+$ and of $[Co(en)_2(ONO)(NO_2)]^+$; these nitrito $\rightarrow$ nitro isomerizations take place without change in configuration at the cobalt. Similarly nitrito $\rightarrow$ nitro isomerization of $(+)$-*cis*-$[Co(en)_2(ONO)(NO_2)]^+$ occurs without loss of chirality. These stereochemical results contrast with normally observed behavior in substitution at cobalt(III).[281] The nitrito $\rightarrow$ nitro isomerization of $[Co(NH_3)_5(ONO)]^{2+}$, in this case generated *in situ* photochemically, has also been studied in poly(vinyl alcohol) media. This very viscous medium, viz. pva sheets, was used to probe the gap between water and solid state studies. The rate of isomerization in pva is similar to that in water, but about ten times faster than in the solid phase. Such a short range of reactivities is, of course, consistent with an intramolecular mechanism in all three media.[282]

Until 1980, all sulfinato– and sulfenato–cobalt(III) complexes thus far characterized were S bonded. In 1980 it was shown that the *O*-sulfinato isomer of $[Co(en)_2(SO_2CH_2CH_2NH_2)]^{2+}$ could readily be generated photochemically from the S-bonded isomer. The O-bonded form was found to be robust. It is not photosensitive, but slowly reverts thermally to the S-bonded form. The half-life for this reversion is about 600 h at room temperature; $\Delta H^{\ddagger} = 22.6$ kcal mol^{-1} and $\Delta S^{\ddagger} = -15.6$ cal deg^{-1} mol^{-1}. Isomerization is a clean first-order process.[283]

5.7.6.3. *Racemization of Dithiocarbamates*

A few years ago, at the end of an assessment of solvent and temperature effects on rates of racemization of dialkyldithiocarbamato complexes $[Co(R_2dtc)_3]$ it proved impossible to reach an unequivocal conclusion as to the precise mechanism.[284] Later work showed that, although rates varied little with solvent, there was a great dependence of activation parameters on the nature of the medium. Thus in the series of solvents ethanol, acetonitrile, dimethylformamide, chloroform, carbon tetrachloride, and toluene, activation enthalpies covered the range 67–130 kJ mol^{-1}, and activation entropies covered the range -124 to $+60$ J K^{-1} mol^{-1}. These activation parameters gave a good isokinetic plot, and it was proposed that racemization was intramolecular in all these solvents[285] despite the uncomfortably large negative limit to the activation entropies. Subsequent determination of activation volumes for racemization of selected complexes of this type has proved a much more convincing guide to mechanism.

Thus $\Delta V^{\ddagger}$ lies between $+5$ and $+10$ cm^3 mol^{-1} for racemization of the pyrrolidyl complex in ethanol, acetonitrile, dimethylformamide, and toluene. This very modest solvent dependence makes a "one-end-off" mechanism unlikely, for such a mechanism should result in a slightly larger range of values which should, moreover, be slightly negative. For a simple twist mechanism $\Delta V^{\ddagger}$ should be very close to zero, so the observed somewhat positive $\Delta V^{\ddagger}$ values can best be rationalized in terms of a twist mechanism "incorporating a low-spin $\rightleftharpoons$ high-spin preequilibrium in the transition state" (quotation marks are provoked by the present author's feeling that it is difficult to accommodate such behavior within the confines of truly authentic transition state theory!). The bond extension which is an unavoidable corollary to the suggested spin change explains the positive $\Delta V^{\ddagger}$ values observed. It is interesting to compare these racemizations with analogous processes of tris(bidentate) iron(II) and chromium(III) complexes. For [Fe(phen)$_3$]$^{2+}$ similarly small positive $\Delta V^{\ddagger}$ values have been explained by the same mechanism, whereas small negative values were observed for [Cr(phen)$_3$]$^+$ where a low-spin $\rightleftharpoons$ high-spin equilibrium is impossible.[286] For racemization of [Co(Ph$_2$dtc)$_3$], activation volumes range between -2 for dimethylformamide and -9.3 cm^3 mol^{-1} for chlorobenzene. So for this dithiocarbamato–cobalt(III) complex it appears that the "one-end-off" mechanism is operating.[287]

5.7.7. *Carbonate and Sulfite Complexes*

The kinetics of formation of [Co(tren)(CO$_3$)]$^+$ from [Co(tren)(OH)$_2$]$^+$ have been investigated over the pH range 10.5–13. The tren system is very suitable for such a study, in a pH range where it is possible to assess the reactivity of CO$_3^{2-}$ as well as of HCO$_3^-$ and CO$_2$, since it is stable both with respect to base hydrolysis and to redox reactions. Rate constants were measured over the temperature range 298–333 K at an ionic strength of 0.5 mol dm^{-3}. It proved necessary to analyze the results in terms of three parallel reactions, since the high reactivity with carbon dioxide meant that reaction with this species had to be included despite its very low concentrations under these conditions.[288]

There has been much treatment of the kinetics and mechanism of decarboxylation of carbonatocobalt(III) complexes. Thus the results for decarboxylation of the [Co(tren)(CO$_3$)]$^+$ cation mentioned above conflict to some extent with earlier results and interpretations in this area.[288] Matters are more straightforward for decarboxylation of monodentate carbonate complexes, for instance, of *trans*-[Co(en)$_2$(CO$_3$)Cl]. The reactivity pattern here,[222] similar to that for other monodentate carbonate complexes, with rates on the stopped-flow time scale, corresponds to carbon–oxygen rather than to cobalt–oxygen bond rupture. Three papers deal with decarboxylate of complexes [Co(LLLL)(CO$_3$)]$^+$, each containing bidentate carbonate. In each case the variation of rate with pH indicates the dependence on acid shown in:

$$k_{obs} = k_1 + k_2[H^+] \tag{34}$$

In each case the acid-catalyzed path is dominant, the k_1 term of minor importance. The solvent isotope effects $k(D_2O)/k(H_2O)$ lie between 2.1 and 2.6. These values, and the activation parameters (Table 5.15),[288–291] show that concerted attack by H_3O^+ is an unlikely mechanism, but indicate rather a two-stage mechanism in which rapid equilibrium protonation of the complex is followed by slow and rate-determining opening of the carbonate chelate ring ($A1$ mechanism with five-coordinate intermediate).[289–291] In the case of the complex of ligand **39**, effects of ligand unsaturation and substituent bulk were also assessed.[289]

Interpretation of kinetic patterns for decarboxylation of the cis-[Co(py)$_2$(CO$_3$)$_2$]$^-$ anion is clouded by a slight uncertainty as to whether both carbonate ligands are indeed bidentate. Kinetic parameters, including the solvent isotope effect, for loss of the first carbonate ligand (Table 5.15) are consistent with decarboxylation of a bidentate group, but the activation parameters do fall within the narrow

Table 5.15. Kinetic Parameters for Decarboxylation of
Cobalt(III)–Tetraazaligand–Carbonate Complexes[a]

Tetraaza ligand	Complex stereo	k_D/k_H	$\Delta H^\ddagger$, kJ mol^{-1}	$\Delta S^\ddagger$, J K^{-1} mol^{-1}	Ref.
Tren, **38**	*cis*		105.9	+15	288
39	*cis*	2.6	82.9	−0.4	289
Cyclen, **34**	*cis*	2.1	100.4	+51	290
Garland, **40**	*cis*-α	2.65	92.9	+34	291
	cis-β	2.56	86.8	−16	291
Cf. py$_2$ [b]	*cis*	5.2	70.7	+10	292

[a] Results from earlier analogous studies are tabulated in Ref. 291.
[b] *bis*-Carbonato complex, viz. *cis*-[Co(py)$_2$(CO$_3$)$_2$]$^-$.

range established for carbon–oxygen bond cleavage in monodentate carbonate complexes of cobalt(III).[292]

Rates of formation and of aquation, i.e., sulfur dioxide loss, of the O-bonded sulfite complex $[Co(NH_3)_5(SO_3)]^+$ are fast, indicating sulfur–oxygen rather than cobalt–oxygen bond breaking and making, here as in the carbonato complexes. Again as with the carbonato complexes, acid catalysis is of major importance in the mechanism of sulfur dioxide loss from the cobalt(III)–sulfite complex. The activation parameters for formation of this $[Co(NH_3)_5(SO_3)]^+$ cation are $\Delta H^\ddagger = 9.8$ kcal mol^{-1} and $\Delta S^\ddagger = +14$ cal deg^{-1} mol^{-1}.[213]

5.7.8. Bis(glyoximato) Complexes

The kinetics and mechanisms of substitution at complexes *trans*-$[Co(dmgH)_2XY]^{n+}$ in nonaqueous media have been reviewed.[293] The kinetics of replacement of the nitro ligand in *trans*-$[Co(dmgH)_2(NO_2)(amine)]$, amine = ammonia, aniline, or pyridine, have been examined in the protic solvent methanol and the dipolar aprotic solvent dimethylformamide, over the temperature ranges 293–323 K and 313–333 K, respectively. The nitro ligand is replaced more rapidly than chloride in these solvolyses. An isokinetic plot ($\Delta H^\ddagger$ vs. $\Delta S^\ddagger$) for three nitro and three chloro complexes of this type undergoing solvolysis in methanol and in dimethylformamide gives a reasonable straight line for all systems except the nitro complexes in methanol. An I_d mechanism is proposed for all the systems on the isokinetic line, but it is felt that there are elements of associative character in the methanolysis of the nitro complexes.[294] To turn to mixed solvents, formation reactions of *trans*-$[Co(dmgH)_2(OH_2)L]^+$ were included in the general analysis of initial state and transition state contributions to reactivity trends for formation at cobalt(III) and nickel(II) centers discussed at length in Section 5.7.4.2 above.[256]

Although the majority of kinetic studies in this area are concerned with dimethylglyoxime complexes, derivatives of other oximes are encountered from time to time. Thus a study of aquation of the methylglyoxime complexes *trans*-$[Co(mgH)_2X_2]^-$, with X = Cl, Br, or I, and of the mixed and parent complexes *trans*-$[Co(mgH)(dmgH)X_2]^-$ and *trans*-$[Co(dmgH)_2X_2]^-$ revealed that both activation energies and activation entropies increase as dmgH is replaced by mgH.[295] Another recent kinetic study involving methylglyoxime complexes dealt with $[Co(mgH)_2(tu)X]$. At pH's above 4 the thiourea is labile, with the labilizing effect of the *trans* halide being in the order Cl < Br ≪ I. However, at a pH of 3.5 it is the halide which leaves, with Cl$^-$ being replaced faster than Br$^-$, and Br$^-$ faster than I$^-$. Presumably these halide replacement reactions go by a D mechanism, in view of the strong *trans*-labilizing effect of thiourea.[296]

Rate constants have been reported for uncatalyzed aquation, and for aquation catalyzed by $HgCl^+$ and by $HgCl_2$, for both steps for the *trans*-$[Co(dmgH)_2Cl_2]^-$ anion; activation parameters were also determined for some of these reactions. It was recognized that the kinetic parameters for the $HgCl^+$ and $HgCl_2$ catalysis reactions may well be composite quantities due to the transient intermediacy of binuclear species. One interesting point to emerge from this study is that the catalytic effect of $HgCl^+$ seems to reside in the log A term rather than in E_a.[297]

Two studies involve linkage isomers. There is a little discussion of mechanism in a primarily preparative paper on the cobalt(III)–bis(dimethylglyoxime)–diphenylguanidine–thiocyanate system.[298] Linkage isomer ratios for *S*- and *N*-bonded thiocyanate *trans* to several ligands, such as substituted pyridines and anilines, and imidazole, give further insight into the *trans* influences of these ligands and thence into the initial states for kinetics of substitution at cobaloxime complexes of this type.[299]

Proton transfer from complexes *trans*-$[Co(dmgH)_2X_2]^{n\pm}$ has been studied by T-jump techniques in solutions of hydroxide in dioxan–water solvent mixtures. Rate constants lie in the range 10^5 to 3×10^6 dm^3 mol^{-1} s^{-1} in water. For the complex with X = CN, the rate constant decreases by a factor of 4 on going from water to mole fraction 0.174 dioxan, but for X = NH_3 the rate constant decreases then increases as the mole fraction of dioxan is increased. These reactivity patterns are discussed with respect to reaction mechanism and the effects of variation of dielectric constant.[300] The replacement of a hydrogen-bonded proton in *trans*-$[Co(dmgH)_2(OH_2)_2]^+$ or in *trans*-$[Co(dmgH)_2(CH_3)(OH_2)]$ by Fe^{3+} is envisaged as proton transfer from cobalt(III) complex to $Fe(OH)_{aq}^{2+}$ with concurrent incorporation of the iron into the oximato–ligand bridging position (**41**). Forward and reverse rate constants are reported for these equilibrium processes, which produce a so-called "pseudomacrocycle."[301]

41

Other recent studies of alkylcobaloximes include those of base-catalyzed cleavage of $[Co(dmgH)_2(Et)(OH_2)]$,[302] methyl transfer between $[Co(dmgH)_2(Me)(SMe_2)]$ and $MeSnCl_3$,[303] and base-catalyzed replacement of water in the diacetylmonoxime complex $[Co(dam)(damH)(R)(OH_2)]^+$.[304] In the last-named reaction, as in the analogous reaction of $[Co(dam)(damH)(R)(NH_3)]^+$ complexes, a conjugate base mechanism operates, involving proton loss from the damH ligand.

5.7.9. μ-Dicobalt(III) Complexes

The fission and formation of a variety of mono-, di, and tri-bridged dicobalt(III) complexes will be discussed in this section, in this order. At the end comparison will be made with a cobalt(III)–iron(II) system, very closely related in that again both metal ions are low-spin d^6 centers.

5.7.9.1. Mono-Bridged Dicobalt(III) Species

The only kinetic investigation in this area has been the establishment of the rate law, rate constants, and activation parameters for aquation of the $[(O_2N)(en)_2Co–O_2–Co(en)_2(NO_2)]^{2+}$ cation. In this complex both nitro ligands are *trans* to the peroxo bridge. This is the first kinetic study of a μ-peroxodicobalt(III) complex that aquates to give hydrogen peroxide as product;[305] aquation of, for instance, the ammine complex $[(H_3N)_5Co–O_2–Co(NH_3)_5]^{4+}$ gives dioxygen via substitution subsequent to intramolecular charge-transfer.[306] In acid solution the μ-peroxo complex $[(NC)_5Co–O_2–Co(CN)_5]^{6-}$ undergoes protonation to give a species with an $–O_2H–$ bridge, but no kinetic study was made of any subsequent bridge rupture.[307]

5.7.9.2. Di-Bridged Dicobalt(III) Species

In acid solution, the *meso* form of the μ-peroxo-μ-hydroxo complex **42** aquates four times more quickly than the racemic form. The rate laws for aquation are given. The observed acid catalysis presumably involves protonation of the peroxo bridge. This is assumed to give a singly bridged bis(aquo)-μ-peroxo intermediate, but as there is no sign of this intermediate it must decompose much more rapidly than the di-μ starting complex.[308] In alkaline solution *meso*-**42** isomerizes to the racemic form.[308] The aquation of **42** in acid solution has also been studied, alongside aquation of its trien analog **43**, in an investigation of

$$\left[(en)_2Co \begin{array}{c} O_2 \\ \diagdown \\ OH \end{array} Co(en)_2 \right]^{3+} \qquad \left[(trien)Co \begin{array}{c} O_2 \\ \diagdown \\ OH \end{array} Co(trien) \right]^{3+}$$

42   **43**

the balance between reduction and dissociation in acidic solutions containing Fe^{2+}. In fact, reduction by the Fe^{2+} is only a minor pathway; acid-catalyzed rupture of the hydroxo bridge is the dominant reaction pathway.[309] However, when a much stronger reducing agent is used, for instance Cd^+, reaction does go wholly via one-electron transfer as rate-determining step. The substrate in the Cd^+ reaction was a μ-amido-μ-superoxo complex, so the comparison is not quite direct.[310] Again electron transfer is a dominant feature of aquation of **44**

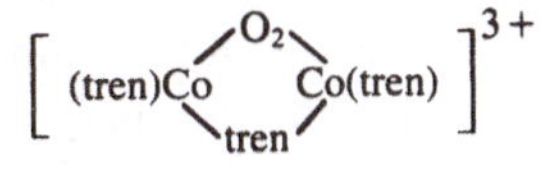
(tren)Co
O2
Co(tren)
tren
3+

44

in acid solution, though preequilibrium bridge protonation and a conformational change in the μ-tren ligand are thought to precede the redox step. In alkaline solution μ-tren is replaced by μ-OH.[311]

5.7.9.3. Tri-Bridged Dicobalt(III) Species

The kinetics of the equilibrium between tri-μ-hydroxo-bis(1,4,7-triazacyclononane)cobalt(III) and *trans*-diaquo-di-μ-hydroxo-bis(1,4,7-triazacyclononane)cobalt(III), 1,4,7-triazacyclononane = **45,** in acid aqueous solution have

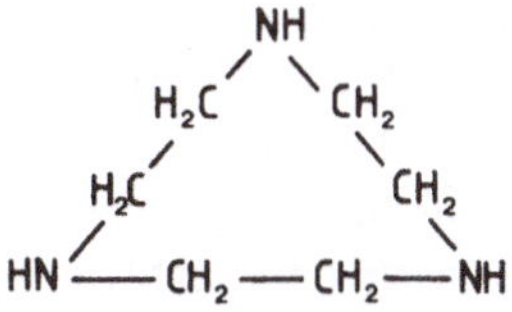
NH
H2C
CH2
H2C
CH2
HN
CH2
CH2
NH

45

been examined. The mechanism involves *cis* ⇌ *trans* interconversion in the diaquo-di-μ-hydroxo-complex as rate-determining step. The kinetic and mechanistic pattern here is consistent with that reported for other tri-μ-hydroxo-complexes.[312]

5.7.9.4. Cobalt(III)–Iron(II)

A study of the kinetics of dissociation of the dinuclear complex $[(edta)Co(III)–NC–Fe(II)(CN)_4(4,4'\text{-bipy})]^{4-}$ forms a key part of an investigation of the reduction of $[Fe(CN)_5(4,4'\text{-bipy})]^{2-}$ by cobalt(II)–edta. The kinetic characteristics of this Co(III)–NC–Fe(II) species in fact show that the redox process cannot involve the intermediacy of this μ-cyano species, and indeed must go by an outer-sphere mechanism.[313]

5.7.10. Photochemistry

5.7.10.1. Ammine and Amine Complexes

Quantum yields have been estimated for photolysis at charge transfer frequencies of $[Co(NH_3)_5L]^{3+}$ cations, with L = ammonia, benzonitrile, propiononitrile, and capronitrile, in water and in binary aqueous solvent mixtures containing ethylene glycol, glycerol, or acetonitrile. Experiments were also conducted in glycol–glycerol mixtures. It was shown that solvent participates in the rate-determining step.$^{(314)}$ Irradiation of neutral aqueous solutions of $[Co(en)_3]^{3+}$ is claimed to give $[Co(en)_2(CH_2NH_2)]^{2+}$, with the CH_2NH_2 ligand coordinated through C and N. An identical product is claimed from $[Co(en)_2(gly)]^{2+}$. In acid solution $[Co(en)_2(CH_2NH_3)(OH_2)]^{3+}$ is said to be obtained both from $[Co(en)_3]^{3+}$ and from $[Co(en)_2(gly)]^{2+}$. The mechanism is thought to involve phototransfer of an electron and fragmentation of the ligand radical thus generated.$^{(315)}$

5.7.10.2. Cyanide Complexes

Photosolvolysis of $[Co(CN)_6]^{3-}$ has been reinvestigated in aqueous ethanol and in ethanol itself. It seems that cyanide is replaced wholly by water in solvent mixtures containing up to at least 60% ethanol. However, in 80% ethanol both $[Co(CN)_5(OH_2)]^{2-}$ and $[Co(CN)_5(EtOH)]^{2-}$ are formed.$^{(316)}$ A study of photosolvolysis of $[Co(CN)_6]^{3-}$ in polyalcohol + water solutions proved an exercise in photophysics rather than photochemistry. The main conclusion was that intersystem crossing efficiencies varied remarkably little as the solvent composition was varied from water to glycerol at mole fraction 0.69. This behavior is similar to that established earlier for some cyano and thiocyanato complexes of chromium(III).$^{(317)}$

A study of photosubstitution in six (substituted) pyridine and pyrazine pentacyanocobaltate(III) complexes $[Co(CN)_5L]^{n-}$ complements an earlier study of similar pentacyanoferrate(II) complexes. Irradiation in ligand field bands results in 100% replacement of L by water; quantum yields range from 0.12 to 0.40. The results are similar to those for the iron(II) complexes, with such differences as are observed assignable to the large difference in importance of π back bonding to cobalt(III) and to iron(II).$^{(318)}$

The complexes *cis*-$[Co(edda)(CN)_2]^-$ and *cis*-$[Co(gly)_2(CN)_2]^-$ are photoreactive only at the charge-transfer-to-metal frequencies ($\lambda \sim 222$ nm); products have been characterized.$^{(319)}$ Both for $[Co(en)_2(CN)Cl]^+$ and for $[Co(en)_2(CN)(OH_2)]^{2+}$ the direction of spontaneous isomerization is *trans* $\rightarrow$ *cis*.$^{(320)}$

5.7.10.3. Sulfinate Complexes

The O-sulfinato isomer of $[Co(en)_2(SO_2CH_2CH_2NH_2)]^{2+}$ is formed on photolysis of the S isomer. The O-bonded form is photoinert, but reverts thermally to the S-bonded form (Section 5.7.6.2).[283]

5.7.11. Reactions of Coordinated Ligands

Such reactions have already cropped up in Sections 5.7.1. and 5.7.3, for both acid-catalyzed aquation and mercury(II)-assisted aquation are reactions in which the coordinated ligand plays a central role. Other manifestations of ligand reactivity are dealt with in this section.

5.7.11.1. Hydrogen and Deuterium Exchange

The kinetics of stereoselective deuteration of malonate hydrogens in bis(malonato)-cobalt(III) complexes $[Co(mal)_2L_2]^-$ containing $L_2 = $ en, pn, N,N'-Me$_2$en, phen, cis-$(NH_3)_2$, or cis-$(py)_2$ have been monitored. Both acid and base hydrolysis are observed, and there is a reversal of stereoselectivity with solution pH. There are some kinetic differences between the amine ligands on the one hand and py and phen on the other, as competition for OD$^-$ between malonate and amine is possible, between malonate and py or phen not.[321] The rate law for deuteration of α hydrogens in α-aminocarboxylato complexes of cobalt(III) containing various combinations of glycine, sarcosine, or alanine with ammonia, ethylenediamine, or diaminopropane is simple second order.

$$\text{rate} = k_2[\text{complex}][\text{OD}^-]. \tag{35}$$

Values of k_2 depend on the geometry and charge of the complex, but are little affected by the nature of the amine or ammine ligands.[322]

An nmr study of the kinetics of proton exchange at cobalt(III)–amine complexes in liquid ammonia has been mentioned in the section on base hydrolysis (Section 5.7.2).[240]

5.7.11.2. Coordinated Amino Acids

In acid solution, $0 \leq \text{pH} \leq 2$, the chelated glycine of $[Co(en)_2(gly)]^{2+}$ exchanges one oxygen atom, at rates proportional to the hydrogen ion concentration. Exchange with the second oxygen is about a thousand times slower. In alkaline solution both oxygens exchange, at rates proportional to hydroxide ion concentration. Oxygen exchange in alkali is considerably faster than in acid, but again exchange of one oxygen is about a thousand times faster than the other in alkali. The slower exchange involves carbon–oxygen bond fission to give *cis*-[Co(en)$_2$(gly)(OH)]$^+$, which subsequently undergoes hydroxide-catalyzed loss of the glycine.$^{(323)}$

Long-range coupling between the imidazole C-4 hydrogen and only one β hydrogen is apparent in the nmr spectrum of [Co(NH$_3$)$_5$(L-hist)]$^{3+}$. This suggests restricted rotation about the β carbon to γ carbon bond.$^{(324)}$ Isomer ratios, Λ-S/ Λ-R, have been used to probe hydroxide-ion-catalyzed epimerization of chelated amino acid anions in cobalt(III) complexes Λ-β$_2$-[CoL$_4$(S-aa)]$^{2+}$ and Λ-β$_2$-[CoL$_4$(R-aa)]$^{2+}$. The relation of product isomer ratios to strain energy minimization was discussed.$^{(325)}$ Product isomer ratios, this time S-alanine/R-alanine, have also been used to probe the effects of methyl substitution in the 2 and 9 positions of tren on acid-catalyzed decomposition of coordinated α,α-aminomethylmalonic acid, LL, in [Co(Me-tren)(LL)]$^+$. There is a small but systematic dependence on ligand bulk; product ratios indicate a difference of only 2 or 3 kJ mol^{-1} between the chemical potentials of the forms.$^{(326)}$

The hydrolysis of 1-acetylimidazole has been studied in the presence of hydroxo–cobalt(III) and –chromium(III) complexes, for example [Co(NH$_3$)$_5$(OH)]$^{2+}$. Catalysis is specific rather than general acid–base, implying strong complex–1-acetylimidazole interaction, presumably complex formation. This result is used to implicate direct involvement of Zn(OH)$^+_{aq}$ in the mechanism of action of carbonic anhydrase.$^{(327)}$

5.7.11.3. Coordinated Phosphates

Two out of the three products of base hydrolysis of the [Co(NH$_3$)$_5$(p-nitrophenylphosphate)]$^{2+}$ cation arise from attack of hydroxide at the coordinated ligand. These products are p-nitrophenolate and hydroxophosphoramidotetraamminecobalt(III). These result from intramolecular attack at the phosphorus by a deprotonated coordinated ammine ligand, which generates a five-coordinate aminophosphorane. Ester hydrolysis is about 10^8 times faster in this complex than in the free p-nitrophenylphosphate—one of the most dramatic examples of the kinetic consequences of coordination to a metal center,$^{(226)}$ comparable with the activation of acetonitrile on coordination to ruthenium(III).$^{(328)}$

In contrast to this striking example, the hydrolysis of bidentate coordinated triphosphate in [Co(NH$_3$)$_5$(H$_2$P$_3$O$_{10}$)] is only about two thirds of the hydrolysis rate for the free ligand.$^{(329)}$ Yet [Co(LL)$_2$(OH$_2$)$_2$]$^{3+}$ cations, with (LL)$_2$ = en$_2$, tn$_2$, or tren, give about a 10^5 times enhancement of the rate of hydrolysis of pyrophosphate at pH 7. The key to the remarkable difference may be that it is possible to have three cobalt(III) cations coordinated to each pyrophosphate.$^{(330)}$

5.7.11.4. *Miscellaneous*

The first step in the reaction of $[Co(NH_3)_5(py)]^{3+}$ with hydroxyl radicals is proposed to be attack at the coordinated pyridine. The absorption spectrum of the $[Co(NH_3)_5(pyOH)]^{3+\bullet}$ radical complex thus produced is given. This intermediate then undergoes redox decomposition. The intermediacy of such a radical complex in the present system[331] is supported by earlier suggestions of the existence of the species $[Co(NH_3)_5(O_2C^\bullet)]^{2+}$ [332] and of this or related species during permanganate oxidation of the formate complex $[Co(NH_3)_5(O_2CH)]^{2+}$.[333]

Iodine oxidation of thiocyanate coordinated to cobalt(III) in *cis*-$[Co(en)_2(S_2O_3)]^-$ gives *trans*-$[Co(en)_2(OH)(S_3O_3)]$. Presumably the tetrathionate complex is a transient intermediate, reacting quickly by sulfur–sulfur and cobalt–sulfur bond breaking to give the characterized product. This product is of considerable interest in its own right, since it represents stabilization of the very unstable species disulfane monosulfonate by coordination to cobalt(III). This ligand is thought to bond to the cobalt through sulfur as depicted in **46**.[334]

$$L_5Co-S \overset{S}{\underset{SO_3}{<}}$$

46

The cobalt(III)–thiolato complexes $[Co(en)_2(LL)]^{n+}$, where LL $=$ $^-SCH_2CH_2NH_2$, $^-SCH_2CO_2^-$, $^-SCMe_2CO_2^-$, $^-SCHMeCO_2^-$, react with hydrogen peroxide to give isolable S-bonded sulfenic acid complexes. This behavior contrasts with that of analogous chromium(III) complexes, where reaction proceeds via chromium–sulfur bond breaking. The mechanism of the reactions of the cobalt(III) complexes is of nucleophilic attack by the peroxide at sulfur. The rate law is:

$$-d[\text{complex}]/dt = \{k_2 + k_2'[\text{H}^+]\}[\text{complex}][\text{H}_2\text{O}_2] \tag{36}$$

Values of $k_2 + k_2'[\text{H}^+]$ are in the region of 1 dm^3 mol^{-1} s^{-1}; activation parameters derived from these composite rate constants are about 10 kcal mol^{-1} ($\Delta H^{\ddagger}$) and -30 cal deg^{-1} mol^{-1} ($\Delta S^{\ddagger}$).[335]

5.8. *Rhodium(III)*

Although there are many fewer references dealing with kinetics and mechanisms of substitution at rhodium(III), there has been enough activity in the period covered by this review to make it worthwhile to arrange topics in roughly the same pattern as used above for cobalt(III).

5.8.1. Aquation

The volume of activation for aquation of $[Rh(NH_3)_5(NO_3)]^{2+}$ is -6.9 cm^3 mol^{-1}. This negative value is consistent with a predominantly associative mechanism.[336] Activation parameters for acid aquation of the *cis*-$[Rh(en)_2Cl_2]^+$ cation are $E_a = 23$ kcal mol^{-1} and $\Delta S^{\ddagger} = +9.6\ (\pm 5.3)$ cal deg^{-1} mol^{-1}. These values were derived from the temperature dependence of rate constants over the range 298–333 K.[337]

5.8.2. Base Hydrolysis

Activation volumes for base hydrolysis in aqueous solution at 313.2 K and an ionic strength of 1.0 mol dm^{-3} of the complexes $[Rh(NH_3)_5X]^{2+}$, with X = Cl, Br, I, and NO$_3$, are $+19.3$, $+20.2$, $+20.4$, and $+22.3$ cm^3 mol^{-1}, respectively. These results, obtained over a pressure range of 1.5 kbar, can be accommodated within the framework of an S_N1CB mechanism.[336]

In a recent publication the effects of stereochemistry on base hydrolysis rates were investigated by measurements on *cis*- and *trans*-$[Rh(cyclam)Cl_2]^+$.[338] This study has now been extended by a comparison of base hydrolysis rates of cyclam (47) and cyclen (34) complexes. For the latter ligand, only the *cis* form

47

of $[Rh(cyclen)Cl_2]^+$ is known. The important difference between these two macrocyclic ligands is the size of the space into which the metal atom is to be fitted. A comparison of activation parameters for base hydrolysis of the two *cis*-$[Rh(LLLL)Cl_2]^+$ complexes suggests that reactivity differences can indeed be attributed to different strains within the complex cations. Various mechanisms for base hydrolysis in these systems are discussed. A dissociative mechanism with a square-pyramidal transition state or transient intermediate seems to be rather tentatively favored, and the authors appear reluctant to rule out some nucleophilic assistance by the incoming hydroxide ion altogether.[339] It has recently been shown that there is a small but significant dependence of rates of base hydrolysis on hydroxide ion concentration for rhodium(III) complexes of this type, though it is often necessary to work at high hydroxide concentrations to observe this term in the rate law.[340] The most recent example of the operation of this hydroxide-dependent path is afforded by the kinetics of reaction of the *trans*-[Rh(en)$_2$ICl]$^+$ cation.[341]

5.8.3. Catalyzed Aquation

The dependence of rates of aquation on concentration of mercury(II) suggest that a binuclear Rh–Cl–Hg intermediate is involved in catalyzed aquation of the *cis*-[Rh(en)$_2$Cl$_2$]$^+$ cation. Activation parameters have been determined from the temperature dependence of rate constants. The activation energy for loss of HgCl$^+$ from the binuclear intermediate appears to be the same as that for unassisted Cl$^-$ loss (cf. Section 5.8.1). Although at first sight the activation entropy, $+1.5 \pm 1.8$ cal deg^{-1} mol^{-1}, is less than the value for unassisted Cl$^-$ loss, the large uncertainty on this latter value makes the difference between the two activation entropies not significant at any reasonable level of confidence.[337] These results are a trifle surprising, but it is possible that some degree of misinterpretation has crept in during the processes of translation, abstracting, and compression required for this review.

5.8.4. Formation

The kinetics of formation of the glycinato product from the reaction of [Rh(NH$_3$)$_5$(OH$_2$)]$^{3+}$ cation with glycine have been monitored in weakly acidic aqueous solution over the temperature range 333–353 K. Activation parameters $\Delta H^{\ddagger} = 22.2$ kcal mol^{-1} and $\Delta S^{\ddagger} = -9.2$ cal deg^{-1} mol^{-1} are reported. Comparison of these with activation parameters for analogous reactions at cobalt(III) suggests that the most likely mechanism is the standard Eigen–Wilkins mechanism, but here with considerable associative character to the interchange step.[342] The kinetics of reaction of [Rh(tpps)(OH$_2$)$_2$]$^{3-}$, where tpps = *meso*- tetrakis(*p*-sulfonatophenyl)porphinato, with chloride, bromide, iodide, and thiocyanate have been measured over the temperature range 288–308 K, in mildly acidic solution. These reactions are first order in entering anion. Rates are about a million times faster than for reaction with Rh$_{aq}$. Activation parameters are $\Delta H^{\ddagger} = 12.8$ to 16.5 kcal mol^{-1} and $\Delta S^{\ddagger} = -10$ to -26 cal deg^{-1} mol^{-1}. All these data and observations, particularly the small but real dependence of $\Delta H^{\ddagger}$ and $\Delta S^{\ddagger}$ on the nature of the incoming group, lead to the tentative proposal of an associative mechanism for these formation reactions. These formation reactions do not go to completion; equilibrium constants are reported in this paper, which also gives an estimate for the pK_a of the coordinated water in this diaquo complex.[343]

Values have also recently been obtained for pK_as of coordinated water in rhodium(III) complexes also containing ammonia and, in some cases, chloride, or bromide.[344] Such pK_a values are valuable in discussion of kinetics of formation reactions of these and related species.

One further reference on formation at rhodium(III), in binary aqueous solvents, will be found in Section 5.8.9 on dioximato complexes.

5.8.5. Solvent and Ligand Exchange

The kinetics and mechanism of acetylacetone exchange with [Rh(acac)₃] have already been discussed in connection with its cobalt(III) analog in Section 5.7.5.2.

Observed first-order rate constants for the reactions of equation (37):

$$trans\text{-}[Rh(en)_2LX]^+ + Y^- \rightarrow [Rh(en)_2LY]^+ + X^- \qquad (37)$$

where X and Y are variously chloride, bromide, iodide, or hydroxide (see also Section 5.8.2) are of the form shown in equation (38).

$$k_{obs} = k_1 + k_2[Y^-] \qquad (38)$$

Values of the activation parameters $\Delta H^\ddagger$ and $\Delta S^\ddagger$ for the k_2 terms in this expression lie on a good isokinetic plot, which covers ranges of 90 kJ mol⁻¹ for the former and 60 J K⁻¹ mol⁻¹ for the latter. These kinetic and parallel stereochemical results are held to show that the mechanisms of these reactions range from S_N1CB to S_N1IP and probably to S_N2.[341]

Kinetic parameters provide evidence for a dissociative mechanism for the ammonia displacement reaction shown in equation (39):

$$[Rh(NH_3)_5X]X_2 \rightarrow [Rh(NH_3)_4X_2]X + NH_3 \qquad (39)$$

when this is carried out in the solid state.[345] It will be interesting to compare this with the mechanism for the analogous process in solution, when this has been unequivocally established.

5.8.6. Ring Opening and Closing

Kinetics of these processes have been studied for rhodium(III)–chelating-arsine–halide complexes in the reactions of equation (40).

$$trans\text{-}[Rh(LL)_2Cl_2]^+ + X^- \rightleftharpoons mer\text{-}[Rh(LL)L'Cl_2X] \qquad (40)$$

Here LL represents the ligand o-dimethylaminophenyldimethylarsine coordinated to the rhodium through one nitrogen and one arsenic atom, L' represents this same ligand bonded only through arsenic, and X = Cl, Br, or I. In the forward direction the reaction is acid catalyzed, with such catalysis most marked for entering chloride, least for entering iodide. In the opposite direction, reaction is inhibited by the presence of acid, with the effectiveness of inhibition increasing in the order I < Br < Cl. Clearly these effects of added acid operate via reversible protonation of the free nitrogen when the chelating arsine is monodentate. Thus for the forward reaction, dissociative ring opening as rate-determining step is coupled with reversible protonation of the no-longer-bound nitrogen, with nucleophilic attack by the entering X^- on the intermediate monodentate–arsine species in both its protonated and nonprotonated forms.[346]

5.8.7. Isomerization

The reader is referred to Section 5.7.6.2 for discussion of nitrito to nitro linkage isomerization kinetics at rhodium(III).[281]

5.8.8. Carbonate and Sulfite Complexes

Kinetic parameters for formation and decarboxylation of carbonato-rhodium(III) complexes are given in Tables 5.16 and 5.17. In acid solution, the cis-$[Rh(en)_2(CO_3)]^+$ cation loses carbon dioxide slowly. Rate constants and activation parameters ($\Delta H^{\ddagger}$ and $\Delta S^{\ddagger}$) are reported for the two paths indicated by:

$$k_{obs} = k_1 + k_2[H^+] \tag{41}$$

Rate constants and activation parameters are also given for loss of carbon dioxide from the intermediate cis-$[Rh(en)_2(CO_3H)(OH_2)]^{2+}$. Here the rhodium(III) system has two advantages over its cobalt(III) counterpart, for not only is it possible to obtain kinetic data for the monodentate bicarbonato intermediate but also there is no concomitant isomerization to complicate the carbon dioxide loss kinetics in the rhodium(III) case. In the reverse direction, kinetic parameters were obtained for the reaction of cis-$[Rh(en)_2(OH_2)(OH)]^{2+}$ and of cis-$[Rh(en)_2(OH_2)_2]^+$ with carbon dioxide–(bi)carbonate. The products are cis-$[Rh(en)_2(CO_3)(OH_2)]^+$ and cis-$[Rh(en)_2(CO_3)(OH)]$; there is no mention of a bidentate carbonato product.[347] Kinetics of decarboxylation of the new complexes $trans$-$[Rh(en)_2(CO_3)(OH_2)]^+$ and $trans$-$[Rh(en)_2(CO_3)_2]^-$ have been reported, with values for rate constants and activation parameters given. The bis(carbonato) complex loses both carbonate ligands at low pHs, but only one carbonate when pH > 6.5. The kinetics of the reverse carbonate-formation reactions, from $trans$-

Table 5.16. Kinetic Parameters for Decarboxylation of Carbonato–Rhodium(III) Complexes (Aqueous Solution; 298.2 K). Subscripts 1 and 2 Referring to the k_1 and k_2 (Acid-Independent and Acid-Dependent) Terms of Equation (41)

Complex	k_1, s⁻¹	$\Delta H_1^\ddagger$, kcal mol⁻¹	$\Delta S_1^\ddagger$, e.u.	k_2, dm³ mol⁻¹ s⁻¹	$\Delta H_2^\ddagger$, kcal mol⁻¹	$\Delta S_2^\ddagger$, e.u.	Ref.
cis-[Rh(en)₂(CO₃)]⁺	9.33×10^{-6}	22.8	−5	1.00×10^{-5}	25.0	+2	347
cis-[Rh(en)₂(OH₂)(CO₃H)]²⁺	0.72	19.4	+5.8				347
trans-[Rh(en)₂(OH₂)(CO₃)]⁺				2.92	10.9	−19.6	348
trans-[Rh(en)(CO₃)₂]⁻, pH < 6	2.26	16.3	−2.1				348
pH > 6.5	1.3						348
trans-[Rh(en)₂Cl(CO₃H)]⁺	1.26	17.4	−0.1				349
trans-[Rh(en)₂Br(CO₃H)]⁺	1.12	18.2	+2.6				349
trans-[Rh(en)₂I(CO₃H)]⁺	0.55	17.2	−2.0				349
(LLL)Rh-μ-(OH)₂(CO₃)-Rh(LLL)²⁺	Negligible path			2.73^a	15.0	−25.4	350

^{a} At 303.2 K.

Table 5.17. Kinetic Parameters for the Reaction of Hydroxo–Rhodium(III) Complexes with Carbon Dioxide (Aqueous Solution; 298.2 K)

Complex	k, dm^3 mol^{-1} s^{-1}	$\Delta H^\ddagger$, kcal mol^{-1}	$\Delta S^\ddagger$, e.u.	Ref.
cis-[Rh(en)$_2$(OH$_2$)(OH)]$^{2+}$	69	15.9	+3.9	347
cis-[Rh(en)$_2$(OH)$_2$]$^+$	215	14.6	+1.1	347
trans-[Rh(en)$_2$(OH$_2$)(OH)]$^{2+}$	81	12.9	−5.5	348
trans-[Rh(en)$_2$(OH)$_2$]$^+$	1.4^a	16.3	+16.4	348
trans-[Rh(en)$_2$(CO$_3$)(OH)]	330	15.4	+4.6	348
trans-[Rh(en)$_2$Cl(OH)]$^+$	260	12.5	−5.3	349
trans-[Rh(en)$_2$Br(OH)]$^+$	395	15.1	+3.9	349
trans-[Rh(en)$_2$I(OH)]$^+$	422	17.6	+11.9	349

^{a} dm^6 mol^{-2} s^{-1}; third-order rate law (cf. text).

[Rh(en)$_2$(OH$_2$)(OH)]$^{2+}$ and from *trans*-[Rh(en)$_2$(CO$_3$)(OH)] have also been examined. The reaction of *trans*-[Rh(en)$_2$(OH)$_2$]$^+$ with carbon dioxide, to give the bis(carbonato) product, exhibits the novel feature of third-order kinetics.$^{(348)}$ Formation and decarboxylation of halide analogs *trans*-[Rh(en)$_2$X(CO$_3$)], where X = Cl, Br, or I, have also been studied kinetically, with rate constants and activation parameters ($\Delta H^\ddagger$ and $\Delta S^\ddagger$) obtained for reaction in both directions. In the discussion of these results, the opportunity is taken to put them into context with those from a variety of earlier kinetic studies of carbonato–rhodium(III) and –cobalt(III) complexes. Thus one can construct remarkably good isokinetic ($\Delta H^\ddagger$ vs. $\Delta S^\ddagger$) plots for these reactions. In these plots, the rhodium(III) and cobalt(III) complexes fall on the same straight line. This result is consistent with, and provides support for, carbon–oxygen rather than metal-ion–oxygen bond breaking being rate determining. There are also some linear relations with certain pK values. Thus a plot of logarithms of rate constants for decarboxylation against pKs of complexes in their bicarbonate form is linear, as is a plot of logarithms of rate constants for reaction with carbon dioxide against pK values for the respective aquo ions. It is also possible to draw some conclusions about *trans*-labilizing effects of various ligands in this context.$^{(349)}$

Kinetics of decarboxylation of μ-carbonato complexes have been studied for the 1,4,7-triazacyclononane (LLL) complexes (45), with M = Rh or Cr. Both components of the right-hand side of the expression $k_{obs} = k_1 + k_2[\text{H}^+]$ are important for the chromium(III) compound, but only the acid-catalyzed path is significant for the rhodium(III) compound. Rate constants and activation parameters are extremely similar for the $k_2[\text{H}^+]$ paths for rhodium(III) and for chromium(III), consistent with the expected carbon–oxygen bond rupture in the rate-determining step. The kinetic parameters here are also, of course, very similar to those for bidentate mononuclear carbonate complexes of rhodium(III), cobalt(III), and chromium(III).$^{(350)}$

The $[Rh(NH_3)_5(OH)]^{2+}$ cation, and its chromium(III) analog, react very rapidly with sulfur dioxide in aqueous solution, to give O-sulfito complexes, which lose sulfur dioxide on addition of acid. Rate constants for reaction of the rhodium(III), chromium(III), and cobalt(III) complexes $[M(NH_3)_5(OH)]^{2+}$ with sulfur dioxide are 1.8×10^8, 2.9×10^8, and 4.7×10^8 dm^3 mol^{-1} s^{-1}, respectively. This great similarity of rates contrasts with the situation in respect of formation of carbonato complexes, where formation rate constants show a marked dependence on the pK_a of the aquo form of the hydroxo substrate.$^{(351)}$

5.8.9. Bis(glyoximato) Complexes

Some kinetic information is given in a paper detailing the chemistry of *trans*-$[Rh(dmgH)_2H(PPh_3)]$ and its conjugate base in aqueous methanol. Forward and reverse rate constants are given for the neutralization reaction, and rate constants are reported for reaction of the conjugate base complex, viz. $[Rh(dmgH)_2(PPh_3)]^-$, with alkyl and benzyl halides. An S_N2 mechanism seems more likely than a radical mechanism for these reactions.$^{(352)}$

A consideration of solvent effects on rates of formation reactions of the *trans*-$[Rh(dmgH)_2(OH_2)_2]^+$ cation in binary aqueous solvent mixtures in terms of initial state and transition state components is included in the paper on formation reactions of aquo–cobalt(III) complexes and of Ni^{2+}_{aq} dealt with fully in Section 5.7.1.3. However it is perhaps proper to point out here the contrast between the D mechanism assumed in this analysis of reactivity trends$^{(209)}$ and the associative mechanism proposed for anation at the very similar species $[Rh(tpps)(OH_2)_2]^{3-}$ (see Section 5.8.4).

5.8.10. Photochemistry

Activity has been concentrated on complexes of the general formula $[Rh(NH_3)_4XY]^{n+}$ and $[Rh(en)_2XY]^{n+}$. There is a slight advantage in studying the tetraammine complexes, in that there can be no complication arising from any steric restraints that might operate in complexes of bi- or polydentate ligands.

Almost a decade ago it was established that *trans*-$[Rh(NH_3)_4X_2]^+$ cations (X = halide) gave *trans*-$[Rh(NH_3)_4(OH_2)X]^{2+}$ photolysis products.$^{(353)}$ Now it has been shown that photoaquation of *cis*-$[Rh(NH_3)_4X_2]^+$, X = Cl or Br, gives *trans*-$[Rh(NH_3)_4(OH_2)X]^{2+}$, and that *cis*-$[Rh(NH_3)_4(OH_2)X]^{2+}$ isomerizes to the *trans* form on irradiation. These reactions go with high quantum yields. There is no indication of a stable intermediate in the photolysis of *cis*-$[Rh(NH_3)_4X_2]^+$; aquation and isomerization are effectively concurrent. All these results on the *cis*- and *trans*-$[Rh(NH_3)_4X_2]^+$ cations are consistent with mechanisms involving five-coordinate square-pyramidal intermediates in which X^- is apical.[354] Photoaquation both of *cis-* and *trans*-$[Rh(NH_3)_4(OH_2)Cl]^{2+}$ gives *trans*-$[Rh(NH_3)_4(OH_2)_2]^{3+}$.[355] However *trans*-$[Rh(NH_3)_4(OH)Cl]^+$ undergoes concurrent aquation and isomerization on irradiation, to give *cis*-$[Rh(NH_3)_4(OH)_2]^+$. This striking example of ligand control of stereochemistry can be combined with earlier stereochemical results on tetraaminerhodium(III) complexes to give the following order of apical preference in the proposed square-pyramidal intermediates in these photoreactions[356]: $OH^- < NH_3 < OH_2 < Cl^- < Br^- < I^-$. A slightly later paper gives further details of photochemistry of this hydroxo-chloro complex, and of its bromo analog. It also points out that *cis-* and *trans*-$[Rh(NH_3)_4(OH)_2]^+$ are inert to photoisomerization, because the hydroxide ligand, the extreme ligand in the above series, is not photolabile.[357]

Studies of the photolytic behavior of complexes $[Rh(en)_2XY]^{n+}$ help to put the ammonia ligand into context when X and Y are taken in various combinations from the ligands NH_3, OH_2, OH^-, Cl^-, and Br^-. Irradiation of *cis-* and of *trans*-$[Rh(en)_2XBr]^{n+}$, with $X = Br$, OH_2, or NH_3, in their ligand field bands gives *trans*-$[Rh(en)_2(OH_2)Br]^{2+}$ in all cases except that of *cis*-$[Rh(en)_2(NH_3)Br]^{2+}$. Irradiation of this complex results in the formation of *cis*-$[Rh(en)_2(NH_3)(OH_2)]^{3+}$ as well as the *trans*-bromoaquo cation. Again these results can be interpreted by a two-stage mechanism with a square-pyramidal intermediate. But here it is necessary to propose that the transformation **48** → **49** is difficult, whereas **50** → **51** is easy, in order to rationalize the mixture of products from photoaquation

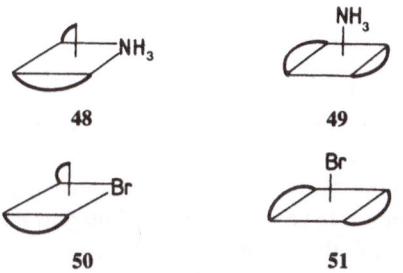

of the *cis*-bromoammine complex. Otherwise all the results conform to the pattern expected from the series set out above; bromide generally prefers to be in the apical position in the square-pyramidal intermediate.[358] Similar remarks and reasoning apply to photolysis of *cis-* and *trans*-$[Rh(en)_2XI]^{n+}$, with $X = I^-$, NH_3, or OH_2. Again all reactions give 100% *trans*-iodoaquo product except that of the *cis*-iodoammine cation, which again gives *cis*-aquoammine and *trans*-iodoaquo products. For the type of complexes covered in this section, the ease of halide liberation decreases in the order $Cl > Br > I$, whereas the order of *trans*-labilizing effects is $I \gg Br > Cl$. From these generalizations one can predict that photolysis of *cis*-$[Rh(NH_3)_4I_2]^+$ should result in loss of ammonia rather than of iodide, and would be the only complex of this class which did not lose halide.[359] The above explanations of the apparently irregular behavior of *cis*-bromoammine and *cis*-iodoammine complexes can also be used for the earlier-reported *cis*-chloroammine complex *cis*-[Rh(en)$_2$(NH$_3$)Cl]$^{2+}$ [360] and cyano–cobalt(III) complexes.[320]

The final photochemical reference concerns *trans* → *cis* isomerization of 4-stilbenecarboxylate coordinated to rhodium(III), or to iridium(III), as a pentaammine derivative. This isomerization has been studied in several solvents. Excitation was at the $S_0 \to S_1$ intraligand absorption band; the metal ions appear to function by inducing $S_1 \to T_1$ intersystem crossing.[361]

5.9. Iridium

The long time scale of most kinetic experiments on iridium substitution kinetics may be presumed to be the reason for the continuity paucity of references in comparison with rhodium(III) and, especially, cobalt(III). There are a few references to substitution at low-spin d^6 iridium(III), and two references to qualitative kinetic observations on the almost equally inert d^5 iridium(IV).

5.9.1. Iridium(III)

5.9.1.1. Hexachloroiridate(III) Aquation

The cations Ga^{3+}, Fe^{3+}, Ce^{3+}, and Eu^{3+} all inhibit aquation of the IrCl$_6^{3-}$ anion. In other words, rate constants for aquation of M^{3+},IrCl$_6^{3-}$ ion pairs are smaller than that for the free anion. The effects of the added cations correlate with their ionic radii.[362] Effects of added Et$_4$N$^+$ salts on aquation rates for the [IrCl$_6$]$^{3-}$ anion have also been attributed to ion pairing. Indeed ion-pairing constants and an enthalpy and entropy for ion pair formation have been calculated from the kinetic results.[363] However one must wonder whether part at least of the effect of the Et$_4$N$^+$ may not be ascribable to its marked effects on water structure, particularly as the Et$_4$N$^+$,IrCl$_6^{3-}$ ion pair is stated to be more reactive than the free anion, in contrast to the M^{3+},IrCl$_6^{3-}$ ion pairs mentioned above. It should also be borne in mind that in kinetic studies where cations have been added with the express intention of altering reactivities via their effects on water structure, effects of M^{n+} and R$_4$N$^+$ have always been found to be opposite to each other.

Another way in which added metal ions might affect aquation rates of the [IrCl$_6$]$^{3-}$ anion is by direct catalysis of the Co(III)–Cl$^-$ plus mercury(II) type (cf. Section 5.7.3). Such metal ion assistance could be involved for Ga^{3+} as it is soft, but seems unlikely in view of the decrease in reactivity on adding this cation. However a kinetic study of the effects of In^{3+} have been interpreted in terms of the usual catalytic scheme (depicted as Scheme 18 here). The

Scheme 18

$$[IrCl_6]^{3-} + In^{3+} \overset{K}{\rightleftharpoons} [Cl_5IrClIn] \overset{k}{\to} [IrCl_5(OH_2)]^{2-} + InCl^{2+}$$

equilibrium constant has the high value of 100 (± 15) at 293 K; the enthalpy and entropy of formation of the binuclear intermediate were also estimated, as were aquation rates. The overall effect was a doubling of the aquation rate on addition of only 10^{-5} mol dm^{-3} indium(III).[364]

5.9.1.2. *Isomerization*

Linkage isomerization of the nitrito ligand to the nitro form [equation (42)] has already been discussed in Section 5.7.6, *q.v.*

$$[Ir(NH_3)_5(ONO)]^{2+} \to [Ir(NH_3)_5(NO_2)]^{2+} \qquad (42)$$

5.9.1.3. *Photochemistry*

An extensive study has been made of photochemical behavior of tetraammine and pentaammine complexes of iridium(III). These complexes were $[Ir(NH_3)_5X]^{n+}$ with X = Cl⁻, Br⁻, I⁻, OH_2, NH_3, MeCN, or PhCN, *trans*-$[Ir(NH_3)_4I_2]^+$, and *trans*-$[Ir(NH_3)_4(OH_2)I]^{2+}$. Photolysis at ligand field frequencies generally resulted solely in photoaquation, with 100% retention of stereochemistry, where relevant. Photolysis at charge transfer bands gave mainly photoaquation, but a small redox contribution was also found. These results in general parallel those for analogous rhodium(III) complexes.[365] The photochemistry of the bis(ethylenediamine) complexes *cis*- and *trans*-$[Ir(en)_2XY]^+$, with X and Y being taken variously from Cl, Br, and OH, has also been established. The *trans*-dihalide complexes give 100% *trans*-halogenoaquo products, but the *cis*(dihalide) complexes give a mixture of *cis*- and *trans*-halogenoaquo products, with the extent of photoisomerization decreasing in the order I > Br > Cl. Moreover, the *cis*-chloroaquo complex photoisomerizes to the *trans* form. As in the case of similar rhodium(III) systems (Section 5.8.10), the introduction of a hydroxo ligand alters the photochemistry, in that it is now the *trans*-hydroxohalogeno complex which photoaquates with concomitant photoisomerization, giving *cis*-$[Ir(en)_2(OH)_2]^+$.[366]

The solvent dependence of chloride substitution rates has been established for excited states of the *cis*-$[Ir(bipy)_2Cl_2]^+$ cation and its d^8-bipy analog. These rates are, of course, very much greater than those for the ground state, but, owing to the dominant role of solvation of the leaving chloride, rates of loss of chloride from the excited and ground states are both very dependent on the nature of the solvent. In this study the solvents used were water, methanol, acetonitrile, and dimethylformamide. Correlation of rates of chloride loss with Gutmann acceptor numbers, but not with donor numbers, is consistent with a dissociative mechanism.[367] Grunwald–Winstein m values can be estimated as 0.48 for the bipy complex and 0.44 for its deuterated analog; compare values of around 0.25 for Ir(III)–chloride complexes undergoing thermal (ground state) solvolysis.[368]

The final reference in this photochemistry section concerns *trans* $\rightarrow$ *cis* isomerization of coordinated stilbenecarboxylate, L, in $[Ir(NH_3)_5L]^{2+}$, as discussed in connection with the analogous rhodium(III) complex (Section 5.8.10).[361]

5.9.2. Iridium(IV)

The effects of pH, temperature, chloride concentration, complex concentration, irradiation, and added cations on the "stability" of hexachloroiridate(IV) solutions have been reviewed, in connection with the use and regeneration of this complex as a catalyst. This chapter includes a large multistep and multi-component scheme. It does not give any rate constants, but gives several references to kinetic data.[369]

Repeat scan spectra of a solution of $[Ir(NO_3)_6]^{2-}$ in concentrated nitric acid give a good isosbestic point, and suggest a half-life of a few hours for the conversion of this hexanitratoiridium(IV) complex into the $[Ir_3O(NO_3)_9]^+$ cation. However, conversion of the starting complex into the trinuclear cation may not quite be complete and thus an estimated rate constant would refer to the approach to equilibrium.[370]

5.10. Platinum(IV)

5.10.1. General

The first step in reactions of $[Pt(S_5)_3]^{2-}$ with sulfide, sulfite, arsenite, hydroxide, or triphenylphosphine is attack of the nucleophile at a sulfur atom α to the metal, **52** $\rightarrow$ **53**.[371] Such attack is presumably greatly facilitated by the $4+$ charge on the platinum, and is reminiscent of Gillard's mechanism for nucleophilic attack at diimines coordinated to iron(II) and ruthenium(II) (see Section 5.4.2).

52 **53**

Some qualitative information germane to the kinetics and mechanism of isomerization and substitution reactions of the compounds [PtMe$_2$X(gly)(OH$_2$)], with X = Br or OH, has been obtained from preparative and nmr studies.[372] Similar information is available on ring opening and closing for bis and tris(glycinato) derivatives of platinum(IV).[373] Quantitative kinetic information is available on ring closure in the [Pt(en)(NH$_3$)(glyH)Br$_2$]$^{2+}$ cation. Rate constants and activation parameters ($\Delta H^{\ddagger}$ and $\Delta S^{\ddagger}$) are reported, the effects of acid concentration described, and catalysis by the platinum(II) complex [Pt(en)(NH$_3$)(glyH)]$^{2+}$ observed and explained.[374] Results are, understandably, very similar to those reported earlier for the chloride analog.

The quantum yield for the replacement of iodide by cyanide:

$$[Pt(CN)_5I]^{2-} + CN^- \rightarrow [Pt(CN)_6]^{2-} + I^- \tag{43}$$

increases as the concentration of cyanide increases, up to a cyanide concentration of 1.8 mol dm^{-3}. At higher cyanide concentrations the quantum yield levels off. A mechanism with a free radical initiation step is proposed.[375]

5.10.2. Inversion at Coordinated Sulfur and Selenium

There have recently been several nmr investigations of the kinetics of pyramidal inversion at sulfur and selenium coordinated to platinum(IV). These studies seem to form an extension of work on inversion at sulfur and selenium coordinated to platinum(II)[376] (cf. Chapter 4). Total band shape fitting methods have been used in the analysis of the temperature dependence of nmr spectra of eleven complexes of the type [PtMe$_3$X(LL)], where X = Cl, Br, or I, and the bidentate ligands LL are MeS(CH$_2$)$_n$SMe with n = 2 or 3 and selenium analogs. For the complexes with ligands where n = 2, the process occuring on the nmr time scale must be inversion at the sulfur or selenium; for compounds with n = 3 it is conceivable that ring reversal might also come within the nmr time scale. However, there do not appear to be any complications from the latter in the present study. Values of E_a, log A, $\Delta G^{\ddagger}$, $\Delta H^{\ddagger}$, and $\Delta S^{\ddagger}$ are tabulated for all 11 complexes, in CDCl$_3$ solution. Observations over a temperature range of about 60 K lead to $\Delta H^{\ddagger}$ values in the range 54–75 kJ mol^{-1} and $\Delta S^{\ddagger}$ values in the range -9 to $+22$ J K^{-1} mol^{-1}. These $\Delta S^{\ddagger}$ values approximate to the value of zero expected for intramolecular processes. Barriers to inversion ($\Delta H^{\ddagger}$ and $\Delta G^{\ddagger}$) are about 10 kJ mol^{-1} higher at selenium than at sulfur.[377] This barrier difference, and the near-zero activation entropies, parallel earlier observations on platinum(II) systems.[378] Barriers also decrease when the chelate ring size increases, which behavior parallels that established for analogous chelates of nitrogen donor ligands. In contrast to the orderly pattern emerging this far, there is no clear-cut trend with variation in halide ligand.[377]

Nmr studies of related dinulear complexes [(PtMe$_3$X)$_2$(MeSCHRSMe)], with X = Cl, Br, or I, and R = H or Me, and selenium analogs, **54**, show a rich variety of intramolecular processes, including inversion at coordinated sulfur or selenium. As in the mononuclear complexes discussed above, activation entropies for inversion approximate to zero. Barriers ($\Delta G^{\ddagger}$ and $\Delta H^{\ddagger}$) to inversion at sulfur are around 40 kJ mol^{-1}, at selenium higher, at around 55 kJ mol^{-1}. These barriers are considerably lower than barriers to ligand switching and methyl scrambling; ligand dissociation, shown by disappearance of ^{195}Pt satellite lines, only becomes significant at temperatures above about 370 K.[379] Three dynamic processes affect nmr spectra of complexes [(PtMe$_3$X)$_2$(MeSSMe)] and [(PtMe$_3$X)$_2$(MeSeSeMe)], **55**, but again it is possible to obtain values for kinetic

54   55

parameters for inversion at sulfur and selenium in these complexes. Barriers here are only 2 or 3 kJ mol^{-1} lower than in the analogous MeSCHRSMe and MeSeCHRSeMe complexes *(vide supra)*. The fact that inversion barriers in the binuclear complexes are some 15 kJ mol^{-1} lower than in the mononuclear complexes is probably simply due to bond strength differences. However, it is possible that synchronous inversion at both sulfur or selenium atoms is important for the binuclear complexes but not for the mononuclear complexes. It is impossible to settle this from the nmr spectra.[380] The amount of kinetic information on intramolecular processes so far deduced from the nmr spectra of this class of compounds is impressive, but clearly more work in this field will be needed to clear up unresolved questions, and will provide yet more detailed insights into the dynamic behavior of these complexes.

Chapter 6

Substitution Reactions of Labile Metal Complexes

6.1. Complex Formation Involving Unsubstituted Metal Ions with Unidentate Ligands and Solvent Exchange

6.1.1. Univalent Ions

Comparatively little work has been done on the kinetics of complex formation between the alkali metal ions and simple ligands in view of the high rate constants and low stability constants involved. Atkinson[1] has recently studied the ultrasonic absorption of the five alkali metal sulfates in water in the frequency range 25–250 MHz, where he found only one relaxation for each salt. The results are analyzed in terms of the normal two-step mechanism (the fast formation of an outer-sphere complex followed by rapid conversion to the inner-sphere complex) in which the rates of the two steps approach each other as the concentration of the solution decreases. (The concentrations were in the range $0.3-1.0$ mol dm^{-3}.) As expected, the reactions are nearly diffusion controlled: the rate constants for inner-sphere complex formation at 0.5 mol dm^{-3} and 25°C are 1.0×10^9 s^{-1} for Li$^+$, Na$^+$, Rb$^+$, and Cs$^+$ sulfates but 2.0×10^9 s^{-1} for the potassium salt.

6.1.2. Bivalent Ions

Merbach and co-workers[2] have provided high-pressure nmr evidence that the same gradual change in mechanism from I_d to I_a occurs for water exchange at the bivalent cations $[M(H_2O)_6]^{2+}$ (for the series Ni, Co, Fe, Mn) as has already been reported[3] for methanol exchange at these cations ($[M(MeOH)_6]^{2+}$). (They had also given preliminary reports of their results with hexaqua-nickel[4] and hexaqua-manganese.[5]) As they point out, from a structural point of view the difference between an I_d and an I_a reaction is in the degree of expansion of the transition state: in the former case there is, by definition, only weak bonding between the metal and the entering (and also the leaving) solvent molecule, whereas in the latter case this bonding is much more significant. This difference, they argue, will be reflected in the volumes of activation and one may envisage a continuous spectrum of transition states ranging from a highly compact one in an associative (I_a or A) mechanism through to a very much more diffuse one in a dissociative (I_d or D) mechanism. Table 6.1 shows the results of these and parallel studies on acetonitrile[6,7] and DMF[6] exchange at Co(II) and Ni(II), and there does seem to be a consistent pattern in which the dissociative activation of Co(II) and Ni(II) (positive $\Delta V^{\ddagger}$) gives way to associative activation at Mn(II) (negative $\Delta V^{\ddagger}$), with a change-over at Fe(II) (zero $\Delta V^{\ddagger}$). Incidentally, some of the assumptions used in the detailed analysis of these results have been questioned by Langford,[8] who has drawn attention to the possible influence exerted by the nonexchanging ligands, but these thoughts have been firmly countered by Merbach[9] and by Swaddle.[10]

Earlier ambiguities associated with the possible deprotonation of a coordinated water molecule in $[Be(H_2O)_4]^{2+}$ have recently been resolved[11] in a pulsed-nmr study of proton and oxygen exchange in concentrated aqueous solutions of beryllium nitrate. The contributions from protolysis and water exchange have been separated and the following rate data obtained for the exchange process: k (per water molecule at 298 K) 1.8×10^3 s^{-1}, $\Delta H^{\ddagger} = 41.5$ kJ mol^{-1}, and $\Delta S^{\ddagger} = -44$ J K^{-1} mol^{-1}. Protolytic dissociation of the aquo-beryllium(II) ion is approximately one order of magnitude faster: $k = 8 \times 10^4$ s^{-1}, $\Delta H^{\ddagger} = 31.2$ kJ mol^{-1}, and $\Delta S^{\ddagger} = -45$ J K^{-1} mol^{-1} for beryllium concentrations up to 0.2 M. The same technique has been used[12] to demonstrate that proton nmr alone may be used to determine the true kinetic parameters for water exchange at the fully hydrated cation provided that a sufficient amount of organic diluent is added

Table 6.1. Volumes of Activation $\Delta V^{\ddagger}$ (cm^3 mol^{-1}) for Solvent Exchange on Divalent Metal Ions

Solvent	Mn(II)	Fe(II)	Co(II)	Ni(II)	Ref.
H_2O	-5.4	$+3.8$	$+6.1$	$+7.2$	2
CH_3OH	-5.0	$+0.4$	$+8.9$	$+11.4$	3
CH_3CN	—	—	$+7.7$	$+9.6$	6
	—	—	$+6.7$	$+7.3$	7
DMF	—	—	$+6.7$	$+9.1$	6

to fully suppress the contribution from the protolytic pathway: water exchange between coordinated sites at the Be(II) ion and the bulk solution dominates the proton exchange rate for DMSO mole fractions greater than 0.4 and for acetone mole fractions greater than 0.6, and the calculated rates are almost independent of the solvent composition and beryllium concentration. For water–DMSO in this concentration region, $k_{ex} = 0.85 \times 10^3$ s^{-1} (per water molecule), $\Delta H^{\ddagger} = 30$ kJ mol^{-1}, and $\Delta S^{\ddagger} = -90$ J K^{-1} mol^{-1}; and for water–acetone, $k_{ex} = 2.1 \times 10^3$ s^{-1}, $\Delta H^{\ddagger} = 34$ kJ mol^{-1}, and $\Delta S^{\ddagger} = -65$ J K^{-1} mol^{-1}.

The kinetics of complex formation between Ni(II) and ortho-phosphate, ribose monophosphate and cytidine monophosphate (CMP) in water have been reported.[13] The results are consistent with an I_d mechanism involving protonated (HL$^-$) and unprotonated (L^{2-}) ligands, and rate constants for reaction with L^{2-} and HL$^-$ are about 1.5×10^5 dm^3 mol^{-1} s^{-1} and 2.3×10^4 dm^3 mol^{-1} s^{-1}, respectively, for all three systems. The cytidine ring appears to exert no effect on the binding of Ni(II) to CMP. The rate parameters have also been reported[14] for the reaction of nickel(II) with 4-phenylpyridine and isoquinoline in water–t-butanol mixtures. Rate parameters for the dissociation of the isoquinoline and thiocyanate complexes of Ni(II) in 1-propanol, and of the former in ethanol and in water, are accommodated[15] within an I_d mechanism, and Tanaka[16] has commented further on the relationships between the activation enthalpies for the dissociation of the same two complexes and the Gutmann donor number of the solvent.

Rorabacher and co-workers[17] have summarized the very considerable problems associated with the measurement of the water-exchange parameters at [Cu(H$_2$O)$_6$]$^{2+}$ and have used the kinetics of this ion reacting with NH$_3$ to form the monoammine complex [Cu(NH$_3$)(H$_2$O)$_5$]$^{2+}$ to estimate them. For the formation and dissociation of the monoammine complex the kinetic parameters are, respectively, k (25°C) $= 2.3 \times 10^8$ dm^3 mol^{-1} s^{-1} and 2.0×10^4 s^{-1}; $\Delta H^{\ddagger} = 4.5$ and 9.5 kcal mol^{-1}; and $\Delta S^{\ddagger} = -5$ and -7 cal K^{-1} mol^{-1}. On the basis of these data and evidence that the reaction proceeds by a dissociative mechanism, the "best" kinetic parameters for the inner-sphere solvent exchange at the aquo-copper(II) ion are calculated to be k_{ex} (25°C) $= 2.0 \times 10^9$ s^{-1}, $\Delta H^{\ddagger}_{ex} = 4.5$ kcal mol^{-1}, and $\Delta S^{\ddagger} = -1$ cal K^{-1} mol^{-1}.

Lincoln[18,19] has reported the rate of ligand exchange at [ZnL$_4$]$^{2+}$ in CD$_2$Cl$_2$, where L $=$ tetramethylthiourea and hexamethylphosphoramide.

6.1.3. Ions of Valency 3 and Higher

High-pressure proton nmr has been used[20] to determine $\Delta V^{\ddagger}$ for ligand exchange at the ions [AlL$_6$]$^{3+}$ and [GaL$_6$]$^{3+}$ in deuterionitromethane solution. For aluminum, $\Delta V^{\ddagger} = +15.6$ (L $=$ DMSO, 358.5 K) and $+13.7$ cm^3 mol^{-1} (L $=$ DMF, 354.5 K), while for gallium, $\Delta V^{\ddagger} = +13.1$ (L $=$ DMSO, 334.6

K) and + 7.9 cm^3 mol^{-1} (L = DMF, 313.8 K). These results, which are consistent with dissociative activation in all cases, were complemented by values of $\Delta H^{\ddagger}$ (88.3 and 85.1 kJ mol^{-1}) and $\Delta S^{\ddagger}$ (28.4 and 45.1 J K^{-1} mol^{-1}) for DMF exchange at [Al(DMF)$_6$]$^{3+}$ and [Ga(DMF)$_6$]$^{3+}$, respectively. Values for DMSO exchange at [Al(DMSO)$_6$]$^{3+}$ and [Ga(DMSO)$_6$]$^{3+}$ in nitromethane are$^{(21)}$ $\Delta H^{\ddagger}$ = 82.6 and 72.5 kJ mol^{-1} and $\Delta S^{\ddagger}$ = 22.3 and 3.5 J K^{-1} mol^{-1}, respectively. The authors are unable to decide between I_d and D mechanisms for these systems.

The rate of trimethyl phosphate (tmp) exchange on [Sc(tmp)$_6$]$^{3+}$ in CD$_3$CN and sym-C$_2$H$_2$Cl$_4$ is$^{(22)}$ virtually independent of tmp concentration (k at 300 K = 65.7 s^{-1} per tmp molecule, $\Delta H^{\ddagger}$ = 29.8 kJ mol^{-1} and $\Delta S^{\ddagger}$ = -111 J K^{-1} mol^{-1}), but in deuterionitromethane the exchange process obeys the rate law as in:

$$\text{rate} = 6k'[\text{Sc(tmp)}_6^{3+}][\text{tmp}] \tag{1}$$

where k' (at 300 K) = 51.3 dm^3 mol^{-1} s^{-1}, $\Delta H^{\ddagger}$ = 26.0 kJ mol^{-1}, and $\Delta S^{\ddagger}$ = -126 J K^{-1} mol^{-1}. The authors favor a D mechanism in the former case but are unable to decide unequivocally between an I_d and an A mechanism for the latter (while, however, preferring an A); they keep an even more open mind in the case of the chlorohydrocarbon as solvent in view of the very limited concentration range available for study here.

The exchange rate of tetramethylurea (tmu) on the ion [Sc(tmu)$_6$]$^{3+}$ is$^{(23)}$ independent of the concentration of free tmu but the kinetic parameters are quite strongly dependent on the nature of the solvent. Thus, in CD$_3$NO$_2$, k (300 K) = 0.26 s^{-1}, $\Delta H^{\ddagger}$ = 91.2 kJ mol^{-1}, and $\Delta S^{\ddagger}$ = 47.8 J K^{-1} mol^{-1}, but in CD$_3$CN the values are 1.08 s^{-1}, 68.6 kJ mol^{-1} and -15.7 J K^{-1} mol^{-1}, respectively. A D mechanism is favored in both cases. In contrast, a two-term rate law was observed$^{(23)}$ for ligand exchange on [Sc(dma)$_6$]$^{3+}$ (dma = N,N'-dimethylacetamide) in the same two solvents:

$$\text{rate} = 6(k_1 + k_2[\text{dma}])[\text{Sc(dma)}_6^{3+}] \tag{2}$$

with k_1 = 4.6 s^{-1}, $\Delta H_1^{\ddagger}$ = 30.3 kJ mol^{-1}, $\Delta S_1^{\ddagger}$ = -132 J K^{-1} mol^{-1}, k_2 = 112 dm^3 mol^{-1} s^{-1}, $\Delta H_2^{\ddagger}$ = 26.0 kJ mol^{-1}, and $\Delta S_2^{\ddagger}$ = -119 J K^{-1} mol^{-1} for the nitromethane system and similar parameters for acetonitrile. D and A mechanisms, respectively, are associated with the two rate terms. A major determinant of the mechanism of exchange at Sc(III) is felt to be steric crowding, including that between first and second solvation spheres, and comparing the rate data available on the four tervalent cations Sc^{3+}, Al^{3+}, Ga^{3+}, and In^{3+}, Pisaniello and Lincoln$^{(23)}$ conclude that at least three factors control the mechanism of ligand exchange at these octahedral species: the ionic radius of M^{3+}, the size of the ligand, and the nature of the environment external to the first coordination sphere.

In the pH range 8–9 the kinetics[24] of the reaction between WO_4^{2-} and *cis*-$[Co(en)_2(H_2O)_2]^{3+}$ [corresponding to substitution at W(VI)] are consistent with a two-path mechanism:

$$[Co(en)_2(OH)_2]^+ + HWO_4^- \rightarrow [Co(en)_2(OH)(WO_4)] + H_2O$$
$$(k_1 = 3.2 \times 10^7 \text{ dm}^3 \text{ mol}^{-1} \text{ s}^{-1}) \qquad (3)$$

$$[Co(en)_2(OH)(H_2O)]^{2+} + HWO_4^- \rightarrow [Co(en)_2(OH)(WO_4)] + H_2O + H^+$$
$$(k_2 = 1.03 \times 10^7 \text{ dm}^3 \text{ mol}^{-1} \text{ s}^{-1}) \qquad (4)$$

These results complement previous data on the exchange of H_2O ($k = 0.44$ s^{-1}) and OH^- ($k = 273$ dm^3 mol^{-1} s^{-1}) at WO_4^{2-} and indicate a rate constant of about 10^7 dm^3 mol^{-1} s^{-1} for substitution at HWO_4^-.

The exchange rate of the equatorial *N*-formylpyrrolidine (fpr) molecules in the dioxouranium ion $[UO_2(fpr)_5]^{2+}$ in CD_2Cl_2 solution is[25] a two-term one:

$$\text{rate} = 5(k_1 + k_2[\text{fpr}])[UO_2(\text{fpr})_5^{2+}] \qquad (5)$$

with k_1 (220 K) $= 42.6$ s^{-1} and k_2 (220 K) $= 369$ dm^3 mol^{-1} s^{-1}. The activation enthalpies and entropies associated with the two rate constants are, respectively, 30.7 and 30.5 kJ mol^{-1}, and -71.6 and -54.9 J K^{-1} mol^{-1}. As with other exchange studies with UO_2^{2+}, it is not possible to assign an unequivocal mechanism. Lincoln[26] has summarized his nmr work with these species and has demonstrated the existence of a linear free energy relationship between $\Delta H^{\ddagger}$ and $\Delta S^{\ddagger}$ for $[UO_2L_5]^{2+}$. Unfortunately, perhaps, there is no obvious connection between the position of a given $[UO_2L_5]^{2+}$ species on the LFER line and the Gutmann donor number of the ligand L.

The rate parameters for the exchange of coordinated water molecules in the equatorial positions of UO_2^{2+} in acetone-d$_6$ are[27] $k(-70°C) = 2.99 \times 10^2$ s^{-1}, $\Delta H^{\ddagger} = 9.9$ kcal mol^{-1}, and $\Delta S^{\ddagger} = 2.1$ cal K^{-1} mol^{-1}.

6.2. Complex Formation Involving Unsubstituted Metal Ions and Multidentate Ligands

6.2.1. Univalent Ions

Rate and equilibrium data for the formation of complexes between the alkali metals and the cryptands (2,2,1) and (2$_B$,2,2) in methanol have been re-

1

ported.$^{(28,29)}$ [Structure 1 shows these ligands, with $m = 1$ and $n = 0$ for $(2,2,1)$, and $m = n = 1$ for $(2,2,2)$; $(2_B,2,2)$ is $(2,2,2)$ with a benzene ring between the two oxygens in one of the bridges.] As with previous work on this type of system, the variation in equilibrium constant is reflected primarily as a variation in the dissociation rate constant k_d, the formation rate constant k_f remaining one or two orders of magnitude below the diffusion-controlled value (which, in these cases, is around 3.5×10^9 dm^3 mol^{-1} s^{-1}). The introduction of the benzene ring into $(2,2,2)$ influences both k_f and k_d for the three heavier members of the series but only k_f for Li$^+$ and Na$^+$ (Table 6.2). K$^+$ fits most effectively into the cavity of $(2_B,2,2)$, while with the smaller cryptand $(2,2,1)$ the preference is for Na$^+$.

The kinetics of dissociation of the thallium cryptates $(2,2,2)$ Tl$^+$ and $(2,2,1)$ Tl$^+$ in water and 90:10 methanol–water have been compared$^{(30)}$ with the previously determined values for the potassium complexes; both direct and acid-catalyzed dissociation rate parameters are reported.

6.2.2. Bivalent Ions

The temperature-jump technique has been used$^{(31)}$ to investigate the interaction between nickel(II) and four adenosine monophosphates (2′-AMP, 3′-AMP, 5′-AMP, and 5′-dAMP) in water. The results are interpreted in terms of a mechanism involving complexation of both protonated and unprotonated ligands (with rate constants similar to those found with the simple phosphates) followed by the formation of a "back-bond" to the adenine moiety. The normal I_d mechanism is thought$^{(32)}$ to operate in the reaction of Ni$_{aq}^{2+}$ with 5,5′-dimethyl-2,2′-bipyridine, but with the 5,5′-dicarboxylate of bipy the rate-determining step is felt to be the chelate-ring closure. Rate-limiting chelate-ring closure is also postulated$^{(33)}$ for the reaction of Co$_{aq}^{2+}$ with malonate, malate, and glycollate and$^{(34)}$ (in contrast to the corresponding reactions in water) for the chelation of nickel(II) with malonic acid, its methyl and n-butyl derivatives, and cyclopropane-1,1-dicarboxylic acid in dioxan–water (25:75 w/w). The rate of formation

Table 6.2. Rate and Equilibrium Parameters for the Formation of Alkali Metal Cryptates in Methanol at 25°C[28,29]

M^+	Cryptand	$10^{-7}k_f$, $dm^3\,mol^{-1}\,s^{-1}$	$\Delta H_f^{\ddagger}$, $kJ\,mol^{-1}$	$\Delta S_f^{\ddagger}$, $J\,K^{-1}\,mol^{-1}$	k_d, s^{-1}	$\Delta H_d^{\ddagger}$, $kJ\,mol^{-1}$	$\Delta S_d^{\ddagger}$, $J\,K^{-1}\,mol^{-1}$	$\log K_s$
Li^+	(2,2,1)	1.88	13.3	−61	78.4	23.8	−129	5.38
	$(2_B,2,2)$	3.3	—	—	2.1×10^5	—	—	2.19
Na^+	(2,2,1)	8.74	15.3	−42	0.0196	64.6	−61	9.65
	$(2_B,2,2)$	8.78	—	—	2.78	55.1	−52	7.50
K^+	(2,2,1)	33.6	10.0	−48	0.969	70.0	−10	8.54
	$(2_B,2,2)$	25.7	—	—	0.158	76.8	−3	9.21
Rb^+	(2,2,1)	30.2	−0.1	−83	60.0	56.3	−22	6.7
	$(2_B,2,2)$	31.5	—	—	20.4	70.1	+7	7.19

of the nickel(II) mono complex with the diethyldithiocarbamate ion ($k = 3 \times 10^3$ dm^3 mol^{-1} s^{-1}) is consistent[35] with an I_d mechanism.

Although the rate constants for the formation of several nickel(II)–diarylthiocarbazone complexes are[36] in general agreement with those of other nickel ligand substitution reactions (while, however, showing an unusually large variation for ligands of the same charge type), the activation enthalpies and entropies (Table 6.3) are considerably different. The very low values of $\Delta H^{\ddagger}$ seem to suggest that the formation of the 1:1 chelate is unusually complex and may involve a preassociation step with a ΔH of -8 to -10 kcal mol^{-1}, which would largely compensate for the expected $\Delta H^{\ddagger}$ of 13–15 kcal mol^{-1}.

The kinetics of the reaction of cobalt(II) and nickel(II) with salicylic acid, and of the latter metal with 5-chloro-, 5-sulfo-, and 5-nitrosalicylic acids (H$_2$L) have been interpreted[37,38] in terms of there being two reactive metal species, the normal ion M_{aq}^{2+} and the deprotonated species MOH$^+$. While MOH$^+$ reacts with rates which are consistent with the I_d mechanism (and the bound OH$^-$ ligand produces no remarkable labilization of the metal-bound waters), the unhydrolyzed cations show rates which are two to three orders of magnitude below the normal values. This is thought to be attributable to the need to break the internal hydrogen bond in the active form of the ligand, HL$^-$.

Analysis of the kinetics of the reaction of nickel(II) with bipy and 4,4'-dimethylbipy in aqueous micellar solutions of sodium dodecylsulfate suggests[39] that complex formation occurs in the region of the micelle surface by the normal I_d mechanism. Thus, the considerable rate enhancement in the presence of micelles may be rationalized quantitatively in terms of the partitioning of the ligand between the micelle surface and the aqueous solution regions. A method for the computation of local concentrations in the vicinity of charged surfaces has been used[40] to analyze the kinetics of the formation of [Ni(pada)]$^{2+}$ in the same micellar solutions.

Table 6.3. Rate Constants (298 K) and Activation Parameters for [NiDz]$^+$ Formation^a

Aryl groups in Dz$^-$	k_f, dm^3 mol^{-1} s^{-1}	$\Delta H^{\ddagger}$, kcal mol^{-1}	$\Delta S^{\ddagger}$, cal K^{-1} mol^{-1}
Phenyl	6.1×10^3	4.4	-27
p-Chlorophenyl	1.2×10^4	4.0	-26
p-Bromophenyl	5.3×10^4	2.8	-27
p-Iodophenyl	3.5×10^5	0.0	-33
p-Tolyl	4.7×10^5	0.6	-31
o-Tolyl	1.0×10^4	4.8	-24

^{a} See Ref. 36 and text.

The activation enthalpies and entropies for the formation of the nickel(II) complexes of bipy, phen and terpy in water–*t*-butanol mixtures pass through distinctive minima[14] (as they do with the complexes of 4-phenylpyridine and isoquinoline). Comparison of kinetic and thermodynamic results for the reaction of Ni(II) with bipy in aqueous methanol shows[41] that the small variation of $\Delta H^{\ddagger}$ with solvent composition represents the difference between large destabilizations of both the nickel(II) (initial state) and the transition state as the proportion of methanol increases.

Kemp, Moore, and Quick[42] have used a rapid-scanning spectrometer to characterize the transient intermediates in the dissociation in water of nickel(II) polyamine complexes. While they were unable to give a complete analysis for $[Ni(trien)(OH_2)_2]^{2+}$ because of the complexity of the spectra involved and the fact that the consecutive stages of dechelation are not significantly separated along the time axis, they could confirm a previous suggestion that the dissociation of $[Ni(dien)(OH_2)_3]^{2+}$ occurred in two stages:

$$[Ni(dien)]^{2+} \xrightarrow{k_1} [Ni(Hdien)]^{3+} \xrightarrow{k_2} Ni_{aq}^{2+} \tag{6}$$

They determined values of 4.7 s^{-1} for k_1 (298 K) and 0.83 s^{-1} for k_2 (for a $[Ni(dien)]^{2+}$ concentration of 0.025 mol dm^{-3} and [HCl] = 0.5 mol dm^{-3}) and found that the spectrum of the intermediate resembled that of $[Ni(en)]^{2+}$. This supports the view that a terminal primary NH_2–Ni bond is broken first rather than the central secondary NH–Ni bond. In the dissociation of $[Ni(dien)_2]^{2+}$, it was established that one ligand is completely removed in two stages before the second starts to disengage.

The kinetics of dissociation is aqueous acid of nickel(II) complexes of a series of 14-17-membered macrocycles containing an O_2N_2-donor set of atoms have been reported.[43,44] The structure of the macrocycle was found to influence strongly the dissociation rate constants, which followed a pattern similar to that of the thermodynamic stabilities (obtained in 95% methanol), the greatest stability occurring with a ring size of 16. Rates of incorporation of Ni(II) into cyclam and other N_4 macrocycles in acetonitrile have also been reported.[45]

Bain-Ackerman and Lavallee[46] have measured the rate parameters for the entry of five bivalent metal ions into *N*-methyltetraphenylporphyrin in dmf solution. All reactions follow a second-order rate law and with the exception of Mn(II), the order of the porphyrin metalation rate constants (Cu(II) > Zn(II) > Co(II) > Ni(II)) coincides with that of solvent exchange at the metal ions. The values of k for this deformed porphyrin (Table 6.4) are all larger than for the corresponding metal ion reacting with planar porphyrins. The authors favor a mechanism which involves solvent dissociation from the metal ion as an important rate-determining factor but they also point out that porphyrin deformation

Table 6.4. Rate Constants (25°C) and Activation Parameters for Complexation with Tetrakis (4-N-methylTPP) in DMF[46]

M	k_f, dm^3 mol^{-1} s^{-1}	$\Delta H^{\ddagger}$, kcal mol^{-1}	$\Delta S^{\ddagger}$, cal K^{-1} mol^{-1}
Cu(II)	289	16.9	9.7
Zn(II)	10.4	14.2	-6.7
Co(II)	0.68	20.4	8.6
Mn(II)	0.010	21.6	4.5
Ni(II)	0.0003	21.4	-3.2

is important. The breaking of the N–H bond during complex formation is of only minor importance in determining the overall rate.

The kinetics of zinc incorporation into cationic[47] and anionic[48] porphyrins in water, and of Mn(II), Co(II), and Ni(II) incorporation into *meso*-tetraphenylporphine in dmf[49] in the presence of acetate have also been reported.

The kinetics and mechanism of the decomposition of the square-planar nickel(II)–tetraglycinamide complex have been presented[50]: the values of the pseudo-first-order rate constants for the solvent and acid dissociation increase from 8×10^{-5} s^{-1} at pH 10.7 to 490 s^{-1} at pH 1.08.

6.2.3. Ions of Valency 3 and Higher

The kinetics of complex formation between Al_{aq}^{3+} and ferron (8-hydroxy-7-iodo-5-quinoline) have been studied[51] by the stopped-flow method. A complex picture involving at least three transients is revealed, and it is thought that the major pathway in the formation of the mono complex involves $Al(OH)_2^+$; assuming a normal I_d mechanism, the rate constant for water exchange at this species is 4.0×10^7 s^{-1}.

The mono complex of Ga(III) with 5-nitrosalicylate in water is formed[52] in two pairs of "proton-ambiguous" paths. Attributing the observed rate exclusively to the two reactions involving the monoprotonated ligand on the one hand and Ga^{3+} and $Ga(OH)^{2+}$ on the other leads to upper limits for the rate constants of 4.1×10^2 and 6.4×10^3 dm^3 mol^{-1} s^{-1}, respectively. Comparison with other results suggests an associative mechanism.

Further evidence is also provided[53] for an associative mechanism in the formation of complexes at vanadium(III). Rate constants are given for the reaction of V^{3+} with HL^- and H_3L^+ (7.0×10^3 and 3.3 dm^3 mol^{-1} s^{-1}, respectively), where the ligand is *p*-aminosalicylic acid (H_2L), but the proton ambiguity precludes values for reaction with the nonpolar H_2L and the zwitterionic $H_2L^{\pm}$.

An associative character to complexation of $[Fe(H_2O)_5OH]^{2+}$ by a series of six hydroxamic acids (Structure **2**) is also suggested by Monzyk and Crumbliss[54]

2

as a result of a recent stopped-flow investigation. Interestingly, these are some of the few systems involving Fe^{3+}_{aq} reacting with weak acid ligands in which an unambiguous assignment of the pH dependence is possible (two pathways involving Fe^{3+} and $FeOH^{2+}$, respectively, and the monoprotonated ligand). The results are shown in Table 6.5.

Rate constants have been determined[55] for the reactions of vanadyl ion with 4,4,4-trifluoro-1-(2-thienyl)butane-1,3-dione (Htftbd), 1,1,1-trifluoropentane-2,4-dione (Htfpd), and pentane-2,4-dione (Hpd) to form the mono complexes. The data are consistent, in the first two cases, with a mechanism involving VO^{2+}_{aq} and both the undissociated enol tautomer ($k = 7.6$ and 3.5 dm^3 mol^{-1} s^{-1}, respectively) and the enolate ion (3.6×10^3 and 2×10^2 dm^3 mol^{-1} s^{-1}); in the other case, the vanadyl ion reacts with the keto tautomer, with a rate constant of 4.4 dm^3 mol^{-1} s^{-1}. Rate constants for the reaction of VO^{2+}_{aq} with the HL$^-$ forms of vanillomandelic, mandelic and thiolactic acids are,[56] respectively, 1.13×10^3, 1.09×10^3 and 6.76×10^2 dm^3 mol^{-1} s^{-1}; those determined for the reaction of the same ligands with the monodeprotonated form of vanadyl (VOOH$^+$) are 4.17×10^4, 5.86×10^4, and 7.93×10^4 dm^3 mol^{-1} s^{-1}, respectively. All these values (except those relating to the keto and enol tautomers of the β-diketones, which are lower, and those involving VOOH$^+$, which are higher) are consistent with the normal dissociative mechanism for vanadyl complex formation.

6.3. The Effects of Bound Ligands

6.3.1. Reactions in Water

The ultrasonic absorption of solutions of the EDTA complexes of Ca(II), Sr(II), Ba(II), and Co(II) has been ascribed[57] to the configurational change of the complex in which the pentacoordinated structure is converted (with loss of the final water molecule from the inner coordination sphere of the metal) into the hexacoordinated structure. The values of k_f (3.6×10^7, 9.2×10^7, 1.45×10^8, and 9×10^6 s^{-1}, respectively) are proportional to, but about one order of magnitude smaller than, the water-exchange rate constants at the unsubstituted

Table 6.5. *Rate Constants (25°C) and Activation Parameters for the Reaction of Fe^{3+} and $FeOH^{2+}$ with Hydroxamic Acids $R_1CO.N(OH)R_2$*[54]

R_1	R_2	Fe^{3+}			FeOH^{2+}		
		k_f, dm^3 mol^{-1} s^{-1}	$\Delta H_f^{\ddagger}$, kcal mol^{-1}	$\Delta S_f^{\ddagger}$, cal K^{-1} mol^{-1}	$10^{-3}k_f$, dm^3 mol^{-1} s^{-1}	$\Delta H_f^{\ddagger}$, kcal mol^{-1}	$\Delta S_f^{\ddagger}$, cal K^{-1} mol^{-1}
CH$_3$	H	1.2	9.2	−24	2.0	4.8	−27
C$_6$H$_5$	H	4.4	14.5	−7	4.3	8.6	−13
CH$_3$	C$_6$H$_5$	2.1	15.3	−6	1.3	6.9	−21
C$_6$H$_5$	C$_6$H$_5$	1	14.5	−9	2.3	7.7	−18
CH$_3$	CH$_3$	2.1	11.9	−17	1.2	7.5	−19
C$_6$H$_5$	CH$_3$	1.4	16.5	−2	0.67	6.6	−23

metal ions, suggesting that water substitution is the rate-determining step in the hexacoordinated-complex formation reaction.

The kinetics of dioxygen uptake in the axial positions of a variety of cobalt(II) complexes containing 12-, 13-, and 14-membered fully saturated macrocyclic tetra-amines have been reported.[58] The presence of the macrocyclic ligand reduces the rate constant by two orders of magnitude compared with linear tetra-amine systems. Axial ligation by the macrocyclic structures, however, significantly promotes[59] the rate of O_2 uptake by a cobalt(II) ion surrounded by a saturated tetra-amine.

The kinetics of the reaction of the nickel(II) tridentate Schiff-base complex

3

3 (L′Ni) with histidine, 3-methylhistidine, and histamine (HL) have been shown[60] to be consistent with the mechanism in equation (7):

$$L'Ni + HL \underset{k_{21}}{\overset{k_{12}}{\rightleftharpoons}} L'Ni-LH \underset{k_{32}[H^+]}{\overset{k_{23}}{\rightleftharpoons}} L'Ni-L \underset{k_{53}}{\overset{k_{35}}{\rightleftharpoons}} L'Ni=L \qquad (7)$$

Values for the rate constants (25°C) are $k_{12} = 3.6 \times 10^3$, 4×10^3, and 1.4×10^3 dm^3 mol^{-1} s^{-1}; and $k_{53} = 0.35$, 0.42, and 1.1 s^{-1}, respectively.

Further evidence has been obtained[61] for the enhancement of the rate of ternary complex formation between octahedral [L′Ni(II)] and L (where L and L′ are both aromatic) through the unusually large outer-sphere association constants attributable to the "stacking" interactions between the incoming and bound ligands. On the other hand, the substitution of the axial water molecule in the five-coordinate aqua[tris{2-(dimethylamino)ethyl}amino]copper(II) by NCO^-, Cl^-, or Br^- is[62] unusually slow for complex formation at Cu(II)—a similar result to that reported previously for the reaction of this complex with NCS^- and N_3^-.

The oxidation of glycine and several other amino acids and carboxylic acids by silver(II) involves the formation of a complex between the substrate and Ag(II). The rate constants for complexation with several substituted Ag(II) species have been reported[63]; they range from 10^6 to 10^8 dm^3 mol^{-1} s^{-1}.

The kinetic parameters for water exchange with the iron(III) complexes of the water-soluble porphyrins *meso*-tetrakis(*N*-methyl-4-pyridyl)porphine (TMpyP) and *meso*-tetrakis(*p*-sulfonatophenyl)porphine (TPPS) have been recorded[64]; they are, respectively, k_1 (per H_2O) at 25°C $= 7.8 \times 10^5 s^{-1}$ and 1.4×10^7 s^{-1}; $\Delta H^{\ddagger} = 13.8$ and 13.7 kcal mol^{-1}; $\Delta S^{\ddagger} = 14.7$ and 20.2 cal K^{-1} mol^{-1}. The exchange rate is enhanced by a factor of 10^4–10^5 over unsubstituted Fe^{3+}_{aq}, $\Delta H^{\ddagger}$ decreasing from 18 to 14 kcal mol^{-1}. Presumably the extra labilization produced by TPPS compared to TMpyP reflects a greater degree of charge donation to the iron.

An interesting report has appeared[65] on the kinetics of substitution at the axial sites of the low-spin d^7 octahedral Ni(III) species.

Rate constants (25°C) and the associated activation parameters for the substitution of a water molecule in (1,4,8,11-tetraazacyclotetradecane)diaquonickel(III) ([Ni(III)cyclam(OH$_2$)$_2$]$^{3+}$) by Cl^-, Br^-, and NCS^- are, respectively, $k = 902, 200, 1160$ dm^3 mol^{-1} s^{-1}; $\Delta H^{\ddagger} = 14.2, 11.3, 11.4$ kcal mol^{-1}; $\Delta S^{\ddagger} = -2.6, 0.0, -6.0$ cal K^{-1} mol^{-1}. The authors favor an I_d mechanism.

The rates of ternary complex formation between the EDTA complexes of the lanthanides and pyridine-2-carboxylate or 8-hydroxyquinoline-5-sulfonate have been reported.[66]

Although the substitution of a single DMSO molecule or Cl^- or Br^- ion in the inner coordination sphere of the uranyl ion UO_2^{2+} has little effect on the rate constant at $-70°C$ for the water-exchange process in the equatorial positions (Table 6.6), the values of $\Delta H^{\ddagger}$ and $\Delta S^{\ddagger}$ are considerably reduced,[27] with the result that the 25°C exchange rate constants are estimated to be an order of magnitude lower in the substituted than in the unsubstituted ion.

Table 6.6. Kinetic Parameters for the Exchange of H$_2$O in Uranyl Complexes[27]

Complex	$\Delta H^{\ddagger}$, kcal mol^{-1}	$\Delta S^{\ddagger}$, cal K^{-1} mol^{-1}	$10^{-2}k_{ex}$, s^{-1} ($-70°C$)	$10^{-4}k_{ex}$, s^{-1} (25°C)
[UO$_2$(H$_2$O)$_4$]$^{2+}$	9.9	2.1	2.99	98.0
[UO$_2$(H$_2$O)$_3$(DMSO)]$^{2+}$	5.9	-16.2	4.55	8.42
[UO$_2$(H$_2$O)$_3$Cl]$^+$	6.1	-17.6	1.63	2.97
[UO$_2$(H$_2$O)$_3$Br]$^+$	6.7	-13.5	2.65	8.49

6.3.2. Reactions in Nonaqueous Solvents

The results of a stopped-flow study of the substitution of axial acetonitrile by imidazole and N-methylimidazole in two complexes of the type $[FeL(AN)_2]^{2+}$ (where L is a 14-membered tetra-aza macrocyclic ligand) in acetonitrile (AN) and acetone are consistent with[67] a dissociative (D) mechanism. The rate-determining step in the reaction between ammineaquonickel(II) complexes and the tridentate ligand 1-(2-hydroxyphenyl)-3,5-diphenylformazan in a 50% (by weight) ethanol–water mixture appears[68] to be ring closure.

Following a temperature-jump study of the tetrahedral ion $[CoCl_4]^{2-}$ solubilized in a reversed micelle system, it is suggested[69] that the rate-determining step in the tetrahedral–octahedral conversion (8) is the final step (9):

$$[CoCl_4]^{2-} + 4H_2O \rightleftharpoons [CoCl_2(H_2O)_4] + 2Cl^- \tag{8}$$

$$[CoCl_3(H_2O)_3]^- + H_2O \rightleftharpoons [CoCl_2(H_2O)_4] + Cl^- \tag{9}$$

The backward bimolecular rate constant for this step is given by $k = k_0 (1 + A[H_2O])$, with $k_0 = 2.5 \times 10^3$ dm^3 mol^{-1} s^{-1} and $A = 1.8$ (at 15°C).

The coordinated dithiocarbamate ion (dtc$^-$) in the mono complex of nickel(II) (in DMSO) exerts[35] a significant labilizing effect on the remaining DMSO molecules in the inner coordination sphere: the rate constant for dtc$^-$ + $[Ni(dtc)]^+$ is $\sim 4 \times 10^5$ dm^3 mol^{-1} s^{-1}, compared with a value of 3×10^3 dm^3 mol^{-1} s^{-1} for dtc$^-$ + Ni^{2+}.

The high-pressure nmr technique has been used[70] to confirm the associative(I_a)–dissociative(D) crossover for the ligand exchange reaction (10):

$$[MX_5L] + L^* \rightleftharpoons [MX_5L^*] + L \tag{10}$$

where M = Nb or Ta, X = Cl or Br, and L is a neutral Lewis base in CH_2Cl_2 or $CHCl_3$ as solvent. Positive values of $\Delta V^{\ddagger}$ (ranging from 15 to 31 cm^3 mol^{-1}) were found for the D reactions [when L = Me_2O, MeCN, Me_3CCN, $(MeO)Cl_2PO$, and $(Me_2N)_3PS$], and negative values (ranging from -10 to -20 cm^3 mol^{-1}) for the I_a reactions (with L = Me_2S, Me_2Se, and Me_2Te).

Acetonitrile solvent exchange at the five-coordinate complex ions $[ML(AN)]^{2+}$ (with M = Co or Ni and L = 1,4,8,11-tetramethyl-1,4,8,11-tetraazacyclotetradecane) appears[71] to follow an I_a mechanism, with large, negative entropies of activation (-69.5 and -47.8 J K^{-1} mol^{-1}, respectively) and low enthalpies of activation (19.5 and 20.3 kJ mol^{-1}). Interestingly, the nickel complexes $[NiL(AN)]^{2+}$ and $[NiL(OH_2)]^{2+}$ are significantly more labile than the cobalt(II) analogs.

The kinetics of substitution of a series of square-planar Ni(II) complexes by dithiocarbamate ion in acetone are[72] generally second order. The variations in rate do not, contrary to earlier reports, appear to reflect the ability of the metal to expand its coordination shell, and it is proposed that there is a significant contribution from back-bonding in these reactions.

The kinetics of ligand substitution at copper(II),

$$[CuL_2] + 2L' \rightleftharpoons [CuL'_2] + 2L \tag{11}$$

(with $L = N$-*tert*-butyl-salicylaldimine and $L' = N$-ethylsalicylaldimine) in various alcohols (ROH) follow[73] a two-term rate law,

$$\text{rate} = (k_0[\text{ROH}] + k_1[\text{L}'])[\text{CuL}_2] \tag{12}$$

although for most alcohols the ligand path (latter term) cannot compete with the solvent path (former term). (In the presence of water, there is an additional term $k'_0 [\text{H}_2\text{O}][\text{CuL}_2]$.) The authors conclude that the mechanism of the solvent path consists of the following steps: (i) the rapid addition of ROH ($[\text{CuL}_2] + \text{ROH} \rightleftharpoons [\text{CuL}_2 \cdot \text{ROH}]$), (ii) fast proton transfer from bound ROH to the phenolic oxygen of the coordinated *tert*-butyl ligand, (iii) rearrangement in the coordination sphere with rupture of the Cu–OH (ligand) bond, and (iv) the rapid, stepwise substitution of the two *tert*-butyl ligands by the ethyl ligands. The rearrangement step (iii) is presumed to be rate determining.

Part 3
Reactions of Organometallic Compounds

Chapter 7

Metal–Alkyl Bond Fission and Formation

7.1. Introduction

This chapter deals chiefly with the breaking and formation of metal–alkyl bonds, though some other topics are included. Section 7.2 reviews the breaking of metal–alkyl bonds, homolytically and heterolytically. Section 7.3 deals with the cleavage of other metal–carbon links and processes involving metal–metal bond fission. Miscellaneous examples of metal–alkyl bond formation not discussed under later headings are included in Section 7.4, while Section 7.5 reviews transalkylation between metals. Other examples of the making and breaking of metal carbon bonds will appear, of course, later under customary headings such as substitution, oxidative addition, and migration in Chapter 8.

During the period under review, two very useful textbooks have appeared, by Kochi[1] and by Collman and Hegedus.[2] The reviewer has tried the latter in graduate teaching with good effect.

7.2. Metal–Alkyl Bond Fission

A review of metal–alkyl bond cleavage (and other topics) has been made by Kochi.[3] Halpern[4] has discussed attempts to obtain both kinetic and thermodynamic data for M–C σ-bond fission. A particular point of interest is the estimation of activation energies for the reverse reaction, equation (1),

$$M^{\cdot} + {}^{\cdot}C \rightarrow M{-}C \tag{1}$$

211

to see how close they are to zero. One such attempt[4] involves the reaction of $[PhCH_2Mn(CO)_5]$ with $[HMn(CO_5)]$ to give $PhCH_3$ and $[Mn_2(CO)_{10}]$, which is first order in the first and zero order in the second reactant. If a mechanism is assumed [equations (2)–(4)]:

$$[PhCH_2Mn(CO)_5] \rightarrow PhCH_2^{\cdot} + [Mn(CO)_5]^{\cdot} \qquad (2)$$

$$PhCH_2^{\cdot} + [HMn(CO)_5] \rightarrow PhCH_3 + [Mn(CO)_5]^{\cdot} \qquad (3)$$

$$2[Mn(CO)_5]^{\cdot} \rightarrow [Mn_2(CO)_{10}] \qquad (4)$$

the first step being slow, then the activation parameters for the overall process can be identified with those for reaction (2); these are $\Delta H^{\ddagger} = 25$ kcal mol^{-1} and $\Delta S^{\ddagger} = 0$ cal deg^{-1} mol^{-1} at 25.0°C. As the thermodynamic bond dissociation energy for CH_3–$Mo(CO)_5$ is ~30 kcal mol^{-1}, this suggests that the activation energy for the radical combination step [the reverse of equation (2)] is close to zero. Further,[5] if the rate of reaction (5)

$$[Ph(Me)CHCo(dmgH)_2py] \rightleftharpoons [Co(dmgH)_2py]$$
$$+ PhCH{=}CH_2 + {}^1\!/_2H_2 \qquad (5)$$

in the forward direction is identified with the individual step (6),

$$[Ph(Me)CHCo(dmgH)_2py] \rightarrow Ph(Me)CH^{\cdot} + [Co(dmgH)_2py]^{\cdot} \qquad (6)$$

then $\Delta H^{\ddagger}$ for C–C bond fission at 25.0°C is 21.2 ± 0.5 kcal mol^{-1}. This implies that $\Delta H^{\ddagger}$ for the reverse reaction involving Co–C bond formation is only 0.9 ± 1.0 kcal mol^{-1} since $\Delta H°$ for reaction (5) is 22.1 ± 0.5 kcal mol^{-1}.

Breaking of the Co–C bond in vitamin B_{12} derivatives and model compounds has attracted attention. Brown[6] has followed the decomposition of $[CH_3Co(dmgH)_2OH]^-$ in alkaline solution to give methane and side products. $d[CH_4]/dt = 1.36 \times 10^{-4}$ [hydroxycomplex] $[OH^-]$ M s^{-1} at 50°C. Methane is not formed from the aquo complex nor from those containing various N- and S-bonding ligands. In the cobalt-containing product one N$=$C link in the ligand system has been hydrolyzed to an NH–C(OH) group. A heterolytic Co–C bond fission is proposed as in:

$$H{-}O{-}H + CH_3{-}Co \rightarrow HO^- + HCH_3 + Co(III) \qquad (7)$$

Grate and Schrauzer[7] have found evidence for steric destabilization of the Co–C bond in alkylcobalamins. The rate of decomposition of compounds containing secondary alkyl groups to alkenes and hydridocobalamin (which occurs spontaneously at room temperature) increases with the bulkiness of substituents at the β carbon, and moreover can be correlated with Taft's steric substituent parameter. Dealkylation occurs through a *syn* β elimination probably involving a concerted heterolytic elimination [equation (8)].

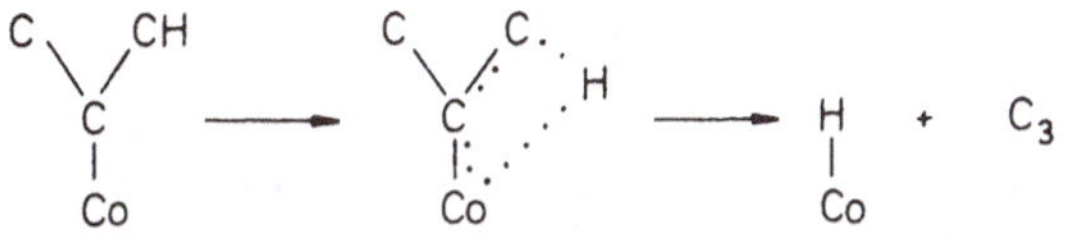

$$ (8) $$

Base-on cobalamins dealkylate considerably faster than base-off, probably owing to the bulk of the base bending the corrin ring toward the alkyl group thereby destabilizing the Co–C bond.

A π-bonded intermediate is postulated by Espenson and Wang[8] to interpret kinetic studies on the decomposition of (β-hydroxyalkyl)cobaloximes to give alkenes. If the formation of the π compex rather than its destruction is rate determining, which the authors consider is probable, k_1 has values at 25°C, of 3.09×10^{-2} and $0.10\ M^{-1}\ s^{-1}$ for R = H and CH$_3$, respectively, while k_2 (R = CH$_3$) lies between 0.7 and 10 $M^{-1}\ s^{-1}$. In Scheme 1, H$^+$ acts as electrophilic

Scheme 1

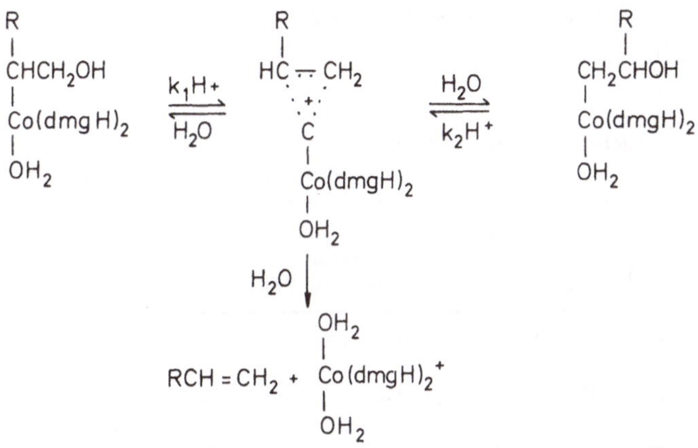

reagent for the σ-alkyl to π-alkene rearrangements, which is illustrated by the enhanced rate when the σ-donor CH$_3$ group is introduced and by a much slower reaction when the (dmgH)$_2$ groups are replaced by (dmgBF$_2$)$_2$.

[(H$_2$O)$_5$CrR]$^{2+}$ complexes in some ways resemble B$_{12}$ compounds. Spreer's group[9] have observed that the aquation of various (α-halo-organo)chromium

Table 7.1. Activation Parameters for the Rate of Disappearance of Organochromium Reactant, $[(H_2O)_5CrR]^{2+}$, during Aquation, with Organic Products[9]

R	$\Delta H^{\ddagger}$, kcal mol^{-1}	$\Delta S^{\ddagger}$, cal K^{-1} mol^{-1}	$k(25°C)$, s^{-1}	Products
CH$_2$Cl	26.1	8.6	5.6×10^{-7}	CH$_3$OH
CH$_2$Br	24.3	-2.7	2.4×10^{-6}	CH$_3$OH
CH$_2$I	25.4	-2.4	4.5×10^{-7}	CH$_3$I
CHCl$_2$	24.4	1.2	4.1×10^{-6}	CO, HCOOH
CHBr$_2$	27.3	6.3	1.4×10^{-6}	CO, HCOOH
CHI$_2$ ^{a}	27.3	0.8	8.5×10^{-8}	CO, HCOOH

^{a} This complex also shows an inverse [H$^+$]-dependent path.

complexes, $[(H_2O)_5CrCH_2(Hal)]^{2+}$ and $[(H_2O)_5CrCH(Hal)_2]^{2+}$, to $[Cr(OH_2)_6]^{3+}$ are first order in reactant. Activation parameters are given in Table 7.1.

Moving from RCo(III) and RCr(III) compounds to RCo(IV), one finds that the metal–carbon bond can break in a variety of ways. Electrochemical studies by an East–West partnership[10] on the attack of pyridine on organocobalt(IV) chelate complexes indicate three, possibly four, main decomposition routes: S_N2, involving a conventional remote side attack on the α-carbon atom; S_Ni, in which there is an initial internal transfer of R from the metal to the chelate ligand; homolytic; and possibly S_N1 (see Table 7.2).

Some observations by Baird's group[11] indicate how steric factors in the

Table 7.2. Reaction Patterns of Organocobalt (IV) Chelates $[RCo(chel)]^+$ in the Presence of Pyridine[10]

Main reaction route	Mechanism of cleavage of Co(IV)–C Bond	$[RCo(chel)]^+$
$[RCo(IV)]^+ \rightarrow R^{\bullet} + [Co(III)]^+$	Homolytic	$[EtCo(acacen)]^+$
		$[MeCo(salen)]^+$
$[RCo(IV)]^+ + Py \rightarrow RPy^+ + [Co(II)]$	S_Ni	$[EtCo(salen)]^+$
		$[EtCo(salphen)]^+$
$[RCo(IV)]^+ + Py \rightarrow RPy^+ + [Co(II)]$	S_N2	$[EtCo(damgH)_2]^+$
		$[EtCo(dpgH)_2]^+$
		$[p\text{-}O_2NC_6H_4CH_2Co(dmgH)_2]^+$
		$[p\text{-}EtOCOC_6H_4CH_2Co(dmgH)_2]^+$
		$[p\text{-}ClC_6H_4CH_2Co(dmgH)_2]^+$
		$[p\text{-}FC_6H_4CH_2Co(dmgH)_2]^+$
		$[C_6H_5CH_2Co(dmgH)_2]^+$
		$[p\text{-}MeC_6H_4CH_2Co(dmgH)_2]^+$
$[RCo(IV)]^+ \rightarrow R^+ + [Co(II)]$	$S_N1(?)$	$[p\text{-}MeOC_6H_4CH_2Co(dmgH)_2]^+$

organo group may influence whether an R–M bond can undergo S_N2 fission. Different types of intermediates are produced in the reaction of $[\eta^5\text{-}C_5H_5Fe(CO)_2R]$ and CuX_2 ($X = Cl$ or Br) to give $[\eta^5\text{-}C_5H_5Fe(CO)_2X]$ (and RX and CuX) depending on R. When R is PhCHD, X is Cl, and the reaction is run in CH_2I_2, $[\eta^5\text{-}C_5H_5Fe(CO)_2I]$ is produced as well as the chloro complex. An S_N2 step [equation (10)] is proposed, preceded by an S_E (oxidative) process, equation (9). ($[\eta^5\text{-}C_5H_5Fe(CO)_2]^•$ abstracts $I^•$ from CH_2I_2.) [The reaction scheme is completed by (11).]

$$[\eta^5\text{-}C_5H_5Fe(CO)_2R] + CuX_2 \rightarrow [\eta^5\text{-}C_5H_5Fe(CO)_2R]^+ + CuX_2^- \quad (9)$$

$$[\eta^5\text{-}C_5H_5Fe(CO)_2R]^+ + CuX_2^- \rightarrow [\eta^5\text{-}C_5H_5Fe(CO)_2]^• + RX + CuX \quad (10)$$

$$[\eta^5\text{-}C_5H_5Fe(CO)_2]^• + CuX_2 \rightarrow [\eta^5\text{-}C_5H_5Fe(CO)_2X] + CuX \quad (11)$$

Evidence for (10) being S_N2 is provided by the inversion of configuration at Ph$\overline{\text{C}}$HD. However, when R is *n*-Bu (which undergoes S_N2 reactions less readily than $PhCH_2$), no iodo product is observed. It is suggested that if R cannot readily undergo S_N2 reactions or if only a poor nucleophile is present, e.g., $[Cu(NO_3)_2(OPPh_3)_2]$, then homolytic fission occurs as in equation (12):

$$[\eta^5\text{-}C_5H_5Fe(CO)_2R]^+ \rightarrow [\eta^5\text{-}C_5H_5Fe(CO)_2]^+ + R^• \quad (12)$$

An interesting example of contrast is provided by the thermal decomposition reaction of two sets of dialkyl complexes. Yamamoto's[12] group has found that the decomposition of various $[R_2Pd(phosphine)_2]$ complexes are first order in reactant and only slightly retarded by free phosphine, suggesting that the major pathway is:

$$[P_2PdR_2] \rightarrow [P_2Pd(R)(H)(ol)] \rightarrow [P_2Pd] + RH + ol \quad (13)$$

Deuterium labeling indicates that the hydride is formed by β elimination. A minor pathway involving $[PPdR_2]$ is also proposed. Activation parameters for the evolution of gases indicate that the phosphines affect the reaction sterically

Table 7.3. Activation Parameters for the Thermolysis of trans-$[PdEt_2(PR_3)_2]$ at 300 K[12]

	PMe_2PH	PEt_3	PEt_2Ph	$PMePh_2$	$PEtPh_2$
$\Delta H^‡$, kcal mol^{-1}	25.8	25.9	26.3	26.4	25.7
$\Delta S^‡$, cal K^{-1} mol^{-1}	9.3	10.6	14.0	14.7	17.5
$\Delta G^‡$, kcal mol^{-1}	23.0	22.7	22.1	22.0	20.5

rather than electronically (Table 7.3). Platinum, in contrast, can form stable metallacyclobutanes. Whitesides' group[13] has followed the thermal decomposition of the dialkyl complex, cis-[Pt(CH$_2$CMe$_3$)$_2$(PEt$_3$)$_2$] in cyclohexane to give a platinacyclobutane and neopentane. As dissociation of a phosphine ligand occurs, as deuterium labeling shows an intramolecular transfer of hydrogen from an Me group of one neopentyl unit to a CH$_2$ group of a second, as neopentyl radicals are not intermediates, and as CH$_3$ and CH$_2$ groups of the neopentyl moieties do not interchange, the following mechanism is proposed (Scheme 2).

Scheme 2

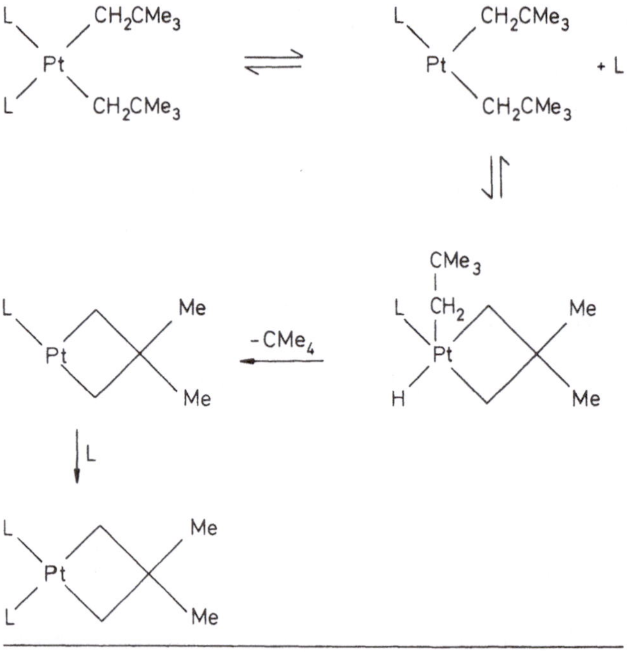

The overall activation parameters are: $E_a = 49$ kcal mol^{-1}, log $A \simeq 20$, while E_a for the initial loss of phosphine is estimated to lie between 27 and 35 kcal mol^{-1}.

7.3. *Miscellaneous Thermal Decompositions*

Several such decompositions involving fission of metal–carbon σ bonds were discussed in the last section. η^5-C$_5$H$_5$–metal fission has been investigated by Razuvaev's group.[14] The rates of thermal decomposition of various metal-

Table 7.4. Activation Parameters for the Thermal Decomposition of [Cp$_2$M], Together with Average Dissociation Energies, Average Free Energies, and Average Entropies of the M–Cp Bond$^{(14)}$

	$10^3 k$, s^{-1} (at 500°C)	E_{act}, kcal mol^{-1}	$\overline{D}$(M–Cp),^{a} kcal mol^{-1}	$\Delta\overline{G}$(M–Cp), kcal mol^{-1}	$\Delta\overline{S}$(M–Cp), cal K^{-1} mol^{-1}
Cp$_2$V	0.14	50.5	88	74	47
Cp$_2$Cr	1.34	27.3	67	55	40
Cp$_2$Mn	7.20	29.3	51	40	35
Cp$_2$Fe	0.01		72	60	41
Cp$_2$Co	3.30	23.1	64	52	41
Cp$_2$Ni	29.5	21.4	60	49	37

^{a} Slightly different values are given for $\overline{D}$(M–Cp) in a more recent reference.$^{(15)}$

locenes are dependent on the surface/volume ratio of the reaction vessel. However, manometry allows the rates and activation energies at one such ratio to be compared for several compounds, although the latter quantity cannot be obtained for ferrocene because of complexities in its kinetics (see Table 7.4).

The thermal decomposition of dimetalcarbonyls continues to be studied by Poë's group. The kinetics of the thermal decomposition of [Re$_2$(CO)$_{10}$] in decalin under oxygen are consistent with an initial, reversible homolytic fission of the Re–Re bond.$^{(16)}$ Activation enthalpies indicate that [Re(CO)$_5$]$^\cdot$ is more stable than [Mn(CO)$_5$]$^\cdot$. The enthalpy of activation for the thermal decomposition of various axially substituted complexes, [Mn$_2$(CO)$_8$L$_2$], L = phosphines and phosphites, in decalin under oxygen is also identified with homolytic fission of the Mn–Mn bonds.$^{(17)}$ There is rough inverse relationship between this enthalpy and the cone angle of L, although phosphines containing mixed groups behave as if they have a larger angle than they actually possess.

The rate of thermal decomposition$^{(18)}$ of [HCo(CO)$_4$] in solution to [CO$_2$(CO)$_8$] and H$_2$ is proportional to [Co$_2$(CO)$_8$]$^{0.5}$ [HCo(CO)$_4$]^2/[CO]^2, which leads to the suggestion that while the reaction cannot be interpreted in detail, the radical [Co(CO)$_3$(PBu$_3$)]$^\cdot$ is involved. [Co(CO)$_4$]$^\cdot$ has also been proposed as an intermediate$^{(19)}$ in the thermal decomposition of [HCo(CO)$_4$]. An analogous radical, [Co(CO)$_3$(PBu$_3$)]$^\cdot$, is postulated$^{(20)}$ in the reaction of [Co$_2$(CO)$_8$] and Bu$_3$P.

7.4. Metal–Alkyl Bond Formation

Although many pieces of work in which metal–alkyl σ bonds are formed will be discussed later, particularly in the sections on *trans*-alkylation and oxidative addition, four sets of experiments do not really fit into these categories.

Two involve formation of Hg–C bonds. Gas phase studies[21] on $MeHg^+$ have shown that the stability of its adduct, $MeHgB^+$, falls along the series of bases, B: $Et_3N > Et_2NH >> i\text{-}PrC_6H_5 > EtC_6H_5 > MeC_6H_5 > C_6H_5I > C_6H_6 > C_6H_5Br > C_6H_5Cl > C_6H_5Br$. In many ways this sequence parallels that for HB^+. A comparison has always been made on the behavior of $MeHg^+$ and H^+ toward bases in solution.[22] Carbon donor ligands show a thermodynamic preference for portonation over methylmercuration, but kinetically $MeHg^+$ is transferred more readily than H^+. A comparison of Brønsted plots for the two transfer processes suggests that less reorganizational energy is involved in the case of $MeHg^+$ because of its better bridging ability.

Cu–C bond formation has also been studied. Rate constants (second order at 25°C) have been measured[23] for the reaction of copper(II) "tetraglycine," $[Cu(II)G_4]$, with various radicals generated by pulse radiolysis:

$$k\ /\ M^{-1}\ s^{-1}$$

$$[Cu(II)G_4] + {}^{\cdot}CH_2COO^- \rightarrow [G_4Cu(III)\text{–}CH_2COO^-] \qquad 9 \times 10^6 \qquad (14)$$

$$[Cu(II)G_4] + {}^{\cdot}CH_2C(Me)_2OH \rightarrow [G_4Cu(III)\text{–}CH_2CMe_2OH] \quad 2 \times 10^7 \qquad (15)$$

$$[Cu(II)G_4] + {}^{\cdot}CH_2CH(OH)CH_3 \rightarrow [G_4Cu(III)\text{–}CH_2CH(OH)CH_3]$$
$$2.7 \times 10^9 \qquad (16)$$

and in contrast,

$$[Cu(II)G_4] + {}^{\cdot}CH_2OH \rightarrow\rightarrow [Cu(I)G_4] + CH_2O \qquad 8.6 \times 10^7 \qquad (17)$$

$$[Cu(II)G_4] + {}^{\cdot}CO_2^- \rightarrow\rightarrow [Cu(I)G_4] + CO_2 \qquad 6.5 \times 10^8 \qquad (18)$$

The decay of the first two organocopper(III) species is pH dependent but that of the third is not, its rate constant being $0.5 \pm 0.2\ s^{-1}$.

Though π-bonded intermediates are often postulated in reactions of cobalamins and model B_{12} compounds, as was done by Espenson and Wang[8] (see above), they remain elusive. Ramasami and Espenson[24] propose reactions (19) and (20) to account for the reaction of a BF_2–cobaloxime derivative and acrylonitrile,

$$[Co(I)(dmgBF_2)_2PBu_3^n] \underset{k_1}{\rightleftharpoons} [Co(I)(dmgBF_2)_2]^- + PBu_3^n \qquad (19)$$

$$[Co(I)(dmgBF_2)_2]^- + CH_2{=}CH_2CN \overset{k_2}{\rightarrow} [Co(I)(dmgBF_2)_2(CH_2{=}CH_2CN)]^-$$
$$(20)$$

where $k_1 = 10.5 \pm 1.8$ s^{-1} and $k_3/k_2 = 0.59 \pm 0.05$ at 25.0°C in methanol, 0.10 M in OH$^-$. Evidence is presented that the acrylonitrile complex is a π-bonded species. It slowly decomposes reforming the starting complex and various organic products.

7.5. Transalkylation

This section is limited to simple systems in which an alkyl group moves from one metal ion to another. Examples in which transalkylation occurs from a nonmetal to a metal can be found in other sections, particularly that on oxidative addition (Chapter 8). Kochi's review[3] mentioned earlier contains a section on transalkylation.

There have been two studies of transalkylation between cobalt centers. The rates of reversible transmethylation[25] between dimethylcobalt (III) systems have been studied by measuring rates of scrambling of Co–CH$_3$ and Co–CD$_3$ groups between compounds of the type [CpCo(phosphine) (CH$_3$)$_2$] (see Table 7.5). The reaction is first order in each reagent.

Endicott's group[26] has measured and made an extremely detailed analysis of rate and equilibrium constants for transmethylation between various cobalt N$_4$-macrocyclic systems (see Table 7.6 and accompanying structures). The reactions are first order in each reagent. Rates can vary by 10^6 (even for the "back" reactions where $\Delta G° < 0$). $\Delta G°$ is analyzed into components, intrinsic free energy barriers to transmethylation being small for cobalt corrin and large for sterically hindered neutral macrocyclic complexes. Estimates for Co–CH$_3$ bond energies are between 33 and 48 kcal mol^{-1}; the bond is stabilized by unsaturation in the N$_4$ macrocycle, but is also very sensitive to stereochemistry.

Trans-alkylation from organochromium systems has been studied by Samuels and Espenson.[27]

Table 7.5. Second-Order Rate Constants for Methyl Exchange between [(η^5-C$_5$H$_4$R)CoL(CH$_3$)$_2$] and [(η^5-C$_5$H$_4$R')CoL'(CD$_3$)(CH$_3$ or CD$_3$)] at 62°C in THF-d$_8$[25]

		k, M^{-1} s^{-1}		
L	L'	R = R' = H	R = H, R' = Me	R = R' = Me
PPh$_3$	PPh$_3$	3.6×10^{-3}	2.2×10^{-4}	8.3×10^{-6}
PPh$_3$	PMe$_3$	3.6×10^{-3}	8.6×10^{-5}	~0
PMe$_3$	PMe$_3$	~0	~0	~0

Table 7.6. Kinetic and Thermodynamic Parameters at 25°C for Transmethylation from $[MeCo[N_4]OH_2]^{2+}$ to $[Co(II)[N_4'](OH_2)_2]^{2+}$ (f and b Denote Forward and Backward Reactions, Respectively)

N_4	N_4'	k_f, M^{-1} s^{-1}	k_b, M^{-1} s^{-1}	$\Delta G^\ddagger_f$, kcal mol^{-1}	$\Delta H^\ddagger_f$, kcal mol^{-1}	$\Delta S^\ddagger_f$, cal K^{-1} mol^{-1}	K_f
Corrin	Me$_4$[14]tetraeneN$_4$	1.07	0.25	15.2	7.5	−22	4.3
Me$_2$pyo[14]trieneN$_4$	Corrin	602		11.4			44
Me$_6$[14]4,11–dieneN$_4$	Corrin	1.55		14.9			1.7 × 10^3
Corrin	(dmgH)$_2$	5.6 × 10^4	~760	8.6			59
[14]aneN$_4$	Corrin						8
(dmgH)$_2$	Me$_4$[14]tetraeneN$_4$	3.96	10.4	14.4			0.38
Me$_2$pyo[14]trieneN$_4$	(dmgH)$_2$	4.6		14.3			29
[14]aneN$_4$	(dmgH)$_2$	2.95		14.5			500
Me$_6$[14]4,11–dieneN$_4$	(dmgH)$_2$	0.059	~0.14	16.9			1.0 × 10^5
Me$_2$pyo[14]trieneN$_4$	Me$_4$[14]tetraeneN$_4$	27.3		13.2			1.1 × 10^2
[14]aneN$_4$	Me$_4$[14]tetraeneN$_4$	0.13		16.4			34
Me$_2$pyo[14]trieneN$_4$	[14]aneN$_4$	0.32		15.9			6
Me$_6$[14]4,11–dieneN$_4$	Me$_4$[14]tetraeneN$_4$	~0.01		18.1	11	−33	7 × 10^3
Me$_6$[14]4,11–dieneN$_4$	Me$_2$pyo[14]trieneN$_4$	~0.01	~3 × 10^{-4}	17.9	13	−23	40

[15]aneN₄   [14]aneN₄   Me₄[14]tetraeneN₄

Me₆[14]4,11-dieneN₄   Me₂[14]4,11-dieneN₄   Me₂pyo[14]trieneN₄

$$[\text{R–Cr}([15]\text{aneN}_4)\text{OH}_2]^{2+} + \text{Hg}^{2+} \rightarrow \text{RHg}^+ + [\text{Cr}([15]\text{aneN}_4)(\text{OH}_2)_2]^{3+} \tag{21}$$

$$[\text{R–Cr}([15]\text{aneN}_4)\text{OH}_2]^{2+} + \text{MeHg}^+ \rightarrow \text{RHgMe} + [\text{Cr}([15]\text{aneN}_4)(\text{OH}_2)_2]^{3+} \tag{22}$$

Reactions (21) and (22) are first order in each reagent and appear to proceed through an S_E2 mechanism, as in many related reactions. Increasing bulkiness in R leads to large decreases in rate constant and entropy of activation (see Table 7.7). Presumably adamantyl has to react by front-side attack, but what of cy-

Table 7.7. Rate Constants at 25°C for Reactions (21) and (22)[27]

R	k_{21}, $M^{-1}\,s^{-1}$	$\Delta H^{\ddagger}_{21}$, kJ mol^{-1}	$\Delta S^{\ddagger}_{21}$, J mol^{-1} K^{-1}	k_{22}, $M^{-1}\,s^{-1}$
CH₃	3.1×10^6			1.63×10^3
CH₂CH₃	2.53×10^3	30.2	−77	9.9
n-C₃H₇	8.21×10^1	48.8	−45	
n-C₄H₉	4.88×10^1	33.5	−100	
n-C₅H₁₁	4.33×10^1	29.7	−114	
CH₂C₆H₅	1.14×10^3	33.4	−75	5.2
i-C₃H₇	4.3×10^{-3}			
c-C₆H₁₁	1.6×10^{-3}			
1-Adamantyl	3.1×10^{-3}			

Table 7.8. Rate Constants for 25°C for Reaction (23) in Various Solvents[28]

R$_4$Sn	k, M^{-1} s^{-1}		
	CH$_3$CN	CH$_2$Cl$_2$	MeOH
Me$_4$Sn	2.4	6.6×10^{-2}	1.6
Et$_4$Sn	1.6×10^{-2}	1.7×10^{-2}	3.3×10^{-3}
n-Pr$_4$Sn	3.7×10^{-3}	9.0×10^{-4}	6.3×10^{-4}
n-Bu$_4$Sn	3.8×10^{-3}	1.4×10^{-3}	6.2×10^{-4}
n-BuSnMe$_3$	1.1	0.10	1.0
n-Bu$_2$SnMe$_2$	0.50	0.15	0.64
i-Pr$_2$SnMe$_2$	2.5	0.13	0.23
t-Bu$_2$SnMe$_2$	4.6×10^{-2}	2.6×10^{-2}	1.4×10^{-3}
i-Bu$_4$Sn	5.6×10^{-4}	3.7×10^{-4}	8.0×10^{-5}
i-Pr$_4$Sn	5×10^{-6}	3×10^{-6}	2×10^{-8}
s-Bu$_4$Sn	7×10^{-7}	8×10^{-7}	

clohexyl which is transferred in a back-side mechanism in the analogous cobaloxime system.

In contrast to the preceding reactions, *trans*-alkylation between tin and mercury in equation (23) does not appear to involve an S_E2 process.[28]

$$HgCl_2 + R_4Sn \rightarrow RHgCl + R_3SnCl \tag{23}$$

Rates of disappearance of HgCl$_2$, which are first order in each reactant, vary somewhat unaccountably with the structure of the tin compound and the polarity of the solvent (see Table 7.8). However, they are equal to the rates of disappearance of the charge transfer bands due to [R$_4$Sn$^+$HgX$_2^-$]. It is proposed that these so-called electrophilic reactions in fact involve a similar intermediate formed by electron transfer.

Chapter 8

Substitution, Oxidative-Addition–Reductive-Elimination, and Migration–Insertion Reactions

8.1. Introduction

The two books mentioned in Chapter 7 by Kochi[1] and by Collman and Hegedus[2] should be included once again in this chapter. The 18 months under survey have produced many interesting papers, so that selection has been difficult. The reviewer has found particularly fascinating the attempts to discover what individual steps occur during oxidative addition processes (namely, those processes which formally involve the loss of two electrons from the metal). The work, in what must be a difficult field experimentally, on the mechanism of formation of Grignard reagents should also be highlighted.

8.2. Substitution Reactions

8.2.1. Substitution Involving Carbonyls or Carbon Monoxide

Entry and departure of carbonyl ligands has attracted considerable attention. Several reactions are found to be dissociative using the criterion that the rate is independent of the concentration of incoming group.

Table 8.1. Kinetic and Thermodynamic Parameters for Reaction (1)

L	$10^6 k$ (130°C), s^{-1}	$\Delta G^{\ddagger}$, kcal mol^{-1}	$\Delta H^{\ddagger}$, kcal mol^{-1}	$\Delta S^{\ddagger}$, cal K^{-1} mol^{-1}	$\Delta G°$, kcal mol^{-1}	$\Delta H°$, kcal mol^{-1}	$\Delta S°$, cal K^{-1} mol^{-1}	Cone angle
AsPh$_3$	11,600	27.4	36.3	22.2	-3.8	-3.9	-3.9	142°
PPh$_3$	99.7	31.3	36.3	12.5	0.1	-3.9	-10.1	145°
P(OPh)$_3$	15.7	32.7	31.9	-2.0	1.5	-8.3	-24.6	121°
CO	130	31.2	40.2	22.6	0	0	0	
Py		24.1	25.4	3.1	-7.1	-14.8	-19.5	

Atwood's group[3] has been studying chromium tetra- and pentacarbonyls. Kinetic parameters for the overall process

$$[Cr(CO)_5L] + CO \rightarrow [Cr(CO)_6] + L \tag{1}$$

which is dissociative, are given in Table 8.1. The activation parameters for the dissociative step show little correlation with cone angle. Excellent correlations with slopes of 1.0 between $\Delta G^{\ddagger}$ and ΔG°, and between $\Delta S^{\ddagger}$ and ΔS° indicate that the least stable point in the reaction profile occurs between $[Cr(CO)_5]$ + CO and $[Cr(CO)_6]$. Reaction (2) is also dissociative; the rate rises as the π-accepting ability of L increases, the variation being much greater than that shown by reaction (1) (cf. Tables 8.1 and 8.2).

$$[Cr(CO)_4L_2] + CO \rightarrow [Cr(CO)_5L] + L \tag{2}$$

The reaction (3) of a molybdenum carbonyl is also dissociative,[4] kinetic parameters being given in Table 8.3.

$$cis\text{-}[Mo(CO)_4L(pip)] + CO \rightarrow cis\text{-}[Mo(CO)_5L] + pip \tag{3}$$

The higher $\Delta H^{\ddagger}$ observed when L is $P(OMe)_3$ is attributed to a stabilizing effect caused by intramolecular hydrogen bonding, N–H $\cdots$ O–P. (Intermolecular hydrogen accelerates the reaction.) Experiments with ^{13}CO demonstrate that the $[Mo(CO)_4P(OMe)_3]$ unit retains its geometry during the reaction, from which it is inferred that the dissociative intermediate is square pyramidal.

The reactions of $[Mn(CO)_5(Hal)]$ with CN^- to give $[Mn(CO)_4(CN)_2]^-$ are, however, somewhat complicated. Rate laws indicate that $[Mn(CO)_5(CN)]$, $[Mn(CO)_4(CN)Br]^-$, and $[Mn(CO)_4(CN)I]^-$ are formed as respective intermediates when Hal is Cl, Br, and I.[5] Activation parameters are strongly dependent on solvent. The reaction of $[Mn(CO)_5Br]$ with β-alanine is complicated by taking

Table 8.2. Kinetic and Thermodynamic Parameters for Reaction (2)

L	$10^6k,$ s^{-1}	$\Delta H^{\ddagger},$ kcal mol^{-1}	$\Delta S^{\ddagger},$ cal K^{-1} mol^{-1}
$P(OPh_3)_3$	397	37.6	18.6
PBu_3	1380	42.5	33.2
$P(OMe)_3$	9.22	43.4	25.5
CO	130	40.2	22.6
PPh_3	9.9×10^7		
$AsPh_3$	Very fast		

Table 8.3. Activation Parameters for Reaction (3)

L	$10^5 k$ (39.7°C), s^{-1}	$\Delta H^{\ddagger}$, kcal mol^{-1}	$\Delta S^{\ddagger}$, cal K^{-1} mol^{-1}
CO	2.77	25.8	2.8
P(OMe)$_3$	5.65	29.4	13.8
PPh$_3$	465	25.5	10.1

taking place through first a thermal then a photochemical step. The thermal step has some, but not all, of the features of a dissociative mechanism.

The rate of isomerization[6] of *trans*-[PhMn(CO)$_4$P(OPh)$_3$] to a *cis–trans* equilibrium mixture is first order in reactant and unaffected by free CO or P(OPh)$_3$. In addition no exchange with labeled d_{15}-P(OPh)$_3$ occurs, from all of which it is concluded that the rearrangement is intramolecular.

The reactions of PPh$_3$, PBu$_3$, and P(OPh)$_3$ (or L) with [MnRe(CO)$_{10}$] lead largely to substitution on rhenium to give [(OC)$_5$MnRe(CO)$_4$L] although [L(OC)$_4$MnRe(CO)$_4$L] and [L(OC)$_4$MnRe(CO)$_5$] can also be formed.[7] Rates are slightly dependent on [L] but this has not been investigated in detail so that the derived $\Delta H^{\ddagger}$ and $\Delta S^{\ddagger}$ could be based on two processes, one dissociative.

Attack[8] of various phosphines, phosphites, and AsPh$_3$ (or L) on [Fe$_2$(CO)$_6$(μ-X)$_2$] (X = S or Se) leads to a variety of products such as [Fe$_2$(CO)$_5$L(μ-X)$_2$], [Fe$_2$(CO)$_4$L$_2$(μ-X)$_2$], and [Fe$_3$(CO)$_{9-n}$L$_n$(μ_3-X)$_2$] (n = 0,1,2). The initial step is not dissociative; the rate law is first order in complex and contains terms in [L] and in [L]^2, suggesting an overall mechanism given by reactions in equations (4)–(6).

$$[Fe_2(CO)_6(\mu\text{-}X)_2] + L \rightleftharpoons [Fe_2(CO)_6 L(\mu\text{-}X)_2] \qquad (4)$$

$$[Fe_2(CO)_6 L(\mu\text{-}X)_2] \rightarrow [Fe_2(CO)_5 L(\mu\text{-}X)_2] + CO \qquad (5)$$

$$[Fe_2(CO)_6 L(\mu\text{-}X)_2] + L \rightarrow \text{other products} \qquad (6)$$

Activation parameters are quoted. There is no dissociative pathway.

It is suggested that the replacement of L, (PPh$_3$, AsPh$_3$, or SbPh$_3$) by CO in [Fe(η^4 − PhCH=CHCH=NCPh) (CO)$_2$L] proceeds[9] as in Scheme 1. The rate of the initial dissociative step, which involves an η^4 to σ rearrangement, falls from L = AsPh$_3$ to L = SbPh$_3$ as expected.

The reaction of carbon monoxide with mixed metal tetranuclear carbonyl hydrides which is first order in each reactant leads to fragmentation, in which H$_2$, tri- and mononuclear products are formed somewhat selectively.[10]For

Scheme 1

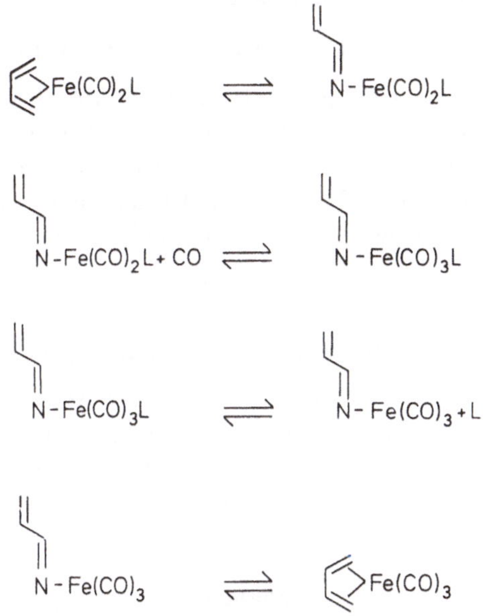

$[H_2FeRu_3(CO)_{13}]$ and $[H_2Ru_4(CO)_{13}]$ $\Delta H^{\ddagger}$ and $\Delta S^{\ddagger}$ are, respectively, 20.0 kcal mol^{-1}, -25.4 cal mol^{-1} K^{-1} and 12.5 kcal mol^{-1}, -36.6 cal mol^{-1} K^{-1}.

In $[Ru(CO)_5]$, replacement of a carbonyl by various phosphines is dissociative.$^{(11)}$ $\Delta H^{\ddagger}$ and $\Delta S^{\ddagger}$ for reaction (7):

$$[Ru(CO)_5] \rightarrow [Ru(CO)_4] + CO \tag{7}$$

are 27.6 kcal mol^{-1} and 15.2 cal K^{-1} mol^{-1}. This carbonyl falls into sequences of increasing lability: $[Fe(CO)_5] < [Ru(CO)_5]$ and $[Mo(CO)_6] < [Ru(CO)_5] < [Pd(CO)_4]$.

The reactions$^{(12)}$ of $[Co_4(CO)_{11}P(OMe)_3]$ and $[Co_4(CO)_{10}P(OMe)_3]$ with PPh$_3$ in which the latter replaces a carbonyl are dissociative, $\Delta H^{\ddagger}$, $\Delta S^{\ddagger}$ being 26.8 kcal mol^{-1}, 8.2 cal K^{-1} mol^{-1} and 27.1 kcal mol^{-1}, 10.7 cal K^{-1} mol^{-1}, respectively. The corresponding reactions with P(OMe)$_3$ are first order in phosphite and retarded by carbon monoxide. If it is supposed that P(OMe)$_3$ is a rather poorer attacking ligand than CO (and hence PPh$_3$ also), the same dissociative mechanism could apply, but the authors seem reluctant to suggest this.

Phosphine exchange between $[CpCo(PPh_3)Me]$ and PMe$_3$ is dissociative.$^{(13)}$ Finally, are two reactions which appear to be simple substitution processes but are not. The apparent substitution process shown in equation (8) occurs by nucleophilic attack by the CH_3 unit on a carbonyl ligand.[14]

$$[Ph_3P(CO)_4ReBr] + CH_3Li \rightarrow [Ph_3P(CO)_4ReCH_3] \tag{8}$$

$[W(CO)_6]$ reacts with the ion pair $[(Ph_3P)_2N^+].[X^-]$ (X = CN, NCO or NCS) to give $[(Ph_3P)_2N]^+[W(CO)_5X]^-$ and $[(Ph_3P)_2N]_2^+[cis-W(CO)_4(CN)_2]^{2-}$, by a process which is first order in each reactant.[15] The data are consistent with an initial attack of X, the nucleophilic group, at a carbonyl carbon, in contrast to the replacement of CO by phosphines and phosphites, which appears to proceed by an I_d mechanism.

8.2.2. *Substitution Involving Di- or Multihapto Ligands*

Exchange reactions of arenes have been reviewed.[16] The exchange of monoalkenes with styrene in $[Fe(CO)_4(CH_2{=}CHPh)]$ is dissociative,[17] $[Fe(CO)_4]$ being formed. However, replacement of α,β-unsaturated ketones by 1,3-dienes (or polyenes) on an $Fe(CO)_3$ center occurs by both a dissociative and an interchange pathway (see Scheme 2). Two pathways[17] also exist for the replacement

Scheme 2

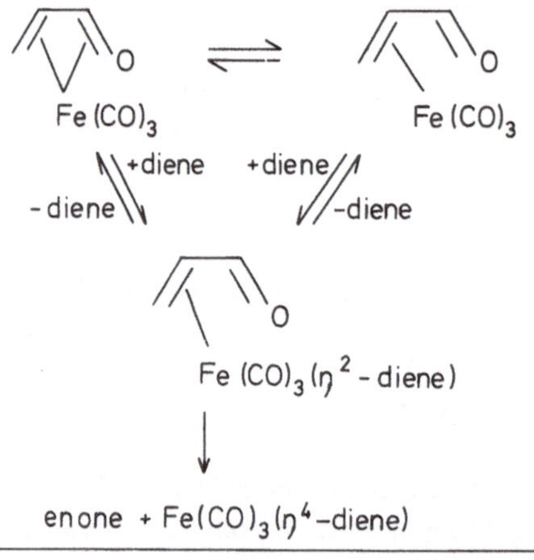

Fe (CO)$_3$ ⇌   Fe (CO)$_3$

‖+diene   +diene⁄⁄
–diene‖   ⁄⁄–diene

Fe (CO)$_3$ (η^2 - diene)

enone + Fe(CO)$_3$(η^4–diene)

of diene in $[Cr(CO)_4(diene)]$ by other dienes or $P(OMe)_3$, the former much preferring the D mechanism, while the latter proceeds by both.

The replacement of 2-methyl-2-butene (mbn) in the optically active *trans*-$[PtCl_2(S\text{-mbn})L]$, by another olefin is first order in complex and olefin when L is a pyridine.[18] Rates fall as the pyridine becomes more donor as expected for an S_N2 process. 2-Methyl-2-butene (mbn) enters more readily than *cis*-1,2-dichloroethene (dce) and with a very much more positive entropy of activation.

Table 8.4. Rate Constants and Activation Parameters for the Reaction of trans-[PtCl$_2$(S-mbn)(4-X-py)] with dmb, dce, or mbn[18]

X	dmb 10^2k (17°C), M^{-1} s^{-1}	dce 10^2k (17°C), M^{-1} s^{-1}	$\Delta H^\ddagger$, kJ mol^{-1}	$\Delta S^\ddagger$, J mol^{-1} K^{-1}	mb 10^2k (17°C), M^{-1} s^{-1}	$\Delta H^\ddagger$, kJ mol^{-1}	$\Delta S^\ddagger$, J mol^{-1} K^{-1}
CN	0.92	20.0	(56)	(−63)	173	86	57
Cl	0.39				30		
H	0.21	3.78	57	−75	18.6	85	37
CH$_3$	0.10	1.44	56	−79	2.98	78	−5
NH$_2$					0.68		

2,3-Dimethyl-2-butene (dmb) reacts the most slowly of the three (see Table 8.4). When L is an aniline, there is evidence for a solvent-dependent pathway also.

A nickel(0) system has been studied in order to look for evidence of nickel(I). The rate of formation[19] of the "sideways on" carbonyl complex, $[Ni(PEt_3)_2(\eta^2\text{-}OCPh_2)]$, is first order in both $[Ni(PEt_3)_4]$ and Ph_2CO, and retarded by free phosphine, indicating steps (9) and (10):

$$[Ni(PEt_3)_4] \rightleftharpoons [Ni(PEt_3)_3] + PEt_3 \qquad (9)$$

$$[Ni(PEt_3)_3] + Ph_2C{=}O \rightarrow [Ni(PEt_3)_2(OCPh_2)] + PEt_3 \qquad (10)$$

Acceptor substituents on the Ph increase the rate of reaction (10), but there is no evidence for the actual formation of a Ni(I) species (as occurs in the oxidative addition reactions of Ni(0) compounds; see later).

The $\eta^6\text{-}C_7H_8$ group is, of course, equivalent to three monodentate σ-bonding ligands. Its displacement from $[\eta^6\text{-}C_7H_8M(CO)_3]$ (M = Cr, Mo, or W) by benzonitrile is first order in complex[20]; the order in entering ligand can be one (Cr), two (Mo), or in between (W). The results can be rationalized in terms of Scheme 3. The rate constant for the first step, which can be obtained in the case

Scheme 3

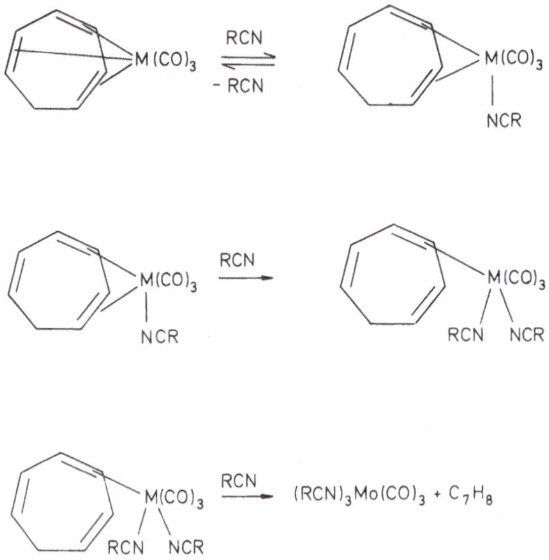

of Cr, is rather insensitive to the nitrile used as entering group in spite of the fact that the large negative values of $\Delta S^{\ddagger}$ suggest an associative process (see Table 8.5).

Table 8.5. Rate and Activation Parameters for Reaction of [η^6-$C_7H_8Cr(CO)_3$] with RCN[20]

R	10^4k (45°C), s^{-1}	$\Delta H^{\ddagger}$, kJ mol^{-1}	$\Delta S^{\ddagger}$, J K^{-1} mol^{-1}
Me	0.09	90.2	−63
Ph	0.068	74.8	−109
o-MeC$_6$H$_4$	0.076		

8.3. Oxidative Addition and Reductive Elimination

The reviewer sympathizes with Crabtree and Alatky,[21] who have highlighted problems in distinguishing between types of "additions," e.g., whether they are "oxidative," "ligand," or even "reductive." They suggest that the addition of a ligand to a metal should be described numerically, e.g., {3,2}, in terms of the number of atoms involved and the number of electrons donated by the ligand to the metal, viz. 3 and 2, respectively, here. Change in oxidation state of the metal, if any, is not included.

First, we will look at reactions in which the oxidation state of the metal in the overall process rises by only 1. Three closely related systems are noteworthy, two involving cobalt and one chromium.

Although it is often thought otherwise, vitamin B$_{12r}$ [the cobalt(II) form] has been shown by Blaser and Halpern[22] to react with organic halides just as model B$_{12r}$ compounds do. All halides in methanol and chlorides and bromides in aqueous solution seem to react by the mechanism followed by the model compounds, as in Scheme 4.

Scheme 4

$$Co(II) + RX \xrightarrow{k_1} Co(III)X + R^{\bullet}$$
$$Co(II) + R^{\bullet} \xrightarrow{fast} R\text{---}Co$$
$$Co(III)X + H_2O \text{ (or MeOH)} \xrightarrow{fast} Co(III)(OH_2) + X^-$$

This gives an overall rate equation, $-d[Co(II)]/dt = 2k_1[Co(II)] [RX]$. B$_{12r}$ reacts more slowly than some but not all of its models (see Table 8.6). The reactions of iodides in aqueous solution are second order in B$_{12r}$ and two possible mechanisms are proposed.

Goh and Goh[23] have shown that the rate of reaction of [Co(CN)$_5$]$^{2-}$ with various bridgehead iodides is first order in each of these two reactants in the presence of acrylonitrile (as is the same reaction with simple organic ions in the

Table 8.6. Rate Constants ($M^{-1} s^{-1}$) for Reaction of Cobalt(II) Complexes with Benzyl Bromides at 25°C[22]

Cobalt(II) complex	Solvent	$p\text{-BrC}_6\text{H}_4\text{CH}_2\text{Br}$	$p\text{-CNC}_6\text{H}_4\text{CH}_2\text{Br}$	$p\text{-NO}_2\text{C}_6\text{H}_4\text{CH}_2\text{Br}$
$[\text{Co(CN)}_5]^{3-}$	Methanol–water (70:30)	7.5		1.0×10^2
$[\text{Co(dmgH)}_2(\text{PPh}_3)]$	Benzene	6.1×10^{-2}	2.2×10^{-1}	3.7×10^{-1}
Vitamin B_{12r}	Methanol	7.5×10^{-3}	1.4×10^{-2}	3.4×10^{-2}
$[\text{Co(saloph)py}]$	Methylene chloride		2.9×10^{-3}	5.5×10^{-3}

absence of this radical trap; see above). They suggest that the first reaction in Scheme 4 still occurs after which R' presumably is scavenged. Table 8.7 shows that k_1 is smaller in some of the bridgehead systems. There is some correlation between this rate constant and others where R' is either produced or destroyed. In the absence of acrylonitrile, the reaction involving bridgehead iodides becomes second order in cobalt complex and is retarded by added $[\text{Co(CN)}_5\text{I}]^{3-}$. The results are compatible with Scheme 4 if the second reaction is the slow step, and the first reaction reversible. The implication, perhaps surprisingly, is that while the introduction of a bridgehead may reduce the rate of the first reaction, it must lower that of the second by a greater amount.

The macrocyclic chromium complex, $[\text{Cr(II)}([15]\text{aneN}_4)]^{2+}$, also reacts[24] in a very similar way to the B_{12r} model compounds with simple organic halides to give *trans*-$[\text{RCr}([15]\text{aneN}_4)\text{OH}_2]^{2+}$ and *trans*-$[(\text{Hal})\text{Cr(III)}([15]\text{aneN}_4)\text{OH}_2]^{2+}$. A mechanism analogous to that in Scheme 4 is proposed. Strong evidence for a radical process is provided by 6-bromo-1-hexene yielding a cyclopentylmethylchromium complex presumably by cyclization of the 5-hexenyl radical. Entropies of activation imply considerable ordering in the transition state. In general the rates of reactions of organic halides with $[\text{Cr(II)}([15]\text{aneN}_4)]^{2+}$, so-called "Cr(II)-en", and $[\text{Co(II)(CN)}_5]^{3-}$ parallel each other (see Table 8.8).

Moving to processes in which the oxidation state changes by 2, one can find examples of all four mechanisms which have been cited for oxidative addition and reductive elimination, namely, S_N2, radical, radical chain, and concerted.

Since cobaloximes$_s$ [i.e., in the Co(I) state] react with organic halides by an S_N2 mechanism, it is not surprising that the rhodium analogs do the same. $[\text{HRh(III)(dmgH)}_2\text{PPh}_3]$, has a p$K_a$ of 9.5 and so is fairly readily deprotonated to give its conjugate base, $[\text{Rh(I)(dmgH)}_2\text{PPh}_3]$. The latter reacts[25] with organic halides to give organorhodium compounds as in:

$$[\text{Rh(dmgH)}_2\text{PPh}_3]^- + \text{RX} \rightarrow [\text{RRh(dmgH)}_2\text{PPh}_3] + \text{X}^- \qquad (11)$$

The large variation in rate constants is compatible with an S_N2 process in which both electronic and steric factors are important (see Table 8.9). The hydride

Table 8.7. Rate Constants for Reaction of $[Co(CN)_5]^{3-}$ with Various Organic Iodides[23]

	t, °C	k, dm^3 mol^{-1} s^{-1}
CH_3I	25.0	0.95×10^{-2}
$CH_3CH_2CH_2I$	25.0	4.3×10^{-2}
$(CH_3)_2CHI$	25.0	1.20
$(CH_3)_3CI$	25.0	9.1
1-Adamantyl iodide	25	2.47×10^{-2}
2,4-Dimethyl-1-bicyclo-[2.2.2]octyl iodide	25	7.5×10^{-3}
7,7-Dimethyl-1-bicyclo-[2.2.1]heptyl iodide	28	$<10^{-6}$
9-Triptycyl iodide	28	$<10^{-6}$

compound yields a hydrocarbon and a rhodium(II) dimer and could react by the steps in equations (12) and (13):

$$[HRh(dmgH)_2PPh_3] + RX + H_2O \rightarrow$$

$$RH + [(H_2O)Rh(dmgH)_2PPh_3]^+ + X^- \quad (12)$$

$$[(H_2O)Rh(dmgH)_2PPh_3]^+ + [HRh(dmgH)_2PPH_3] \rightarrow$$

$$H_3O^+ + [Rh(dmgH)_2PPh_3]_2 \quad (13)$$

A radical mechanism is also thought to be possible, but the limited data are considered to be more in keeping with a nucleophilic type of process involving hydride donation to R as in equation (12).

Table 8.8. Rate Constants, k_1 ($M^{-1}s^{-1}$) for the Reaction of Typical Organic Halides with Various Cr(II) and Co(II) Complexes as in Scheme 4[23]

Halide	$[Cr(II)([15]aneN_4)]^{2+}$	["Cr(II)en"]	$[Co(II)(CN)_5]^{3-}$
Primary aliphatic bromide	0.082^a	0.1	
Secondary aliphatic bromide	0.9	1	
Tertiary aliphatic bromide	5	5	
Primary aliphatic iodide	0.21^b		0.043
Secondary aliphatic iodide	2.5	50	1.2
$PhCH_2Cl$	1.6×10^{2c}		5×10^{-4}
$PhCH_2Br$	1.0×10^4		2.3

^{a} EtBr, $\Delta H^{\ddagger} = 44.2$ kJ mol^{-1}, $\Delta S^{\ddagger} = -118$ J mol^{-1} K^{-1}.
^{b} EtI, $\Delta H^{\ddagger} = 44.1$ kJ mol^{-1}, $\Delta S^{\ddagger} = -110$ J mol^{-1} K^{-1}.
^{c} $PhCH_2Cl$, $\Delta H^{\ddagger} = 30.6$ kJ mol^{-1}, $\Delta S^{\ddagger} = -100$ J mol^{-1} K^{-1}.

Table 8.9. Rate Constants (M^{-1} s^{-1}) at 25°C for Reactions (11) and (12)

RX	k_{11}	k_{12}
PhCH$_2$Br	16,600	960
PhCH$_2$Cl	1040	54
Ph(Me)CHCl	96	18

Osborn's group[26] has continued its studies on whether alkyl halides add oxidatively to iridium(I) compounds by a radical chain or an S_N2 mechanism by using more reactive metal complexes such as [Ir(PMe$_3$)$_2$(CO)Cl]. For simple alkyl (methyl excepted), vinyl, and aryl halides and α-halo-esters, evidence based on the effect of radical initiators and inhibitors, structure–reactivity relationships, the trapping of radicals by acrylonitrile, and the loss of stereospecificity at the reacting carbon atom all indicate a radical chain process, perhaps as in equations (14) and (15):

$$R^{\bullet} + Ir(I) \rightarrow R\text{-}Ir(II) \qquad (14)$$

$$R\text{-}Ir(II) + RX \rightarrow R\text{-}Ir(III)\text{-}X + R^{\bullet} \qquad (15)$$

The reactions of methyl, allyl, and benzyl halides, as well as chloromethyl methyl ether, do not show features of chain reactions. While this does not conclusively exclude a radical chain process, it is consistent with the S_N2 pathway proposed several years ago on the basis of kinetic studies. However, a possible objection is that the rate of such an S_N2 process for methyl bromide would have to be at least 10^5 times faster than that for ethyl bromide, which is an unusually high factor.

Lappert's group[27] report more work on platinum-containing systems. Spin trapping esr experiments have demonstrated the existence of free organic radicals (as relevant intermediates, not artifacts) in oxidative addition reactions of the following type: platinum(0) phosphines + alkyl halides and platinum(II) (alkyl)phosphines + sulfonyl or acyl halides. If Ph$_3$CCl is used in the first reaction, Ph$_3$C$^{\bullet}$ can be detected directly. Galvinoxyl inhibits the addition of p-Me-C$_6$H$_4$SO$_2$Cl, but not the corresponding bromide, to [PtMe$_2$(PMe$_2$Ph)$_2$]. The trapping of a platinum(I) compound is reported: [Pt(trap)I(PPh$_3$)$_2$]. It is concluded that for more reactive alkyl halides, there is a radical, nonchain process as shown in equations (16) and (17):

$$Pt + RX \rightleftharpoons PtX + R^{\bullet} \qquad (16)$$

$$R^{\bullet} + PtX \rightarrow Pt(R)X \qquad (17)$$

However, for less reactive halides such as $R'SO_2Cl$, there is probably a radical, chain mechanism as in equations (18) and (19)($R = R'SO_2$):

$$Pt + R^\bullet \rightarrow PtR^\bullet \tag{18}$$

$$PtR^\bullet + RX \rightarrow Pt(R)X + R^\bullet \tag{19}$$

Bolsman and Van Doorn[28] have exploited the readiness of *gem*-halonitroalkanes to give radical anions to compare the oxidative addition behavior of Ir(I) and Pt(0). Contrasting mechanisms are shown by the reaction of 1-chloro-1-nitroethane with *trans*-[IrCl(CO)L$_2$] (L = PMe$_2$Ph) and with [Pt(PPh$_3$)$_4$]. The first reaction appears to be an oxidative addition forming as an initial intermediate a compound which is assigned the following structure:

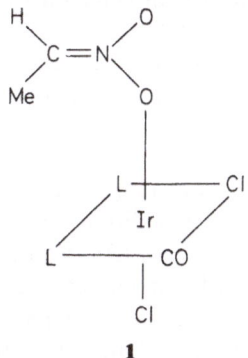

1

The reaction is retarded by 2-methyl-2-nitrosopropane and by 1,4-dinitrobenzene, which scavenge radicals and radical anions, respectively. (The radical anion of the latter can be detected.) It is proposed that the process is initiated by reactions (20) and (21):

$$[Ir(I)Cl(CO)L_2] + MeCHCl(NO_2) \rightarrow [Ir(II)Cl(CO)L_2]^+ + MeCHCl(NO_2)^\cdot \tag{20}$$

$$MeCHCl(NO_2)^\cdot \rightarrow Me\dot{C}H(NO_2) + Cl^- \tag{21}$$

Then reactions (22) and (23) maintain the radical chain:

$$Me\dot{C}H(NO_2) + [IrCl(CO)L_2] \rightarrow [IrCl(CO)L_2(O_2NCHMe)] \tag{22}$$

$$[IrCl(CO)L_2(O_2NCHMe)] + MeCHCl(NO_2) \rightarrow [IrCl_2(CO)L_2(O_2NCHMe)]$$
$$+ MeCH(NO_2) \tag{23}$$

1-Chloro-1-nitro anions are known to decompose as in equation (21). In contrast the reaction of [Pt(PPh$_3$)$_4$] is inhibited by neither of the scavengers used in the Ir(I) example. A radical cage step, as in equation (25), is proposed as part of the overall process, equations (24)–(26):

$$[Pt(PPh_3)_4] \rightarrow [Pt(PPh_3)_2] + 2PPh_3 \qquad (24)$$

$$[Pt(0)(PPh_3)_2] + MeCH(Cl)NO_2 \rightarrow [Pt(I)(PPh_3)_2Cl \cdot MeĊ(H)NO_2]$$
$$\rightarrow [Pt(II)(PPh_3)_2Cl]^+ + MeC(H)NO_2^- \qquad (25)$$

$$[Pt(II)(PPh_3)_2Cl]^+ + Cl^- \rightarrow [Pt(PPh_3)_2Cl_2] \qquad (26)$$

While there is some formal similarity between the first part of reactions (25) and (16), the platinum product formed in equation (26) is nòt that expected in a conventional oxidative addition process.

In contrast to the preceding examples, a reaction of a platinum(0) compound can be quoted which may undergo oxidative addition in yet another way. During the oxidative addition of various tin compounds, SnR_1Cl_3 or SnR_2Cl_2, to[Pt(C$_2$H$_4$)(PPh$_3$)$_2$] to give *cis*-[PtPh(SnPhCl$_2$)(PPh$_3$)$_2$] or *cis*- and *trans*-[PtCl(SnCl$_{1,2}$R$_{2,1}$)(PPh$_3$)$_2$] (R = Me or Ph), no evidence was found for radicals.[29]

If the law of microscopic reversibility is invoked, an example can be given of a concerted oxidative addition since [RhMeCl$_2$CO(PPh$_3$)$_2$] reductively eliminates methylchloride in a first-order, intramolecular process[30] with an activation energy of 95.0 kJ mol^{-1}. However, the example is more complicated than it seems at first. The formation of [Ph$_3$PMe]$^+$Cl$^-$ in the presence of free triphenylphosphine suggests that there may be a second pathway for the production of *trans*-[RhCl(CO)(PPh$_3$)$_2$] which involves an S_N2 attack by Ph$_3$P on the carbon of the Rh–Me group.

[NiMe(Ph)(PEt$_3$)$_2$] provides an example in which reductive elimination proceeds through both an intramolecular and a radical chain mechanism.[31] The first pathway yields MePh and [Ni(PEt$_2$)], is inversely related to concentration of free phosphine, is first order in reactant and involves no scrambling of organic groups between the compound above and, for example, [Ni(CD$_3$)(σ-tolyl)(PEt$_3$)$_2$]. It therefore probably involves an intramolecular elimination from [NiMe(Ph)(PEt$_3$)]. The second reaction is induced by free aryl halide. It yields MePh and [Ni(Hal)(Ph)(PEt$_3$)$_2$], has an induction period, can be retarded or accelerated by radical inhibitors or promoters, and results in scrambled arylmethane products. For it, a radical chain mechanism is proposed as in equations (27)–(29) (phosphines are omitted):

$$[PhNi(I)] + PhX \rightarrow [Ph_2Ni(III)X] \qquad (27)$$

$$[Ph_2Ni(III)X] + [PhNi(II)Me] \rightarrow [PhNi(II)X] + [Ph_2Ni(III)Me] \qquad (28)$$

$$[Ph_2Ni(III)Me] \rightarrow PhMe + [PhNi(I)] \quad (29)$$

The second step involves a *trans*-arylation between Ni(III) and Ni(II). A closely related oxidative addition will be mentioned shortly.

In an attempt to see if the mechanism of oxidative addition can be correlated with relative reactivities, Pearson and Figdore[32] have measured rate constants and activation parameters for the reactions of methyl iodide and methyl tosylate (MeOTs) with a number of transition metal nucleophiles. k_{MeI} and k_{MeOTs} both vary enormously (see Table 8.10). It is curious that the $\Delta H^{\ddagger}$ and $\Delta S^{\ddagger}$ for the

Table 8.10. Nucleophilic Reactivities at 25°C for Various Transition Metal Nucleophiles with Methyl Iodide and Methyl Tosylate[32]

	$k_{MeI}, M^{-1} s^{-1}$	k_{MeI}/k_{MeOTs}
$[CpFe(CO)_2]^-$	2.8×10^6	1.2×10^4
$[Ni(PEt_3)_4]$	$> 8 \times 10^5$	3×10^5
$[CpRu(CO)_2]^-$	3×10^5	
$[CpNi(CO)]^-$	2.2×10^5	$< 1 \times 10^4$
$[Co(dmgH)_2pyr]^-$	1.8×10^4	1.2×10^3
$[Pd(PEt_3)_3]$	1.3×10^3	8.6×10^5
$[Rh(CN)_4]^{3-}$	7.8×10^2	300
$[Re(CO)_5]^-$	1×10^3	1×10^3
$[Fe(CO)_4]^{2-}$	5×10^2	88
$[Ni(PPh_3)_3]$	2.1×10^2	Large
$[Rh(I)(C_2DOBF_2)]$	$> 1 \times 10^2$	$> 2 \times 10^3$
$[Pt(PEt_3)_3]$	9.7	2.2×10^4
$[Mn(CO)_5]^-$	7.4	430
$[CpW(CO)_3]^-$	2.4	1.5×10^3
$[CpMo(CO)_3]^-$	1.5	940
$[LiCuMe_2]_2$	1.46	8.1×10^{-3}
$Li_2[PtMe_4]$	1.6×10^{-1}	< 0.3
$Li[AuMe_2]$	1.3×10^{-1}	$< 5 \times 10^{-3}$
$[CpCr(CO)_3]^-$	7.5×10^{-2}	29
$[Co(CO)_4]^-$	4.37×10^{-2}	> 15
$[Pd(PPh_3)_3]$	3.09×10^{-2}	Large
$[Co(CN)_5]^{3-}$	2.52×10^{-2}	$> 10^8$
$[CpIr(CO)PPh_3]$	4.8×10^{-3}	
$[Pt(PPh_3)_3]$	1.1×10^{-2}	3.7×10^3
$[CpRh(CO)PPh_3]$	8×10^{-4}	
$[IrCl(CO)(PPh_3)_2]$	1.0×10^{-3}	Large
$[CpCo(CO)PPh_3]$	6×10^{-4}	
$[RhCl(CO)(PPh_3)_2]$	3.6×10^{-5}	
PhS^-	1×10^5	7.7
PPh_3	6×10^{-5}	7.7

reaction of the nucleophiles in this table with both MeI and MeOTs are jointly isokinetic (provided various methyl complexes, such as LiAuMe$_2$, are ignored) in spite of the fact that some of the reactions proceed through S_N2 substitution and others by one or other of the radical mechanisms.

Several pieces of work suggest that organic halide radical anions are involved in some oxidative additions, either as individual species, RX$^\tau$, or as radical pairs, M·RX$^\tau$.

Connor and Riley$^{(33)}$ have inferred the formation of RX$^\tau$ in the oxidative addition to *cis*-[Mo(CO)$_2$(dmpe)$_2$] (dmpe is Me$_2$PCH$_2$CH$_2$PMe$_2$) of a variety of R–X such as Cl$_3$C–Cl, Ph$_3$C–Cl, and PhCH$_2$–Br. This yields first *trans*-[Mo(CO)$_2$(dmpe)$_2$]X, which can be detected by ir, visible–uv, and esr spectroscopy, and then either *cis*-[Mo(CO)$_3$(dmpe)$_2$X]X (or *trans*-[MoH(CO)$_2$(dmpe)$_2$]X in some instances). (The analogous reaction involving chromium yields only the initial product.) Radicals such as Ph$_3$C$^\cdot$ and PhCH$_2^\cdot$ can be detected either directly or by trapping using esr. The detection of the intermediate containing the *trans*-[Mo(CO)$_2$(dmpe)$_2$]$^+$ cation implies steps (30) and (31):

$$\textit{cis-}[Mo(CO)_2(dmpe)_2] + RX \rightarrow \textit{cis-}[Mo(CO)_2(dmpe)_2]^+ + RX^\tau \quad (30)$$

$$\textit{cis-}[Mo(CO)_2(dmpe)_2]^+ \rightarrow \textit{trans-}[Mo(CO)_2(dmpe)_2]^+ \quad (31)$$

the overall reaction being completed by reactions (32)–(34):

$$RX^\tau \rightarrow R^\cdot + X^- \quad (32)$$

$$\textit{trans-}[Mo(CO)_2(dmpe)_2]X + RX \rightarrow \textit{cis-}[Mo(CO)_2(dmpe)_2X]X + R^\cdot \quad (33)$$

$$R^\cdot + R^\cdot \rightarrow R\text{---}R \quad (34)$$

For CCl$_4$, the disappearance of complex in equations (30) and (33) is first order in both Mo and RX. Reaction (31) seems to be too fast to detect.

Tsou and Kochi$^{(34)}$ have got evidence for an intermediate of the type M.RX$^\tau$ by studying the oxidative addition of aryl halides, ArX, to [Ni(PEt$_3$)$_4$] using a combination of techniques such as rate data, product ratios, Hammett relationships, esr, and cyclic voltametry. Both Ni(I) and Ni(II) products are formed, [NiX(PEt$_3$)$_3$] and *trans*-[NiX(Ar)(PEt$_3$)$_2$], respectively. Rate constants for the disappearance of [Ni(PEt$_3$)$_4$] correlate with changes in ArX and with half-wave reduction potentials of ArX, but not with Ni(I)/Ni(II) product ratios formed in the parallel reactions (37) and (38), from which it is concluded that reaction (36) must be rate determining (L = PEt$_3$). The rate constant for the overall reaction has an inverse first-order term in free phosphine, which provides evidence for equilibrium (35):

$$[Ni(O)L_4 \rightleftharpoons [Ni(O)L_3] + L \qquad (35)$$

$$[Ni(O)L_3] + ArX \rightarrow [Ni(I)L_3 \cdot ArX]^\tau \qquad (36)$$

$$[Ni(I)L_3 \cdot ArX]^\tau \underset{\longrightarrow}{\overset{\longrightarrow}{}} \begin{array}{l} [ArNi(II)XL_2] + L \qquad\qquad (37) \\[2mm] [Ni(I)L_3]^+ + X^- + Ar^\cdot, \text{ etc.} \qquad (38) \end{array}$$

A reductive elimination involving a similar nickel system was mentioned earlier, while (36) might be compared with reaction (10).

Costa's group[35] has detected an intermediate in the reaction of [Co(I)(salen)]⁻ and t-BuBr, which takes place in two steps. The first, followed by cyclic voltametry, is rapid, first order in each reactant, and involves the disappearance of [Co(I)(salen)]⁻. In contrast, the second step, studied by polarography, is slow, corresponding to the appearance of [Co(II)(salen)]. Equations (39) and (40) are proposed:

$$[Co(I)(salen)]^- + t\text{-BuBr} \rightarrow [Co\text{-}t\text{-Bu}(Br)(salen)]^- \qquad (39)$$

$$[Co\text{-}t\text{-Bu}(Br)(salen)]^- \rightarrow [Co(II)(salen)] + C_4H_8 + \tfrac{1}{2}H_2 + Br^- \qquad (40)$$

It would be interesting to know more about the distribution of electrons in [CoBu(Br)(salen)]⁻.

At this point it is worth mentioning other reactions in which radical pairs, not dissimilar to $M \cdot RX^\tau$, seem to be formed. One process involving the participation of a species, $[R_4Sn^+HgX_2^-]$, had already been discussed in Chapter 7. Kochi's group[36] also postulates that intermediates such as $[R_3SnH \overset{+}{\cdot}(TCNE^\tau)]$ are formed in the production of the 1:1 adducts, $R_3M(TCNE)H$, from trialkyl-hydrides, R_3MH ($M = Sn$, Ge, and Si), and tetracyanoethylene (TCNE). Use of a variety of techniques including kinetics, Taft σ^* values, reduction potentials and ionization potentials suggests the formation of a charge transfer complex as in equation (41), followed by formation of the radical pair in equation (42):

$$R_3SnH + TCNE \rightarrow [R_3SnH(TCNE)] \qquad (41)$$

$$[R_3SnH(TCNE)] \rightarrow [R_3Sn\overset{+}{H}{}^\bullet(TCNE^\tau)] \qquad (42)$$

In some instances the charge transfer and the radical pair intermediates can be detected. Hydrogen transfer follows electron transfer [equation (43)]:

$$[R_3SnH^{+\cdot}(TCNE^{\cdot})] \rightarrow R_3Sn(TCNE)H \tag{43}$$

Something with $ArX^{\delta-}$ character seems to be involved in the reduction of tri-n-butylstannane by benzyl halides. Relative relativities have been shown by Blackburn and Turner[37] to follow Hammett relations in which σ^- (as opposed to σ^+ or σ) substituent constants are used. The authors point out that this is a novelty for a free radical chain reaction, and suggest a transition state in which the organic halide has anionic character. Two mechanisms are suggested for chlorides and bromides, while iodides appear to react differently. Fluorides are very inert.

Four papers from Whitesides' group[38–40] build up the picture of the mechanism of formation of Grignard reagents. The rate of reaction of organic halides with solid magnesium is proportional to the concentration of the former and the surface area of the latter. All the iodides examined, and many secondary alkyl bromides, react at mass transport or diffusion-controlled rates in the diethylether. This is shown by a detailed study of cyclopentylbromide, whose rate of reaction is inversely proportional to solution viscosity and proportional to stirring rate and whose activation energy is 2.3 kcal mol^{-1}, and also by the observation that the rate of reaction of all the halides in this category are equal. The reaction of most organic chlorides in ether is chemically controlled; structure–reactivity profiles indicate chlorine atom abstraction or single-electron transfer may be with magnesium insertion to be the most probable rate-determining step giving the following as possible transition states:

$$(RX).Mg, \ RX^{\cdot}, \ R^{\cdot}, \ or \ R \cdots X$$
$$\cdot Mg_s \cdot$$

Organic bromides fall between iodides and chlorides, the rate-controlling step containing contributions from mass transfer and chemical reaction. Hammett ρ constants for bromobenzene support any of the last three transition states. In the last of the four papers[41] rate–structure profiles for the formation of Grignard reagents are shown to correlate, with coefficients between 0.73 and 0.99, with other processes involving the electrochemical and chemical reduction of alkyl halides.

Esr studies on the analogous reaction in the solid state[42] between atomic magnesium and various mono- and polychloroalkanes reveal signals due to the radicals formed by loss of one chlorine atom (e.g., $BuCl \rightarrow Bu^{\cdot}$, $ClCH_2CH_2Cl \rightarrow {}^{\cdot}CH_2CH_2Cl$, etc.). From monochloroalkanes a second signal is produced which is attributed to the ion–radical pair, $Mg^{+\cdot}RCl^{\cdot}$, an interesting assignment since the resonance is a narrow singlet.

On the subject of Grignard reagents, a review of their mechanism of reaction with ketones has been made by Ashby.[43]

Systems involving oxidative cleavage of metal–metal bonds have been studied by Poë's group.[44,45] The order of reactions of bromine and iodine with dimetal carbonyls (M = Mn and/or Re) containing axial phosphines and phosphites, L, indicate the initial formation of adducts which may contain several molecules of Hal_2 as in, e.g., equations (44) and (45):

$$[M_2(CO)_8L_2] + n(Hal)_2 \overset{K_n}{\rightleftharpoons} [M_2(CO)_8L_2 \cdot n(Hal)_2] \tag{44}$$

$$[M_2(CO)_8L_2 \cdot n(Hal)_2] \overset{k_n}{\rightarrow} [M(CO)_4L(Hal)] \tag{45}$$

n, which can be as high as 4 for iodine, is affected by solvent. Rates of reaction tend to be slower for I_2 than for Br_2. Though there is some evidence of steric retardation, rates generally rise with increasing basicity of L and fall with increasing carbonyl stretching frequency, which suggests that attack is electrophilic at an oxygen of a carbonyl ligand. This proposal is reinforced by the fact that k_{ii}, when L is PPh_3, falls in the sequence Re_2, MnRe, Mn_2, which is the opposite to the trend for M–M homolytic fission. In Table 8.11 $k_i = k_n$ ($n = 1$), $k_{ii} = K_n k_n$ ($n = 2$), $k_{iii} = K_n k_n$ ($n = 3$).

A somewhat similar adduct to those just mentioned may be formed in the oxidative elimination process[11] in which [Ru(CO)₅] reacts with I_2 to give *cis*-

Table 8.11. Kinetic Parameters for Reactions of I_2 with $[M_2(CO)_8L_2]$ in Cyclohexane at 25°C[45]

M_2	L	$10^3 k_i$, s^{-1}	k_{ii}, $M^{-1} s^{-1}$	$10^{-4}k_{iii}$, $M^{-2} s^{-1}$	ν_{CO}, cm^{-1}
Mn_2	$P(C_6H_{11})_3$	1.2	6.67	<0.01	1946
	$P(p\text{-}MeOC_6H_4)_3$	~0.1	3.80	1.61	1955
	PEt_3			2.48	1950
	$PPhEt_2$			10.1	1952
	PPh_3	0.07		0.120	1960
MnRe	PPh_3	0.06		0.185	1965
Re_2	PPh_3	~0		11.8	1960
Mn_2	PBu_3			9.35	1950
	$P(OMe)_3$		1.52		1967
	PPh_2Et		0.21	0.37	1958
	$P(OPh)_3$			6.7×10^{-4}	1985

[Ru(CO)$_4$I$_2$]. The reaction has an induction period after which the rate equation contains both zero- and first-order terms in I$_2$. The rate constant for the zero- and first-order term is too large to be ascribed to the dissociative step given in equation (7). Formation of an adduct, [Ru(CO)$_5$.I$_2$] is suggested.

This section closes with some reductive elimination reactions in which C–C bonds are formed. The elimination of C$_2$H$_6$ from [NiMe$_2${Ph$_2$P(CH$_2$)$_n$PPh$_2$}] is unimolecular[46]; when $n = 2$, $\Delta H^{\ddagger} = 26.8$ kcal mol^{-1}, $\Delta S^{\ddagger} = 1.9$ cal deg^{-1} mol^{-1}, and for $n = 3$, $\Delta H^{\ddagger} = 25.1$ kcal mol^{-1}, $\Delta S^{\ddagger} = 4.8$ cal deg^{-1} mol^{-1}.

The Puddephatt–Tipper team[47] have shown that reductive elimination involving the formation of cyclopropanes from platinacyclopropanes appears to involve a concerted process rather than the production of carbene–alkene intermediates (as does also the oxidative addition involving the reverse reaction, and the skeletal isomerization of platinacyclopropanes). They[48] have also proposed a similar concerted behavior for a reaction which could be looked upon either as a reductive elimination or a substitution, namely, the overall process in equation (46).

$$[\text{PtX}_2(\overline{\text{CHRCH}_2\text{CH}_2})(\text{THF})] + \text{ol} \rightarrow [\text{PtX}_2(\text{ol})\text{THF}] + \overline{\text{RCHCH}_2\text{CH}_2} \quad (46)$$

It is first order in each reactant and its rate is lowered when R is Ph and when the alkene is cyclic (see Table 8.12).

Table 8.12. Rate Parameters for the Overall Reaction of [PtX$_2$(CHRCH$_2$CH$_2$)(THF)] with Olefins[48]

Platinum compound	Olefin	$10^3 k$, liters mol^{-1} s^{-1}	E, kJ mol^{-1}	A, liters mol^{-1} s^{-1}	$\Delta S^{\ddagger}$, J K^{-1} mol^{-1}
[PtCl$_2$·C$_3$H$_6$]	*cis*-Pent-2-ene Hex-1-ene 3-Methylpent-1-ene Dec-1-ene	176	68	2.3×10^{11}	-27
	Penta-1,3-diene	69	77	3.6×10^{12}	-4
	Cyclopentene	7.4	80	1.3×10^{12}	-13
	Cyclohexene	1.4	89	1.0×10^{13}	4
[PtCl$_2$·C$_3$H$_5$Ph]	*cis*-Pent-2-ene	57.6	65	2.3×10^{10}	-46
	Cyclopentene	1.7	88	8.7×10^{12}	3
[PtBr$_2$·C$_3$H$_6$]	*cis*-Pent-2-ene	153			
	Penta-1,3-diene	119	73	1.2×10^{12}	-14

8.4. *Migration–Insertion Reactions*

A useful review has been written by Kuhlmann and Alexander[49] on the migratory insertion of carbon monoxide into metal–carbon σ bonds. It includes a section on kinetics and mechanism.

The acetyl(benzoyl)rhenium complex, $[Re(CO)_4(COMe)(COPh)]^-$, shown in Scheme 5, has provided a useful way of comparing rates of migration and

Scheme 5

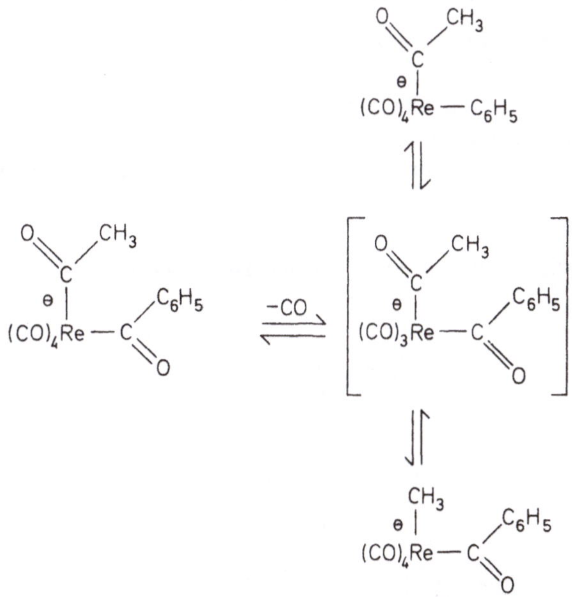

stabilities of methyl and phenyl groups.[50] Rate studies include decarbonylation and addition of phosphine and phosphite to the tricarbonyl intermediate. The phenylrhenium product is more stable (thermodynamically) than the methyl-rhenium by a factor of 50, while methyl migration is 28–29 times faster than phenyl.

Extended CNDO calculations[51] support the alkyl migration mechanism for the well-known process in:

$$[RMn(CO)_5] + CO \rightleftharpoons [RCOMn(CO)_5] \tag{47}$$

The observed sequence of increasing reaction rate for the forward reaction, R $= CF_3 < CH_3 < C_2H_5$, is calculated. It is further shown that a mechanism involving β-H elimination is reasonable for the side reaction:

$$[C_2H_5Mn(CO)_5] \rightarrow [HMn(CO)_5] + C_2H_4 \tag{48}$$

The kinetics of the water–organic-solvent biphasic reaction of $PhCH_2Br$, $Na^+[Co(CO)_4]^-$, $[Bu_4N]^+Cl^-$, and CO to give $[PhCH_2COCo(CO)_4]$ have been analyzed in terms of reactions (49) and (50), which occur in the organic phase.[52]

$$PhCH_2Br + [Bu_4N]^+[Co(CO)_4]^- \rightarrow [PhCH_2Co(CO)_4] + [Bu_4N]^+Br^- \tag{49}$$

$$[PhCH_2Co(CO)_4] + CO \rightleftharpoons [PhCH_2COCo(CO)_4] \tag{50}$$

The first step is first order in each reactant, rate determining, and slightly dependent on the organic solvent.

Schriver's group[53] has shown that even in the absence of free carbon monoxide, the Lewis acids such as $AlBr_3$ induce an alkyl migration in various complexes such as $[MnMe(CO)_5]$, $[Mn(CH_2Ph)(CO)_5]$, $[CpFeMe(CO)_2]$, and $[CpMoMe(CO)_3]$ to give products of the type in structure **2.**

2

These cyclic adducts react with one equivalent of CO to give the normal product of the migratory insertion reaction, e.g., $[Mn(COMe)(CO)_5]$, but they do so more rapidly than the parent alkyl complexes. The Lewis acid serves three functions in forming the cyclic adduct, namely, increasing the rate of alkyl migration, stabilizing the acyl group, and providing an electron-rich atom to fill a coordination site. Rates of CO uptake are given in Table 8.13. The fact that the stability of an adduct is roughly inversely related to its ability to take up carbon monoxide suggests that the rate-determining step in the latter is the breaking of the metal–Lewis-acid linkage, namely, the Mn–Br bond in structure **2.** The Lewis acid can catalyze the formation of acyl complexes, not normally

Table 8.13. Rate of Carbon Monoxide Uptake of Alkylmetal Complexes and Acylmetal–Lewis-Acid Complexes in Toluene[(53)]

Complex	10^6 (initial rate), $M\ s^{-1}$
$[Mn(CH_3)(CO)_5]$	0.18
$[Mn(CH_3)(CO)_5] + Mg(oleate)_2$	0.16
$[Mn(CH_3)(CO)_5 + GeCl_4$	0.22
$[Mn(CH_3)(CO)_5 + B(C_6H_5)F_2$	0.32
$[Mn(C(OAlBrBr_2)CH_3)(CO)_4]$	0.72
$[Mn(C(OAlClCl_2)CH_3)(CO)_4]$	4.2
$[Mn(C(OBFF_2)CH_3(CO)_4]$	7.8
$[Mn(CH_2C_6H_5)(CO)_5]$	0
$[Mn(C(OAlBrBr_2)CH_2C_6H_5)(CO)_4]$	1.6
$[(\eta^5-C_6H_5)Fe(CH_3)(CO)_2]$	0
$[(\eta^5-C_5H_5)Fe(C(OAlBrBr_2)CH_3)(CO)]$	2.5
$[(\eta^5-C_5H_5)Mo(CH_3)(CO)_3]$	0
$[(\eta^5-C_5H_5)Mo(C(OAlBrBr_2)CH_3)(CO)_2]$	35

[Metal complex] $= 1.00\ M$; $t = 20.0 \pm 0.1°C$; $P_{CO} = 383 \pm 17$ torr. [Lewis acid] $= 0.10\ M$.

stable, such as $[CpFe(COMe)(CO)_2]$, but not that of the elusive formyl complex, $[Mn(COH)(CO)_5]$. In this context it is interesting that surfaces of alumina and silica catalyze alkyl migration in related metal carbonyls.[(54)]

Trans effects can be important in migratory insertion reactions. Thus *trans* labilization appears to determine the product distribution[(55)] in the reaction of $[Ru(CO)_2X(Me)L_2]$ and L' (where $X = $ Cl, Br, or I; $L = PMe_2Ph$ or $AsMe_2Ph$, and $L' = $ CO or PMe_2Ph). Of the products, $[RuXL_2L'(COMe)]$, that in which L' and the acyl group are *trans,* is formed quickly and reversibly, owing to the *trans* effect exerted by the MeCO ligand, while the more stable isomer with the *cis* configuration is produced slowly. Scheme 6 is proposed.

A *trans*-labilizing ligand is also important in carbonyl insertion in $[PtCl(Ph)(CO)(PMePh_2)]$ complexes.[(56)] Of the three isomers only that with Ph *trans* to $PMePh_2$ yields an aroyl complex, namely, $[Pt_2(\mu-Cl)_2(COPh)_2(PMePh_2)_2]$. The position of equilibrium between carbonyl and acyl moves toward the former with variation of the organic ligand: Et, Ph, Me, CH_2Ph. When $PMePh_2$ is replaced by other phosphines and arsines, a similar change in equilibrium accompanies the fall in *trans* influence.[(57)]

Scheme 6

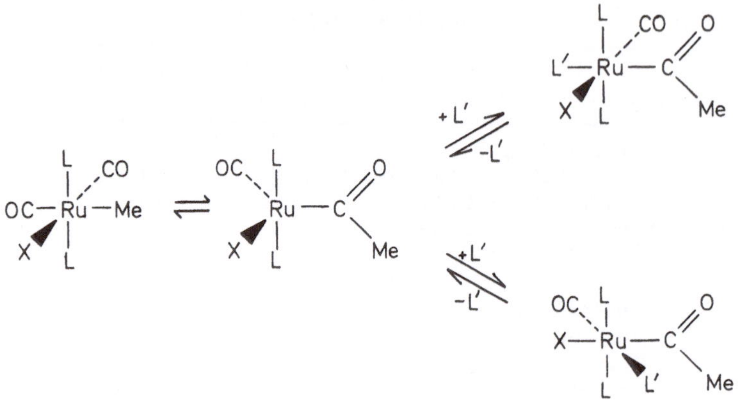

SO$_2$ has been shown to insert[58] into the M–Me rather than M–Ph bond in reactions (51) and (52):

$$cis\text{-}[PtMe(Ph)(PMePh_2)_2] + SO_2 \rightarrow cis\text{-}[Pt(SO_2Me)Ph(PMePh_2)_2] \qquad (51)$$

$$cis\text{-}[AuMe_2(Ph)(PMe_3)] + SO_2 \rightarrow abcd\text{-}[AuMe(SO_2Me)(PMe_3)(Ph)] \qquad (52)$$

In both cases the geometry at the metal is retained (which is shown in the second case by replacing one CH$_3$ group by CD$_3$). The reactions probably involve coordination of SO$_2$ followed by Me migration. Thus there is a contrast with insertion into Me$_n$Ph$_{4-n}$Sn complexes where the Ph–Sn bond is cleaved and where the attack of the SO$_2$ is probably electrophilic at carbon.

In several Fe–alkyl systems, SO$_2$ insertion is also electrophilic, involving backside attack on the α carbon. The same behavior is shown in the insertion of N-sulfinylsulfonamides and disulfonylsulfur diimides[59] into [CpFe(CO)$_2$CH$_2$R].

Two carbene systems have been studied by H. Fischer. The carbene [(OC)$_5$Cr(=C(SnPh$_3$)NEt$_2$)] rearranges[60] with loss of CO to the carbyne, [(Ph$_3$Sn)(OC)$_4$Cr($\equiv$CNEt$_2$)], by a first-order process with $\Delta H^{\ddagger} = 102$ kJ mol and $\Delta S^{\ddagger} = 34$ J mol^{-1} K^{-1}. As free carbon monoxide has no effect on the rate and solvent very little, the reaction may involve an intermediate in which Sn bridges the Cr=C carbene link. The insertion[61] of dimethylcyanamide into a tungsten carbene is first order in each reactant [see equation (53)].

$$[(OC)_5W{=}C(p\text{-}C_6H_4R)_2]$$
$$+ \ Me_2NCN \rightarrow [(OC)_5WC(NMe_2)N{=}C(p\text{-}C_6H_4R)_2] \quad (53)$$

Since the rate constant rises and correlates well with Hammett σ constants as R becomes increasingly donor, the reaction may involve nucleophilic attack by the terminal N of cyanamide on the C of the carbene.

Ugo's group[62] has studied the insertion of ketones into a dioxygen system, namely, $[L_2PtO_2]$ (L = PPh$_3$) (see Scheme 7). Rate studies indicate two path-

Scheme 7

ways, A and B, (i) and (ii) being rate-determining steps. Reaction A is associative since k_A falls as the ketone becomes less nucleophilic and more bulky, and since

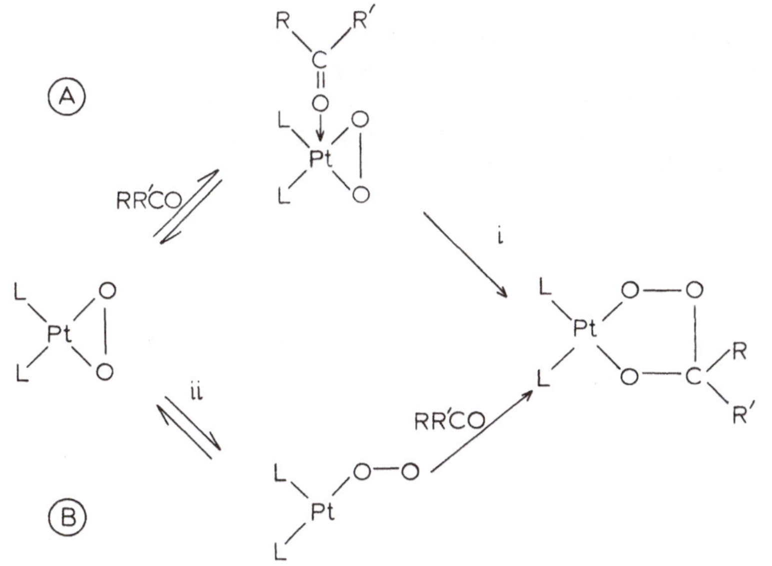

Table 8.14. Activation Parameters for Scheme 7 (Ketone = Acetone)[62]

Solvent	$\Delta H^{\ddagger}_A$, kcal mol^{-1}	$\Delta S^{\ddagger}_A$, cal K^{-1} mol^{-1}	$\Delta H^{\ddagger}_B$, kcal mol^{-1}	$\Delta S^{\ddagger}_B$, cal K^{-1} mol^{-1}
Benzene	8.8	−42.4	17.0	−20.3
Acetonitrile	11.9	−36.6	29.0	15.4
THF	12.7	−30.7	13.8	−32.4

Table 8.15. Influence of Solvent on Rate Constants at 30°C in Scheme 7 (Ketone = Acetone)[62]

Solvent	$10^4\, k_A$, $M^{-1}\, s^{-1}$	$10^5 k_B$, s^{-1}
Benzene	14.0	12.7
DMF	1.69	1.36
Propylene carbonate	1.84	0.266
Benzene + 0.195 M MeOH	133	494

$\Delta S_A^{\ddagger}$ is significantly negative (see Table 8.14). $\Delta H_B^{\ddagger}$ is larger than $\Delta H_A^{\ddagger}$, $\Delta S_B^{\ddagger}$ is somewhat variable, and k_B rather dependent on solvent, so that the product of step B could be the end-on dioxygen complex shown in the scheme. The fact that methanol causes k_B to rise and that hydrogen bonding stabilizes monohapto dioxygen systems supports this supposition (see Table 8.15).

Chapter 9

Rearrangements, Intramolecular Exchanges, and Isomerization of Organometallic Compounds

9.1. *Mononuclear Compounds*

9.1.1. *Isomerizations and Intramolecular Exchanges*

The mechanisms of *cis–trans* isomerization of Pd(II) and Pt(II) complexes (except organometallics) have been considered in Chapter 4. The uncatalyzed *cis–trans* isomerization of [PtXR(PEt$_3$)$_2$] (X = halide, R = aryl) occurs by an initial release of halide ion. The dependence of rate on the solvent (acetonitrile or alcohols) has been attributed to hydrogen bonding which assists the dissociation of halide to give T-shaped three-coordinate intermediates, rather than to nucleophilic displacement of halide by solvent.[1a] However, others have shown that the first step has a $\Delta V^{\ddagger}$ value of -12 cm^3 mol^{-1} (X = Br, R = mesityl, solvent = MeOH) and is hence believed to be an associative solvent displacement of halide. The isomerization step which follows ($\Delta V^{\neq}$ = + 7.7 cm^3 mol^{-1}) could be dissociative or intramolecular.[1b,c] The photoisomerization of [PtClPh(PEt$_3$)$_2$] is unaffected by [Cl$^-$] and must have a different mechanism to the thermal process. The *cis*-to-*trans* mechanism, possibly an intramolecular rearrangement of a low-lying ligand field state, is also believed to be different from the reverse which occurs by charge transfer excitation.[2] The rate of interconversion of *cis–trans* isomers of [Pt(η^3-crotyl)(C$_6$Cl$_5$)(PPh$_3$)] is suppressed by added PPh$_3$ and a dissociative path via a three-coordinate intermediate is considered. Substitution of PPh$_3$ by P(C$_6$H$_4$OMe-4)$_3$ occurs faster than isomerization and with retention of stereochemistry.[3]

The rate of exchange of the ligand L which is *trans* to hydride with those *cis* in [RhHL$_3$] increases in the order L = PPr$_3^i$ > PPh$_3$ > PEt$_3$. This process is clearly intramolecular [unlike those of Pt(II)] and involves intermediates of threefold symmetry. Since the P–Rh–P angles in the planar PPr$_3^i$ complex are already 109.3°, an intermediate close to trigonal pyramidal (planar RhP$_3$) is preferred to a ψ-tetrahedral one for which there would be no enlargement of these angles.[4]

^{31}P nmr line-shape analysis has been used to obtain the rates of intramolecular exchange of coordinated with noncoordinated phosphorus nuclei in the 16-electron complexes [PtMe$_2$(polyphosphine)][5] and in the four-coordinate 18-electron complex [Rh(NO){P(CH$_2$CH$_2$PPh$_2$)$_3$}].[6] Negative $\Delta S^{\neq}$ values indicate associative mechanisms even though the rhodium complex would require then a 20-electron intermediate, unless its formation is associated with a bending of the Rh–N–O unit.

The isotopically enriched complexes [W(CO)$_4$(^{13}CO){P(OMe)$_3$}] and [W(CO)$_4$(C^{18}O){P(OMe)$_3$}] were synthesized with preferential enrichment in the equatorial position (*cis* to phosphite). At 78°C there is a slow equilibration between the axial and equatorial sites to give a statistical distribution but this occurs without any exchange of coordinated with free CO; the reaction was carried out under ^{13}CO to establish this. Therefore the rearrangement is intramolecular.[7] The *cis–trans* isomerization of [W(CO)$_4$(PR$_3$)$_2$] is also believed to occur intramolecularly but this has not been so well established.[8] As the number of phosphine ligands in such molecules increases there is a greater tendency to have dissociative exchange. In *cis*-[Mo(CO)$_2$(PMe$_2$Ph)$_4$] there is a rapid exchange of only two phosphines with free phosphine leading to coalescence of ^{1}H nmr signals. Although not confirmed, it is likely to be the PMe$_2$Ph ligands *trans* to CO that undergo dissociative exchange.[9] Finally, for octahedral complexes, the photoisomerization of [RuCl$_2$(PPh$_3$)$_2$(Bu^{t}NC)$_2$] involves dissociation of PPh$_3$. The dimer formed by condensation of the five-coordinate intermediates was detected.[10]

The intramolecular exchange behavior of several seven-coordinate hydride compounds has been examined.[11–14] The pentagonal bipyramidal compound $[CrH_2\{P(OMe)_3\}_5]$ (1) has the hydrides ligands *cis* and in the equatorial plane

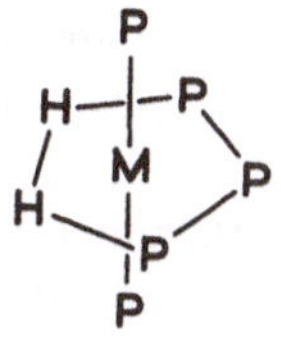

1

consistent with the observed $AB_2CC'XX'$ spin system. Nmr coalescence leads to a ^{31}P nmr triplet and a ^1H (hydride) nmr sextet. Of the eight basic permutational sets for the exchange, only one (C_2) agrees with the observed line shapes. This corresponds closely with a Berry pseudorotation (BPR) mechanism appropriate to this geometry, that is, a simultaneous exchange of two axial and two equatorial $P(OMe)_3$ ligands.[11] The tungsten analog appears even more fluxional (barrier < 32 kJ mol^{-1}), while $[WH_4\{P(OMe)_3\}_4]$ also undergoes intramolecular exchange.[12] Assuming that there is no exchange intermediate of any appreciable concentration, an analysis of the nmr line shapes at different temperatures for $[MoH(CF_3CO_2)\{P(OMe)_3\}_4]$ (2) does not lead to any mechanism involving any

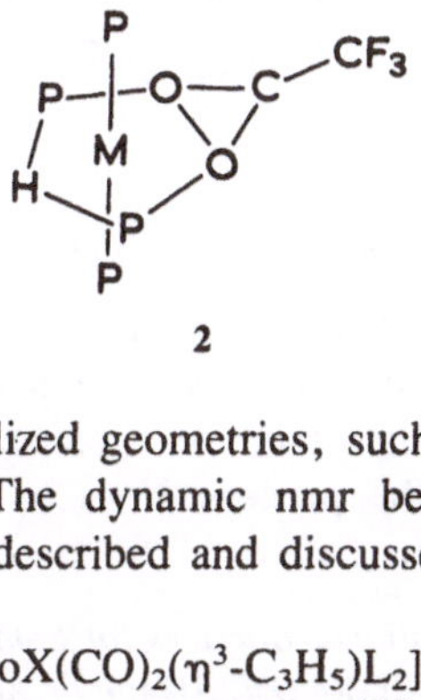

2

simple paths involving idealized geometries, such as a capped octahedron or capped trigonal prism.[13] The dynamic nmr behavior of $[VH(CO)_3L]$ and $[VH(CO)_4L]$ has also been described and discussed where L is a polydentate phosphine ligand.[14]

The allyl complexes $[MoX(CO)_2(\eta^3\text{-}C_3H_5)L_2]$ where X = halide, L = Py or substituted Py[15] or X = Cl, L = $P(OMe)_3$[16] are also dynamic. Structures and intramolecular behaviour are discussed.

9.1.2. Simple Ligand Rotation at a Metal Center

There continue to be reports on rotations of coordinated alkenes, alkynes, etc. about the metal–ligand axis rapid enough to coalesce nmr signals. In square planar complexes the alkene is preferentially orientated perpendicularly. Orientations for other coordination geometries have also been examined experimentally and theoretically. The two observed rotamers of *cis*-[PtCl$_2$(PMe$_2$Ph)-(CH$_2$=CHOCOMe)] in ratio 1.4:1.0 interconvert with $\Delta G\ddagger = 62.7 \pm 0.8$ kJ mol^{-1}, higher than in the corresponding ethene and propene compounds (51.4 and 56.4 kJ mol^{-1}, respectively).[17] Although vinylacetate should be a better π acceptor, the difference could be of steric origin. Calculations indicate that alkene orientations in linear two-coordinate, in square planar, and in octahedral complexes are predominantly under steric rather than electronic control and that small rotational barriers are to be expected if bulky ligands are not employed.[18] In trigonal bipyramidal and trigonal planar molecules, π-bonding effects are more important and favor the alkene oriented in the trigonal plane.[18,19] For example, the conformation with the alkenes perpendicular to this plane in [Ni(C$_2$H$_4$)$_3$] is less stable than the in-plane conformation, whereas in [Ni(C$_2$H$_4$)$_2$] there is little angular preference.[19]

Coupling of changes of coordination geometry with alkene rotation is possible. An ethene rotation barrier of 134 kJ mol^{-1} in [Fe(CO)$_4$(C$_2$H$_4$)] is calculated if the Fe(CO)$_4$ group is maintained rigidly C_{2v} but coupling of Berry pseudo-rotation (BPR) with alkene rotation reduces this to 50 kJ mol^{-1}. Scheme 1 shows

Scheme 1

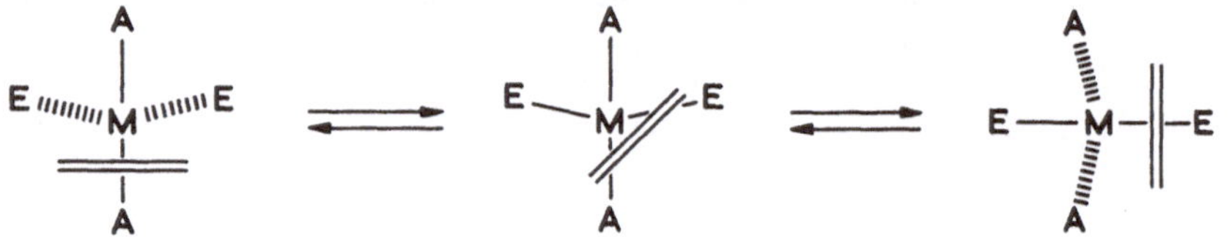

these coupled motions.[18] The tetracyanoethene compounds [M(Bu^{t}NC)$_4$(TCNE)]$^+$ (M = Co or Rh) are believed to behave in this way, but in CH$_2$Cl$_2$ solution, for example, ion pairing is important. As a consequence, the counterion has a marked effect on the rate of this process. For Co, E_a is 54.4, 67.8, and 77.4 kJ

Table 9.1. Rotational Barriers for Some Complexes of Type [Fe(C$_5$H$_5$)(CO)L(Alkene or Alkyne)]$^+$ [22]

Alkene/alkyne	L	$\Delta G^{\ddagger}$, kJ mol^{-1}	Temperature, °C
C$_2$H$_4$	P(OPh)$_3$	33	−95
PhC≡CPh	P(OPh)$_3$	58	+25
EtC≡CEt	p(OPh)$_3$	55	−5
C$_2$H$_4$	PPh$_3$	42	−40
MeC≡CMe	PPh$_3$	60	+1
EtC≡CEt	PPh$_3$	63	+18

mol^{-1} with counterions ClO_4^-, PF_6^-, and BPh_4^-, respectively, but such variations are not easily interpreted.[20,21]

For the cationic cyclopentadienyl compounds given in Table 9.1, dependence of the rotational barrier on the unsaturated ligand has been studied. Alkynes have higher barriers than ethene and the use of $P(OPh)_3$ gives lower barriers than PPh_3. Although steric effects can account for this, it does seem that metal-to-ligand π bonding is important. In particular, $PhC{\equiv}CPh$ has only a slightly higher rotational barrier than $EtC{\equiv}CEt$ [L = $P(OPh)_3$], which suggests that the steric contribution to the barrier is not large.[22]

The alkene ligands in *trans*-[M(CO)$_4$(methylacrylate)$_2$] (M = Mo or W) are mutually perpendicular and the two diastereomers **3** and **4** are not interconvertible by alkene rotation. Rotational barriers are somewhat higher for W than Mo and are different for the two diastereomers: 69.4 ± 2.0 (**3**; M = W) and 81.5 ± 2.0 kJ mol^{-1} (**4**; M = W). This illustrates the dangers in accounting for differences in barriers in simple terms; in this case steric effects must be essentially the same in **3** and **4** while electronic differences must be rather subtle.[23] Alkyne rotation is observed in *cis*-[W(CO)(Me$_2$NCS$_2$)$_2$(alkyne)] (**5**) to occur

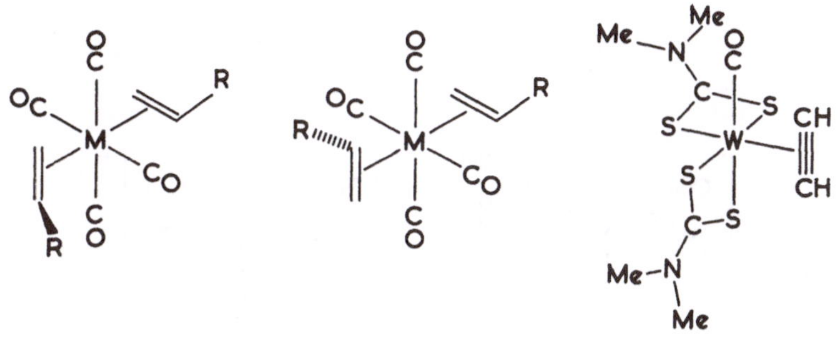

3   4   5

without exchange of the four nonequivalent Me groups and must occur without any complications involving monodentate dithiocarbamate ligands, etc.[24]

Rotation of an η^2 ligand is also required in the isomerization of the η^2-acyl compound **6** to **7**. Considerably slower isomerizations occur when R = aryl, for example, $\Delta G^{\ddagger}$ (214 K) = 63.5 kJ mol^{-1} (R = Ph) compared with $\Delta G^{\ddagger}$

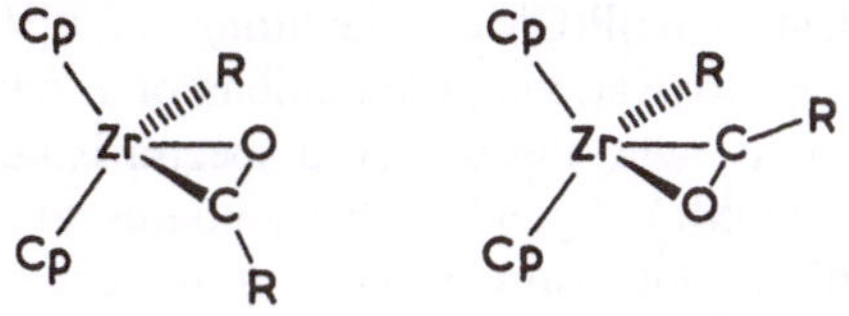

6   7

$(150 \text{ K}) = 47.7 \text{ kJ mol}^{-1}$ (R = Me). Whether an η^1-acyl intermediate is involved or not seems unknown.[25]

The plane of the CH_2 ligand in $[Fe(C_5H_5)(Ph_2PCH_2CH_2PPh_2)(CH_2)]^+$ is perpendicular to the FeP_2 plane to maximize Fe–C π bonding. The two CH_2 hydrogen atoms are therefore nonequivalent (^1H nmr signals at $\delta 17.29$ and $\delta 13.89$) but there is restricted rotation about the Fe–CH_2 bond ($T_c \sim -30°C$). Compared with $[Fe(C_5H_5)(CO)_2(CH_2)]^+$ the diphosphine complex would have greater multiple iron–carbon bonding giving a greater rotational barrier and reduced electrophilicity of the CH_2 ligand.[26]

Dynamic behavior of η^3-allyl complexes occurring without *syn–anti* hydrogen exchange is normally attributed to rotation about the allyl–metal bond. This can normally be detected only where different geometric isomers are present in equilibrium as for $[FeBr(C_3H_5)(CO)_3]$ or where the allyl termini are different as in $[Fe(NO)_2(PPh_3)(\eta^3\text{-}C_3H_5)]^+$ [27] and $[Co(CO)_2(PPh_3)(\eta^3\text{-}C_3H_5)]$.[28] The intermolecular interconversion of enantiomers 8 and 9 and of 10 and 11 is detected

by the exchange of termini a and b. η^1-Allyl intermediates are not involved as in other cases involving dynamic allyl (see Section 9.1.3).

Iron and ruthenium compounds of type $[M(\eta^4\text{-}1,3\text{-diene})L_3]$, where L_3 can be various combinations of CO, tertiary phosphine, or isocyanide, have square-pyramidal geometries with one ligand L in a unique axial position. It is universally observed that there is an intramolecular exchange of axial and basal ligands L.[29-33] For $[Fe(\eta^4\text{-pentadiene})P(OMe)_3]$ the fitting of $^{31}P\{^1H\}$ nmr spectra (ABC spin system), calculated for various permutations of P nuclei and linear combinations of permutations, with the observed spectra is best for a cyclical exchange of the three $P(OMe)_3$ ligands. This amounts to a rotation about the Fe–diene axis but this is not distinguishable from sequential BPR steps (the square-pyramidal form passing through trigonal bipyramidal forms) (Scheme 2).[29] Activation data have been obtained for variously substituted complexes

Scheme 2

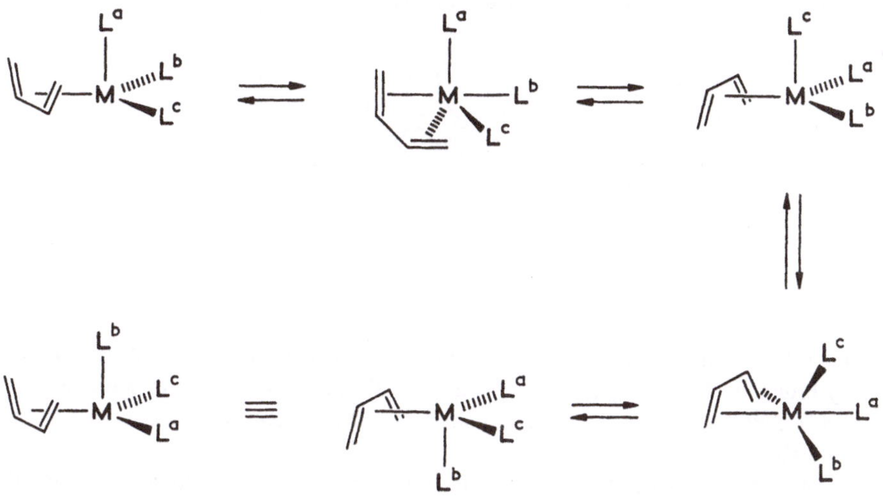

with $\Delta G^{\ddagger}$ varying between 28 and 46 kJ mol^{-1} depending upon L and the diene. Higher barriers are found for Ru than Fe,[30] although for BPR of [M(PF$_3$)$_5$] (M = Fe or Ru) the reverse is true. A rather high value of $\Delta G^{\ddagger}$ (68 ± 1 kJ mol^{-1} with T_c 120°C) is found for [Fe(η^4-Bu^{t}N=C=CPhCPh=C=NBu^t)(Bu^{t}NC)$_3$] (12).[33]

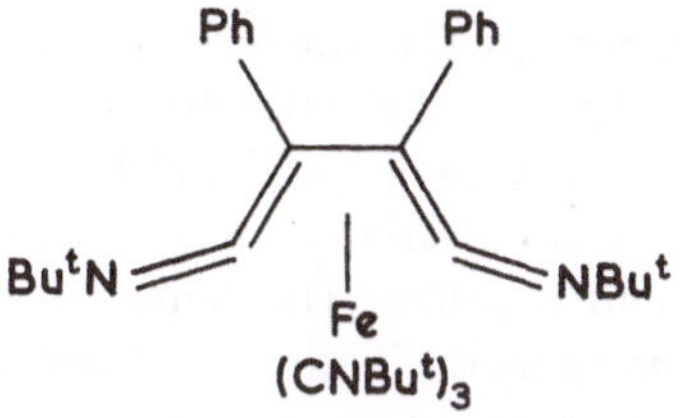

12

Some structures of compounds [Fe(cyclobutadiene)L$_3$] close to one of the idealized forms **13** or **14** are found but an analysis of known structures has shown no real angular preferences with structures being found with various conformations between **13** and **14**. No case of a frozen-out nmr spectrum has been

13 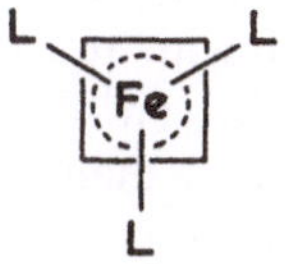   **14**

reported so rotational barriers must be exceedingly low.[34] In contrast the $C_4(CF_3)_4$ complex **15** shows four different CF_3 signals in the ^{19}F spectrum which coalesce

15

above 25°C. The orientation shown is preferred because replacement of CO and I by bidentate Me_2NCS_2 gives a spectrum with three ^{13}F signals which coalesce at 70°C.[35] The origin of these unusually high rotational barriers is still to be established.

A calculation of the energy differences between eclipsed and staggered metallocenes has given a value of 2.78 kJ mol^{-1} for ferrocene (exptl. value 3.8 ± 1.3 kJ mol^{-1}) and 4.66 kJ mol^{-1} for ruthenocene.[36] Recent X-ray and neutron diffraction data for ferrocene have shown that the triclinic low-temperature form has cyclopentadienyl rings rotated only 9° out of the eclipsed conformation. The structure of the high-temperature monoclinic form is disordered while the gas phase structure is eclipsed.[37-39] A gas phase electron diffraction structure of $[Fe(C_5Me_5)_2]$ is also eclipsed with a rotational barrier of 4.2 ± 1.3 kJ mol^{-1}.[40] Rotational barriers in 1,1′-dihaloferrocenes have been determined from the temperature dependence of dielectric measurements.[41] A method (mechanical spectroscopy) of determining rotational barriers has been applied to $[PtMe_3(C_5H_5)]$. A solution with polystyrene in benzene is evaporated to dryness and the residue pressed into a pellet. The temperature at which the energy dispersion is a maximum when an oscillating mechanical stress is applied to the pellet corresponds to a coincidence of the applied frequency with the rotational frequency.[42]

9.1.3. Ligand Motion Involving Changes in Hapticity

Syn–anti hydrogen exchange at a terminal CH_2 group of an η^3-allyl or η^4-diene is established in some cases to result from having an intermediate with a M–C σ bond at this carbon. No evidence has been given for the unlikely rotation about the CH_2–C bond with retention of hapticity. The molecules $[Mo(\eta^3$-$C_3H_5)(CO)_2(C_5H_5)]$ and $[Mo(\eta^3$-$C_3H_5)(NO)I(C_5H_5)]$ have been compared structurally and with respect to their *exo–endo* isomerization mechanisms. Allyl rotation in the dicarbonyl (no *syn–anti* exchange) interchanges *exo* and *endo* isomers. The crystallized *exo* isomers contain an entirely symmetrical η^3-allyl unlike the crystallized *endo* form of the nitrosyl compound, which has a very unsymmetrical allyl. The allyl terminus *cis* to the high *trans* influence NO ligand is closer to Mo and related to this the *exo–endo* interconversion is via a η^1-allyl (Scheme 3). In the η^1-allyl intermediate the σ-bonded carbon recognizes its place

Scheme 3

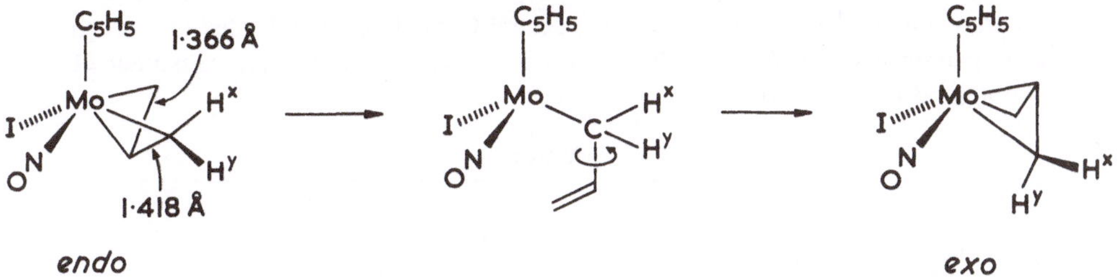

endo   *exo*

of origin since it returns *cis* to NO in the *exo* product. Spin-saturation transfer techniques have shown that the *syn* hydrogen H^y in the *endo* isomer becomes *anti* in the *exo* isomer. Allyl rotation is slower than η^3 to η^1 transformation in this case.[43,44]

Activation data for *syn–anti* hydrogen exchange in $[M(C_3H_5)_4]$ (M = Zr or Hf), $[Zr(cot)(C_3H_5)_2]$, and $[Hf(cot)(C_3H_5)_2(thf)]$ have been reported.[45] Although *syn–anti* exchange is observed for $[Ir(NO)(\eta^3-C_3H_5)(PPh_3)_2]^+$ above $-17°C$, below $-83°C$ signal broadening may be due to a slowing of exchange between isomers with linear and bent M–N–O groups, respectively (there is infrared evidence for both forms in solution).[46]

Addition of PMe_3 to $[Ni(\eta^3\text{-allyl})_2]$ gives a 1:1 adduct for which the 18-electron form (16) is found in the crystal.[47,48] This form exists exclusively in solution but *syn–anti* exchange ($\Delta G^{\ddagger} = 40 \pm 6$ kJ mol^{-1}) occurs with T_c at $-70°C$ without loss of ^{31}P coupling by an intramolecular path via the 16-electron compound (17) (Scheme 4). Above $-40°C$ the loss of ^{31}P coupling implies a rapid reversible dissociation to 18. With $[Ni(CH_2CHCMe_2)_2(PMe_3)]$ or $[M(C_3H_5)_2(PMe_3)]$

Scheme 4

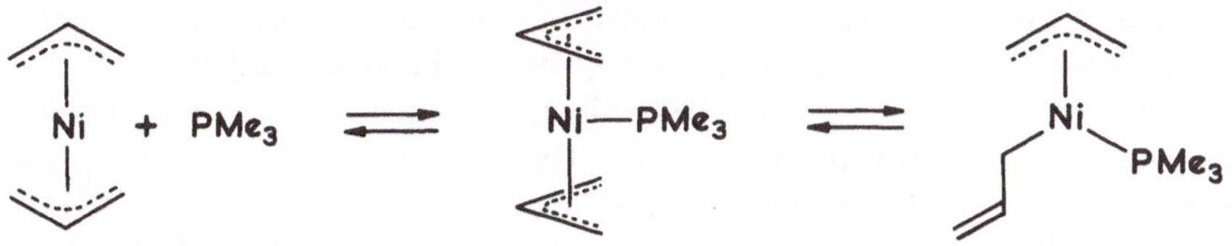

18   16   17

(M = Pd or Pt) the 16e forms like **17** predominate in solution. In these cases there are rapid interconversions of the η^1- and η^3-allyl ligands with $\Delta G^{\ddagger}$ 40 ± 6 (M = Pd) and 88 ± 6 (M = Pt) kJ mol^{-1}. Presumably this exchange is via an 18-electron species like **16**. The two noninterconverting isomers of **19** are believed to differ only in which face of the central double bond is coordinated. These isomers react at different rates with PMe$_3$ to give the same compound [Ni(C$_{12}$H$_{18}$)(PMe$_3$)] (**20**) for which there is a rapid interchange (T_c = $-5°C$) between the η^1 and the η^3 ends of the C$_{12}$ chain. It is not known whether this occurs via a di-η^1-intermediate with the central C=C bond coordinated or through a di-η^3-intermediate. At 15°C the *trans,trans,trans*-cyclododecatriene isomer **21** is obtained (Scheme 5).[48]

Scheme 5

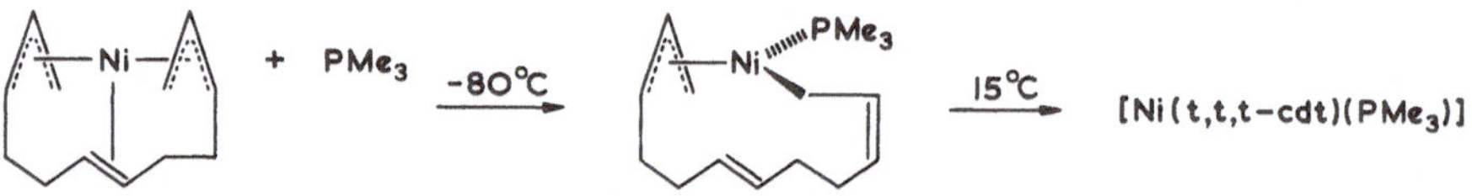

Similar changes in hapticity occur when the η^5 and η^1 ligands in [Pd(η^5-C$_5$H$_5$)(η^1-C$_5$H$_5$)(PR$_3$)] exchange intramolecularly; migration of Pd about the η^1 ring occurs even more rapidly. There is a marked rate dependence of ligand interchange on the tertiary phosphine with rates in the order PMe$_3$ > PPh$_3$ > PPr$_3^i$.[49] In the above example ligand interchange is a degenerate process, but for related cyclopentadienyl–allyl compounds the corresponding process is an exchange of isomers as in equation (1) (R = Pr^i or cyclohexyl). In this equilibrium between 18-electron and 16-electron species, it is not known whether

$$[\text{Pd}(\eta^5\text{-C}_5\text{H}_5)(\eta^1\text{-2-MeC}_3\text{H}_4)(\text{PR}_3)] \rightleftharpoons [\text{Pd}(\eta^1\text{-C}_5\text{H}_5)(\eta^3\text{-2-MeC}_3\text{H}_4)(\text{PR}_3)]$$

(1)

there is an associative or dissociative mechanism.[50] The reversible gain and loss of η^2-styrene from [Rh(η^2-styrene)(η^3-C$_6$H$_5$CHMe)(cod)] seems to be fundamentally associated with the exchange of the two *ortho*-hydrogen atoms of the η^3-benzyl ligand, presumably via a η^1-benzyl intermediate.[51]

The complex [Ni(η^3-C$_3$H$_5$)(SH)(PMe$_3$)] undergoes *syn–syn* and *anti–anti* exchange more rapidly (observed above $-100°C$) than *syn–anti* exchange, which is only observed in the ^{1}H nmr spectrum above $-20°C$.[52]

A very interesting *syn–anti* exchange is observed for the complex [Zr(C$_5$H$_5$)$_2$(s-*cis*-C$_4$H$_6$)] (**22**), formed along with the unusual *s-trans*-isomer on photolysis of a mixture of butadiene and $[Zr(C_5H_5)_2Ph_2]$. Within experimental error the rate of *syn–anti* hydrogen exchange is equal to the rate of exchange of cyclopentadienyl rings A and B (Scheme 6) which is compatible with the di-σ-bonded intermediate (23) shown. This is formally a Zr(II)–Zr(IV) interconversion and

Scheme 6

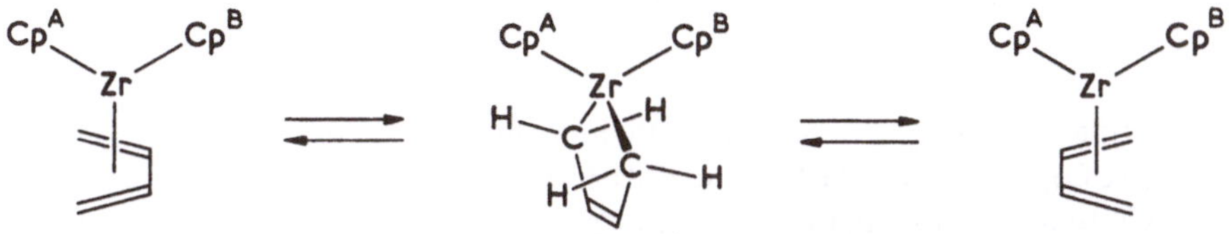

might only be accessible for 1,3-diene complexes having a highly favorable oxidation state 2 higher.[53] There is no evidence for an intramolecular diene flip of this sort with well-known η^4-diene complexes of the later transition metals. For example, *syn–anti* isomerization of $[Fe(CO)_3(CH_2{=}CHCH{=}CHCOMe)]$ occurs only when catalyzed by base.[54]

9.1.4. *Metal Migration between Different Ligand Sites*

Ring whizzing and related behavior is frequently observed but has been little studied recently from a mechanistic viewpoint. However, the nmr spectra relating to the fluxional behavior of $[M(CO)_3(\eta^6\text{-cot})]$ (M = Cr or W) have been reexamined and interpreted in terms of combinations of 1,2 and 1,3 shifts.[55]

The η^4-cot ligand in $[Ru(\eta^6\text{-}C_6Me_6)(\eta^4\text{-cot})]$ is closer to the metal than in $[Ru(CO)_3(\eta^4\text{-cot})]$. Although there is greater metal to cot π donation in the hexamethylbenzene compound, the activation energy for 1,2-metal atom shifts at the cot ligand is around 8 kJ mol^{-1} lower. A transition state for a 1,2 shift requiring the ligand to approach $[cot]^{2-}$, favored by increased π backing–bonding, was suggested to explain this unexpected result.[56]

The dynamic behavior of η^1-, η^3-, and η^5-tropylium complexes has also been studied.[57-59] The iron cations $[Fe(CO)_3(\eta^5\text{-}C_7H_6R)]^+$ (R = Me, Ph, cyclohexyl, Pr^i) contain various interconverting isomers which give frozen-out nmr spectra at − 100°C showing the 2-isomer 25 and the 6-isomer 27 predominantly. The 1-isomer 24 is not observed. As the temperature is raised coalescences show that in order of increasing activation energy there is a degenerate enantiomerization of the 2-isomer 25 via the 3-isomer 26, then a degenerate enantiomerization of the 6-isomer 27, and only at still higher temperatures do the 2- and 3-isomers 25 and 26 interconvert with the 6-isomer 27. The unobserved 1-isomer 24 must intervene in this last reaction.[57]

24   25   26   27

Ring whizzing of the η^3-tropylium ligand in the complex $[M(\eta^3\text{-}C_7H_7)(CO)_2(Me_2NCH_2CH_2OGaMe_2C_3H_3N_2)]$ has been identified,[58] while in $[Re(CO)_5(\eta^1\text{-}C_7H_7)]$ dnmr (dynamic nmr) behavior was analyzed in terms of 1,2 shifts $(\Delta G^{\ddagger} = 83 \pm 4 \text{ kJ mol}^{-1})$.[59] The η^3 form of $[Pt(\text{cyclohexadienyl})(PH_3)_2]^+$ was calculated to be only slightly more stable than the η^5 form and of the two possible orientations of the latter, the one with the PtP_2 plane coinciding with the plane of symmetry through the C_6H_7 ligand is just the more stable. These quite small energy differences are consistent with the observed dynamic behavior of compounds such as $[Pd(\eta^3\text{-cycloheptadienyl})(PR_3)_2]^+$.[60]

The trityl complex $[Pd(acac)(\eta^3\text{-}CPh_3)]$ **28** shows three distinct dynamic

28

processes. The lowest energy process is the transfer of Pd between the two *ortho* sites of the π-bound Ph ring which can be distinguished from a transfer between all six *ortho*-Ph sites which occurs more slowly. These processes retain an unsymmetrical acac ligand and the rate of the slower exchange of the acac termini is slightly concentration dependent and probably not intramolecular.[61] Transfer of Fe between the two six-membered rings in $[Fe(C_5H_5)(\text{naphthalene})]$ was used to explain the Jahn–Teller-effect-induced transition in the esr spectrum from anisotropic to isotropic. Isomers are observed when unsymmetrically substituted naphthalenes are used.[62]

9.1.5. *Hydrogen Migrations Involving Hydrogen Bound to Carbon*

Intramolecular hydrogen migration between carbon and metal atoms is, of course, important to various catalytic reactions and may be sufficiently fast to give nmr coalescence. Treatment of compound **29** with D_2O in CD_3NO_3 at

$-20°C$ leads to complete exchange with the ethene hydrogen atoms as well as the hydride.

$$[RhH(C_2H_4)(C_5H_5)(PMe_3)]^+ \rightleftharpoons [Rh(C_2H_5)(C_5H_5)(PMe_3)]^+ \qquad (2)$$
$$\text{29 (18e)} \qquad\qquad\qquad \text{30 (16e)}$$

In addition to reversible deprotonation there must be exchange between the hydride and the C_2H_4 ligand. Although **29** shows sharp hydride and C_2H_4 ^1H nmr signals at $-20°C$, at room temperature these were broadened due to the increasingly rapid equilibrium with **30**.[63] A similarly rapid hydrogen transfer is found for [IrH(cod)(η^4-cyclohexadiene)], which is the predominant species in a rapid tautomeric equilibrium with [Ir(cod)(η^3-cyclohexenyl)]. A separate hydride nmr signal is observed at $-70°C$ but by $57°C$ this has coalesced with the two *endo*-hydrogen signals and there is a rapid exchange of all the carbon atoms of the six-membered ring. The cod ligand is not involved.[64] Protonation of [Fe(diene)L$_3$] [L = P(OMe)$_3$] gives a closely related system except that the incoming proton resides at carbon. The protonated butadiene compound (**31**) contains a 2e-3c Fe–H–C bridge and, although H^A, H^B, and H^C gives a coalesced signal at δ -5.92 at $22°C$, three separated signals are observed at $-95°C$: δ -15.2, -2.4, and -2.0. There is successive contact of the methyl hydrogen atoms with iron and the Fe–H bond is believed to be strongly covalent and like that established by neutron and X-ray diffraction in [Fe(η^3-cyclooctenyl)L$_3$]$^+$ (**32**). The exchange of L^A with L^B is very rapid even at $-70°C$ because of the

31   32

oscillatory exchange of the hydrogen bonded to iron with the corresponding hydrogen atom on the other side of the allyl. Exchange of L^A and L^B with L^C is slower. One-electron reduction of **33** gives a structurally related compound but with a much longer and weaker Fe–H bond; dynamic processes are correspondingly faster. For example, E_a for the permutation of all three phosphorus nuclei is 21 kJ mol^{-1} for the neutral compound compared with 60 kJ mol^{-1} for the cation.[65]

Scheme 7

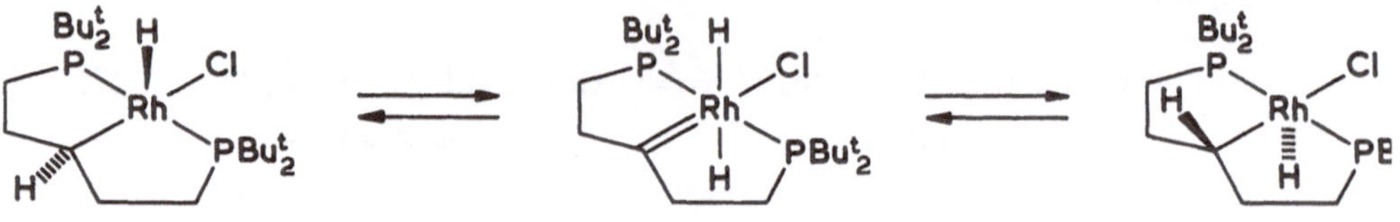

33

A $^1H[^{31}P, ^{103}Rh]$ INDOR experiment on **33** (Scheme 7) gave the ^{103}Rh resonance as a triplet with equivalent coupling to two protons consistent with the rapid process shown. A concerted exchange of the two hydrogen atoms is possible but nonconcerted pathways seem more reasonable for a coordinatively unsaturated species.[66] Complex **34** at 145°C undergoes intramolecular H-atom

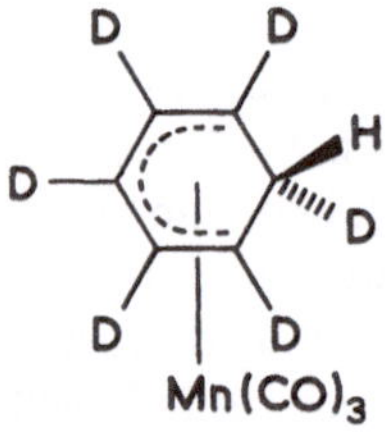

34

scrambling into the vinylic sites ($\Delta G^{\ddagger} = 142$ kJ mol^{-1}) with relative initial rates into the 1- (or 5-), 2- (or 4-), and 3-positions of 1.1, 1.2, and 1.0, respectively. Observations are consistent with the formation of [MnD(CO)$_3$(η^4-C$_6$D$_5$H)] which can immediately transfer D back to carbon or undergo 1,2-manganese shifts at the η^4-arene, allowing D migration back to other sites.[67] All the cases so far considered involved transfer of an *endo*-hydrogen between carbon and metal, but complex **35** undergoes slow exchange of D with the ring hydrogen atoms at 20°C implying an *exo*-hydrogen sigmatropic shift in **36** formed by D transfer.[68]

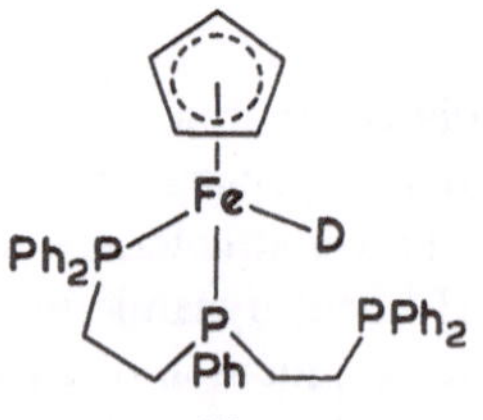

35 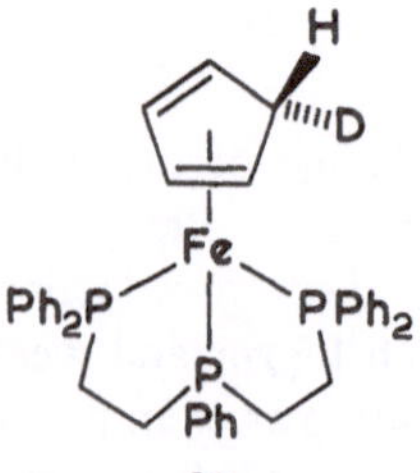   **36**

9.1.6. Alkyl Migration Reactions

Intramolecular transfer of Me to CO in $[MnMe(CO)_5]$ is the first step in the CO insertion reaction, but the 16-electron species so formed is extremely reactive toward the reverse Me migration or to coordination of donor molecules so has not been detected. Two cases have been reported recently where the an acetyl species is in equilibrium with its methyl isomer.[69,70] The crystals of acyl complex shown in equation (3) contains an η^2-acetyl ligand albeit with a

$$[Ru(\eta^2\text{-}COMe)I(CO)(PPh_3)_2] \rightleftharpoons [RuMeI(CO)_2(PPh_3)_2] \tag{3}$$

fairly long Ru–O bond so that an 18-electron rather than 16-electron formulation is appropriate. In solution the methyl complex is the main species.[69] In contrast the intramolecular isomerization between **37** and **38** shown in Scheme 8 (L = $PMePh_2$) involves a five-coordinate η^1-acyl ligand. When crystalline samples of

Scheme 8

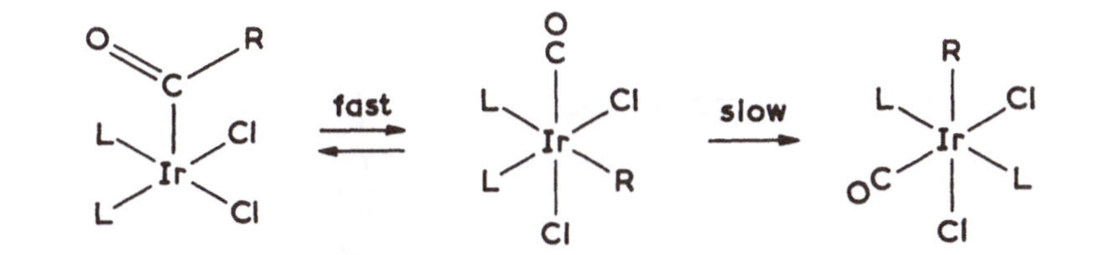

37 are dissolved, an equilibrium mixture with **38** (R = Me or Pr^n) is established within 1 min as established by ^1H nmr and infrared spectra. Results are consistent with an intramolecular isomerization with the ligands L remaining *cis*. The first-order *cis*-to-*trans* isomerization giving **39** (R = Et) is very slow ($t_{1/2} \sim 1$ week in $CDCl_3$) but much faster with methanol present ($t_{1/2} = 7.7$ min in 2:1 $CDCl_3$/ CD_3OD). The rates are also increased by adding $LiClO_4$ and decreased with LiCl. Dissociation of Cl^- allows a *cis–trans* interconversion of a five-coordinate cationic intermediate.[70]

9.1.7. Intraligand Rotations and Rearrangements

The barriers for rotation about the C(carbene)–C(aryl) bond in $[Fe(C_5H_5)(CO)_2(CHC_6H_4R\text{-}4)]^+$ are 38 (R = H) and 44 (R = Me) kJ mol^{-1} and were determined from coalescence of the two *ortho*-hydrogen signals. The arene ring is in the plane of the Fe–CH–C(aryl) atoms bisecting the CO ligands and there must be a strong conjugative interaction (stronger when $R = Me$) between the arene and the carbene carbon atom.[71] There is also restricted rotation about the C(cyclobutadiene)–mesityl bonds in $[Co(C_5H_5)(\eta^4\text{-}1,3\text{-}Ph_2\text{-}2,4\text{-}Mes_2\text{-cyclobutadiene})]$.[72] The interesting intramolecular interconversions of **40** and **41** were originally thought to pass through the tricyclic intermediate (**42**) or the acyclic one (**43**), but a theoretical treatment has indicated that **42** and **43** are much higher in energy than **40** or **41** and unlikely to be intermediates. The cyclooctadienediyne intermediate **44** is another possibility so far unexplored.[73]

40   **41**   **42**

43   **44**

The *exo–endo* interconversion of $[Fe(CO)_3(C_6H_7OMe)]$ formed by methoxide addition at the cyclohexadienyl complex $[Fe(CO)_3(\eta^5\text{-}C_6H_7)]^+$ has been shown to be possible. *Exo* addition is faster than *endo* but the *endo* product may be formed slowly if the *exo* addition is reversible and the *endo* product thermodynamically favored.[74]

9.2. Dinuclear Compounds

9.2.1. Migration of CO and Related Ligands

The exchange of bridging with terminal CO ligands in compounds such as $[Co_2(CO)_8]$ and $[Fe_2(C_5H_5)_2(CO)_4]$ is believed to occur in pairs. In solution cobalt carbonyl has isomers with and without bridging CO in equilibrium, but the only isomer observed for $[Co_2(Bu^tNC)_8]$ in the crystal and in solution (^{13}C nmr analysis) has two bridging isocyanide ligands. Exchange with free Bu^tNC is slow, and there is a much faster exchange of terminal with bridging Bu^{t}NC probably via the unobserved nonbridged isomer.[75] The mechanisms of interconversion of isomers of $[M_2(C_5H_5)_2(CO)_3(RNC)]$ (M = Fe or Ru) and of $[Fe_2(C_5H_5)_2(CO)_2(RNC)_2]$ appear to be similar to that established for bridging-terminal CO exchange in the parent tetracarbonyl compound.[76] The complex $[Fe_2(CH_2)(CO)_8]$ has a structure in the crystal like that of $[Fe_2(CO)_9]$ but with a bridging CH$_2$. The mechanism of conversion into the non-carbonyl-bridged form found in solution is probably similar.[77] A very specific exchange occurs with the alkylidene-bridged compound (**45**). Scheme 9 shows how there can be exchange of the C$_5$H$_5$ ligands without exchange of the mobile CO ligands when R^1 and R^2 are different.[78]

Scheme 9

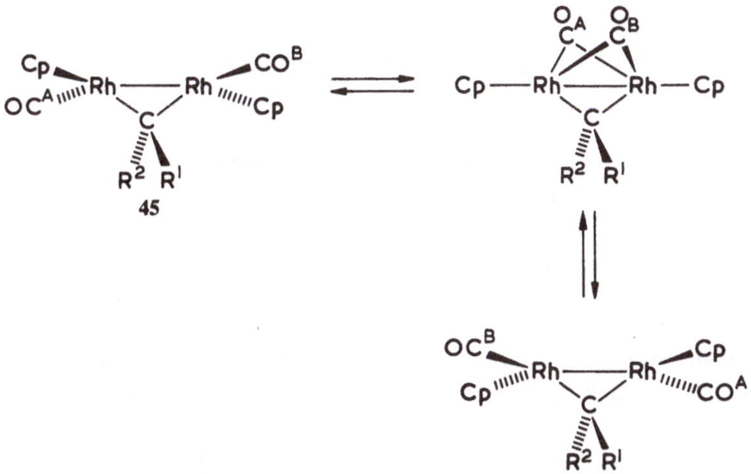

9.2.2. *Hydrogen Migration Reactions*

Compound **46** (phosphine ligand = Ph$_2$PCH$_2$CH$_2$CH$_2$PPh$_2$) maintains the same structure in solution as in the crystal but undergoes a rapid exchange of the bridging hydride ligands (H^b with H$^{b'}$) without exchange of either with the

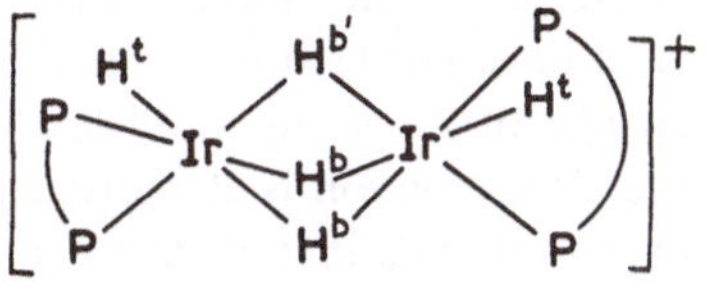

46

terminal hydrides.$^{(79)}$ In [(PPh$_3$)$_2$Rh(μ-H)$_2$W(C$_5$H$_5$)$_2$] the hydrido, ^{31}P, and ^{103}Rh nuclei form a AA′MXX′ spin system. The low-temperature ^{1}H nmr spectrum can be interpreted as such but there is exchange of the hydrides (or phosphines) to give a doublet of triplets for the hydride ligands at room temperature. Again the mechanism is unknown but must be intramolecular.$^{(80)}$

9.2.3. *Motion Involving Bridging Organic Ligands*

An interesting range of dinuclear complexes has been reported containing unsymmetrical ligands which rapidly oscillate between the metal atoms. The six compounds 47,$^{(81)}$ 48,$^{(82)}$ 49,$^{(83)}$ 51,$^{(84)}$ 52,$^{(85)}$ and 53$^{(86)}$ all contain σ-M–C bonds to only one of the bridged metal atoms but there is nevertheless a rapid interchange of the metal atoms. All would seem to pass through a transition state (or intermediate) with a plane of symmetry bisecting and perpendicular to the M–M axis. The oscillation of the four-electron-donating cyanide in 47 was detected by

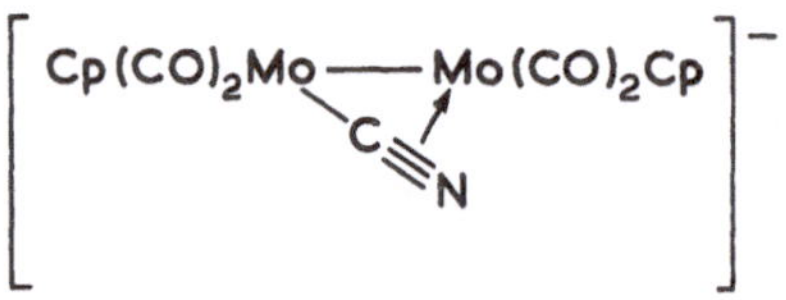

$$\left[\text{Cp(CO)}_2\text{Mo} \text{—} \text{Mo(CO)}_2\text{Cp} \right]^-$$

47

coalescence of the C$_5$H$_5$ nmr signals.$^{(81)}$ As well as the coalescence of the C$_5$H$_5$ signals for 48 there is an associated coalescence of the H^b and H^c signals; the

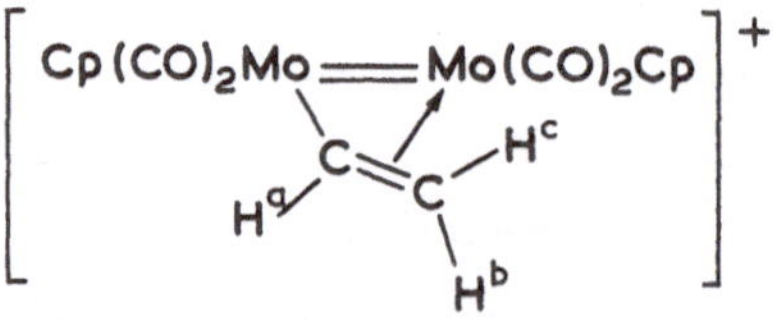

$$\left[\text{Cp(CO)}_2\text{Mo} \text{=} \text{Mo(CO)}_2\text{Cp} \right]^+$$

48

H^a nmr signal changes from a double doublet to a triplet. The transition state for vinyl oscillation must be quite unlike that for the oscillation of [Os$_3$H(C$_2$H$_3$)(CO)$_{10}$], which does not lead to exchange of the vinylic hydrogen atoms.$^{(82)}$ The rapid exchange of Zr atoms in 49 leads to exchange of ligands A with C and of B with D. The reaction is stereospecific both at Zr (retention) and at carbon atom E (inversion). The inversion at carbon was established from

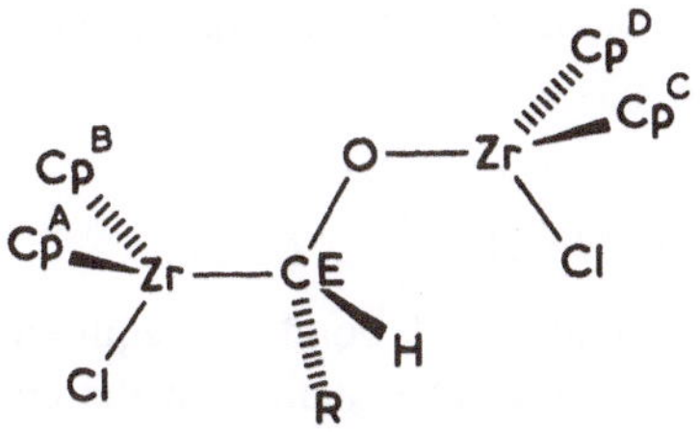

49

the exchange of the diastereotopic benzylic hydrogen atoms when $R = CH_2Ph$ and probably requires a transition state (or intermediate) (**50**).[83] There is an

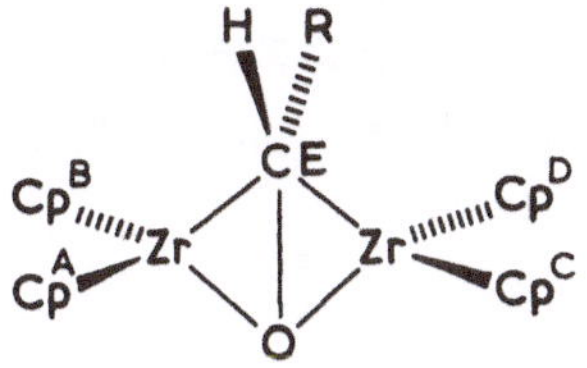

50

intramolecular substitution at the carbon atom of the CHRO group which becomes planar in **50**. The exchange of iron atoms in compound **51** amounts to the oscillation of a μ-vinyl,[84] while compound (**52**) is quite remarkable in not only

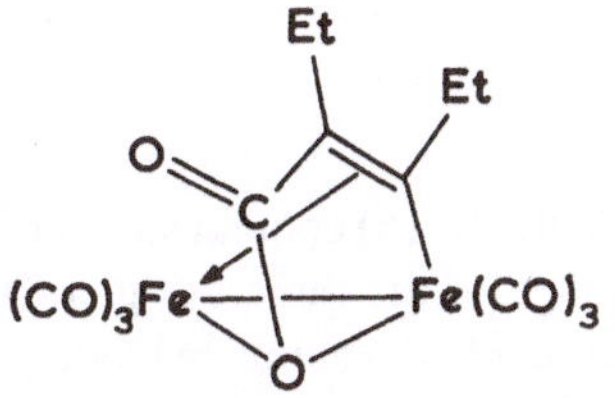

51

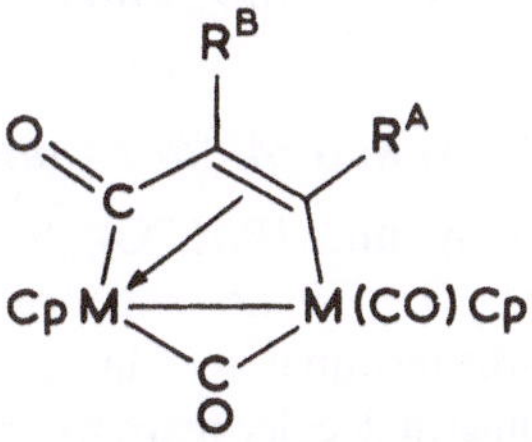

52

exchanging the C_5H_5 ligands but also R^A with R^B. This results from a reversible cleavage of the CR^ACR^B–CO bond allowing the alkyne and CO fragments to recouple in the opposite direction.[85] The μ^2-CPh_2 ligand in **53**, best described

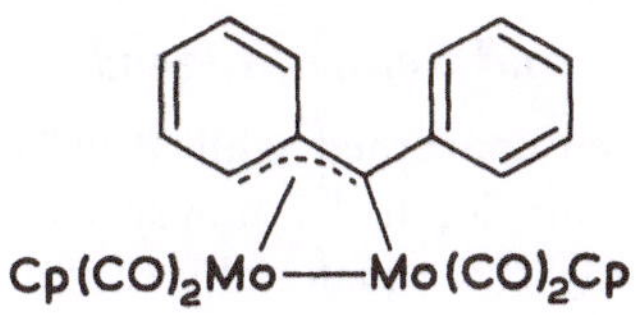

53

as an η^3-allyl as shown, oscillates between the Mo atoms exchanging the C_5H_5 rings as well as the two aryl groups.[86]

9.2.4.   *Rotation about Metal–Metal Bonds*

Restricted rotation about M–M bonds is required for certain nonbridged systems such as the nonbridged intermediate in the *cis–trans* isomerization of $[Fe_2(C_5H_5)_2(CO)_4]$. Rotation about a triple metal–metal bond has now been described. The compound $1,1\text{-}[Mo_2(NMe_2)_2(CH_2SiMe_3)_4]$ (**54**) (R $=$ CH_2SiMe_3)

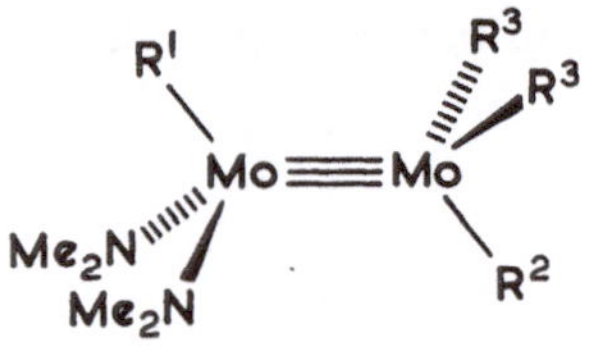

54

adopts the staggered ethane-type structure shown but undergoes conformational changes equilibrating R^2 with R^3. The signals for the diastereotopic CH_2 protons of R^3 coalesce with those for R^2 in the ^1H nmr spectrum.[87]

9.3.   *Cluster Compounds*

9.3.1.   *Fluxional Metal Frameworks*

The compounds $[Pt_n(CO)_{2n}]^{2-}$ ($n = 3, 6, 9, 12, 15$) contain stacks of eclipsed Pt triangles. There is no evidence for exchange of the three bridged CO groups with the three terminal ones in each layer but there is dynamic behavior involving intra- and intermolecular rearrangement of the layers. For example, in $[Pt_9(CO)_{18}]^{2-}$ ^{195}Pt nmr shows that there is rapid intramolecular rotation of the layers with respect to each other even at $-85°C$. Addition of $[Pt_{12}(CO)_{18}]^{2-}$, however, leads to collapse of nmr multiplets and the exchange of Pt triangles between the clusters is given as an explanation.[88]

9.3.2.   *Migration of CO and Related Ligands*

Most reports have been concerned with identification of particular intramolecular exchanges. For example, the ^{13}C nmr spectrum of $[Os_3(CO)_{11}(Bu'NC)]$ at $-60°C$ shows that axially and equatorially substituted isomers are present. Between $-60°C$ and $-30°C$ localized axial–equatorial exchange at the $Os(CO)_4$ groups of the axial isomer can be observed, but at $50°C$ there is a rapid exchange of axial and equatorial isomers, but whether by a localized exchange at $Os(CO)_3(Bu^tNC)$ groups is not clear.[89] Similarly there are localized CO exchanges at $Fe(CO)_3$ groups in $[Fe_3(CO)_9X_2]$ (X = S, Se, Te, or NMe).[90] In contrast the observation of a ^{13}C nmr quartet for the μ^3-CH ligand of $[Rh_3(C_5H_5)_3(CO)_2(CH)]^+$ (^{103}Rh coupling) must be due to a nonlocalized CO migration about the Rh_3 triangle.[91,92] The structure of $[Rh_4(CO)_8L_4]$ (L = $P(OPh)_3$) (55) in the crystal persists in solution; only one of the ligands L at the

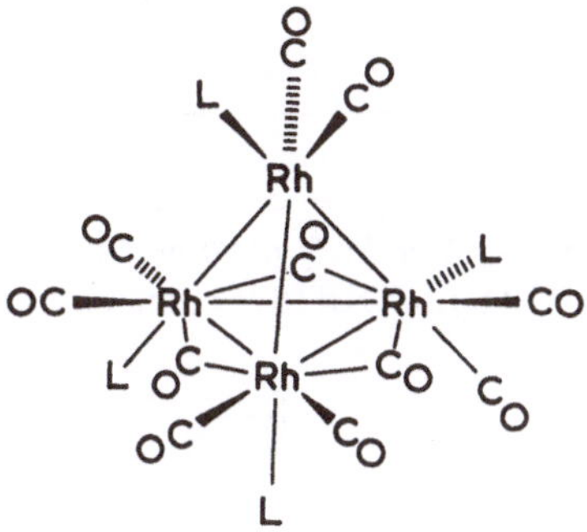

55

basal Rh atoms occupies an equatorial position. Specific CO exchanges in the lower temperature range have been interpreted in terms of a Cotton mechanism involving a nonbridged intermediate together with a degenerate rocking motion of the unique $Rh(CO)_2L$ group in this intermediate.[93] The important anion $[Rh_5(CO)_{15}]^-$ (56) shows ^{13}C nmr signals with relative intensities 2.6:6:6 at

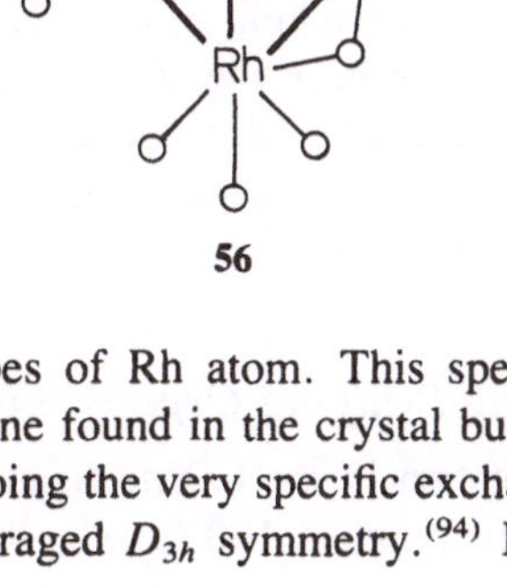

56

$-80°C$ with only two types of Rh atom. This spectrum seems to fit a D_{3h} structure rather than a C_2 one found in the crystal but would also be consistent with a C_2 molecule undergoing the very specific exchange illustrated in Scheme 10 that generates time-averaged D_{3h} symmetry.[94] Fluxional behavior of the

Scheme 10

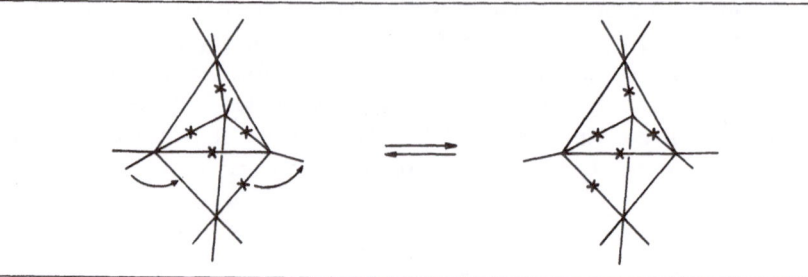

large clusters $[Rh_{14}(CO)_{25}H_{4-n}]^{n-}$ (n = 3 or 4),[95] $[Ni_6(CO)_{12}]^{2-}$, $[Ni_9(CO)_{18}]^{2-}$, and $[Ni_{12}(CO)_{21}H_{4-x}]^{x-}$ (x = 2 or 3)[96] have also been described and discussed. The complex $[Nb_3(C_5H_5)_3(CO)_7]$ (**57**) contains a unique $\mu^3:\eta^2$-CO regarded

$$
\begin{array}{c}
\text{structure of compound 57}
\end{array}
$$

57

as a six-electron donor.[97] Dynamic behavior leads to equilibration of the cyclopentadienyl ligands which occurs by motion of the bridging CO around the metal ring and of the $(CO)_2(C_5H_5)$ units locally at each niobium atom.[98]

9.3.3. Hydrogen Migration Reactions

A new isomeric form of $[Ru_4H_4(CO)_{10}(Ph_2PCH_2CH_2PPh_2)]$ contains a bridging rather than a chelating diphosphine ligand. Three separate intramolecular processes of this molecule have been identified, including a specific migration of one hydride ligand between two edges of the metal tetrahedron, a conformational change in the diphosphine bridge, and finally total equilibration of the four nonequivalent hydride ligands.[99]

9.3.4. Motion Involving Bridging Organic Ligands

An internal ligand rotation in the COMe ligand of compound **58** leads to ^{13}C nmr coalescences in the range $-100°C$ to $-40°C$ and the equilibration of

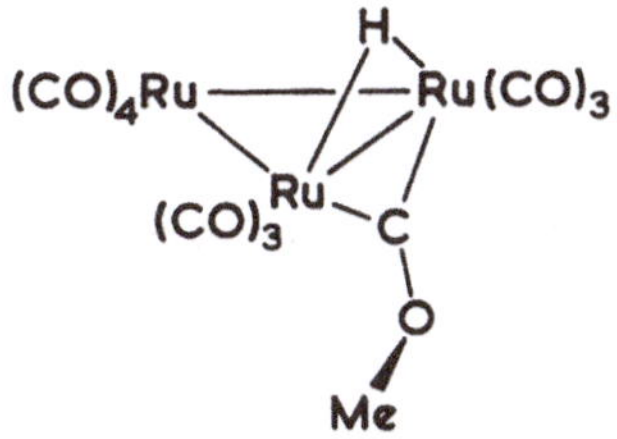

58

the two bridged Ru atoms. There is clearly significant multiple bond character in the C–O bond of this ligand (length 1.299 Å). At higher temperatures there is localized exchange in the $Ru(CO)_4$ group and finally complete CO scrambling.[100] Compound **59** has a unique mode of coordination for the μ^3-C_6H_4

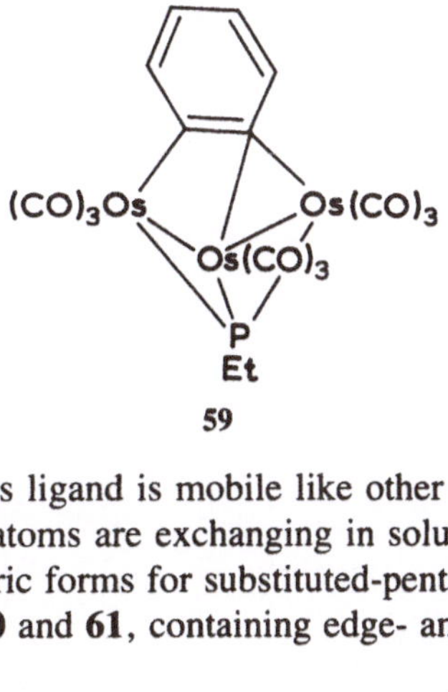

59

ligand but nonetheless this ligand is mobile like other μ^3-C_6H_4 ligands and the two metal-bound carbon atoms are exchanging in solution.[101]

There are two isomeric forms for substituted-pentalene compounds of type $[Ru_3(CO)_8(pentalene)]$, **60** and **61**, containing edge- and face-bridged pentalene

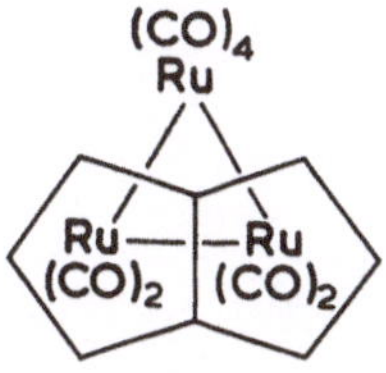

60

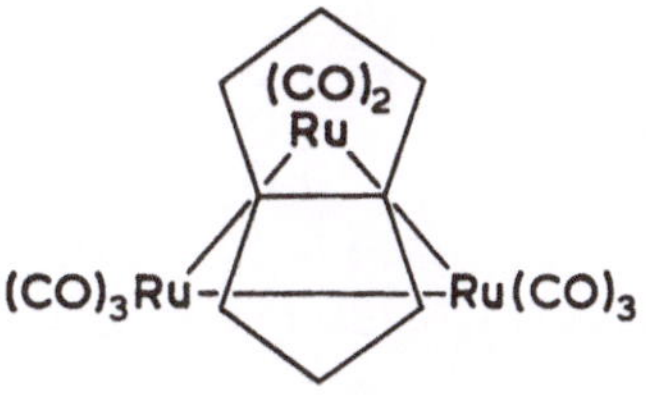

61

ligands respectively. Although these isomers are in equilibrium in solution, their interchange does not give nmr coalescences and its mechanism is unknown. Isomer **61** has a temperature-invariable nmr spectrum but **60** shows a rapid oscillation of the pentalene ligand from above to below the metal plane while remaining bridging the same two ruthenium atoms.[102]

Chapter 10

Reactivity of Coordinated Hydrocarbons

10.1. Introduction

This chapter reviews mechanistic studies on the stoichiometric reactions of co-ordinated hydrocarbons with nucleophiles and electrophiles, together with some related processes. Kinetic investigations of such reactions, especially nucleo-philic additions, have increased considerably in recent years, greatly increasing our knowledge of the intimate mechanisms involved and of the factors important in controlling ligand reactivity.

Fundamentally related, but catalytic, processes are discussed elsewhere (Chapter 11), as are intramolecular ligand rearrangements (Chapter 9).

10.2. Nucleophilic Addition and Substitution

10.2.1. σ-Bonded Hydrocarbons

A number of quantitative studies have appeared concerning addition to coordinated carbenes. Formation of the dimethylamino(imino)carbene complexes (1) from the reactions of $[W(CO)_5\{C(4\text{-}RC_6H_4)Ph\}]$ (2) (R = H, Me, CF_3, or

Scheme 1

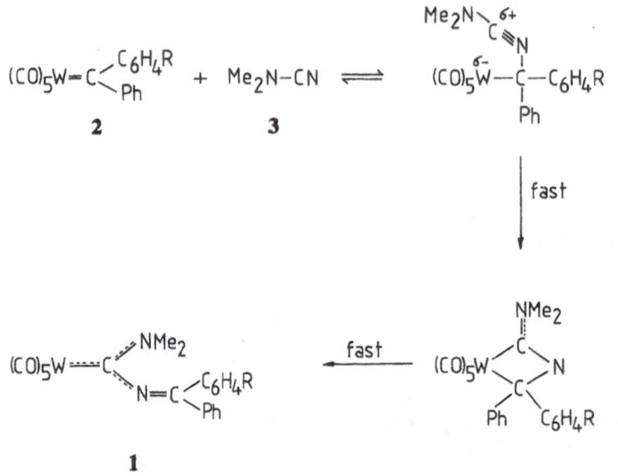

Br) with dimethylcyanamide (3) in methylcyclohexane has been shown to follow the rate law given in equation (1).[1]

$$\frac{d(1)}{dt} = k[2][3] \tag{1}$$

The second-order rate constants, k, correlate well with Hammett σ constants, with a positive ρ value. These observations, together with the low $\Delta H^{\ddagger}$ (37.3–41.6 kJ mol^{-1}) and strongly negative $\Delta S^{\ddagger}$ values (-119 to -133 J K^{-1} mol^{-1}), were rationalized by the associative stepwise mechanism shown in Scheme 1. This involves initial nucleophilic attack by the $C\equiv N$ group of **3** on the carbene carbon of **2**.

    Similar kinetic behavior was observed for the related insertion reactions of $[M(CO)_5\{C(4\text{-}RC_6H_4)OMe\}]$ (M = Cr or W; R = H, Me, OMe, CF$_3$, or Br) with Et_2N—$C\equiv CMe$ to produce **4**.[2] Once again the low activation enthalpies

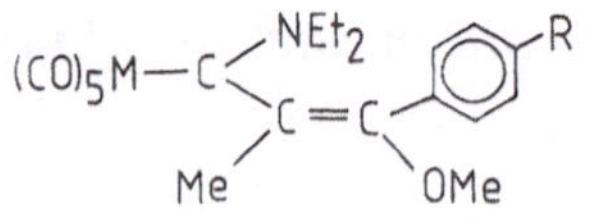

($\Delta H^{\ddagger} = 25.1 - 39.4$ kJ mol^{-1}) and strongly negative $\Delta S^{\ddagger}$ values (-129 to -145 J K^{-1} mol^{-1}) were discussed in terms of nucleophilic addition of the

ynamine at the carbene carbon in the first step (cf. Scheme 1). The tungsten carbene complexes were about three times as reactive as their chromium analogs.

A report[3] of relevance to asymmetric synthesis has shown that methoxide ion adds stereospecifically to only one face of the carbene ligand in the chiral complex **5** to give the diastereoisomer **6** in high yield {equation (2)}. Similar addition of MeO⁻ to the other geometric isomer of **5** also generates a new chiral center with high stereoselectivity (9:1). Reaction of **6** with $Ph_3C^+BF_4^-$ results in chemospecific abstraction of the methoxy group and the stereospecific regeneration of **5**.

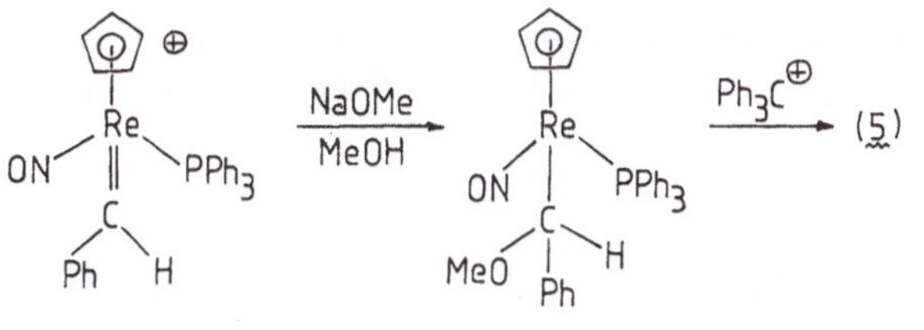

$$(2)$$

Numerous studies into the role of metal carbene species as reactive intermediates in homogeneous metal-catalyzed olefin methathesis are generally outside the scope of this chapter.[4] A related report of considerable mechanistic significance is the demonstration[5] that isolable metal carbene complexes such as $[W(CO)_5(CPhX)]$ (X = Ph or MeO) can act as initiators of acetylene polymerization. This provides experimental support for an earlier hypothesis[6] that metal-catalyzed polymerizations of acetylenes are propagated as shown in Scheme 2.

Scheme 2

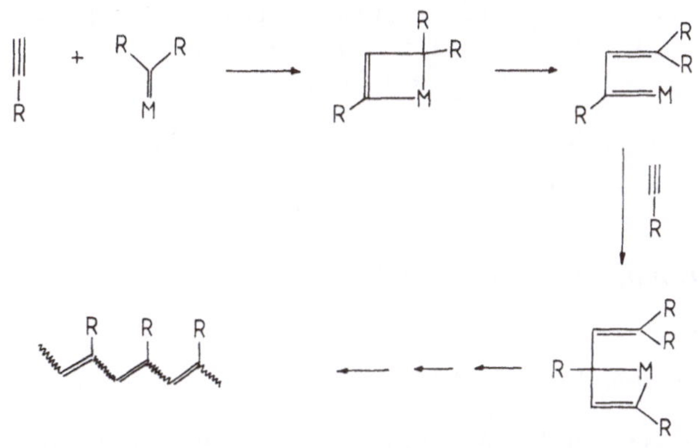

As part of an investigation of the reductive elimination of methyl chloride from [RhMeCl$_2$(CO)(Ph$_3$P)$_2$], kinetic and spectroscopic evidence has been obtained [7] for the unprecedented S_N2 attack by free Ph$_3$P on the coordinated methyl group to yield *trans*-[RhCl(CO)(Ph$_3$P)$_2$] and the phosphonium salt Ph$_3$MeP$^+$Cl$^-$. Activation of the methyl group apparently arises solely from the strongly electrophilic rhodium(III), although prior dissociation of a chloride ligand to give a cationic intermediate could not be excluded. A detailed kinetic investigation of the auto-oxidation of dimethylcadmium in *n*-decane to give MeOOCdMe and (MeOO)$_2$Cd has also been reported.[8] A chain free radical mechanism was confirmed and the role of peroxide products clarified. Conversion of the cobaltacyclopentadiene phosphite complexes **7** into the 1-alkoxyphosphole oxide species **8** is believed from kinetic studies (rate $= k$[complex]) and the

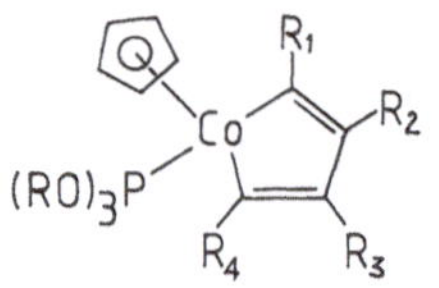

7 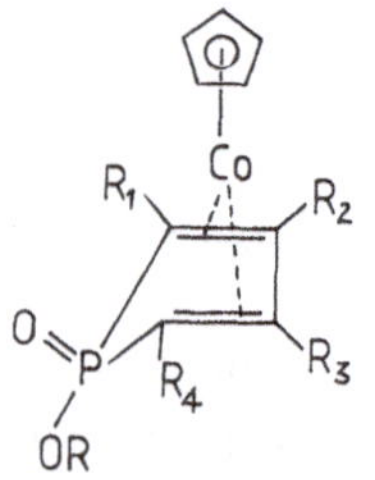   8

absence of solvent effects to involve 1,4-cycloaddition of the phosphite to the α,α'-carbons of the metallacycle.[9] The negative $\Delta S^{\ddagger}$ values (~ -17 J K^{-1} mol^{-1}) found for the conversions with P(OMe)$_3$ are consistent with such a concerted nonionic intramolecular transformation involving a pentacoordinated phosphorus in the transition state, as is the reactivity order:

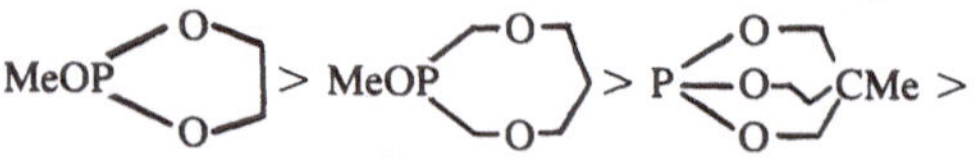

An alternative mechanism involving insertion of the coordinated phosphite into the Co–carbon bond could not be eliminated.

10.2.2. π-Bonded Hydrocarbons

10.2.2.1. Addition at Mono-Olefins

Processes related to the Wacker process, namely, addition of nucleophiles to olefins coordinated to Pd(II) and Pt(II), continue to attract attention. Equilibrium constants, K, have been measured in acetone for the formation of a wide range of σ-bonded 2-ammonioethanide complexes **(9)** via equation (3):

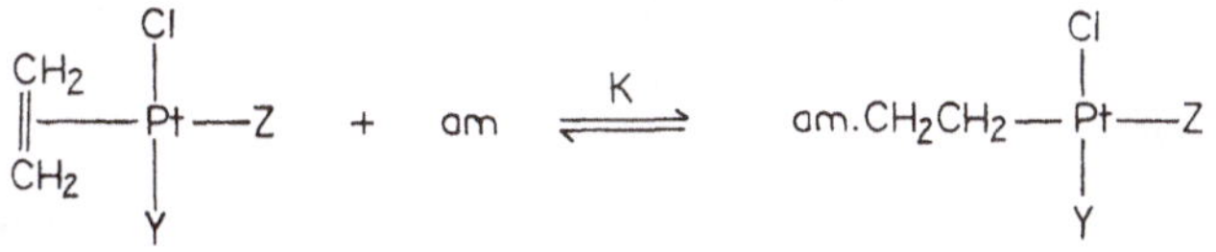

$$\text{(3)}$$

9

where Y = Cl, Z = am, and Y–Z = acac; am = primary, secondary, tertiary aliphatic and aromatic amines.[10] The basicity of the amines is important in determining the magnitude of K, no σ adducts being formed with very weak bases ($pK_a < 5$). π-Effects are also relevant. However, steric factors have the major influence on K, as shown by the decrease in K along the series primary > secondary > tertiary, linear > branched, and alicyclic > aliphatic amines. Interestingly, with am = pyridine or imidazole, competitive reversible formation of the five-coordinated species **(10)** was also observed at low temperature [equation (4):

$$\text{(4)}$$

10

where Y = Cl, Z = py, and Y–Y = acac]. The preference between these competing pathways shown in equations (3) and (4) was rationalized by considering the Pt(II) center as a soft acid and the η-C_2H_4 unit as a hard one.

A novel alternative to the Wacker process for the catalytic oxidation of olefins has been demonstrated, involving oxygen transfer from the nitro ligand of cobalt(III)–nitro complexes such as [Co(py)(saloph)(NO$_2$)] [saloph = N,N'-bis(alicylidene-o-phenylene) diamino] and [Co(py)(TPP)(NO$_2$)] (TPP = tetraphenylporphyrin) to Pd(II)-bound olefins, followed by reoxidation of the reduced nitrosyl ligand by molecular oxygen.[11] Preliminary results are consistent with the mechanism outlined in Scheme 3, in which reoxidation of the nitrosyl to the nitro ligand appears to be the rate-determining step (in oxygen-lean atmosphere). Unlike the Wacker process, the palladium remains in the divalent state throughout the catalytic cycle and serves exclusively as a cocatalyst. In another related

Scheme 3

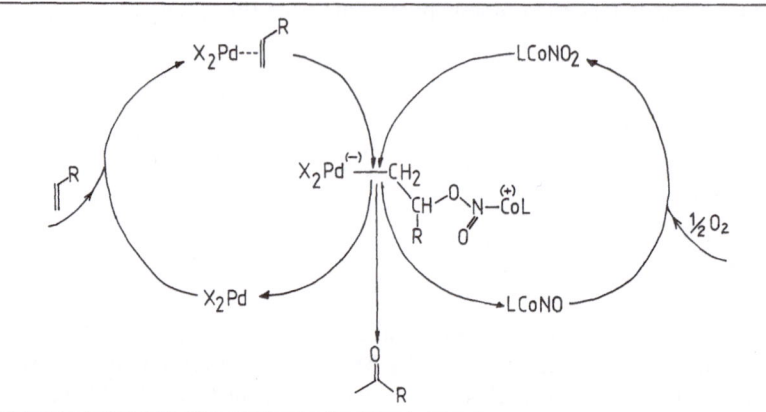

study, the oxidation of $[PtCl_3(C_2H_4)]^-$ by molecular chlorine in aqueous solution, to yield CH_2ClCH_2OH and $[PtCl_4]^{2-}$, has been shown[12] to proceed via the stepwise mechanism in Scheme 4. The intermediates **11** and **12,** formed via

Scheme 4

$$[Pt(II)Cl_3(C_2H_4)]^- \xrightarrow{Cl_2, \ Cl-} \underset{\mathbf{11}}{[Pt(IV)Cl_5(CH_2CH_2Cl)]^{2-}}$$

$$\downarrow H_2O(-HCl)$$

$$ClCH_2CH_2OH + [PtCl_4]^{2-} \longleftarrow \underset{\mathbf{12}}{[Pt(IV)Cl_5(CH_2CH_2OH)]^{2-}}$$

nucleophilic attack on the coordinated hydrocarbon [highly activated by the Pt(IV) center], were intercepted and characterized both in solution and as stable salts.

The first kinetic investigation of nucleophilic addition to coordinated mono-olefins other than in the Wacker process has been reported[13] for the reactions depicted in equation (5) (R = Ph, Bu^n):

$$[Fe(CO)_2(\eta\text{-}Cp)(\eta\text{-}C_2H_4)]^+ + R_3P \rightarrow [Fe(CO)_2(\eta\text{-}Cp)(\eta^1\text{-}C_2H_4.PR_3)]^+ \quad (5)$$

These rapid processes obey the rate law

$$-\frac{d[\text{Fe}]}{dt} = k[\text{Fe}][R_3P] \quad (6)$$

in acetone. This is consistent with direct addition of the tertiary phosphines to the ethylene ligand, as is the large negative $\Delta S^{\ddagger}$ value of $-63(5)$ J K^{-1} mol^{-1} found for the Ph$_3$P reaction. Comparison with previous kinetic data on Bu$_3^n$P addition to other (π-hydrocarbon) metal complexes revealed the electrophilicity order [Fe(CO)$_3$(1-5-η-C$_6$H$_7$)]$^+$ > [Co(η-C$_5$H$_5$)(η^3-C$_8$H$_{11}$)]$^+$ > [Cr(CO)$_3$(η-C$_7$H$_7$)]$^+$ > [Fe(CO)$_2$(η-C$_5$H$_5$)(η-C$_2$H$_4$)]$^+$ (13) > [Mn(CO)$_3$(η-C$_6$H$_6$)]$^+$ (relative rates ~ 400*:140:59:2.3:1). The comparatively low reactivity of (13) arises from

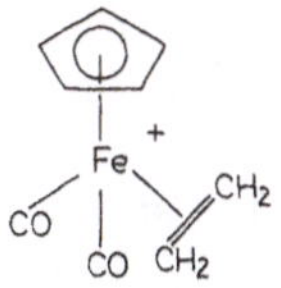

13

its relatively high $\Delta H^{\ddagger}$ value {42.4 (1.4) kJ mol^{-1} for reaction with Ph$_3$P}. The electrophilicity of (13) and related substituted olefin complexes is also shown[14] by their reactions with the η^1-allyl complex [Fe(CO)$_2$(η^1-C$_3$H$_5$)(η-C$_5$H$_5$)] to give the dinuclear condensation products (14.) Regioselectivity of addition is low for

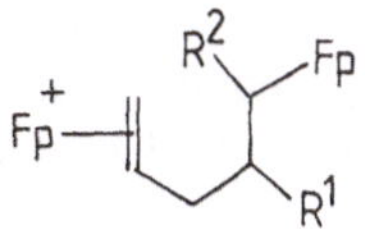

14

olefins without a strong polarizing substituent. However, when the coordinated olefin contains an electron-withdrawing group, C–C bond formation is not only more facile but regiospecificity is complete.

10.2.2.2. Addition at η^3-Enyls

Recent stereochemical studies have established two principal modes of addition of nucleophiles to π-allyl ligands coordinated to Pd(II), as shown in Scheme 5.[15,16] Dimethylmalonate and amines add *trans* (path A), whereas hydride adds *cis* (path B). The first unequivocal example of the *cis* migration

---

*The preliminary value of 160 given for [Fe(CO)$_3$(1-5-η-C$_6$H$_7$)]$^+$ in Ref. 13 arose from partial oxidation of the Bu$_3^n$P nucleophile.

Scheme 5

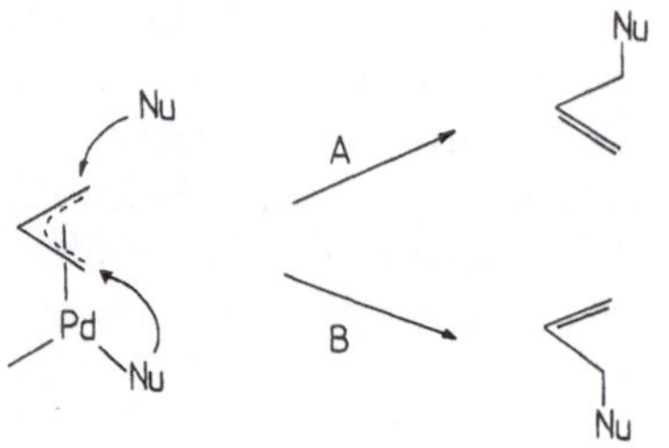

of a heteronucleophile from Pd(II) to an unsaturated carbon ligand (π-olefin or π-allyl) has now been demonstrated[17] [equation (7)]. The *trans* configuration in the allylic acetate product **15** was confirmed by hydrolysis and hydrogenation

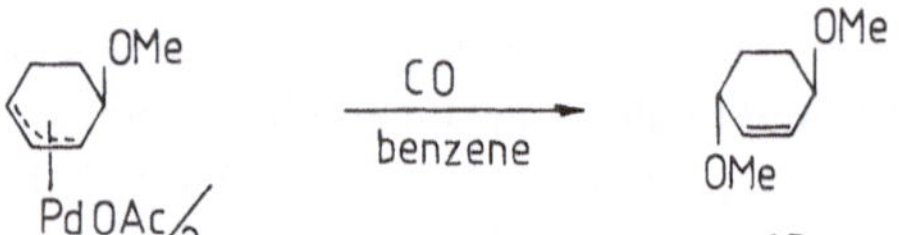

(7)

to give the known *trans*-4-methoxycyclohexanol. The acetate migration may occur either directly in the π-allyl complex, or, more likely, after rearrangement to a σ-allyl complex **(16)**, which would resemble an S_N2' substitution (S_N2'

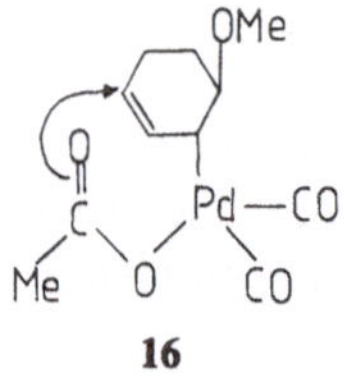

16

substitutions often occur with *syn*-stereospecificity). Reversible *cis* addition (as well as *trans* addition) of an acetate nucleophile to a Pd(II)–allyl intermediate has been previously postulated to explain the isomerization of *cis*- and *trans*-3-acetoxy-5-methoxycarbonylcyclohex-1-ene **(17a,b)** catalyzed by [Pd(Ph$_3$P)$_4$].[18]

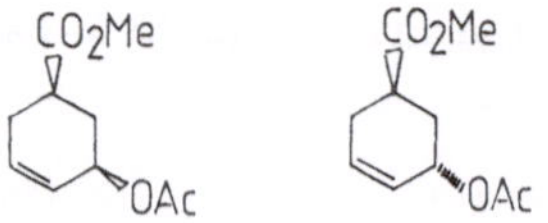

17a   **17b**

Similarly, the presence of both *cis*- and *trans*-attack paths is indicated in the Pd(0)-catalyzed amination of isomers **17a,b,** which in both cases gave a *cis–trans* mixture of amine products (path A:path B, 1.9:1).[15b] However, *trans* addition by amines predominates (20:1) in related reactions on **18.**[17]

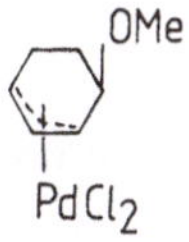

18

Kinetic ($k_{obs} = k[Me_2PhP]$) and spectroscopic results for reaction (8):

$$[MoCl(CO)_2(Me_2PhP)_2(\eta\text{-}C_3H_5)]$$
$$+ 3Me_2PhP \rightarrow [Mo(CO)_2(Me_2PhP)_4] + C_3H_5PMe_2Ph^+Cl^- \quad (8)$$

have been rationalized[19] in terms of initial nucleophilic attack on the η-allyl ligand to give the intermediate phosphonium adduct **19.** This route was favored

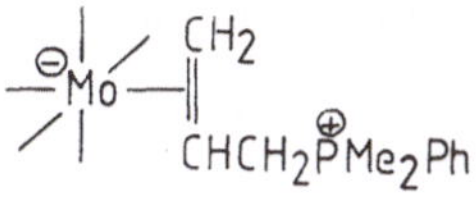

19

over alternative initial nucleophilic attack at the metal because of the marked decrease in rate with decreasing solvent dielectric constant (100-fold drop in k between ethanol and benzene), and the much slower reaction with the 2-methylallyl complex. The intermediacy of species such as **19** may be fairly general in the reduction of π-allyl metal complexes by strongly basic phosphine ligands, since intermediates of the type $[Fe(CO)_2(NO)(R_3P.C_3H_4X)]$ (R = Me, Et, Bu^n; X = H, 1-Cl, 2-Br) have been isolated and characterized in the related reactions of $[Fe(CO)_2(NO)(\eta\text{-}C_3H_4X)]$ with R_3P to yield ultimately $[Fe(CO)_3(NO)]^-$ and allyltrialkylphosphonium ions (in acetone or CH_3CN).[20]

The factors controlling the reactivity of nucleophiles toward coordinated π-allyl groups have also been explored by means of CNDO-type SCF–MO calculations.[21] The calculations suggest that as a nucleophile, X, approaches the π-allyl ligand in $[Pd(II)YZ(\eta\text{-}C_3H_5)]^{n+/-}[Y,Z = Cl_2, Cl(H_3P),$ or $(H_3P)_2;$ n = 0 or 1] complexes, the Pd(II) is reduced to Pd(0), and the allyl product $CH_2{=}CHCH_2X$ leaves the Pd(0) via an intermediate π-olefin Pd(0) complex. Attack on the cationic complexes containing Y,Z = H_3P is predicted to be much more facile than with the anionic chloro substrates, in keeping with experimental observations.[15a] Orbital mixing and changes in the Pd–Cl and Pd–P bond strengths were shown to be important factors in this ligand effect upon reactivity. On the other hand, the electron density on the C-3 atom of π-allyl ligand seems unimportant.

10.2.2.3. Addition at η^5-Dienyls

There is considerable synthetic interest in the reactions of $[Fe(CO)_3(\eta^5$-cyclohexadienyl)]$^+$ cations with nucleophiles, and the high chemo-, regio-, and stereospecificity of such processes have recently been reviewed.[22] The reactions are generally chemospecific for the dienyl ligand, there being very few established examples of nucleophilic addition at other sites such as the metal or a carbonyl ligand. Similarly, addition often occurs regiospecifically at one terminus of the dienyl group, and almost always in a stereospecific fashion at the M-*exo* face of the ligand (opposite to the metal).

It is therefore interesting that the site of nucleophilic attack on the osmium cation **20** (M = Os) varies depending on the nature of the nucleophile.[23] Thus,

$$\left[\bigcirc\hspace{-0.2em}-\hspace{-0.2em}M(CO)_3 \right]^+$$

20

whereas amines, alcohols, thiols, phosphines, and carbon nucleophiles other than cyanide give exclusively 1,3-diene derivatives **(21)** (M = Os), hydride and cyanide give mixtures of **21** and the σ, allyl-bonded products **(22)** (M = Os).

21

22

This behavior contrasts with the iron analog **(20)** (M = Fe) where all nucleophiles add at the dienyl terminus to give **21** (M = Fe). The osmium results were rationalized in terms of a common mechanism involving initial kinetically favored attack at the inner 2- or 4-positions of the dienyl system to give **22** (M = Os). Except for Nu = H$^-$ or CN$^-$, where formation of **22** is considered irreversible, subsequent dissociation to regenerate the cation **20** followed by nucleophilic attack at the dienyl terminus then leads to **21** as the thermodynamically favored product.

Also unusual is the formation of both *exo*- and *endo*-substituted diene products **21** (M = Fe, Nu = OMe) and **23** (R = OMe), respectively, upon refluxing

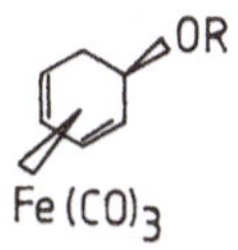

23

cation **20** (M = Fe) in acidified methanol.[(24)] This reaction was shown to proceed via initial rapid formation of the *exo* product. However, this *exo* addition is reversible owing to the presence of acid, and sufficient dienyl cation **(20)** (M = Fe) is therefore always available to allow the much slower *endo* addition to proceed until an equilibrium between **21** (M = Fe, Nu = OMe) and **23** (R = OMe) is achieved. Preliminary kinetic studies of the reaction of **23** (R = OMe) with trifluoroacetic acid were also reported, release of the methoxy substituent being much slower than with the corresponding *exo* isomer. In view of the isolation and characterization of the ester complexes [Fe(CO)$_2$(COOR)(1-5-η-dienyl)] (dienyl = C$_6$H$_7$ or C$_7$H$_9$) from the reactions of **20** (M = Fe) and [Fe(CO)$_3$ (1-5-η-C$_7$H$_9$)]$^+$ with NaOR in the corresponding alcohol at low temperature,[(25)] the most likely mechanism for the formation of the *endo* derivatives **23** from **20** (M = Fe) is shown in Scheme 7.

Scheme 7

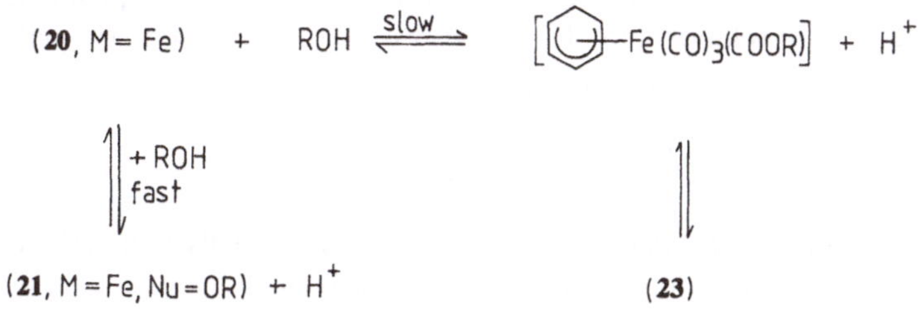

Initial nucleophilic addition at a carbonyl ligand has also been invoked in the reductions of the acyclic dienyl cations **24** (R = H, Me, Et, Pr^i) with lithium triethylborohydride, which give mainly diene products of inverted configuration:[(26)]

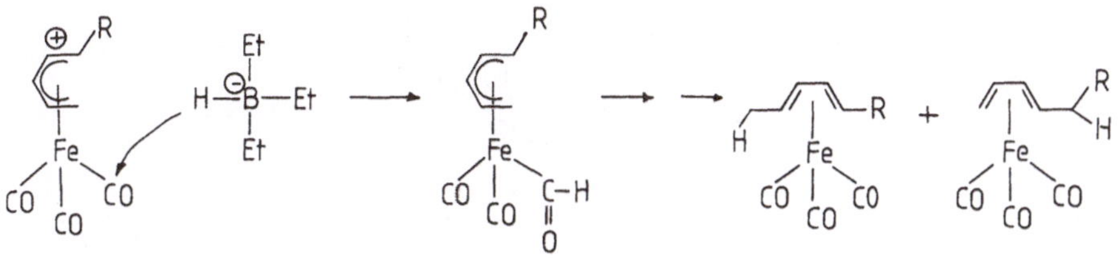

24

(9)

In contrast, reductions with the much less reactive $LiBH_4$ and $NaBH_3CN$ proceed largely (70%–100%) with retention of configuration.

In contrast to organic chemistry, where extensive studies have attempted to correlate nucleophilicity with various parameters (e.g., the Swain–Scott, Edwards, and Ritchie equations), there is currently little quantitative information concerning nucleophilic reactivities toward coordinated π-hydrocarbons. Recent kinetic studies[27] of reactions shown in equations (10) and (11) have established

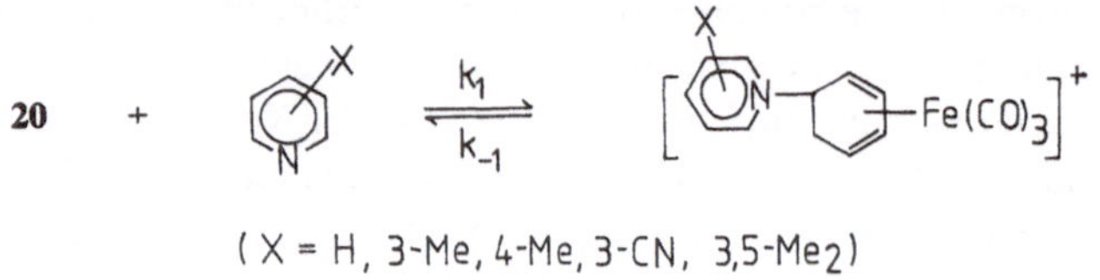

$$20 \quad + \quad \underset{}{\text{pyridine-}X} \quad \underset{k_{-1}}{\overset{k_1}{\rightleftharpoons}} \quad \left[\text{complex} -Fe(CO)_3 \right]^+ \tag{10}$$

$$(X = H,\ 3\text{-Me},\ 4\text{-Me},\ 3\text{-CN},\ 3,5\text{-Me}_2)$$

$$\left[\underset{OMe}{\text{complex}} -Fe(CO)_3 \right]^+ + 2\ \underset{}{\text{aniline-}X}-NH_2 \longrightarrow \left[\underset{OMe}{\text{complex}} -Fe(CO)_3 \right] + \left[\text{aniline-}X-NH_3 \right]^+ \tag{11}$$

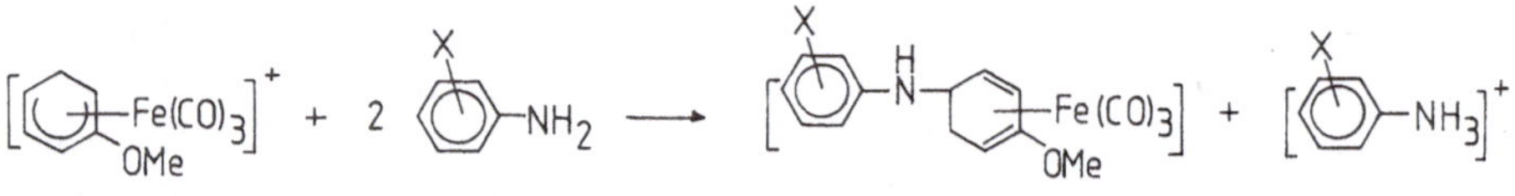

$$\mathbf{25} \qquad\qquad\qquad\qquad\qquad\qquad \mathbf{26}$$

for the first time the importance of basicity in controlling amine nucleophilicity. Reactions (10) obey the rate expression in equation (12), except for the X = 3-CN case where relationship (13) is followed. For the more basic anilines

$$k_{obs} = k_1 \text{[amine]} \tag{12}$$

$$k_{obs} = k_1 \text{[amine]} + k_{-1} \tag{13}$$

(X = 4-Me, 4-MeO) equation (12) is again obeyed. This is most readily explained in terms of the stepwise mechanism in Scheme 8, involving rapid amine-

Scheme 8

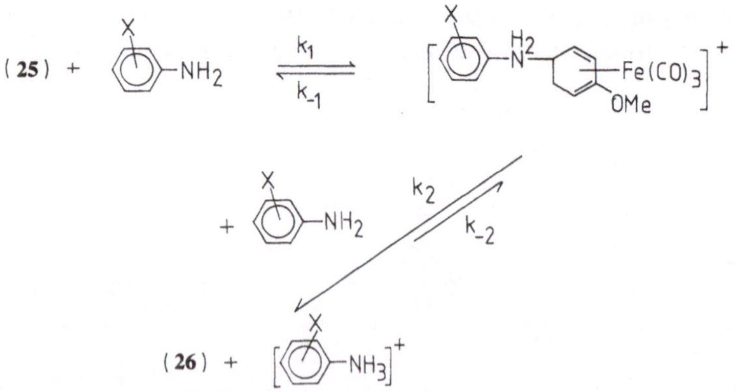

$$(\mathbf{25}) + \underset{}{\text{aniline-}X}-NH_2 \quad \underset{k_{-1}}{\overset{k_1}{\rightleftharpoons}} \quad \left[\underset{OMe}{\text{complex}} -Fe(CO)_3 \right]^+$$

$$+ \underset{}{\text{aniline-}X}-NH_2 \quad \overset{k_2}{\underset{k_{-2}}{\text{\Large/}}}$$

$$(\mathbf{26}) + \left[\text{aniline-}X-NH_3 \right]^+$$

assisted proton removal in the second step ($k_2 \gg k_1, k_{-1}, k_{-2}$). In contrast, with the less basic anilines the two-term equation (14) is obeyed:

$$k_{obs} = k' \text{ [amine]} + k'' \tag{14}$$

Assuming very rapid reversible deprotonation, k' and k'' can be equated to k_1 and $k_{-1}[\text{H}^+]/([\text{H}^+] + K_2K_a)$, respectively. Most significantly, for both reactions (10) and (11) the Brønsted relationship [equation (15)] is obeyed:

$$\log k_1 = \alpha pK_a + \text{const} \tag{15}$$

The large slopes ($\alpha = 0.7\text{--}1.0$) indicate a very strong dependence of k_1 on amine basicity, and considerable carbon–nitrogen bond formation in the transition states. These results suggest a "hard" character for the dienyl rings in cations **20** (M = Fe) and **25**.

Detailed kinetic studies have also been reported[28,29] for the reactions of **20** (M = Fe, Ru, or Os) with N,N-dimethylaniline yielding the substituted diene products **27**. These processes may best be regarded as unusual examples of electrophilic aromatic substitution, as shown in Scheme 9. Observation of the

Scheme 9

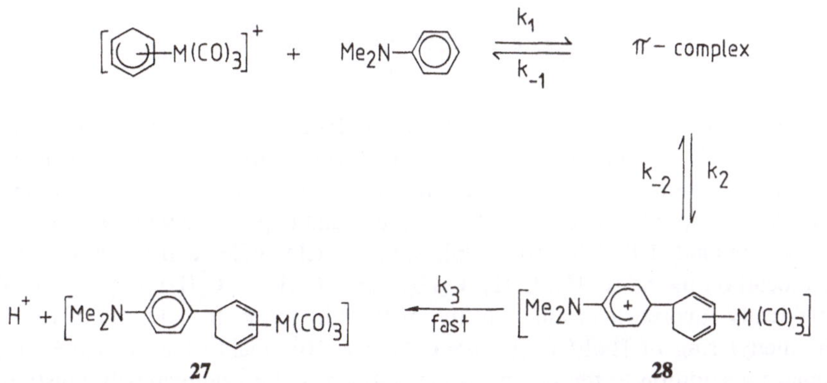

rate law in equation (16) with both the Fe and Ru cations in CH_3NO_2 indicates that either π-complex formation

$$\text{rate} = k[\text{complex}][\text{Me}_2\text{NC}_6\text{H}_5] \tag{16}$$

is rate determining (i.e., $k = k_1$), or there is no significant contribution from a π-complex intermediate and formation of the Wheland-type σ-complex (**28**) is rate limiting. The results provide the first quantitative comparison of the reactivity of cyclohexadienyl rings when coordinated to each of the iron triad metals, namely, Fe > Os > Ru (44:3.3:1 at 45°C). The large k_{Fe}/k_{Ru} ratio arises almost entirely from the lower enthalpy of activation in the iron case ($\Delta H^{\ddagger}_{Fe} = 53.3 \pm 1.3$ kJ mol^{-1}, $\Delta H^{\ddagger}_{Ru} = 62.8 \pm 0.7$ kJ mol^{-1}), since both processes have the same $\Delta S^{\ddagger}$ value within experimental error (~ -102 J K^{-1} mol^{-1}).

The reactions of [FeI(CO)$_2$(1-5-η-dienyl)] (dienyl = C$_6$H$_7$ or C$_7$H$_9$) with Bu$_3^n$P have been shown to proceed via the stepwise sequence in Scheme 10.$^{(30)}$

Scheme 10

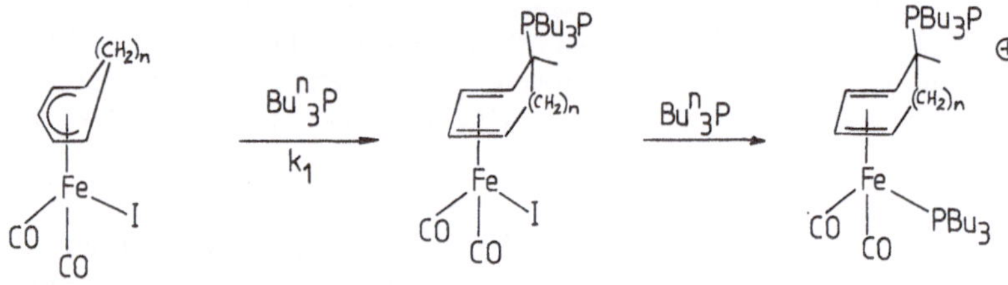

The rapid first step obeys the rate law in equation (17) for both complexes.

$$\text{rate} = k_1[\text{complex}][\text{Bu}_3^n\text{P}] \tag{17}$$

The much faster (~ 80 times) rate for the C$_6$H$_7$ complex, and its large negative $\Delta S^{\ddagger}$ value of $-130(3)$ kJ mol^{-1}, support direct addition to the dienyl ligands. In contrast, a dissociative mechanism was confirmed for the slower iodide replacement step. On the basis of these results and experience with other systems it was concluded that for trialkylphosphine nucleophiles *exo* addition at a π-hydrocarbon ligand (C$_3$H$_5$, C$_4$H$_4$, C$_6$H$_6$, C$_6$H$_7$, C$_7$H$_7$, or C$_7$H$_9$) will be generally kinetically favored over attack at the metal. Finally, k_1 for addition of Bu$_3^n$P to the dienyl ring of [FeI(CO)$_2$(1-5-η-C$_6$H$_7$)] is 10^5 smaller than that previously found for addition to the parent cation (20) (M = Fe), dramatically illustrating the effect of a substrate positive charge.

A development which should be of value in future mechanistic studies on such systems is the isolation of optically active [Fe(CO)$_3$(1-5-η-dienyl)]$^+$ and [Fe(CO)$_3$(1-4-η-diene)] species, and the determination of the absolute configurations in some cases.$^{(31-33)}$ Of particular relevance to the present chapter is the reaction of 25 with the enol trimethylsilyl ether of (+)-camphor. Some diastereoselectivity occurs during this nucleophilic addition since the recovered unreacted cation was found to be optically active.$^{(31)}$

10.2.2.4. Addition and Substitution at η^6-Arenes

Rate and equilibrium constants for the rapid reversible processes in equation (18)

$$[M(\eta\text{-}C_6H_6)_2]^{2+} + Ph_3P \underset{k_{-1}}{\overset{k_1}{\rightleftharpoons}} [M(\eta\text{-}C_6H_6)(1\text{-}5\text{-}\eta\text{-}C_6H_6.PPh_3)]^{2+} \quad (18)$$

29

(M = Fe, Ru, or Os) have been obtained via variable temperature ^{1}H and ^{31}P nmr spectroscopy.$^{(34)}$ The rate constants, k_1, for triphenylphosphine addition to the benzene ring of **29** vary markedly with the nature of the metal (Fe $\gg$ Ru $>$ Os; relative order 390:7:1). This reactivity order is similar to that found$^{(28,29)}$ for reactions of cations **20** (M = Fe, Ru, or Os) with *N,N*-dimethylaniline, and suggests that it may be a general phenomenon. It was rationalized$^{(34)}$ in terms of decreasing metal → arene π-backbonding in the order Os, Ru $>$ Fe. A LFER (Linear Free Energy Relationship) was observed between $\log k_1$ and $\log K_{eq}$, the slope of ~ 0.6 indicating considerable C–PPh$_3$ bond formation in the transition state. The assignment of an *exo* stereochemistry to the phosphonium adducts and the absence of spectroscopic evidence for an intermediate supported direct addition of Ph$_3$P to the benzene ligand. Preliminary kinetic studies of the related additions of (Bu^{n}O)$_3$P to cations **29** in CH$_3$CN to give adducts **30** (M = Fe, Ru, or Os; R = Bu^n) reveal a similar trend in k_1 (Fe $\gg$ Ru $>$ Os; relative rates 64:3:1).$^{(35)}$ With (MeO)$_3$P these phosphonium adducts undergo Arbusov elimination to yield the corresponding phosphonates (**31**). Interestingly, in the pres-

Scheme 11

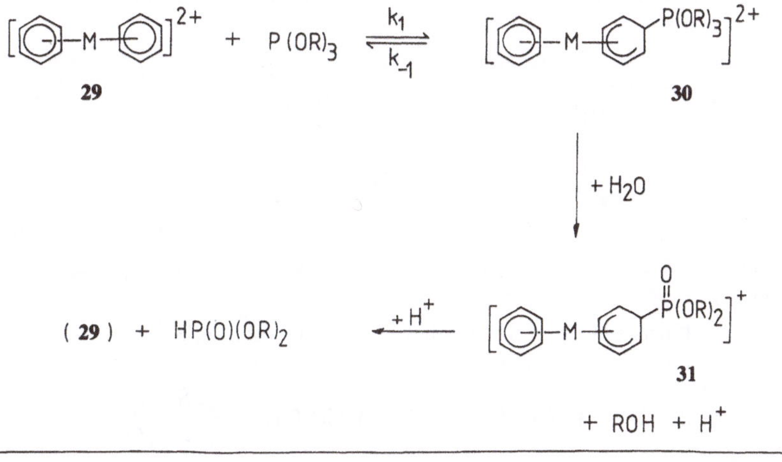

ence of a little water the cations **29** (M = Fe or Ru) were found to catalyze the conversion of excess phosphite into $HP(O)(OR)_2$ according to the mechanism suggested in Scheme 11.

Kinetic studies in methanol of the displacement of halide by MeO^- from the complexes **32** [X = F or Cl; ML_n = $Cr(CO)_3$, $Mo(CO)_3$, $Fe(\eta\text{-}C_5H_5)^+$,

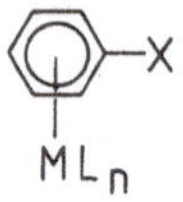

$$\text{(arene)}-X$$
$$\underset{|}{ML_n}$$

32

and $Mn(CO)_3^+$] have shown that activation of the halobenzene ligands by the metal units decreases in the order $Mn(CO)_3^+ > Fe(\eta\text{-}C_5H_5)^+ \gg Mo(CO)_3 \sim Cr(CO)_3$.[36] As noted previously for the ML_n = $Cr(CO)_3$ substrates, the fluorobenzene complexes are more reactive than chlorobenzene analogs. This is in accord with a mechanism involving rate-determining nucleophilic addition to form a steady state concentration of a Meisenheimer-type anionic intermediate. In contrast to the $M(CO)_3$ (M = Cr or Mo) complexes which give linear k_{obs} vs. $[MeO^-]$ plots, such plots are distinctly curved for the ML_n = $Fe(C_5H_5)^+$ and $Mn(CO)_3^+$ compounds. Although other explanations are possible, these observations suggest that ion pairs are rapidly formed between these latter cations and MeO^-, resulting in a reduction in their reactivity toward nucleophiles (e.g., Scheme 12, where $k_2^{ip} < k_2$).

Scheme 12

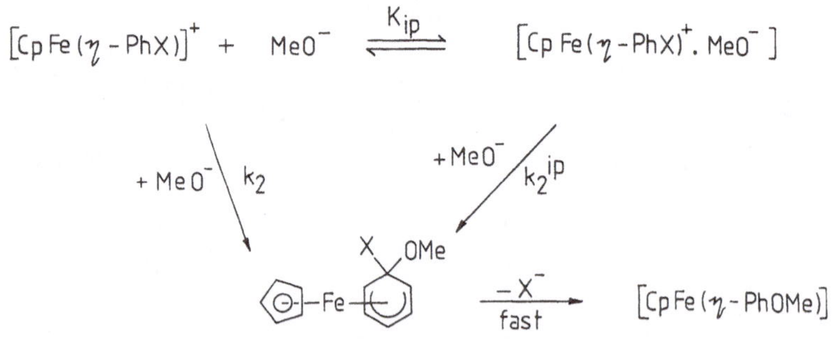

$$[CpFe(\eta\text{-}PhX)]^+ + MeO^- \underset{}{\overset{K_{ip}}{\rightleftharpoons}} [CpFe(\eta\text{-}PhX)^+ \cdot MeO^-]$$

$$+MeO^- \;\; k_2 \qquad\qquad +MeO^- \;\; k_2^{ip}$$

$$\text{(Cp)}-Fe-\text{(cyclohexadienyl, X, OMe)} \xrightarrow[\text{fast}]{-X^-} [CpFe(\eta\text{-}PhOMe)]$$

Meisenheimer-type cyclohexadienyl adducts such as **33**

$$R = CH_2CN,\; CMe_2CN,\; CH(SPh)_2,\; \text{or} \begin{array}{c} S \\ S \end{array}\!\!\!\!\text{>}$$

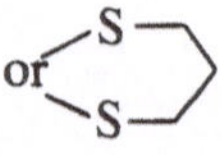

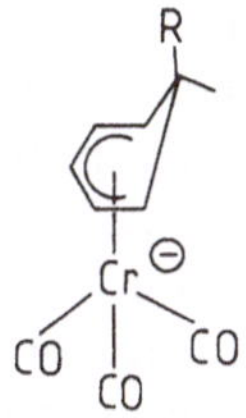

33

have been isolated and characterized from the reactions of reactive carbanions with $[Cr(CO)_3(\eta\text{-}C_6H_6)]$.[37] With the 1,3-dithiane derivative an X-ray structure confirmed *exo* addition to the benzene ring. Extended Hückel MO calculations suggest that the regioselectivity of nucleophilic substitution on the arene rings of $[Cr(CO)_3(\eta\text{-arene})]$ complexes is controlled not only by the ring substituent but also by the conformation of the $Cr(CO)_3$ unit.[38] Arene carbons which are eclipsed with respect to the carbonyl groups are predicted to be preferentially attacked by nucleophiles. In a related study, INDO molecular orbital calculations on the cations **34** (R = H, Me, OMe, or COOMe) have shown that the position

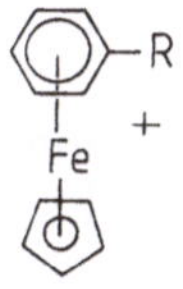

34

of attack on the arene ligand is neither charge controlled nor influenced by frontier orbital interactions with the complex LUMO.[39] However, charge considerations do explain the preferential attack of nucleophiles upon the arene rather than the cyclopentadienyl ligand of **34.** Initial nucleophilic attack at the iron is also predicted to be unlikely, since the only vacant metal $3d$ orbitals (the $3d_{xz}$, $3d_{yz}$ pair) are not only considerably higher in energy than the LUMO (~ 1.5 eV), but are also unfavorably aligned spatially with respect to an approaching nucleophile.

The degree of asymmetric induction obtained in the reactions of Grignard reagents with chiral chromium tricarbonyl complexes of various substituted diarylimines has been used to define the geometry of the transition state for these processes.[40]

10.2.2.5. *Addition at the Tropylium Ligand*

The stereospecificity, regioselectivity, and general mechanisms of nucleophilic addition to η-tropylium complexes of the group VI metals have been reviewed.[41] Interestingly, in contrast to the generally held view that nucleophilic addition to coordinated π-hydrocarbons occurs almost invariably from an *exo* direction, reduction of $[Mo(CO)_3(\eta\text{-}C_7H_7)]^+$ with $NaBD_4$ and $NaBD_3CN$ has been unequivocally shown (from ^1H and ^2H nmr) to give mixtures of *exo*- and *endo*-$[Mo(CO)_3(1\text{-}6\text{-}\eta\text{-}C_7H_7D)]$.[42] For example, with $NaBD_3CN$ a 60/40 ratio of *exo*/*endo* products was obtained. Doubt was cast on the validity of stereochemical assignments based on ir $\nu_{(C-H)}$ stretching frequencies. Formation of *endo*-$[Mo(CO)_3(1\text{-}6\text{-}\eta\text{-}C_7H_7D)]$ was considered to involve initial addition at the metal or a CO ligand (to give metal hydride or metal formyl intermediates, respectively), followed by hydride migration to the tropylium ring. A reexamination of deuterium migration in *exo*-$[Mo(CO)_3(1\text{-}6\text{-}\eta\text{-}C_7H_7D)]$ at higher temperatures using ^2H nmr similarly suggested the formation of a molybdenum–deuteride intermediate with a fluxional $\eta^5\text{-}C_7H_7$ ring, and random migration of the deuterium back to the ring.

The reaction of $[Cr(CO)_3(\eta\text{-}C_7H_7)]^+$ with *N,N*-dimethylaniline to yield **35**

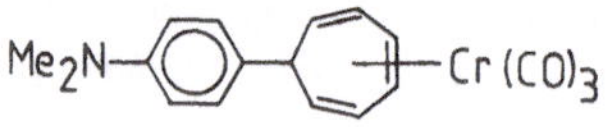

35

follows a two-term rate law of the form in equation (19).[43] This is consistent with electrophilic aromatic substitution involving the steady state formation

$$\text{rate} = k_a[\text{complex}][Me_2NC_6H_5] + k_b[\text{complex}][Me_2NC_6H_5]^2 \ldots \quad (19)$$

of a Wheland-type σ-complex intermediate **36**, followed by competing amine- and solvent-assisted proton removal. The markedly different kinetic behavior

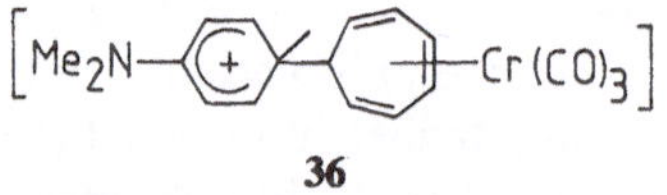

36

noted above [equation (16)] for the analogous reaction of *N,N*-dimethylaniline with cation **20** (M = Fe) may simply arise from much more rapid proton removal in the iron case, or it may reflect fundamental differences in mechanism.

10.2.3. *Reactions at Side Chains and Exocyclic Carbocations*

Rate constants for the addition of a range of nucleophiles to ferrocenylalkylium ions (37) [equation (20):

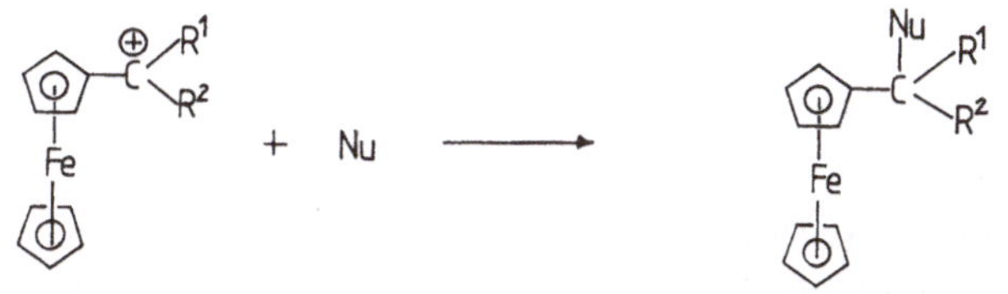

$$\tag{20}$$

37

where, R^1, R^2 = H, alkyl, aryl, or ferrocenyl] in water and water–CH_3CN mixtures have been shown[44] to correlate reasonably well with Ritchie's N_+ scale established earlier for addition to triarylmethyl and tropylium cations. Also in keeping with Ritchie's observations was the nonapplicability of the reactivity–selectivity principle to reaction (20) with anionic and amine nucleophiles. The absence of a secondary α-hydrogen kinetic isotope effect (k^H/k^D = 1.0) for the addition of anions to cation $37(R^1$ = H, R^2 = Fc) showed that there is little or no change in hybridization at the reaction center (Cα) of the carbocation on going to the transition state. That is, an "early" transition state resembling an ion pair was indicated, with encounter and desolvation of the nucleophile being the major contributors to the free energy of activation. In contrast, appreciable rehybridization at Cα from sp^2 to sp^3 (and considerable C–O bond formation) occurs in forming the transition state for the reaction with water, since k^H/k^D was $\sim$ 0.90. An intermediate situation apparently applies to the reactions with amines ($k^H/k^D \sim 0.96$). Detailed kinetic and equilibrium investigations of the forward and reverse processes with water, i.e.,

$$37 + H_2O \underset{k_{-1}}{\overset{k_1}{\rightleftharpoons}} FeR^1R^2OH + H^+ \tag{21}$$

reveal no correlation between k_1 and the equilibrium constants K_1.[45] Medium, salt, and solvent isotope effects on rates and equilibria are consistent with a carbocationlike transition state.

In a related study log k_{rel} values for the solvolysis of the acetates **38** (X = H, Me, OMe, or CF_3) correlate linearly with Hammett σ^+ substituent con-

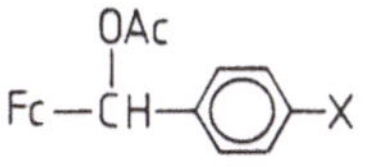

38

stants.[46] The derived ρ value of -1.39 is much less negative than those previously found for the S_N1 solvolyses of benzyl and benzhydryl derivatives. However, a carbocation-type transition state is supported by the LFER between log k_{rel} and pK_R^+ for the corresponding intermediate carbocations **39**. The small

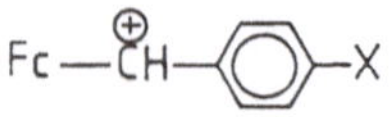

$$Fc - \overset{\oplus}{C}H - \langle \bigcirc \rangle - X$$

39

sensitivity of the ferrocenyl substrates to the aryl substituent effects was ascribed to swamping by the pronounced electron-releasing capacity of the ferrocenyl group. Rate constants have also been determined for the base-catalyzed hydrolyzes of the m- and p-carbomethoxymethyl ligands of the cobaloxime complexes $[Co(oxime)L(MeOOCC_6H_4)]$ ($L = CN^-$, SCN^-, NO_2^-, N_3^-, or primary amines).[47] Correlation via the Taft dual-substituent parameter equation shows that substantial resonance interaction can occur between the $-Co(oxime)L$ substituents and the carbomethoxyphenyl ligands when L is an unsaturated inorganic ligand capable of π-interaction with the cobalt atom.

10.2.4. Attack at Carbonyl Ligands

The reactions of $[W(CO)_6]$ with the pseudohalide salts $[(Ph_3P)_2N]^+$ $[X]^-$ ($X = CN$, OCN, or SCN) in chlorobenzene to give $[W(CO)_5X]^-$ obey the rate law in equation (22)[48]:

$$-\frac{d[W(CO)_6]}{dt} = k[W(CO)_6][PPN^+X^-] \tag{22}$$

An associative or I_d mechanism involving attack by the anions at the tungsten center was considered unlikely, and a concerted process involving initial attack at a carbonyl carbon was proposed (Scheme 13). The alternative initial addition of X^- to a carbonyl ligand followed by loss of COX and rapid pick up of X^- was eliminated by the failure to trap the five-coordinate $[W(CO)_5]$. On the basis of these and earlier studies it seems probable that most reactions of anionic nucleophiles (N_3^-, halides, pseudohalides) with $[M(CO)_6]$ ($M = Cr$, Mo, or W) complexes involve interaction at a carbonyl carbon.

Scheme 13

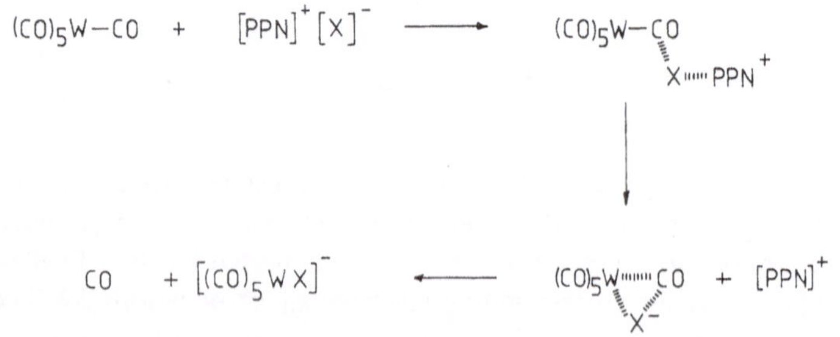

10.3. Electrophilic Attack

Competitive studies of the cycloaddition reactions of the complexes

$$\left\{ \textbf{40:} \text{ a, L = CO; b, L = P} \underset{O}{\overset{O}{\diagdown}} O \text{ ; c, L = P(OPh)}_3 \right\}$$

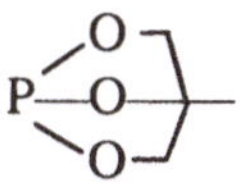

with the electrophilic olefins β,β-dicyanostyrene or ethoxymethy-lene–malononitrile to give **41** (R = Ph or OEt) have revealed the reactivity order

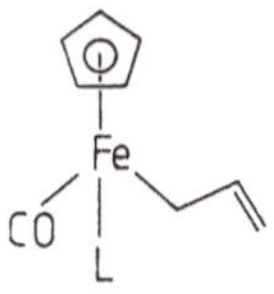

40

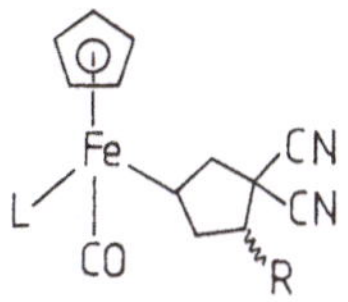

41

(40b) > **(40c)** > **(40a)** (relative rates 900:180:1).[49] This order follows that of decreasing basicity of the complexes, which arises from the increasing σ-donating ability of L along the series

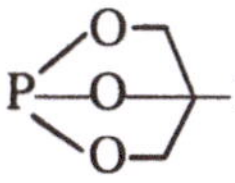

$$\text{P} \underset{O}{\overset{O}{\diagdown}} O \text{->P(OPh)}_3 > \text{CO}$$

In contrast, replacement of a CO group in [Cr(CO)$_3$(η-mesitylene)] by the stronger electron accepting ligand, maleic anhydride, leads to an approximately 20-fold decrease in the rate of arene hydrogen isotope exchange in CF$_3$COOD.[50] However, exchange is still very much faster than that with free mesitylene, confirming the increased acidity of the arene upon coordination.

Extended Hückel M.O. calculations suggest that the regioselectivity of electrophilic attack upon the arene ligand in [Cr(CO)$_3$(η-arene)] complexes is controlled not only by the ring substituent, but also by the conformation of the Cr(CO)$_3$ unit.[38] Electrophilic substitution is preferred at arene carbons which are staggered with respect to the carbonyl groups.

Kinetic studies have been made of the reactions of dimetal carbonyls of the type [M$_2$(CO)$_{10-n}$L$_n$] (M$_2$ = Mn$_2$, MnRe, Re$_2$; n = 1 or 2; L = range of tertiary phosphines and phosphites) with bromine and iodine in cyclohexane or decalin.[51,52] Complex rate laws were obtained, usually in the iodine case containing a predominant k[I$_2$] term, but a term as high as k'[I$_2$]^4 was obtained in one case. The influence of L on the rate constants for both I$_2$ and Br$_2$ reactions was best interpreted in terms of initial electrophilic attack at the O atom of the CO ligands. Rates increased in the order Mn$_2$ < MnRe < Re$_2$.

Chapter 11

Homogeneous Catalysis of Organic Reactions by Complexes of Metal Ions

11.1. Introduction

11.1.1. Scope of the Review

The literature continues to provide many examples of the application of homogeneous catalysis in organic synthesis and despite their difficulties these methods continue to attract attention becase of possible industrial applications. A volume of review papers has appeared recently.[1] Although the number of papers published on this subject is large the number that present detailed studies of kinetics and mechanism is quite small. This review concentrates on such reports, although numerous references are given to other papers. The reactions on which a number of detailed studies have appeared are metathesis of alkenes, carbonylation reactions, the water gas shift reaction, and hydrogenation, especially asymmetric hydrogenation.

11.1.2. Elementary Steps in Homogeneous Catalysis

Since homogeneous catalytic reactions are kinetically complex, detailed understanding requires prior investigation of various steps in the catalytic cycles. A few papers likely to be of value in the elucidation of catalytic kinetics are reviewed in this section. A novel approach to the study of steps after the rate-determining step is to irradiate the precursor at low temperature. Thus the irradiation of $[W(CO)_3(\eta^5\text{-}C_5H_5)(n\text{-}C_5H_{11})]$ at 25°C in *iso*-octane shows that $[WH(CO)_3\,(\eta^5\text{-}C_5H_5)]$ is a secondary photoproduct, while the first product is $[WH(CO)_2\,(\eta^5\text{-}C_5H_5)(n\text{-pentene})]$.[2] It has been shown that the photochemical cleavage of metal–metal bonds in species such as $[Co_2(CO)_6(PBu_3)_2]$ generates odd-electron molecules that are not isomerization catalysts for alkenes but they become active in the presence of a hydrogen source such as $[SiHEt_3]$ which yields a hydridocobalt complex that then promotes isomerization.[3] There have been many discussions of the role of α-hydrogen migrations and carbene reactions in catalysis.[4] However, $[Co(PPh_3)(\eta^5\text{-}C_5H_5)(CD_3)_2]$ reacts with C_2H_4 to yield CD_3H and $CD_3CH{=}CH_2$ by the classical alkyl migration route.[5,6] Reductive elimination from hydrido(alkyl) metal complexes has often been proposed as a step in hydrogenation. Using a Rh(I) species and (Z)-(α)-acetamidocinnamate a hydrido(alkyl) intermediate has actually been isolated at $-78°C$.[7] However, $[CoH(CO)_2(\eta^5\text{-}C_5H_5)]$ reacts with $[Co(CO)_2(\eta^5\text{-}C_5H_5)R]$ to yield RH and RCHO, suggesting a bimolecular process in this case.[8] $[Ir(PMe_3)_4]^+[PF_6]^-$ reacts with paraformaldehyde to yield a hydrido(formyl) complex of iridium.[9] Such complexes have attracted attention as possible intermediates in the Fischer–Tropsch reaction but it now seems more likely that the mechanism involves dissociative adsorption of CO.[10] The formation of a C–C bond by reaction of CO with a cluster carbide has been demonstrated.[11] A CO ligand in $[Fe_4(CO)_{13}]^{2-}$ can be protonated using $SO_2(OH)(CF_3)$ producing, ultimately, methane.[12] Hydridocarbonyl clusters of several metals are active catalysts in solution in polar solvents where deprotonation is feasible. Stopped-flow kinetic studies of the reaction with MeO^- reveal first-order behavior for the iron group complexes (e.g., $k_1 = 1.3 \times 10^6\ s^{-1}$ for $[FeH_2(CO)_4]$) and reactions which are fast, but not exceptionally so, for acid–base processes involving MeO^-. $[OsH_2(CO)_4]$ and $[Os_4H_4(CO)_{12}]$ have similar deprotonation rates even though the Os_4 cluster has only bridging hydride ligands. Thus deprotonation must involve substantial structural and electronic rearrangements.[13] Exchange experiments between H_2 and D_2O catalyzed by $[Co(CN)_5]^{3-}$ confirm the heterolytic mode of dissociation for H_2.[14]

11.2.   Reactions Catalyzed by Carbene Complexes

11.2.1.   Alkene Metathesis

Many of the leading workers in this field participated in the Lyon symposium on the subject.[15] The mechanism of alkene metathesis proceeds through a metallocyclobutane derivative formed by reaction of an alkene with a metal–carbene complex. Retro-carbene addition yields a product molecule and a new carbene

Scheme 1

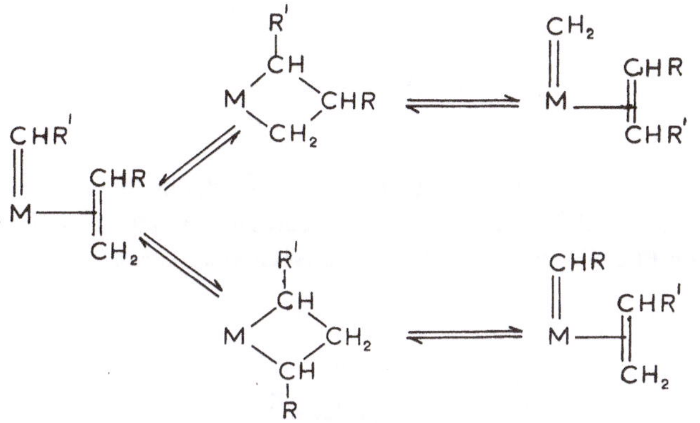

complex (Scheme 1). Many metathesis reactions are very fast and the carbene complex intermediates are not readily detected at least in their catalytically active forms. Thus the carbene complex is to be regarded as a typical chain-carrying intermediate, but the structural relationships to the precursors are far from certain in many cases. Reaction with a cocatalyst, dissociation of a ligand, dimerization, and change of oxidation state of the catalytic metal are all possible steps which may be involved in initiation of the chain reaction. Possible modes of termination are bimolecular reaction of two metal–carbene complexes or the decomposition of a metallocyclobutane to an alkene or cyclopropane. Such steps have been studied in model compounds which are themselves too unreactive to sustain a long kinetic chain length. The most effective catalytic metals are Mo, W, Re, and Ti. Slower reactions are seen with Nb and Ta which are useful in model systems.

In general Fischer-type carbene complexes are catalytically inactive but there are various means of converting them into active carbenes. Thus nucleophilic attack by CH_3Li in **1** at the carbene carbon followed by removal of the

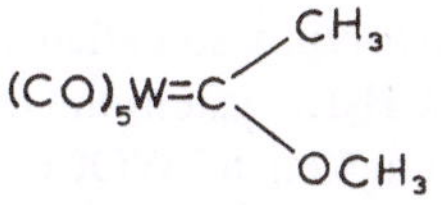

1

methoxide by H^+ should yield a dialkylcarbene complex. In fact, the dialkylcarbene complex reacts at a second metal center and the product was a μ-carbeneditungsten complex (**2**).[16] Such a complex is an alkyne metathesis catalyst.[17] Attempts were made to trap the dialkylcarbene complex using norbornene enol ether and cyclopentene enol ether. The latter undergoes a reversible

$$2$$

chain polymerization, suggesting that the terminal double bond remains attached to the metal as in **3**.[16] The same methoxycarbene complex can be activated by reaction with PhCCH, which yields a phenylvinylcarbene complex of tungsten,

$$3$$

and this was used for polymerization of cyclopentene.[18,19] The formation of a metallocyclobutane from a metal–carbene complex can be either a direct dipolar addition to a double bond or else prior coordination of the alkene may be necessary. A model system for the latter type of process has been described.[20] Formation of the stable tungsten carbene alkene complex **5** involves prior dissociation of a CO ligand from **4**. Decomposition yields a cyclopropane derivative (**6**) in what would be a termination step in metathesis. This reaction is autocatalytic, suggesting that a $W(CO)_4$ fragment promotes further reaction of **4** by abstraction of CO.

$$4 \qquad\qquad 5 \qquad\qquad 6$$

The activation process involving so-called catalyst and cocatalyst can be very complex. In the $WOCl_4$–CH_3Li system the true catalyst is $[WOCl_2(CH)_2)]$. Methyl lithium reduces the tungsten to $WOCl_2$ via $[WOCl_3(CH_3)]$ and these species react further to give $[WOCl_2(CH_2)]$ and $[WOCl_3H]$.[21] Reaction of $Mg(CH_2C(CH_3)_3)_2$ with $WOCl_4$ also yields a metathesis catalyst containing $[WO(CH_2C(CH_3)_3)_4]$ and $[WOCl(CH_2C(CH_3)_3)]$. Neither is active for metathesis and hence the mixture must contain yet another component, perhaps more $WOCl_4$. It should be noted that many active catalysts contain ligands with oxygen as a "hard" donor atom.[22] An *ab initio* calculation of the energy of the intermediate in the reaction of $[CrOCl_2(CH_2)]$ with C_2H_4 has been made.[23]

A series of alkylidenetantalum(V) complexes has provided considerable insight into another type of activation process in metathesis catalysis. Thus benzyl and neopentyl complexes undergo α abstraction, the more hindered neopentyl complex the more readily. Treatment with an excess of phosphine $[L = P(CH_3)_3]$ gives α abstraction in a very hindered seven-coordinate species $[TaL_2(CH_2C(CH_3)_3)_2Cl_3]$. Nmr studies show a complicated series of dissociation and dimerization equilibria among carbene complexes of the type $[TaL_n(CHC(CH_3)_3)Cl_3]$ with $n = 1–3^{(24)}$ and a full account of metathesis catalysis will have to take these equilibria into account.$^{(25)}$ Such effects have already been demonstrated in the oxotungsten series. Thus metathesis of 1-butene or 2-pentene by $[WO(PEt_3)_2(CHC(CH_3)_3)Cl_2]$ requires prior removal of chloride by complexation with $AlCl_3$ or a palladium complex. The 16-electron five-coordinate complex is very active and has a W–C bond order between 2 and 3.$^{(26)}$

Some possible termination steps are easily identified in the slow reactions of tantalum complexes such as $[Ta(C_5H_5)(CHC(CH_3)_3)Cl_2]$ and $[TaL_2(CHC(CH_3)_3)Cl_3]$ (7a) and (7b) $(L = P(CH_3)_3)$. Addition of $RCH{=}CH_2$

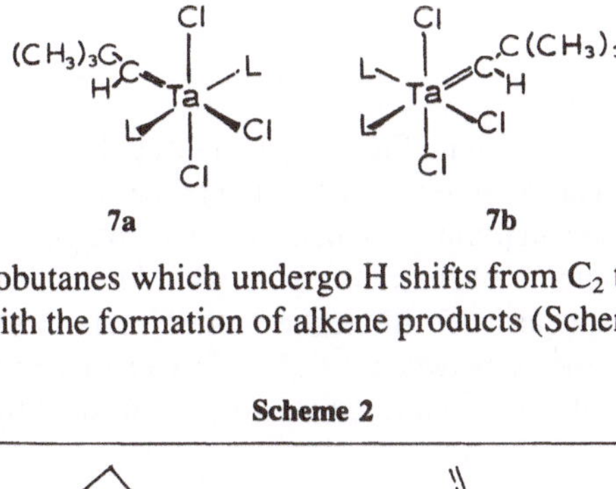

7a   7b

yields metallocyclobutanes which undergo H shifts from C_2 to C_1 thus removing the active center with the formation of alkene products (Scheme 2). Replacement

Scheme 2

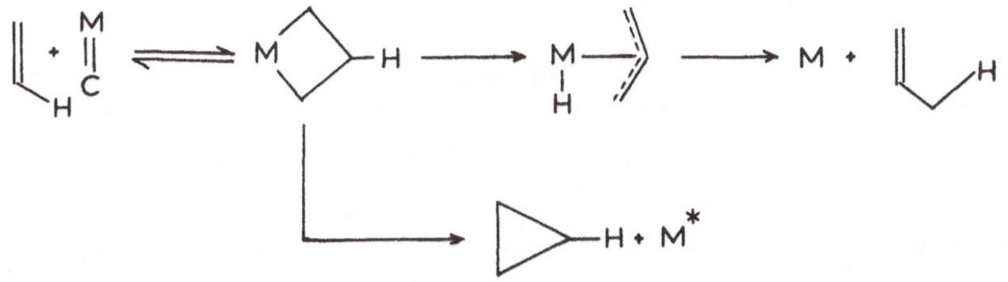

of $P(CH_3)_3$ by tetrahydrofuran in **7** promotes some metathesis, which can be further enhanced by substitution of the hard ligand CH_3O^- for Cl^- and addition of $P(CH_3)_3$. Presumably these conditions suppress the hydride shift. $[Ta(P(CH_3)_3)(CHC(CH_3)_3)(OC_4H_9)_2Cl]$ reacts with styrene by transalkylindenation and then formation of stilbene by a bimolecular coupling reaction, which also terminates the chain.$^{(25–27)}$ The possibility of termination by formation of cyclopropanes was studied in a series of tungstacycles which were formed by reaction of η^3-allyls with H^- followed by thermal decomposition (Scheme 3).[28]

Scheme 3

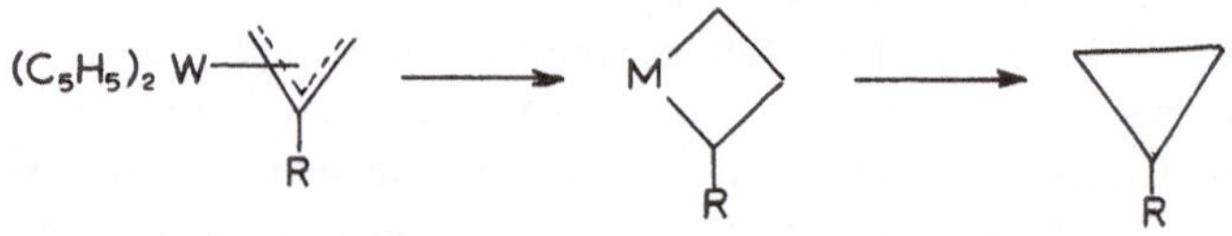

Exchange of methylene groups (degenerate metathesis) has been studied using methylenecyclohexane and $^{13}CH_2{=}C(CH_3)_2$ with $[\eta^5\text{-}C_5H_5)Ti(CH_2)AlCl(CH_3)_2]$

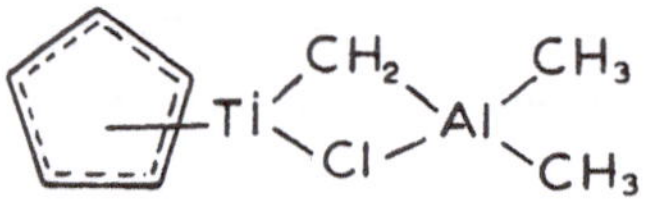

8

(**8**) as catalyst. Reaction of **8** with $PhC{\equiv}CPh$ yields a titanacyclobutene complex and $AlCl(CH_3)_2$.[29] The saturated titanacycle can be otained by treating **8** with 3,3-dimethylbut-1-ene and pyridine in benzene. The role of the base is to remove the $AlCl(CH_3)_2$. Treatment of the titanacycle with $AlCl(CH_3)_2$ regenerates **8** so that the "cocatalyst" appears to facilitate the carbene metallacycle equilibrium.[30] The degenerate metathesis between $^{13}CH_2{=}CHCO_2Et$ and $CH_2{=}CHCN$ has been used to show that cyclic intermediates can be formed by polar addition in a Wittig-type reaction (Scheme 4).[31]

Scheme 4

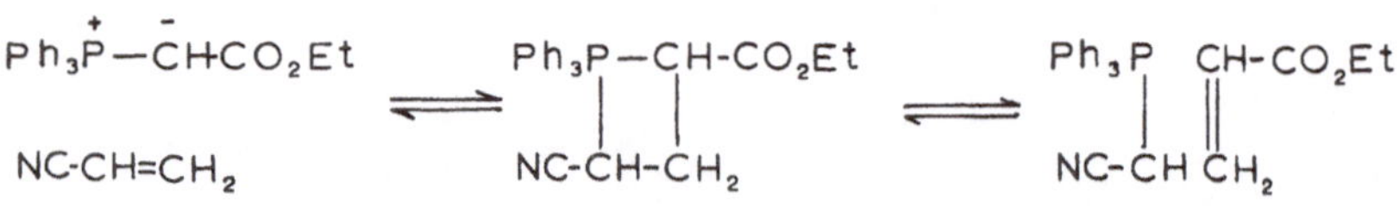

The reactivity and stereoselectivity of metathesis are not well understood at present. Conjugated carbene complexes ($-CR{=}CH-CR{=}M$) react faster with alkynes than with alkenes and conversely for nonconjugated types.[32] Using $[W(CO)_2(AsPh_3)_2Cl_2]$, degenerate metathesis of 3,3-dimethylbut-1-ene is much faster than productive metathesis. In cross-metathesis with norbornadiene the main chain carrier seems to be $W{=}CHC(CH_3)_3$.[33] Stereoselectivity is generally greater in the reactions of strained cyclic systems than other types. However, in the metathesis of terminal alkenes the higher-molecular-weight product gives a higher *trans/cis* ratio than the lower-molecular-weight one consistent with a metallacyclobutane in which the 1-2-*e-a* or 1-2-*e-e* configurations are more favored than 1-2-*a-a* in disubstitution.[34]

11.2.2. Ring-Opening Polymerization by Metallocarbene Complexes

Ring-opening polymerization of cycloheptene gives 98% *cis*-polymer which is largely translationally invariant and which has a number average molecular weight of 2.1×10^5.[35] Using 5-substituted norbornenes the m and r configurational diads can be determined by ^{13}C nmr. If the catalytic complex is octahedral there are four relative positions of the polymer chain and the vacancy (structures **9–12**); **9** is unreactive; **10** and **12** are mirror images. Formation of the metallacyclobutane can be *cis*-**13** or *trans*-**14**. The sequence **11** → **13** → **12** then takes place highly specifically, thus giving a block polymer.[36]

9   10   11   12

13   14

11.3. Hydrogenation

11.3.1. Introduction

The literature of homogeneous catalytic hydrogenation is very extensive.[37] There have been few wholly new developments in alkene hydrogenation where the emphasis has been on asymmetric syntheses. However, there have been some novel reports in the field of hydrogenation of arenes.

11.3.2. Hydrogenation of Alkenes

A very novel approach to hydrogenation is to use syngas (i.e., mixtures of CO and H_2) in place of H_2 for the reduction, thus eliminating a separate shift step in converting CO to H_2. It is necessary to achieve satisfactory rates in the two cycles that reduce CO and the unsaturate, respectively. In the case of the reduction of nitrobenzene to aniline, $[Ir_4(CO)_{12}]$ or $[Rh_6(CO)_{16}]$ allows the combination of the hydrogenation and the shift cycles (Scheme 5). The titanium

Scheme 5

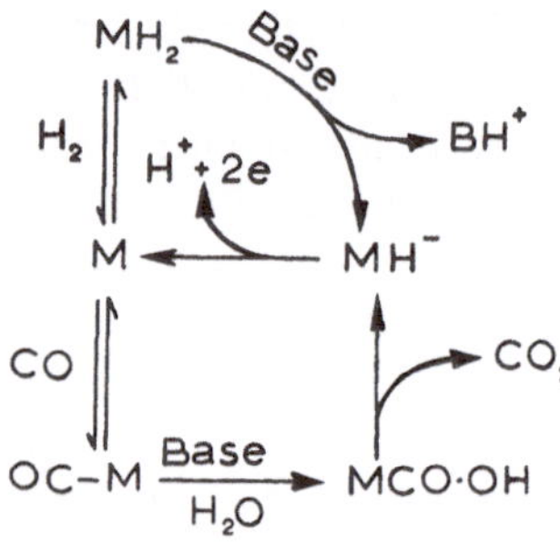

complexes $[Ti(\eta^5\text{-}C_5H_5)_2R_2]$ are interesting catalyst precursors which require photochemical activation under H_2 which causes evolution of RH (R = CH_3, $CH_2C_6H_5$, or C_6H_5). The alkene to be reduced can then be added in a solvent such as heptane.[39]

Numerous complexes of group VIII metals are active for hydrogenation. The mechanism generally involves migration of a hydride ligand onto the co-ordinated "unsaturate" followed by elimination of product from the hydrido(alkyl) intermediate. Triphenylphosphine complexes of ruthenium are unusual in that $[Ru(PPh_3)_3(styrene)]$ and $[RuH_2(PPh_3)_4]$ can both be isolated and thus the "unsaturate" and "hydride" routes can be established for a single system.[40]

There have been several attempts to raise the activity of homogeneous catalysts by use of solvents of low coordinating power. $[Ir(PMePH_2)_2(COD)]^+[PF_6]^-$ is very active in dichloromethane at 0°C.[41] Reduction of the (dicarbollide)Rh complex (15) gives a catalyst 30 times as active as $[Rh(PPh_3)_3Cl]$ for hydrogenation of trimethylvinylsilane.[42] The solvates $[Pd(diphos)S_2]^+$ (S = acetone, water, acetonitrile, but not benzene or dichloromethane) are active for styrene hydrogenation at 30°C/1 atm.[43] Several catalysts have good selectivity for partial hydrogenation of polyenes or alkynes to alkenes. $[Fe(\eta^5\text{-}C_5H_5)(\mu_3\text{-}CO)]_4$ catalyzes the formation of 1-pentene from 1-pentyne without isomerization. The cluster is very stable and can be recovered without fragmentation.[44] $[Ru(\eta^4\text{-}COD)(\eta^6\text{-}C_7H_8)]$ is an efficient catalyst for hydrogenation of cycloheptatriene to

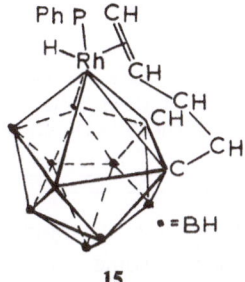

15

cycloheptene.[45] Removal of chloride from [Co(PPh$_3$)$_3$Cl] by AgClO$_4$ yields a complex which is also a selective catalyst for hydrogenation of butadiene to 1-butene.[46]

There are rather few water-soluble hydrogenation catalysts. A number of rhodium complexes derived from ligands such as (PH$_2$PCH$_2$CH$_2$)$_2$NCOC$_6$H$_4$SO$_3^-$ are active in this way.[47] Catalysts have been developed for hydrogenation of ketones,[48,49] carboxylic acids,[50] esters,[51] nitriles,[52] and nitro compounds.[53]

11.3.3. Asymmetric Hydrogenation

Numerous investigations of asymmetric hydrogenation have been published since the early disclosure of the commercial production of *l*-DOPA. Most complexes are based on complexes of rhodium(I) with asymmetric chelate phosphane ligands. However, reduction catalyzed by Ru(II) as exemplified by [RuH(PPh$_3$)$_3$Cl] has been extended by use of the ligand (4R,5R)-(−)-DIOP (**16**). The precursor

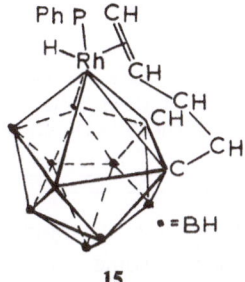

16

is [Ru$_2$(DIOP)$_3$Cl$_4$], which reacts with H$_2$ to give [RuH(DIOP)$_2$Cl]. This hydrido-chloride is unusually reactive for such a complex of *trans* stereochemistry. Using *N,N*-dimethylacetamide solvent, acrylamide is reduced in good optical yield. Kinetic data fit Scheme 6, which is similar to that established for [RuH(PPh$_3$)$_3$Cl].[54]

Scheme 6

$$\rightarrow \text{HRuCl(DIOP)}_2 + \text{acrylamide} \overset{K_1}{\rightleftharpoons} \text{RuCl(alkyl)(DIOP)} + \text{DIOP}$$

$$\text{RuCl(alkyl)(DIOP)} + \text{H}_2 \rightarrow \text{H}_2\text{RuCl(alkyl)(DIOP)}$$

$$\text{H}_2\text{RuCl(alkyl)(DIOP)} \rightarrow \text{HRuCl(DIOP)} + \text{propionamide}$$

$$\text{DIOP (fast)}$$

Before giving a general review of asymmetric hydrogenation catalyzed by rhodium it is worth noting some results obtained with $(2S,3S)$-bis(diphenylphosphino)butane. Reduction of ethyl-(Z)-α-acrylamidocinnamate gives up to 100% stereoselectivity. The rate is proportional to the concentration of **17**, the structure of which was confirmed by X-ray crystallography. The H_2

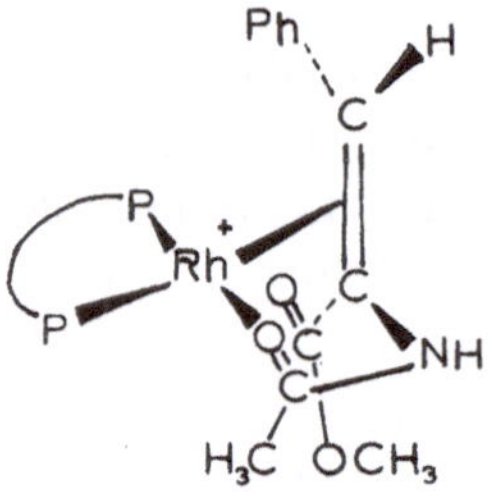

17

should become attached to the face coordinated to Rh, which is not so. It is proposed that the catalyst is actually a highly reactive but less abundant diastereomer and that it is relative rates and not mode of binding that determines the enantioselectivity.[55]

The fact that eneamides bind to chiral complexes of Rh(I) and then undergo hydrogenation with high enantiomeric selectivity has allowed determination of intermediate structures by nmr methods since often the single diastereomer present is easily characterized. $[\text{Rh(dioxop)(CH}_3\text{OH)}_2]^+$ (**18**) adds H_2 reversibly and models show that one ring oxygen coordinates without strain in $[\text{RhH}_2(\text{Dioxop)(CH}_3\text{OH)}]^+$ (**19**). When α-Benzamidocinnamic acid is added to

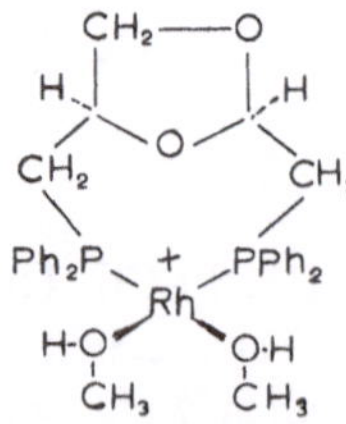

18

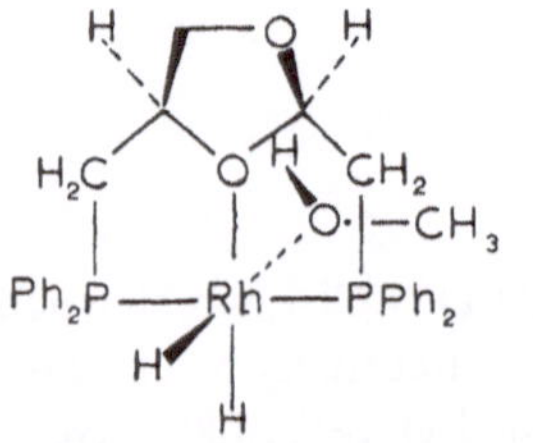

19

18 the ^{31}P nmr spectrum shows that only one of the four possible isomers is formed (**20**) (R = H). This highly selective reaction requires the presence of NEt_3 and does not take place selectively for the ester (**20**) (R = CH_3). Further nmr studies show that an intermediate chelate carboxylate (**21**) is formed before **20**

20   **21**

(R = H) and it is this step which is catalyzed by NEt_3. The tridentate binding gives an optically efficient hydrogenation route.[56] Single-carbon enrichment with ^{13}C in (Z)-enamides shows that the C=C and the amide C=O coordinate to Rh(I) in a bidentate mode when **16** is the chiral ligand. The corresponding (E)-enamides complex through the C=C and the carboxylate C=O. Only the (Z)-isomers undergo stereoselective hydrogenation.[57] The full characterization of these enamide complexes is a useful confirmation of the "unsaturate route" in asymmetric hydrogenation.[58] Also, using (R,R)-1,2-bis(o-methoxyphenyl-phenylphosphino)ethanerhodium(I), the α-benzamidocinnamic acid complex can be formed and treated with H_2 at low temperature. On warming the hydrido(alkyl) intermediate is formed (**22**) and this can be characterized by its nmr spectrum.[59]

22

Asymmetric syntheses with the ligand **16** have also been extended to α,β-unsaturated acids such as atropic acid [CH_2=C(Ph)COOH]. Again, bases such as NEt_3 must be added to promote formation of carboxylate anions which then act as ligands which give rise to a single diastereomer. In these asymmetric syntheses chiral five-ring chelates are effective for potentially tridentate enamides, while seven-ring chelates are effective with bidentate substrates.[61] By contrast the structures of chiral complexes of itaconic acid $[CH_2{=}C(CO_2H)(CH_2CO_2H)]$ seem to be determined by hydrogen bonding. The disodium salt gives poor asymmetric synthesis.[60] Some caution is appropriate in interpreting these results because there seems to be little correlation between the configurations of the diastereomeric structures determined by nmr methods and the final chiral hydrogenation products.[61]

A comparison of rhodium-catalyzed hydrogenation of α-acylaminocinnamic acids with the ligands **23** (X=NH and X=NCH₃) which have the same config-

23

uration shows that the products have opposite configurations. Presumably the N-methylation must change the helical configuration of the phenyl groups.[62] At least 150 chiral diphosphines have now been used in hydrogenation[63] and new examples continue to appear.[64-66]

An attempt has been made to use the chiral center in $[Cr(CO)_2(PPh_3)(C_{10}H_8)]$ for hydrogenation.[67] The use of optically active ligands in catalysts such as reduced $[Ti(C_5H_5)_2(menthyl)_2]$ gives a small enantiomeric excess.[68] A DIOP rhodium complex promoted by NEt_3 gives asymmetric reduction of methyl ketones.[69]

11.3.4. Hydrogenation of Aromatic Hydrocarbons

While the mechanism of action of a number of catalysts for alkene hydrogenation is well understood, the same is not true for arenes. The available catalysts seem to show widely different modes of action, including free radical hydrogenation.[70] Work using allyl–cobalt complexes has brought to light some interesting new principles. The catalyst precursor is $[Co(\eta^3\text{-}C_3H_5)L_3]$ $[L = P(OCH_3)_3]$ and its reaction with H_2 is promoted by a $\eta^3 \rightarrow \eta^1$ interconversion. The reverse $\eta^1 \rightarrow \eta^3$ interconversion then promotes loss of $P(OCH_3)_3$, yielding $[CoH_2(\eta^3\text{-}C_3H_5)L_2]$. A further displacement of $P(OCH_3)_3$ by benzene generates the true catalyst and the proposed hydrogenation cycle is shown in Scheme 7. In this case it appears that by means of $\eta^3 \rightleftharpoons \eta^1$ interconversions the allyl ligand is (a) able to free a coordination site for oxidative addition and (b) promotes a change in the number of coordination sites used by the cyclic fragment during reduction. The structural chemistry of arene complexes reveals that the η^2 and η^6 types are most stable and most common. η^4 Complexes are rare and involve loss of aromatic energy with formation of a nonplanar ligand. The unusual role of this allyl–cobalt catalyst seems to be to promote hydrogenation by formation of a noncoplanar ring.[71]

Scheme 7

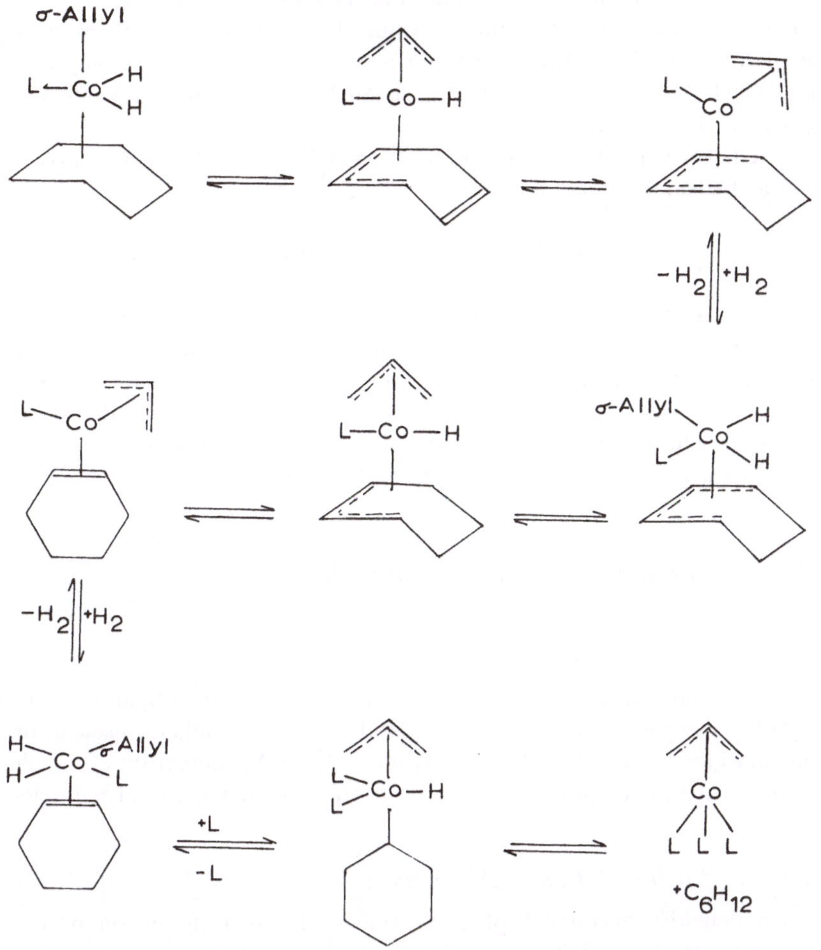

Ruthenium has a marked tendency to interaction with phenyl groups in PPh$_3$ by *ortho*-metallation. Thus it is not surprising that [RuH$_2$(PPh$_3$)$_3$] is an effective catalyst for the hydrogenation of naphthalene to tetrahydronaphthalene.$^{(72)}$

11.3.5. Hydrogen Transfer Hydrogenation

Several catalysts are available that effect the reduction of secondary alcohols to ketones. Typical acceptors are cyclohexene or various ketones.$^{(73-76)}$ Diphenylacetylene has been used as an acceptor for dehydrogenation of primary alcohols.$^{(77)}$ Attempts to activate alkanes in this way have been attempted and indeed

$[Ir(COE)_2L_2]^+$ ($L = PPh_3$, COE = cyclo-octene) can be used to catalyze the disproportionation of COE, cyclohexene and cyclopentene, etc.[78] A useful extension of transfer hydrogenation would be an asymmetric synthesis. Using $[Ru_4H_4(CO)_8\{(-)-DIOP\}_2]$ as catalyst alkylmethyl ketones gave only poor optical yields but the reaction of $PhCOCH_2CH(CH_3)_2$ with 2-propanol gave an optical yield of 9.8%.[79]

A novel variant is the reaction of acetaldehyde with water to give acetic acid and ethanol [equation (1)]:

$$CH_3CHO + 2H_2O \rightarrow CH_3CH_2OH + CH_3COOH \qquad (1)$$

The catalysts are various half-sandwich complexes such as $[Rh_2(C_5Me_5)_2(OH)_3]^+[PF_6]^-$, which implicates a hydride intermediate. The rate is first order in acetaldehyde and half-order in the catalyst, which must therefore dissociate in water giving $[Rh(C_5Me_5)(OH)_2(OH_2)]$. Oxidative addition of the aldehyde yields a hydrido(hydroxo)(acyl) complex which can eliminate acetic acid. The metal hydride then reduces another molecule of acetaldehyde.[80]

11.4. Reactions of Carbon Monoxide

11.4.1. Hydroformylation

There have been relatively few studies of conventional hydroformylation. Rhodium complexes of some of the chiral phosphines hitherto used in hydrogenation give asymmetric hydroformylation.[81-83] A comparison of cobalt and rhodium from the point of view of concurrent isomerization has been made.[84]

11.4.2. Carbonylation of Alcohols

A definitive account[85] of acetic acid synthesis using an iridium catalyst has appeared. Its activity is similar to rhodium but the mechanism is more complex (Scheme 8). The catalyst is usually introduced as $[Ir(H_2O)_3Cl_3]$ with CH_3I as promoter. At low I^- concentration the cycle A generates acetyl iodide, the main species in solution being $[Ir(CO)_3I]$. At much higher I^- concentration the main species is $[Ir(CO)_2(CH_3)I_3]^-$. The methyl migration step is common to cycles A and B, the difference being whether five-coordinate $[Ir(CO_2)(COCH_3)I_2]$ is trapped by CO (cycle A) or by I^- (cycle B). Cycle A is important at low methyl iodide concentration because otherwise methyliridium species predominate. Also, cycle A is characterized by the slow reaction of $[Ir(CO)_2I]_n$ with CH_3I, which requires predissociation of $[Ir(CO)_3I]$. Hence this regime is subject to CO inhibition. By contrast, in cycle B oxidative addition of CH_3I occurs

Scheme 8

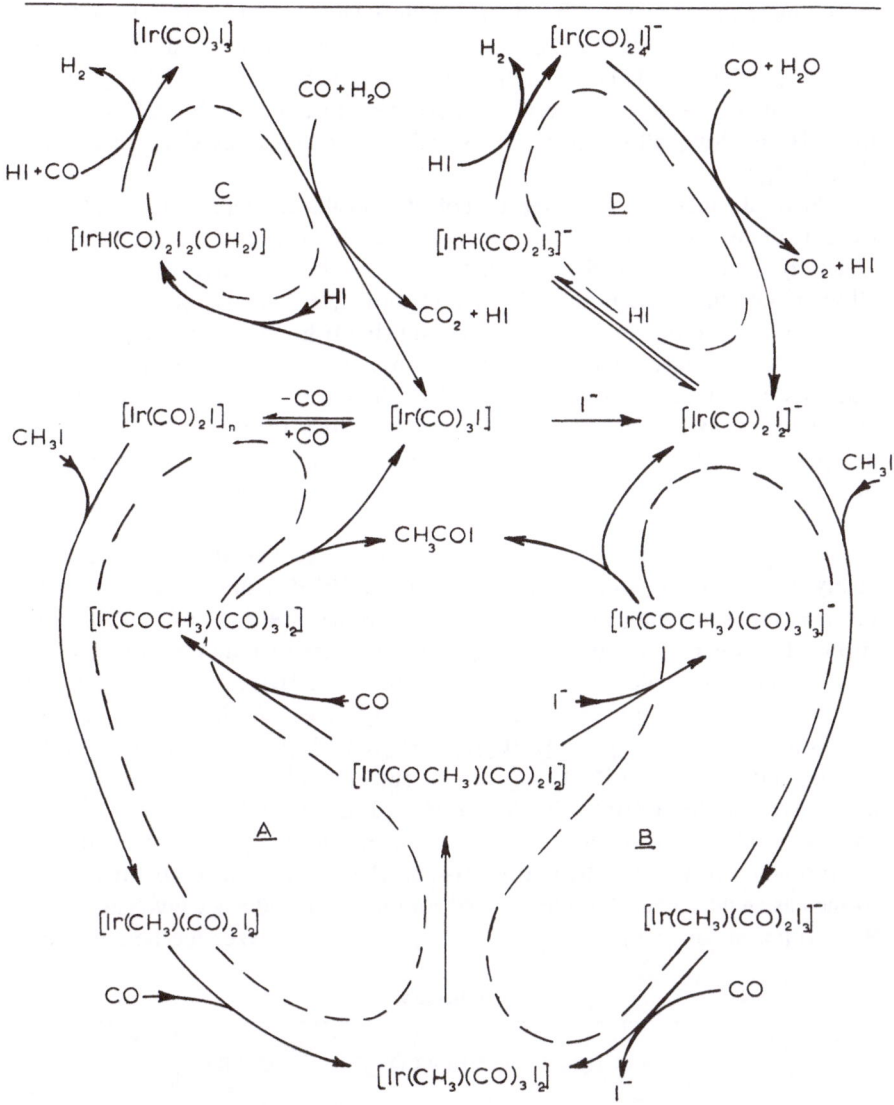

readily because $[Ir(CO_2I_2]^-$ is so much more nucleophilic than $[Ir(CO)_2I]$ and the rate is first order in CO because of the displacement of I^- by CO in the predominant $[Ir(CO)_2(CH_3)I_3]^-$.

Water gas shift reactions compete with both carbonylation cycles (cycles C and D). In essence this is competition between CH_3I and HI for oxidative addition to iridium(I). The ensuing hydridoiridium species react with more HI, liberating H_2, after which iridium(III) is reduced by CO. At high H_2O or CH_3OH concentration this leads to a third kinetic regime, which is first order in methanol and independent of CO concentration. The reaction between $[Ir(CO)_2I_2]^-$ and HI is fast but reversible. However, the next step, reaction of $[IrH(CO)_2I_3]^-$ with more HI is slow, allowing cycles B and D to compete as CH_3I reacts with $[Ir(CO)_2I_2]^-$.[85]

Several papers report the reaction of CO and alkenes with alcohols catalyzed by palladium complexes to give saturated and unsaturated esters.[86–88] It is generally supposed that these reactions proceed by formation of a palladium alkoxide and then insertion of CO. Alkyl carbonates can also be formed.[89] In conventional hydroformylation of ethylene using $[Rh_4(CO)_{12}]$ it has been shown that the cluster catalyst dissociates.[90] The problems of recovery of these rhodium catalysts have received some attention.[91] The use of a heterogeneous catalyst and syngas (mixtures of CO and H_2) as a feedstock for production of organic compounds has been of interest industrially for many decades. There is a report of extensive screening tests using a large number of soluble metal–complex catalysts with syngas at 2000 atm. Products include esters, alcohols, alkanes, ethylene glycol, etc.[92] Care must be taken to see that these homogeneous catalysts are actually free of precipitated metal.[93] It is now accepted that the Fischer–Tropsch synthesis involves dissociative adsorption of CO on a catalyst surface but it is possible that a homogeneous synthesis would involve the attack of coordinated hydride on coordinate CO. $[NbH_3(C_5H_5)]$ reacts with carbonyls such as $[Co_2(CO)_8]$ to give ethane. $[Cr(^{13}CO)_6]$ gives $^{13}C_2H_6$.[94] The reduction of CO in a η^2-acyl complex, $[Zr(C_5H_5)(CH_3)(COCH_3)]$, by $[MoH_2(C_5H_5)]$ has also been demonstrated.[95] Using various ruthenium-based precursors in acetic acid solvent at 200°C/300–400 atms of H_2/CO, the actual catalyst is $[Ru(CO)_5]$ and the products are methyl acetate, ethylene, ethylene glycol, etc. Although the formation of formaldehyde from H_2 and CO is very unfavorable it could be formed as an intermediate undissociated complex. The sequence given in Scheme 9 is proposed for methanol synthesis.[96] Rhodium catalyzes the homologation

Scheme 9

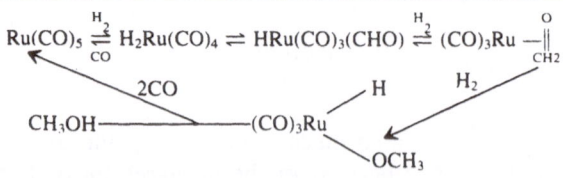

of methanol by H_2/CO.[97] Using CoI_2/PBu_3 as catalyst for homologation of methanol the addition of borates improves the selectivity for ethanol possibly by increasing the carbene character of the acyl ligand which would promote hydrogenation (Scheme 10).[98] Model syntheses of various C_1 and C_2 organic molecules have been achieved with syngas and various manganese carbonyl complexes.[99]

Scheme 10

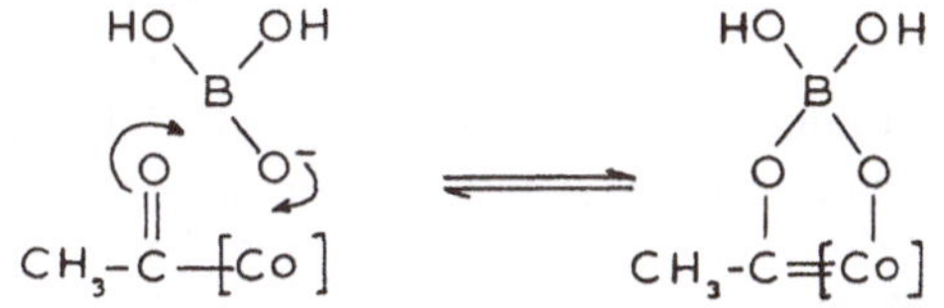

11.4.3. The Homogeneous Water Gas Shift Reaction (WGSR)

For many years there has been some interest in carrying out the WGSR by homogeneous catalysis. It has been presumed to occur as a side reaction in various carbonylation processes such as the acetic acid synthesis described above[85] and hydroformylation catalyzed by $[Co_2(CO)_8]/Ph_2PCH_2CH_2PPh_2$.[100] Only within the last five years has the reaction been studied in detail. Several recent papers have good reviews of early work.[101] The gas phase reaction is exothermic and is carried out at high temperature over mixed-oxide catalysts. When the reactant water is in the liquid phase the reaction becomes slightly endothermic but the equilibrium is still favorable because the entropy change becomes positive:

$$CO(g) + H_2O(g) = H_2(g) + CO_2(g),$$
$$\Delta H° = -9.84 \text{ kcal mol}^{-1}, \Delta S° = -10.1 \text{ cal deg}^{-1} \text{ mol}^{-1} \quad (2)$$

$$CO(g) + H_2O(l) = H_2(g) + CO_2(g),$$
$$\Delta H° = 0.68 \text{ kcal mol}^{-1}, \Delta S° = 18.3 \text{ cal deg}^{-1} \text{ mol}^{-1} \quad (3)$$

Thus under some circumstances it may be desirable to operate a low temperature liquid phase reactor. It has been shown[102] that several metal carbonyl WGSR catalysts are tolerant of sulfur in the CO feed, unlike their heterogeneous counterparts. It now seems that a wide range of complexes (e.g., carbonyls of Fe, Ir, Re, Rh, and Ru[101]) have WGSR activity and while most systems involve the familiar nucleophilic attack by OH⁻ on coordinated CO,[103] others operate in acidic media where the oxidative addition of HX to a low-valent metal is followed by further reaction of the hydrido complex to generate H_2.

Several catalytic systems use mononuclear or cluster carbonyls in the presence of a base such as KOH or NEt_3. The solvent is usually aqueous methanol or ethoxyethanol since pure water gives low activity. Temperatures are in the range 100–200°C with CO pressures of 1–40 atm. The CO_2 product is normally very soluble in the reaction medium, but despite this the base is often present as a formate under steady state operation:

$$OH^- + CO \rightleftharpoons HCO_2^- \tag{4}$$

Using $[Fe(CO)_5]$ the rate is first order in the catalyst but zero order in CO. However, at low CO pressures the catalyst deactivates by formation of $FeCO_3$ through oxidation of both metal and the CO ligands. Infrared studies show that the only iron species present in the catalyst solutions are $[Fe(CO)_5]$ and $[FeH(CO)_4]^-$. It is proposed that CO_2 results from the known reaction of $[Fe(CO)_5]$ with OH^-. As reaction proceeds the pH decreases from the initial value of 8.6 when the base is present as formate to a lower value of 7.4 as a result of the presence of CO_2 product and hence bicarbonate.

$$CO_2 + H_2O + HCO_2^- \rightleftharpoons HCO_2H + HCO_3^- \tag{5}$$

Under these conditions the hydridocarbonyl, $[FeH(CO)_4]^-$, can be protonated[104]:

$$[Fe(CO)_5] + OH^- \rightarrow [FeH(CO)_4]^- + CO_2 \tag{6}$$

$$[FeH(CO)_4]^- + H_2O \rightleftharpoons [FeH_2(CO)_4] + OH^- \tag{7}$$

$$[FeH_2(CO)_4] \rightarrow [Fe(CO)_4] + H_2 \tag{8}$$

$$CO + [Fe(CO)_4] \rightarrow [Fe(CO)_5] \tag{9}$$

The group VI carbonyls are also active WGSR catalysts and infrared studies show that the $M(CO)_6$ species are the only significant ones in solution. Under conditions similar to those for catalysis by $[Fe(CO)_5]$ the rate is first order in CO. Hence the first step must be dissociation of the catalytic complex. Since the base is present as formate and the anion $[W(CO)_5(OCHO)]^-$ has been synthesized independently it is proposed that this is the next formed species. The mechanism given below is consistent with the known thermal decomposition properties of the anion.[105]

$$[M(CO)_6] \rightleftharpoons [M(CO)_5] + CO \tag{10}$$

$$[M(CO)_5] + HCO_2^- \rightarrow [M(CO)_5(OCHO)] \tag{11}$$

$$[M(CO)_5(OCHO)] \rightarrow [MH(CO)_5] + CO_2 \tag{12}$$

$$[MH(CO)_5]^- + H_2O \rightarrow [MH_2(CO)_5] + OH^- \tag{13}$$

$$[MH_2(CO)_5] \rightarrow [M(CO)_5] + H_2 \tag{14}$$

$$M = Cr, Mo, \text{ or } W$$

By contrast with the catalysts based on Fe, Cr, Mo, and W which are mononuclear, the cluster carbonyls of Ru seem to remain intact in solution under reaction conditions.[101] However, three precursors, $[Ru_3(CO)_{12}]$, $[Ru_4H_2(CO)_{13}]$, and $[Ru_4H_4(CO)_{12}]$, all give identical catalyst solutions in the presence of base and in which there are probably at least three carbonylate anions present. The main one is $[Ru_4H_3(CO)_{12}]^-$ together with some $[Ru_3H(CO)_{11}]^-$. The concentrations of water and base have little effect on the rate which is first order in catalyst precursor and is consistent with the clusters remaining undissociated. The rate is first order in P_{CO}. Reaction of OH^- with a cluster carbonyl is unlikely to be rate determining since such reactions are known to be fast for Ru. A sequence involving loss of H_2 and then reaction with CO can also be ruled out on the basis of the rate equation. In the tetranuclear series it is proposed that Ru–Ru bond scission may be involved (Scheme 11), and in the trinuclear series individual steps of this type have actually been demonstrated.

Scheme 11

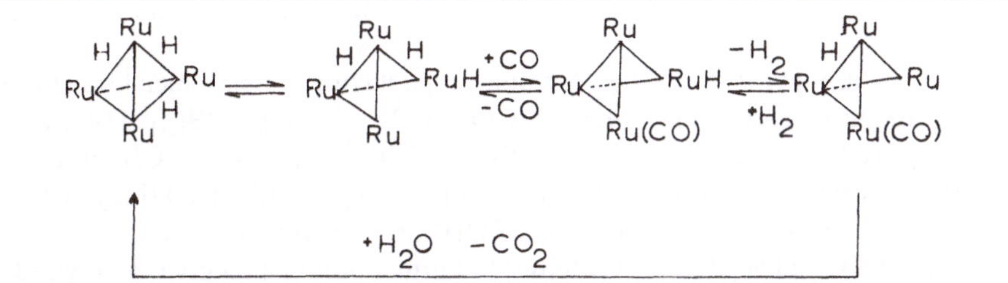

Some Ni(II) complexes of bidentate phosphines are very active WGSR catalysts under basic conditions.[106]

A WGSR catalyst that is typical of those that are active in acidic medium is $[Rh(CO)_2I_2]^-$. The reaction is carried out in the presence of HI in aqueous acetic acid. Such conditions are typical of those that favor oxidative addition of a species HX to Rh(I). In this case the generation of CO_2 will involve nucleophilic attack by water rather than OH^-. The reaction conditions are exceptionally mild, requiring less than 1 atm of CO at 80–100°C. Kinetic measurements show two distinct reaction regimes at 55–66°C and 72–97°C, and curiously the lower temperature range gives rise to the higher activation energy. Quench experiments show that at low temperature the catalyst is present as Rh(I) in $[Rh(CO)_2I_2]$, while at high temperature the main species is $[Rh(CO)_2I_4]^-$. At low temperature the rate is independent of P_{CO} and dependent on $[H^+][I^-]^2$. This implies attack on $[Rh(CO)_2I_2]^-$ by I^- and then H^+ after which further reaction with HI yields H_2 [steps (i) and (ii) in Scheme 12]. At high temperature, CO displaces I^- in $[Rh(CO)_2I_4]^-$ and $[Rh(CO)_3I_3]$ reacts with H_2O, giving H^+ and $[Rh(CO)_2I_3(COOH)]$. Thus the rate is dependent on $P_{CO}/[H^+][I^-]$ [steps (iii) and (iv) in Scheme 12].[103] Using a series of rhodium complexes $[RhH_2(py)_2L_2]^+$ it has been shown that elimination of H_2 is favored by electron-withdrawing ligands such as $(CH_3)_3CNC$ or CO.[107]

Scheme 12

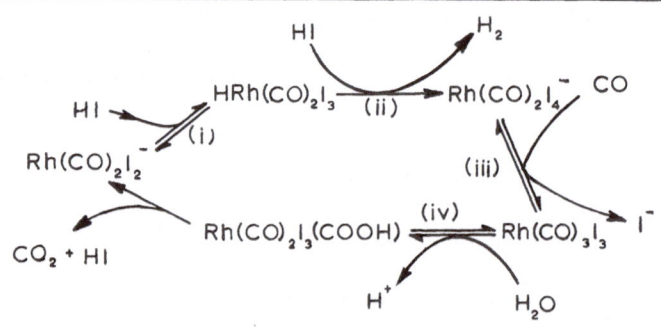

A rhodium complex of the "A-frame" type having a metal–metal bond can be protonated and then treated with CO to give $[Rh_2(\mu\text{-}H)(\mu\text{-}CO)(Ph_2CH_2CH_2PPh_2)]^+$, which is an active WGSR catalyst at 90°C/1 atm.[108] Another novel system is the bipyridylruthenium complex $[Rh(H_2O)(bipy)_2Cl]^+$. Displacement of Cl^- by CO, reaction of OH^- at the coordinated CO, and elimination of CO_2 yields $[RuH(H_2O)(bipy)]^+$. Further protonation gives a dihydride which eliminates H_2 photochemically. This completes the cycle for a photoassisted catalytic reaction.[109]

11.5. Oxidation

Mechanisms of oxidation are extremely diverse and may be of several free radical or charge transfer types as well as organometallic or Wacker-type reaction. Most of the last type depend on the reaction of a substrate with an oxidized form of the catalyst. An alternative would be to form a complex with the reduced form of the catalyst and then oxidize it, e.g.,

$$[PtCl_3(C_2H_4)]^- + Cl_2 + H_2O \rightarrow [PtCl_4]^{2-} + ClCH_2CH_2OH + H^+ \quad (15)$$

The mechanism involves formation of $[PtCl_5(CH_2CH_2Cl)]^{2-}$, which hydrolyzes to $[PtCl_5(CH_2CH_2OH)]^{2-}$ (24). The rate-determining step is the second-order reaction of 24 with Cl^-.[110]

$$\begin{array}{c} CH_2CH_2OH \\ Cl \diagdown \overset{|}{\underset{|}{Pt}} = Cl \\ Cl \diagup \quad \diagdown Cl \\ Cl \end{array}$$

24

Peroxide mechanisms are common in hydrocarbon oxidation. *tert*-Butyl hydroperoxide (BuOOH) can be used as an oxidizing agent with a palladium trifluoroacetate catalyst. The "catalyst" is $[Pd(OCOCF_3)(OOBu)]$, which adds to the alkene linkage to give 25. Decomposition of this intermediate to ketone

$$CF_3COOPd \diagup \overset{CH_2}{\underset{O \diagdown O}{\diagdown}} \overset{H}{\underset{R}{\diagup}} C \diagup$$
$$Bu$$

25

and $[Pd(OCOCF_3)(OBu)]$ requires a H shift. The catalytic cycle generates 1 mol of allyl complex, $[Pd(\eta^3-C_6H_{11})(OCOCF_3)]$, for every oxidation cycle.[111] The pathway in the formation of H_2O_2 from $[Pt(PPh_3)_2O_2]$ has been investigated.[112] There have been several investigations of the cooxidation of PPh_3 and alkenes using rhodium catalysts and the participation of $[Rh(PPh_3)Cl(O_2)]$ is assumed. The reaction by which this species generates triphenylphosphine oxide has a rate inversely dependent on the PPh_3 concentration and the oxidation step is assumed to involve $[Rh(PPh_3)_2Cl(O_2)]_2$.[113]

There have been further investigations of oxygen atom transfer reactions. Thus $[Rh(PPh_3)_3Cl]$ catalyzes the reaction of CO_2 and PPh_3 to give $POPh_3$ and CO_2[114] and aqueous $[PdCl_2-CuCl_2]_2$ catalyzes the reaction between CO and NO, yielding CO_2 and N_2O.[115] An interesting new system is $[CO(Saloph)(pyridine)NO_2]$, which oxidizes PPh_3, SPh_2, and RCH_2OH by O-transfer. Alkenes can be made reactive towards $Co—NO_2$, which is a weak nucleophile, by coordination to Pd. Thus $[Pd(C_2H_4)Cl_2]_2$ is oxidized to acetaldehyde in a dry solvent and the cobalt species is reduced to $[Co(Saloph)-NO]$.[116]

A study has been made of the role of copper in the oxidation of toluene to $PhCH_2OAc$ and $PhCH_2Br$ catalyzed by $Co—Cu—Br^-$.[117] $[Rh_6(CO)_{16}]$ and

Scheme 13

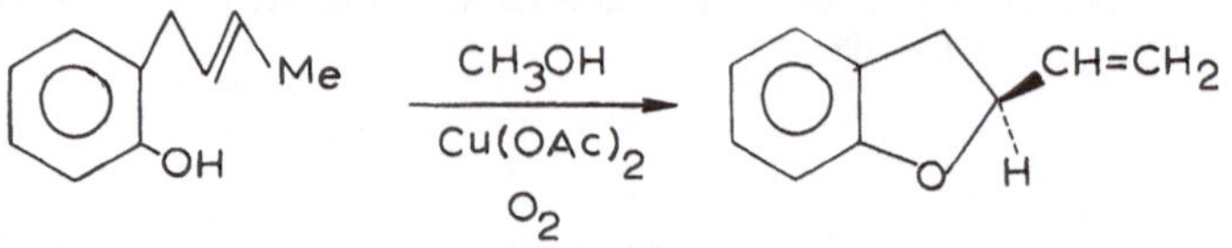

$[Re_2(CO)_{10}]$ are useful catalysts for free radical oxidation of alcohols and ketones such as cyclohexanol and cyclohexanone which yield adipic acid.[118]

Pinanylpalladium acetate is a chiral catalyst for oxidative cyclization (Scheme 13). The allylic palladium fragment remains intact during reaction which is presumed to involve a bridged Pd–Cu species.[119]

References

References for Chapter 1

1. R. D. Cannon, in *Inorganic Reaction Mechanisms* (A. G. Sykes, ed.), Vol. 7, The Chemical Society, London (1981), pp. 3–60.
2. R. D. Cannon, *Electron Transfer Reactions*, Butterworths, London (1980).
3. R. G. Linck, in *Treatise of Analytical Chemistry* (I. M. Kolthoff, ed.), 2nd ed., Part 1, Vol. 2, Wiley, New York (1980), pp. 645–698.
4. E. Buhks, M. Bixon, J. Jortner, and G. Navon, *Inorg. Chem.* **18**, 2014 (1979).
5. M. A. Ratner and R. D. Levine, *J. Am. Chem. Soc.* **102**, 4898 (1980).
6. E. L. Yee and M. J. Weaver, *Inorg. Chem.* **19**, 1077 (1980).
7. N. Sutin, M. J. Weaver, and E. L. Yee, *Inorg. Chem.* **19**, 1096 (1980).
8. M. J. Weaver and E. L. Yee, *Inorg. Chem.* **19**, 1936 (1980).
9. H. Fischer, G. M. Tom, and H. Taube, *J. Am. Chem. Soc.* **98**, 5512 (1976).
10. M. J. Weaver, *J. Phys. Chem.* **84**, 568 (1980).
11. T. J. Meyer, *Chem. Phys. Lett.* **64**, 417 (1979).
12. J. C. Curtis, B. P. Sullivan, and T. J. Meyer, *Inorg. Chem.* **19**, 3833 (1980).
13. B. P. Sullivan and T. J. Meyer, *Inorg. Chem.*, **19**, 752 (1980).
14. M. J. Powers and T. J. Meyer, *J. Am. Chem. Soc.*, 102, 1289 (1980).
15. H. D. Abruna, J. L. Walsh, T. J. Meyer, and R. W. Murray, *J. Am. Chem. Soc.* **102**, 3272 (1980).
16. H. E. Toma, *J. Chem. Soc. Dalton Trans.*, 471 (1980).
17. A. J. Miralles, R. E. Armstrong, and A. Haim, *J. Am. Chem. Soc.* **99**, 1416 (1977).
18. C. L. Wong and J. K. Kochi, *J. Am. Chem. Soc.* **101**, 5593 (1979).
19. S. Fukuzumi, C. L. Wong, and J. K. Kochi, *J. Am. Chem. Soc.* **102**, 2928 (1980).
20. C. L. Wong, R. J. Klinger, and J. K. Kochi, *Inorg. Chem.* **19**, 423 (1980).
21. R. J. Klinger and J. K. Kochi, *J. Am. Chem. Soc.* **102**, 4790 (1980).
22. H. M. Huck and K. Wieghardt, *Angew. Chem. Int. Ed.* **19**, 558 (1980).
23. H. M. Huck and K. Wieghardt, *Inorg. Chem.* **19**, 3688 (1980).
24. H. Bertram and K. Wieghardt, *Inorg. Chem.* **18**, 1799 (1979).
25. J. V. Beitz, J. R. Miller, H. Cohen, K. Wieghardt, and D. Meyerstein, *Inorg. Chem.* **19**, 966 (1980).
26. H. Bertram and K. Wieghardt, *Chem. Ber.* **111**, 832 (1978).

27. K. D. Whitburn, M. Z. Hoffman, M. G. Simic, and N. V. Brezniak, *Inorg. Chem.* **19,** 3180 (1980).
28. N. V. Brezniak and M. Z. Hoffman, *Inorg. Chem.* **18,** 2935 (1979).
29. J. P. Candlin and J. Halpern, *J. Am. Chem. Soc.* **85,** 2518 (1963).
30. G. McLendon and W. F. Mooney, *Inorg. Chem.* **19,** 12 (1980).
31. P. Natarajan and N. V. Raghavan, *J. Chem. Soc. Chem. Commun.* 268 (1980).
32. G. McLendon and A. E. Martell, *Inorg. Chem.* **15,** 2662 (1976).
33. N. Al-Shatti, M. Ferrer, and A. G. Sykes, *J. Chem. Soc. Dalton Trans.*, 2533 (1980).
34. M. Ferrer, T. D. Hand, and A. G. Sykes, *J. Chem. Soc. Dalton Trans.*, 14 (1980).
35. T. Shibahara, H. Kuroya, and M. Mori, *Bull. Chem. Soc. Japan* **53,** 2834 (1980).
36. J. H. Timmons, R. H. Niswander, A. Clearfield, and A. E. Martell, *Inorg. Chem.* **18,** 2977 (1979).
37. W. R. Harris, G. L. McLendon, A. E. Martell, R. C. Bess, and M. Mason, *Inorg. Chem.* **19,** 21 (1980).
38. S. R. Pickens and A. E. Martell, *Inorg. Chem.* **19,** 15 (1980).
39. A. B. Soares, R. C. Taylor, and A. G. Sykes, *J. Chem. Soc. Dalton,* 1101 (1980).
40. B. S. Brunschwig and N. Sutin, *Inorg. Chem.* **18,** 1731 (1979).
41. R. Marčec and M. Orhanović, *Inorg. Chim. Acta* **37,** 67 (1979).
42. A. H. Martin and E. S. Gould, *Inorg. Chem.* **14,** 873 (1975).
43. G. A. K. Thompson and A. G. Sykes, *Inorg. Chem.* **18,** 2025 (1979).
44. V. S. Srinvasan, Y.-R. Hu, and E. S. Gould, *Inorg. Chem.* **19,** 3470 (1980).
45. C. A. Radlowski, P.-W. Chum, L. Hua, J. Heh, and E. S. Gould, *Inorg. Chem.* **19,** 401 (1980).
46. M. G. Segal and R. M. Sellers, *J. Chem. Soc. Chem. Comm.*, 991 (1980).
47. C. S. Glennon, T. D. Hand, and A. G. Sykes, *J. Chem. Soc. Dalton Trans.*, 19 (1980).
48. A. Peloso, *J. Chem. Soc. Dalton Trans.*, 1160 (1979).
49. G. M. Summa and B. A. Scott, *Inorg. Chem.* **19,** 1079 (1980).
50. W. L. Waltz, J. Lilie, R. T. Walters, and R. J. Woods, *Inorg. Chem.* **19,** 3284 (1980).
51. G. J. Ferraudi and J. F. Endicott, *Inorg. Chim. Acta* **37,** 219 (1979).
52. Y. Narusawa and K. Nakano, *J. Inorg. Nucl. Chem.* **42,** 875 (1980).
53. D. E. Pennington, *J. Coord. Chem.* **10,** 135 (1980).
54. R. J. Balahura and N. A. Lewis, *Can. J. Chem.* **57,** 1765 (1979).
55. H. Ogino, K. Tsukahara, and N. Tanaka, *Inorg. Chem.* **19,** 255 (1980).
56. D. H. Huchital and J. Lepore, *Inorg. Chim. Acta* **38,** 131 (1980).
57. J. Phillips and A. Haim, *Inorg. Chem.* **19,** 1616 (1980).
58. J. Phillips and A. Haim, *Inorg. Chem.* **19,** 76 (1980).
59. P. L. Gauss and J. L. Villaneuva, *J. Am. Chem. Soc.* **102,** 1934 (1980).
60. J. J. Jwo, P. L. Gauss, and A. Haim, *J. Am. Chem. Soc.* **101,** 6189 (1979).
61. G. M. Brown, H. J. Krentzien, M. Abe, and H. Taube, *Inorg. Chem.* **18,** 3374 (1979).
62. R. Berkoff, K. Kristand, and H. D. Gafney, *Inorg. Chem.* **19,** 1 (1980).
63. K. W. Hicks and G. A. Chappelle, *Inorg. Chem.* **19,** 1623 (1980).
64. M. A. Rawoff and J. R. Sutter, *J. Phys. Chem.* **71,** 2767 (1967).
65. P. N. Balsubramanian and V. R. Vijayaraghavan, *Inorg. Chim. Acta* **38,** 49 (1980).
66. A. Bakac and J. H. Espenson, *Inorg. Chim. Acta* **30,** L329 (1978).
67. D. R. Rosseinsky and G. A. Jauregui, *J. Chem. Soc. Faraday Trans. 1* **75,** 473 (1979).
68. C. W. Mowforth, D. R. Rosseinsky and K. Stead, *J. Chem. Soc. Faraday Trans. 1* **75,** 1268 (1979).
69. G. Davies, *Inorg. Chem.* **10,** 1155 (1971).
70. R. C. Thompson, *Inorg. Chem.* **18,** 2379 (1979).

71. K. A. Rubinson, J. Cagin, R. W. Hurst, E. Habashi, T. M. Kenyherz, W. R. Heineman, and H. B. Mark, *J. Chem. Soc. Chem. Commun.*, 47 (1980).
72. J. H. Espenson and H. B. Gjerde, *Inorg. Chem.* **19**, 3549 (1980).
73. P. Worthington and P. Hambright, *Inorg. Chim. Acta* **46**, L87 (1980).
74. J. S. Rybka, J. L. Kurtz, T. A. Neubecker, and D. W. Margerum, *Inorg. Chem.* **19**, 2791 (1980).
75. A. G. Lappin, M. P. Youngblood, and D. W. Margerum, *Inorg. Chem.* **19**, 407 (1980).
76. G. S. Yoneda, G. L. Blackmer, and R. A. Holwerda, *Inorg. Chem.* **16**, 3376 (1977).
77. J. K. Yandell and M. A. Augustin, *Inorg. Chem.* **18**, 577 (1979).
78. M. P. Youngblood and D. W. Margerum, *Inorg. Chem.* **19**, 3068 (1980).
79. G. D. Owens, R. W. Taylor, T. Y. Ridley, and D. W. Margerum, *Anal. Chem.* **52**, 130 (1980).
80. H. Bruhn, J. Westerhausen, J. F. Holzworth, and J. H. Fuhrhop, in NATO Advanced Study Institute Series, Series C, Vol. 50: *Techniques and Applications of Fast Reactions in Solution* (W. J. Gettins and E. Wyn-Jones, eds.), Reidel, Dordrecht, The Netherlands (1979), pp. 523–534.
81. E. Bjergbakke, *Nukleonika* **24**, 825 (1979).
82. A. G. Lappin and R. D. Peacock, *Inorg. Chim. Acta* **46**, L71 (1980).
83. P. G. Rasmussen and C. H. Brubaker, *Inorg. Chem.* **3**, 977 (1964).
84. P. V. Subba Rao, M. Suryanarayana, K. V. Subbaiah, and P. S. N. Murty, *J. Inorg. Nucl. Chem.* **42**, 1181 (1980).
85. M. Woods and J. C. Sullivan, *Inorg. Chem.* **18**, 3317 (1979).
86. M. J. Blandamer and J. Burgess, *Coord. Chem. Rev.* **31**, 93 (1980).
87. A. M. Kjaer and J. Ulstrup, *Inorg. Chem.* **18**, 3624 (1979).
88. J. L. Walsh, J. A. Bauman, and T. J. Meyer, *Inorg. Chem.* **19**, 2145 (1980).
89. E. S. Yang, M.-S. Chan, and A. C. Wahl, *J. Phys. Chem.* **84**, 3094 (1980).
90. E. Pelizzetti and E. Pramauro, *Inorg. Chim. Acta* **40**, X152 (1980).
91. G. Wada, Y. Ohta, S. Hayashida, H. Sugino, T. Mizui, and Y. Aihara, *Bull. Chem. Soc. Japan* **53**, 1278 (1980).
92. G. Wada and A. Kawauchi, *Bull Chem. Soc. Japan* **53**, 3023 (1980).
93. J. G. H. Du Preez, H. E. Rohwer, and D. C. Morris, *Inorg. Chim. Acta* **43**, 65 (1980).
94. K. R. Beckham and D. W. Watts, *Aust. J. Chem.* **32**, 1425 (1979).
95. T. J. Westcott and D. W. Watts, *Aust. J. Chem.* **32**, 2139 (1979).
96. L. Spiccia and D. W. Watts, *Aust. J. Chem.* **32**, 2275 (1979).
97. R. W. Soukop, V. N. Sapunov, R. Schmid, and V. Gutman, *E. Phys. Chem.* **119**, 129 (1980).
98. E. Pelizzetti and E. Pramauro, *Inorg. Chim. Acta* **46**, L29 (1980).
99. E. Pelizzetti and E. Pramauro, *Inorg. Chem.* **19**, 1407 (1980).
100. A. W. Maverick and H. B. Gray, *Pure Appl. Chem.* **52**, 2339 (1980).
101. I. S. Sigal, K. R. Mann, and H. B. Gray, *J. Am. Chem. Soc.* **102**, 7252 (1980).
102. G. Giro, G. Casalbore, and P. G. di Marco, *Chem. Phys. Lett.* **71**, 476 (1980).
103. O. Johansen, A. Launikonis, A. W.-H. Mau, and W. H. F. Sasse, *Aust. J. Chem.* **33**, 1643 (1980).
104. K. Kalyanasundaram and M. Grätzel, *Angew. Chem. Int. Ed. Engl.* **18**, 701 (1979).
105. J. Kiwi, E. Bargarello, E. Pelizzetti, M. Visca, and M. Grätzel, *Angew. Chem. Int. Ed. Engl.* **19**, 646 (1980).
106. N. Sutin and C. Creutz, *Pure Appl. Chem.* **52**, 2717 (1980).
107. C. Creutz, M. Chou, T. L. Netzel, M. Okumura, and N. Sutin, *J. Am. Chem. Soc.* **102**, 1309 (1980).
108. J. M. Malin, *Inorg. Chim. Acta* **45**, L87 (1980).

109. J. E. Baggott and M. J. Pilling, *J. Phys. Chem.* **84**, 3012 (1980).

110. T. K. Foreman, C. Griannotti, and D. G. Whitten, *J. Am. Chem. Soc.* **102**, 1170 (1980).

111. K. Chandrasekaran and P. Natarajan, *Inorg. Chem.* **19**, 1714 (1980).

112. W. Böttcher and A. Haim, *J. Am. Chem. Soc.* **102**, 1564 (1980).

113. R. E. Sassoon and J. Rabani, *J. Phys. Chem.* **84**, 1319 (1980).

114. M. W. Blaskie and D. R. McMillin, *Inorg. Chem.* **19**, 3519 (1980).

115. A. Kirsch-de-Mesmaeker, L. Wilputte-Steinert, and J. Nasielski, *Inorg. Chim. Acta* **45**, L37 (1980).

116. L. Ballardini, G. Varani, and V. Balzani, *J. Am. Chem. Soc.* **102**, 1719 (1980).

117. R. P. Ashbury, G. S. Hammond, P. H. P. Lee, and A. T. Poulos, *Inorg. Chem.* **19**, 3461 (1980).

118. D. A. Geselowitz and H. Taube, *J. Am. Chem. Soc.* **102**, 4525 (1980).

119. P. J. Heaney, A. G. Lappin, R. D. Peacock, and B. Stewart, *J. Chem. Soc. Chem. Comm.*, 769 (1980).

120. G. B. Porter and R. H. Sparks, *J. Chem. Soc. Chem. Commun.*, 1094 (1979).

121. E. T. Adman, *Biochim. Biophys. Acta* **549**, 107 (1979).

122. V. T. Taniguchi, N. Sailasuta-Scott, F. C. Anson, and H. B. Gray, *Pure Appl. Chem.* **52**, 2275 (1980).

123. R. A. Holwerda, D. B. Knaff, H. B. Gray, J. D. Clemmer, R. Crowley, J. M. Smith, and A. G. Mauk, *J. Am. Chem. Soc.* **102**, 1142 (1980).

124. A. G. Mauk, R. A. Scott, and H. B. Gray, *J. Am. Chem. Soc.* **102**, 4360 (1980).

125. D. J. Cookson, M. T. Hayes, and P. E. Wright, *Biochim. Biophys. Acta* **591**, 162 (1980).

126. D. J. Cookson, M. T. Hayes, and P. E. Wright, *Nature (London)* **283**, 682 (1980).

127. M. G. Segal and A. G. Sykes, *J. Am. Chem. Soc.* **100**, 4585 (1978).

128. A. G. Lappin, M. G. Segal, D. C. Weatherburn, and A. G. Sykes, *J. Am. Chem. Soc.* **101**, 2297 (1979).

129. J. Butler, R. A. Henderson, F. A. Armstrong, and A. G. Sykes, *Biochem. J.* **183**, 471 (1979).

130. F. A. Armstrong, R. A. Henderson, and A. G. Sykes, *J. Am. Chem. Soc.* **102**, 6545 (1980).

131. J. G. Reynolds, C. L. Coyle, and R. H. Holm, *J. Am. Chem. Soc.* **102**, 4350 (1980).

132. R. A. Henderson and A. G. Sykes, *Inorg. Chem.* **19**, 3103 (1980).

133. B. A. Feinberg and W. V. Johnson, *Biochem. Biophys. Res. Commun.* **93**, 100 (1980).

134. I. A. Mizrahi, F. E. Wood, and M. A. Cusanovich, *Biochemistry* **15**, 343 (1976).

135. I. A. Mizrahi, T. E. Meyer, and M. A. Cusanovich, *Biochemistry* **19**, 4727 (1980).

136. I. A. Mizrahi and M. A. Cusanovich, *Biochemistry* **19**, 4733 (1980).

137. T. Goldkorn and A. Schejter, *J. Biol. Chem.* **254**, 12562 (1979).

138. Y. Ilan, A. Shafferman, B. A. Feinberg, and Y.-K. Lau, *Biochim. Biophys. Acta* **548**, 565 (1979).

139. Y. Ilan and A. Shafferman, *Biochim. Biophys. Acta* **548**, 161 (1979).

140. A. Adegite and M. I. Okpanachi, *J. Am. Chem. Soc.* **102**, 2832 (1980).

141. R. A. Scott and H. B. Gray, *J. Am. Chem. Soc.* **102**, 3219 (1980).

142. M. A. Augustin and J. K. Yandell, *Inorg. Chim. Acta* **37**, 11 (1979).

References for Chapter 2

1. *Comprehensive Chemical Kinetics*, Sec. 6, Vol. 16, *Liquid Phase Oxidation* (C. H. Bamford and C. F. H. Tippen, eds.), Elsevier, Amsterdam (1980).

2. G. W. Parshall, *Homogeneous Catalysis: The Applications and Chemistry of Catalysis by Soluble Transition Metal Complexes*, Wiley, New York, (1980).

3. J. F. Endicott and B. Durham, in *Coordination Chemistry of Macrocyclic Compounds*, (G. A. Melson, ed.), Plenum, New York (1979), p. 393.
4. R. M. Noyes, *Ber. Bunsenges. Phys. Chem.* **84**, 295 (1980).
5. A. E. Shilov, *Kinet. Katal.* **21**, 26 (1980); *Chem. Abs.* **92**, 186454d (1980).
6. M. L. Khidekel, *Kinet. Katal.* **21**, 53 (1980); *Chem. Abs.* **93**, 25385n (1980).
7. T. L. Brown, *Ann. N.Y. Acad. Sci.* **333**, 80 (1980).
8. J. K. Kochi, *Pure Appl. Chem.* **52**, 571 (1980).
9. R. G. Linck, *Treatise on Analytical Chemistry*, Part I, Vol. 2, (I. M. Kolthoff and P. J. Elving, eds.), Wiley, New York (1979), p. 645.
10. G. Stedman, *Adv. Inorg. Chem. Radiochem.* **22**, 113 (1979).
11. NATO Advanced Study Institutes, Series C, Vol. 50: *Techniques and Applications of Fast Reactions in Solution* (W. J. Gettins and E. Wyn-Jones, eds.), Reidel, Dordrecht, The Netherlands (1979).
12. G. D. Owens, R. W. Taylor, T. Y. Ridley, and D. W. Margerum, *Anal. Chem.* **52**, *130 (1980)*.
13. J. F. Holzworth, Ref. 11, p. 13.
14. J. P. Maher, Ref. 11, p. 29.
15. D. F. C. Morris and T. J. Ritter, *J. Chem. Soc. Dalton Trans*, 216 (1980).
16. K. K. Sengupta and P. K. Sar, *Inorg. Chem.* **18**, 979 (1979).
17. M. S. Frank, A. K. Ramaiah, and P. V. K. Rao, *Ind. J. Chem. Sec. A* **18A**, 369 (1979).
18. V. S. Koltunov and G. I. Zhuravlev, *Radiokhimiya* **22**, 57 (1980); *Chem. Abs.* **92**, 169926s (1980).
19. A. K. Banerjee, A. K. Basak, and D. Banerjee, *Ind. J. Chem. Sec. A* **18A**, 332 (1979).
20. A. F. M. Nazer and C. F. Wells, *J. Chem. Soc. Dalton Trans.* 1532 (1980).
21. P. K. Sen, S. Maiti, and K. K. S. Gupta, *Ind. J. Chem. Sect. A* **19A**, 865 (1980).
22. P. Collings and G. Stedman, *Inorg. Chem. Acta* 41, 5 (1980).
23. Wu-H. Ho, *Proc. Natl. Sci. Council Rep. China* 4, 1 (1980); *Chem. Abs.* **92**, 204190w (1980).
24. V. S. Koltunov and A. A. Rayabova, *Radiokhimya* **22**, 635 (1980); *Chem. Abs.* **94**, 8180v (1981).
25. D. M. Stanbury, W. K. Wilmarth, S. Khalaf, H. N. Po, and J. E. Byrd, *Inorg. Chem.* **19**, 2715 (1980).
26. U. Muralikrishna, K. V. Bappanayya, and P. S. R. Sarma, *Ind. Chem. J.* **14**, 21 (1980).
27. C. W. J. Scaife and R. G. Wilkins, *Inorg. Chem.* **19**, 3244 (1980).
28. R. N. Mehrotra and R. G. Wilkins, *Inorg. Chem.* **19**, 2177 (1980).
29. D. Pinnell and R. B. Jordan, *Inorg. Chem.* **18**, 3191 (1979).
30. M. Giurgiu, *Stud. Univ. Babes-Bolyai (Ser.) Chem.* **24**, 8 (1979); *Chem. Abs.* 92, 169987n (1980).
31. G. J. Lamprecht, J. G. Leipoldt, C. R. Dennis, S. S. Basson, *React. Kinet. Catal. Lett.* **13**, 269 (1980); *Chem. Abs.* **93**, 13786f (1980).
32. C. O. Adedinsewo and A. Adegite, *Inorg. Chem.* **18**, 3597 (1979).
33. G. Nord, B. Pedٓmٓon, and O. Farver, *Inorg. Chem.* **17**, 2233 (1978).
34. R. I. Haines and A. McAuley, *Inorg. Chem.* **19**, 719 (1980).
35. L. S. A. Dikshitulu, V. H. Rao, and S. N. Dindi, *Ind. J. Chem. Sec. A* **19**, 283 (1980).
36. C. L. Wong and J. K. Kochi, *J. Am. Chem. Soc.* **102**, 2928 (1980).
37. C. L. Wong, R. J. Klinger, and J. K. Kochi, *Inorg. Chem.* **19**, 423 (1980).
38. R. S. Drago, *Coord. Chem. Rev.* **32**, 97 (1980).
39. J. A. Streeky, D. G. Pillsbury, and D. H. Busch, *Inorg. Chem.* **19**, 3148 (1980).
40. K. Nag and A. Chakravorty, *Coord. Chem. Rev.* **33**, 87 (1980).
41. D. Ip and J. Rocek, *J. Am. Chem. Soc.* **101**, 6311 (1979).
42. S. N. Mahapalro, M. Krumpolc, and J. Rocek, *J. Am. Chem. Soc.* **102**, 3799 (1980).

43. F. Hasan and J. Rocek, *J. Am. Chem. Soc.* **97,** 1444; 3762 (1975).
44. K. Nagarajan, S. Sundaram, and N. Ventakasubramanian, *Ind. J. Chem. Sec. A* **18A,** 335 (1979); **19A,** 81 (1980).
45. J. F. Harrod and A. Pathak, *Can. J. Chem.* **58,** 686 (1980).
46. K. K. Banerji, *Ind. J. Chem. Sec. A.* **17A,** 300 (1979).
47. F. Freeman, C. R. Armstead, M. G. Essig, E. M. Karchevski, C. J. Kojima, V. C. Manopoli, and A. H. Wickman, *J. Chem. Soc. Chem. Commun.,* 65 (1980).
48. A. F. M. Nazer and C. F. Wells, *J. Chem. Soc. Dalton Trans.,* 2143 (1980).
49. V. S. Koltunov, M. F. Tikhonov, K. M. Frolov, and O. V. Lebedev, *Zh. Fiz. Khim.* **54,** 363 (1980).
50. P. V. S. Rao, R. V. S. Murty, K. V. Subbaiah, *Ind. J. Chem. Sec. A.* **18A,** 228 (1979).
51. E. Pelizzetti, *J. Chem. Soc. Dalton Trans.,* 484 (1980).
52. R. Varadarajan and M. Joseph, *Ind. J. Chem. Sec. A* **19A,** 977 (1980).
53. V. V. Rao, B. Sethuram, and T. N. Rao, *Z. Phys. Chem. Leipzig* **261,** 1171 (1980).
54. T. Ernst, M. Wawrzenczyk, and M. Wronska, *Z. Phys. Chem. Leipzig* **216,** 306 (1980).
55. M. Jaky and L. Simandi, *Magy. Kem. Foly.* **86,** 127 (1980); *Chem. Abs.* **93,** 113574d (1980); *Magy. Kem. Foly.* **86,** 13b (1980); *Chem. Abs.* **93,** 185441f (1980).
56. Z. Szeverenyi, L. Simandi, and M. Jaky, *Magy. Kem. Foly.* **86,** 132 (1980); *Chem. Abs.* **93,** 113575e (1980).
57. K. W. Hicks and G. Chappelle, *Inorg. Chem.* **19,** 1623 (1980).
58. T. J. Kemp, P. Moore, and G. R. Quick, *J. Chem. Soc. Perkin Trans.* 2, 291 (1980).
59. E. Pelizzetti and E. Mentasti, *Inorg. Chem.* **18,** 583 (1979).
60. M. J. Frank and P. V. K. Rao, *Ind. J. Chem. Sec. A* **19A,** 538, 632 (1980).
61. T. Okamoto, A. Ohno, and S. Oka, *Inorg. Chem.* **19,** 3176 (1980).
62. H. S. Singh, V. P. Singh, and D. P. Pandey, *Monatsh Chem.* **110,** 1455 (1979).
63. S. S. Srivastava, S. Kumar, and P. C. Mathur, *J. Chim. Phys. Phys-Chim. Biol.* **76,** 667 (1979).
64. A. F. Godfrey and J. K. Beattie, *Aust. J. Chem.* **32,** 1905 (1979).
65. F. Ahmad, S. Kumar, and V. K. Biswani, *J. Inorg. Nucl. Chem.* **42,** 999 (1980).
66. L. Treindl and M. Melichercik, *Collect. Czech. Chem. Commun.* **45,** 1173 (1980).
67. R. Lorbie and M. Zador, *Can. J. Chem.* **58,** 1305 (1980).
68. H. N. Po., C-F. Lo, N. Jones, and R. W. Lee, *Inorg. Chim Acta* **46,** 185 (1980).
69. J. S. Littler, G. R. Quick, and D. Wozniak, *J. Chem. Soc. Perkin Trans.* 2, 657 (1980).
70. D. F. C. Morris and T. J. Ritter, *J. Less Comm. Met.* **65,** 49 (1979).
71. L. N. Arzmaskovska, A. V. Romanenko, and Y. I. Ermakov, *React. Kinet. Catal. Lett.* **13,** 395 (1980); *Chem. Abs.* **93,** 238312y (1980).
72. D. Soria, M. L. De Castro, and H. L. Chum, *Inorg. Chim. Acta.* **42,** 121 (1980).
73. H. L. Chum and M. E. M. Helene, *Inorg. Chem.* **19,** 876 (1980).
74. E. Baciocchi, C. Rol, and L. Mandolini, *J. Am. Chem. Soc.* **102,** 7597 (1980).
75. J. Dziegiec, *Pol. J. Chem.* **53,** 1821 (1979).
76. M. Ignaczak and M. Deka, *Pol. J. Chem.* **54** 259 (1980).
77. R. Varadarajan and C. F. Wells, *J. Chem. Soc. Trans. Faraday Trans. 1,* 2017 (1980).
78. E. Pelizetti and R. Giordano, *J. Chem. Soc. Dalton Trans.,* 1516 (1979).
79. R. Varadarajan, M. D. A. Hossain, and M. Joseph, *Ind. J. Chem. Sec. A* **19A,** 117 (1980).
80. L. Eberson and L. G. Wistrand, *Acta Chem. Scand. B* **34,** 349 (1980).
81. M. Hirano, E. Kitagawa, and T. Morimoto, *J. Chem. Soc. Perkin Trans.* 2, 569 (1980).
82. M. Burger and E. Koros, *J. Phys. Chem.* **84,** 496 (1980).
83. A. M. Zhabotinskii, *Ber. Bunsenges. Phys. Chem.* **84,** 303 (1980).
84. A. M. Zhabotinskii and A. B. Rovinskii, *Teor. Esp. Khim.* **16,** 386 (1980).*Chem. Abs.* **93,** 102031y (1980).

85. K. Prasad and M. S. Prasad, *Ind. J. Chem. Sec. A* **18A,** 398 (1979).
86. G. Rabai, C. Bazsa, and M. Beck, *Magy. Kem. Foly.* **86,** 112 (1980). Chem. Abs. *93,* 7055y (1980).
87. E. Pelizzetti, M. Woods, and J. C. Sullivan, *Inorg. Chem.* **19,** 524 (1980).
88. V. S. Koltunov, M. F. Tzikhonov, and K. M. Frolov, *Radiokhimiya* **22,** 483 (1980).
89. K. H. Schmidt, S. Gordon, R. C. Thompson, and J. C. Sullivan, *J. Inorg. Nucl. Chem.* **42,** 611 (1980).
90. V. S. Koltunov, M. F. Tzikhonov, K. M. Frolov, and M. P. Shapovalov, *Radiokhimiya* **22,** 65 (1980).
91. C. Baiocchi and E. Pelizzetti, *Transition Met. Chem.* **5,** 259 (1980).
92. A. Kumar and P. Neta, *J. Am. Chem. Soc.* **102,** 7284 (1980).
93. T. Raviprasad, B. Sethuram, and T. N. Rao, *Ind. J. Chem. Sec. A* **19A,** 261 (1980).
94. G. Annibale, L. Canovese, L. Cattalini, and G. Natile, *J. Chem. Soc. Dalton Trans.,* 1017 (1980).
95. L. S. A. Dikshitulu, H. V. Rao and P. Vani, *Ind. J. Chem. Sec. A* **19A,** 974 (1980).
96. G. Marcu, G. Pop, I. Pop, and C. Nascu, *Rev. Chim. (Bucharest)* **30,** 1101 (1979); *Chem. Abs.* **92,** 136058g (1980).
97. I. R. Epsrein, K. Kustin, and L. J. Warshaw, *J. Am. Chem. Soc.* **102,** 3751(1980).
98. W. C. E. Higginson and D. A. Macarthy, *J. Chem. Soc. Dalton Trans.* 797 (1980).
99. K. Prasad, G. S. Verma, and M. S. Prasad, *Ind. J. Chem. Sec. A* **19A,** 695 (1980).
100. J. Ige, J. F. Ojo, and O. Olubuyide, *Canad. J. Chem.* 54, 2065 (1979).
101. M. J. Blandamer, J. Burgess, P. P. Duce, and R. I. Haines, *J. Chem. Soc. Dalton Trans.* 2442 (1980).
102. J. C. Bates, P. Reveco, and G. Stedman, *J. Chem. Soc. Dalton Trans.,* 1487 (1980).
103. L. M. Bharadwaj, D. N. Sharma, and Y. K. Gupta, *J. Chem. Soc. Dalton Trans.,* 1526 (1980).
104. T. Ohno and N. N. Lichtin, *J. Am. Chem. Soc.* **102,** 4836 (1980).
105. J. C. Curtis, B. P. Sullivan, and T. J. Meyer, *Inorg. Chem.* **19,** 3833 (1980).
106. A. J. Miralles and A. Haim, *Inorg. Chem.* **19,** 1158 (1980).
107. T. G. Dunne and J. K. Hurst, *Inorg. Chem.* **19,** 1152 (1980).
108. R. A. Holwerda and J. S. Petersen, *Inorg. Chem.* **19,** 1775 (1980).
109. P. Sevcik and P. Cifra, *Collect. Czech. Chem. Commun.* **45,** 2481 (1980).
110. P. Sevcik and M. Khir, *Collect. Czech. Chem. Commun.* **45,** 21 (1980).
111. J. Konstantatos, E. Vrachnou-Astra, N. Katsaros, and D. Katakis, *J. Am. Chem. Soc.* **102,** 3035 (1980).
112. V. S. Srinivasan, Y. R. Hu, and E. S. Gould, *Inorg. Chem.* **19,** 3470 (1980).
113. L. Adamcikova, *Collect. Czech. Chem. Commun.* **45,** 3287 (1980).
114. N. P. Luneva, L. A. Nikonova, and A. E. Shilov, *Kinet. Katal.* **21,** 1458 (1980); *Chem. Abs.* 94, 53610w (1981).
115. N. T. Denisov, S. I. Koboleva, A. E. Shilov, and N. I. Shuvalova, *Kinet. Katal.* **21,** 1527 (1980); *Chem. Abs.* 93, 246272a (1980).
116. H. O. Blaser and J. Halpern, *J. Am. Chem. Soc.* **102,** 1684 (1980).
117. J. Fiala and A. A. Vlcek, *Inorg. Chim. Acta* **42,** 85 (1980).
118. J. Fiala and A. A. Vlcek, *Inorg. Chim. Acta* **40,** 123 (1980).
119. S. L. Kessel, R. M. Emberson, P. G. Debrunner, and D. N. Hendrickson, *Inorg. Chem.* **19,** 1170 (1980).
120. L. Adamcikova and L. Treindl, *Chem. Zvesti* **34,** 145 (1980); *Chem. Abs.* **93,** 81217w (1980).
121. L. Treindl and L. Adamcikova, *Collect. Czech. Chem. Commun.* **45,** 3266 (1980).
122. L. Adamcikova and L. Treindl., *Collect. Czech. Chem. Commun.* **45,** 26 (1980).
123. F. Vierling, *Bull. Soc. Chim. France,* 144 (1980).

124. J. Halpern and R. A. Jewsbury, *J. Organomet. Chem.* **181,** 223 (1979).
125. J. S. Rybka, J. L. Kurtz, T. A. Neubecker, and D. W. Margerum, *Inorg. Chem.* **19,** 2791 (1980).
126. L. J. Kirschenbaum and D. Meyerstein, *Inorg. Chem.* **19,** 1373 (1980).
127. T. Ramasami and J. H. Espenson, *Inorg. Chem.* **19,** 1846 (1980).
128. R. van Eldik and G. M. Harris, *Inorg. Chem.* **19,** 880 (1980).
129. N. Selvarajan and N. V. Raghavan, *J. Chem. Soc. Chem. Commun.* **336** (1980).

References for Chapter 3

1. R. F. Modler and M. M. Kreevoy, *J. Am. Chem. Soc.* **99,** 2271 (1977).
2. B. S. Meeks and M. M. Kreevoy, *J. Am. Chem. Soc.* **101,** 4918 (1979).
3. B. S. Meeks and M. M. Kreevoy, *Inorg. Chem.* **18,** 2187 (1979).
4. D. Rehorek, R. Herzschuh, and H. Hennig, *Inorg. Chim. Acta* **44,** L75 (1980).
5. P. C. Keller and J. V. Rund, *Inorg. Chem.* **18,** 3197 (1979).
6. H. C. Kelly and V. B. Marriott, *Inorg. Chem.* **18,** 2875 (1979).
7. M. A. Mathur and G. E. Ryschkewitsh, *Inorg. Chem.* **19,** 887, 3054 (1980).
8. M. G. Hu and R. A. Geanangel, *Inorg. Chem.* **18,** 3297 (1979).
9. J. D. Odom, T. F. Moore, W. H. Dawson, A. R. Garber, and E. J. Stampf, *Inorg. Chem.* **18,** 2179 (1979).
10. J. D. Odom and T. F. Moore, *Inorg. Chem.* **19,** 2651 (1980).
11. D. A. Kleier, J. Bicerano, and W. N. Lipscomb, *Inorg. Chem.* **19,** 216 (1980).
12. M. M. Midland, J. E. Petre, and S. A. Zderic, *J. Organomet. Chem.* **182,** C53 (1979).
13. S. Nagase, N. K. Ray and K. Morokuma, *J. Am. Chem. Soc.* **102,** 4536 (1980).
14. K. Yoshino, M. Kotaka, M. Okamoto, and H. Kakihana, *Bull. Chem. Soc. Japan.* **52,** 3005 (1979).
15. S. A. Ter-Grigoryan, G. N. Kononova, and A. I. Martyushin, *Russ. J. Phys. Chem.* **54,** 727, 728 (1980).
16. L. Babcock and R. Pizer, *Inorg. Chem.* **19,** 56 (1980).
17. R. B. Moodie and J. P. Sansom, *J. Chem. Res. (S),* **390** (1979).
18. R. J. P. Corriu and C. Guerin, *J. Organomet. Chem.* **198,** 231 (1980).
19. R. J. P. Corriu, G. Dabosi, and M. Martineau, *J. Organomet. Chem.* **186,** 25 (1980).
20. M. H. Novice, H. R. Seikaly, A. D. Seiz, and T. T. Tidwell, *J. Am. Chem. Soc.* **102,** 5835 (1980).
21. G. Seconi, C. Eaborn, and A. Fischer, *J. Organomet. Chem.* **177,** 129 (1979).
22. D. Pietropaolo, M. Fiorenza, A. Ricci, and M. Taddei, *J. Organomet. Chem.* **197,** 7 (1980).
23. G. Baruch and A. Horowitz, *J. Phys. Chem.* **84,** 2535 (1980).
24. H. Sakurai, Y. Nakadaira, M. Kira, H. Sugiyama, K. Yoshida, and T. Takiguchi, *J. Organomet. Chem.* **184,** C36 (1980).
25. R. N. Haszeldine, R. V. Parish, and B. F. Riley, *J. Chem. Soc. Dalton Trans.* **705** (1980).
26. C. A. Tsipis, *J. Organomet. Chem.* **188,** 53 (1980).
27. C. Eaborn, D. A. R. Happer, S. P. Hopper, and K. D. Safa, *J. Organomet. Chem.* **188,** 179 (1980).
28. H. Meyer, J. Klein, and A. Weiss, *J. Organomet. Chem.* **177,** 323 (1979).
29. R. K. Harris, C. T. G. Knight, and D. N. Smith, *J. Chem. Soc. Chem. Commun.,* 726 (1980).
30. L. S. Dent Glasser and D. N. Smith, *J. Chem. Soc. Chem. Commun.,* 727 (1980).
31. C. Eaborn and B. Singh, *J. Organomet. Chem.* **177,** 333 (1979).

32. E. L. J. Breet and R. Van Eldik, *Inorg. Chim. Acta* **40**, 153 (1980).
33. J. R. Pembridge and G. Stedman, *J. Chem. Soc. Dalton Trans.* 1657 (1979).
34. P. Collings and G. Stedman, *Inorg. Chim. Acta* **41**, 1 (1980).
35. R. B. Moodie, K. Schofield, and P. G. Taylor, *J. Chem. Soc. Perkin Trans.* 2, 133 (1979).
36. S. A. Andreev, B. A. Lebedev, I. V. Tselinskii, and I. N. Shokhor, *Zh. Org. Khim.* **16**, 1353, 1360, 1370, 1374 (1980).
37. R. D. Bach, J. W. Holubka, R. C. Badger, and S. J. Rajan, *J. Am. Chem. Soc.* **101**, 4416 (1979).
38. G. A. Olah, B. G. B. Gupta, and S. C. Narang, *J. Am. Chem. Soc.* **101**, 5317 (1979).
39. S. S. Emeish and K. E. Howlett, *Can. J. Chem.* **58**, 159 (1980).
40. S. S. Emeish, *Can. J. Chem.* **58**, 902 (1980).
41. J. D. Buhr and H. Taube, *Inorg. Chem.* **19**, 2425 (1980).
42. J. C. Bates, P. Reveco, and G. Stedman, *J. Chem. Soc. Dalton Trans.*, 1487 (1980).
43. S. E. Aldred and D. L. H. Williams, *J. Chem. Soc. Chem. Commun.*, 73 (1980).
44. V. Napoleone and Z. A. Schelly, *J. Phys. Chem.* **84**, 17 (1980).
45. B. C. Challis and D. E. G. Shuker, *J. Chem. Soc. Perkin Trans.* 2, 1020 (1979).
46. B. C. Challis and D. E. G. Shuker, *Food Cosmet. Toxicol.* **18**, 283 (1980).
47. Y. L. Chow, K. S. Pillay, and H. Richard, *Can. J. Chem.* 57, 2923 (1979).
48. B. C. Challis and J. R. Outram, *J. Chem. Soc. Perkin Trans.* 1, 2678 (1979).
49. J. R. Blackborrow, D. P. Clifford, I. M. Hollinshead, T. A. Modro, J. H. Ridd, and M. C. Worley, *J. Chem. Soc. Perkin Trans.* 2, 632 (1980).
50. D. J. Mills and J. H. Ridd, *J. Chem. Soc. Perkin Trans.* 2, 637 (1980).
51. R. Bonnett and P. Nicolaidou, *J. Chem. Soc. Perkin Trans.* 1, 1969 (1979).
52. A. P. Gosney and M. I. Page, *J. Chem. Soc. Perkin Trans.* 2, 1783 (1980).
53. P. M. G. Bavin, G. J. Durant, P. D. Miles, R. C. Mitchell, and E. S. Pepper, *J. Chem. Res. (S)*, 212 (1980).
54. B. G. Gowenlock, R. J. Hutchison, J. Little, and J. Pfab, *J. Chem. Soc. Perkin Trans.* 2, 1110 (1979).
55. G. Hallett and D. L. H. Williams, *J. Chem. Soc. Perkin Trans.* 2, 1372 (1980).
56. J. Casado, A. Castro, MA. L. Quintel, and J. M. Cachaza, *Monatsh. Chem.* **110**, 1331 (1979).
57. S. S. Johal, D. L. H. Williams, and E. Buncel, *J. Chem. Soc. Perkin Trans.* 2, 165 (1980).
58. G. Hallett and D. L. H. Williams, *J. Chem. Soc. Perkin Trans.* 2, 624 (1980).
59. S. S. Singer, G. M. Singer, and B. B. Cole, *J. Org. Chem.* **45**, 4931 (1980).
60. M. J. Akhtar, C. A. Lutz, and F. T. Bonner, *Inorg. Chem.* **18**, 2369 (1979).
61. A. E. Smolyar, N. P. Zaretskii, and O. P. Charkin, *Russ. J. Inorg. Chem.* **24**, 1758 (1979).
62. A. E. Smolyar, N. P. Zaretskii, N. M. Klimenko, and O. P. Charkin, *Russ. J. Inorg. Chem.* **24**, 1761 (1979).
63. A. J. Kresge and Y. C. Tang, *J. Chem. Soc. Chem. Commun.*, 309 (1980).
64. A. J. Kresge, Y. C. Tang, A. Awwal, and D. P. Onwood, *J. Chem. Soc. Chem. Commun.*, 310 (1980).
65. A. Fattah, M. Nazer, and C. F. Wells, *J. Chem. Soc. Dalton Trans.*, 1533 (1980).
66. D. J. Pasto, *J. Am. Chem. Soc.* **101**, 6852 (1979).
67. W. Kremers and A. Singh, *Can. J. Chem.* **58**, 1592 (1980).
68. E. G. Janzen, H. J. Stronks, D. E. Nutter, E. R. Davis, H. N. Blount, J. L. Poyer, and P. B. McCay, *Can. J. Chem.* **58**, 1596 (1980).
69. A. Vogler, R. E. Wright, and H. Kunkely, *Angew. Chem. Int. Ed. Engl.* **19**, 717 (1980).
70. D. F. C. Morris and T. J. Ritter, *J. Chem. Soc. Dalton Trans.*, 216 (1980).
71. M. F. Tikhonov and V. S. Koltunov, *Russ. J. Phys. Chem.* **53**, 1143 (1979).
72. D. H. Cuatecontzis and J. D. Miller, *Inorg. Chim. Acta* **38**, 157 (1979).
73. *J. Phys. Chem.* **84**(10), (1980).

74. S. Balt, *Inorg. Chim. Acta* **45**, L241 (1980).
75. N. Lau and R. R. Dewald, *J. Phys. Chem.* **84**, 2348 (1980).
76. R. R. Holmes (ed.), *Pentacoordinated Phosphorus:* Vol. 1, *Structure and Spectroscopy;* Vol. II, *Reaction Mechanisms,* American Chemical Society, Washington, D.C. (1980).
77. K. B. Dillon, M. P. Nisbet, and T. C. Waddington, *J. Inorg. Nucl. Chem.* **41**, 1273 (1979).
78. P. W. A. Hubner and R. M. Milburn, *Inorg. Chem.* **19**, 1267 (1980).
79. R. D. Cornelius, *Inorg. Chem.* **19**, 1286 (1980).
80. J. M. Harrowfield, D. R. Jones, L. F. Lindoy, and A. M. Sargeson, *J. Am. Chem. Soc.* **102**, 7733 (1980).
81. R. A. Lazarus, P. A. Benkovic, and S. J. Benkovic, *J. Chem. Soc. Perkin Trans. 2,* 373 (1980).
82. I. S. Sigal and F. H. Westheimer, *J. Am. Chem. Soc.* **101**, 5329 (1979).
83. I. S. Sigal and F. H. Westheimer, *J. Am. Chem. Soc.* **101**, 5334 (1979).
84. I. Granoth and J. C. Martin, *J. Am. Chem. Soc.* **101**, 4618 (1979).
85. A. Skowronska, M. Pakulski, and J. Michalski, *J. Am. Chem. Soc.* **101**, 7412 (1979).
86. J. J. H. M. Font Freide and S. Trippett, *J. Chem. Soc. Chem. Commun.,* 157 (1980).
87. M. J. P. Harger, *J. Chem. Soc. Perkin Trans. 2,* 154 (1980).
88. K. Schäfer and K.-D. Asmus, *J. Phys. Chem.* **84**, 2156 (1980).
89. C. Srinivasan, P. Kuthalingam, and N. Arumugam, *J. Chem. Soc. Perkin Trans. 2,* 170 (1980).
90. C. Srinivasin and S. Rajagopal, *React. Kin. Catal. Lett.* **12**, 45 (1979).
91. L. M. Bharadwaj, D. N. Sharma, and Y. K. Gupta, *J. Chem. Soc. Dalton Trans.,* 1526 (1980).
92. G. P. Panigrahi and R. Panda, *Bull. Chem. Soc. Japan* **53**, 2366 (1980).
93. H. Eshtiagh-Hosseini, H. W. Kroto, J. F. Nixon, and O. Ohashi, *J. Organomet. Chem.* **181**, C1 (1979).
94. H. Estiagh-Hosseini, H. W. Kroto, J. F. Nixon, S. Brownstein, J. R. Morton, and K. F. Preston, *J. Chem. Soc. Chem. Commun.,* 653 (1979).
95. H. R. Allcock and P. J. Harris, *J. Am. Chem. Soc.* **101**, 6221 (1979).
96. S. S. Krishnamurthy and P. M. Sundaram, *Inorg. Nucl. Chem. Lett.* **15**, 367 (1979).
97. S. S. Krishnamurthy, K. Ramachandran, A. C. Sau, R. A. Shaw, A. R. V. Murthy, and M. Woods, *Inorg. Chem.* **18**, 2010 (1979).
98. T. L. Evans and H. R. Allcock, *Inorg. Chem.* **18**, 2342 (1979).
99. A. Okumura, S. Watanabe, M. Sakaue, and N. Okazaki, *Bull. Chem. Soc. Japan* **52**, 2783 (1979).
100. C. D. Baer, J. O. Edwards, M. J. Kaus, T. G. Richmond, and P. H. Rieger, *J. Am. Chem. Soc.* **102**, 5793 (1980).
101. T. Antonio, A. K. Chopra, W. R. Cullen, and D. Dolphin, *J. Inorg. Nucl. Chem.* **41**, 1220 (1979).
102. Y. Kurimura and R. Onimura, *Inorg. Chem.* **19**, 3516 (1980).
103. M. E. Landis and D. C. Madoux, *J. Am. Chem. Soc.* **101**, 5107 (1979).
104. A. D. Nadezhdin and H. B. Dunford, *Can. J. Chem.* **57**, 3017 (1979).
105. E. J. Nanni, M. D. Stallings, and D. T. Sawyer, *J. Am. Chem. Soc.* **102**, 4481 (1980); D. T. Sawyer and M. J. Gibian, *Tetrahedron* **35**, 1471 (1979).
106. E. J. Nanni and D. T. Sawyer, *J. Am. Chem. Soc.* **102**, 7591 (1980).
107. H. Sagae, M. Fujihira, H. Lund, and T. Osa, *Bull. Chem. Soc. Japan* **53**, 1537 (1980).
108. Yu. N. Kozlov, A. P. Purmal, and A. M. Uskov, *Russ. J. Phys. Chem.* **54**, 992 (1980).
109. A. M. Held, D. J. Halko, and J. K. Hurst, *J. Am. Chem. Soc.* **100**, 5732 (1978).
110. B. Mani, C. R. Mohan, and V. S. Rao, *React. Kin. Catal. Lett.* **13**, 277 (1980).
111. I. I. Vasilenko, *Russ. J. Phys. Chem.* **53**, 1611 (1979).

112. I. I. Vasilenko, *Russ. J. Phys. Chem.* **54**, 1002 (1980); I. I. Vasilenko and V. G. Babich, *Russ. J. Phys. Chem.* **54**, 1046 (1980).
113. M. M. Morrison, J. L. Roberts, and D. T. Sawyer, *Inorg. Chem.* **18**, 1971 (1979).
114. M. Kimura, T. Kawajiri, and M. Tanida, *J. Chem. Soc. Dalton Trans.*, 726 (1980).
115. P. S. N. Murty, K. V. Subbaiah, and P. V. Subba Rao, *React. Kin. Catal. Lett.* **11**, 79 (1979).
116. N. A. McAskill and D. F. Sangster, *Aust. J. Chem.* **32**, 2611 (1979).
117. M. Kimura, T. Akazome, K. Takenaka, and A. Kobayashi, *Bull. Chem. Soc. Japan* **53**, 1270 (1980).
118. M. J. Blandamer, J. Burgess, P. P. Duce, and R. I. Haines, *J. Chem. Soc. Dalton Trans.*, 2442 (1980).
119. V. Holba and O. Volarova, *Collect. Czech. Chem. Commun.* **44**, 3588 (1979).
120. R. C. Thompson, P. Wieland, and E. H. Appelman, *Inorg. Chem.* **18**, 1974 (1979).
121. R. Somuthevan, R. Renganathan, and P. Maruthamuthu, *Inorg. Chim. Acta* **45**, L165 (1980).
122. G. J. Lamprecht, J. G. Leipoldt, C. R. Dennis, and S. S. Basson, *React. Kin. Catal. Lett.* **13**, 269 (1980).
123. M. Kimura and M. Shukutani, *Bull. Chem. Soc. Japan* **52**, 2535 (1979).
124. R. Brodzinsky, S. G. Chang, S. S. Markowitz, and T. Novakov, *J. Phys. Chem.* **84**, 3354 (1980).
125. S. B. Sant, *React. Kin. Catal. Lett.* **12**, 195 (1979).
126. C. F. Shaw, M. P. Cancro, P. L. Witkiewicz, and J. E. Eldridge, *Inorg. Chem.* **19**, 3198 (1980).
127. A. R. Butler, C. Glidewell, and J. Needham, *J. Chem. Res. (S)*, 47 (1980).
128. S. S. -T. Chu and E. T. Kaiser, *J. Chem. Soc. Chem. Commun.*, 636 (1979).
129. H. Asefi and J. G. Tillett, *J. Chem. Soc. Perkin Trans. 2*, 1579 (1979).
130. D. Veltwisch, E. Janata, and K. -D. Asmus, *J. Chem. Soc. Perkin Trans. 2*, 146 (1980).
131. K. -D. Asmus, *Acc. Chem. Res.* **12**, 436 (1979).
132. M. Boniface and K. -D. Asmus, *J. Chem. Soc. Perkin Trans. 2*, 758 (1980).
133. J. Milne and P. La Haie, *Inorg. Chem.* **18**, 3181 (1979).
134. J. G. Leipoldt, C. R. Dennis, A. J. van Wyk, and L. D. C. Bok, *Inorg. Chim. Acta* **34**, 237 (1979).
135. E. Gebert, E. H. Appelman, and A. H. Reis, *Inorg. Chem.* **18**, 2465 (1979).
136. R. C. Thompson and E. H. Appelman, *Inorg. Chem.* **19**, 3248 (1980).
137. C. J. Schack and K. O. Christe, *Inorg. Chem.* **18**, 2618 (1979).
138. E. J. Hart and W. G. Brown, *J. Phys. Chem.* **84**, 2237 (1980).
139. M. Kimura, T. Suzuki, and Y. Ogata, *Bull. Chem. Soc. Japan* **53**, 3198 (1980).
140. S. K. Sharma and V. P. Kudesia, *React. Kin. Catal. Lett.* **13**, 55 (1980).
141. D. Hass, O. Bechstein, and S. Melenk, *Z. Chem.* **20**, 71 (1980).
142. D. Hass, O. Bechstein, and S. Melenk, *Z. Chem.* **20**, 113 (1980).
143. *Ber. Bunsenges. Phys. Chem.* **84**, 295–369 (1980).
144. R. M. Noyes, *J. Am. Chem. Soc.* **102**, 4644 (1980).
145. C. Vidal and A. Noyau, *J. Am. Chem. Soc.* **102**, 6666 (1980).
146. M. Orbán, *J. Am. Chem. Soc.* **102**, 4311 (1980).
147. M. Burger and E. Körös, *J. Phys. Chem.* **84**, 496 (1980).
148. K. Bar-Eli and S. Haddad, *J. Phys. Chem.* **83**, 2945 (1979).
149. J. Boissonade and P. De Kepper, *J. Phys. Chem.* **84**, 501 (1980).
150. L. Treindl and P. Fabian, *Collect. Czech. Chem. Commun.* **45**, 1168 (1980); K. Bar-Eli and S. Haddad, *J. Phys. Chem.* **83**, 2952 (1979).
151. P. Herbine and R. J. Freed, *J. Phys. Chem.* **84**, 1330 (1980).
152. M. Orbán, E. Körös, and R. M. Noyes, *J. Phys. Chem.* **83**, 3056 (1979).
153. Z. Noszticzius, *J. Am. Chem. Soc.* **101**, 3660 (1979).

154. E. Körös, M. Orbán, and I. Habon, *J. Phys. Chem.* **84**, 559 (1980).
155. J. C. Sullivan and R. C. Thompson, *Inorg. Chem.* **18**, 2376 (1979).
156. S. P. Srivastava, G. Bhattacharjee, V. K. Gupta, and S. Pal, *React. Kin. Catal. Lett.* **13**, 231 (1980).
157. T. R. Thomas, D. T. Ponce, and R. A. Hasty, *J. Inorg. Nucl. Chem.* **42**, 183 (1980).
158. A. Manglik, S. K. Sharma, and V. P. Kudesia, *React. Kin. Catal. Lett.* **15**, 467 (1980).
159. U. Nickel, M. Borchardt, M. R. Bapat, and W. Jaenicke, *Ber. Bunsenges.* Phys. Chem. **83**, 877 (1979).

References for Chapter 4

1. D. W. Watts, *Pure Appl. Chem.* **51**, 1713 (1979).
2. E. F. Caldin, *Pure Appl. Chem.* **51**, 2067 (1979).
3. M. J. Blandamer and J. Burgess, *Pure Appl. Chem.* **51**, 2087 (1979).
4. M. J. Blandamer and J. Burgess, *Coord. Chem. Rev.* **31**, 93 (1980).
5. R. van Eldik, *CHEMSA,* 46 (1980).
6. W. P. Jencks, *Acc. Chem. Res.* **13**, 161 (1980).
7. R. J. Mureinik, *Rev. Inorg. Chem.* **1**, 1 (1979).
8. A. Vlcek and A. A. Vicek, *Proc. 8th Conf. Coord. Chem.*, 445 (1980).
9. H. B. Gray and R. J. Olcott, *Inorg. Chem.* **1**, 481 (1962).
10. M. Kotowski, D. A. Palmer, and H. Kelm, *Inorg. Chim. Acta* **44**, L113 (1980).
11. W. Rindermann, D. A. Palmer, and H. Kelm, *Inorg. Chim. Acta* **40**, 179 (1980).
12. D. A. Palmer and H. Kelm, *Inorg. Chim. Acta* **19**, 117 (1976).
13. M. Kotowski, D. A. Palmer, and H. Kelm, *Inorg. Chem.* **18**, 2555 (1979).
14. C. H. Langford, *Inorg. Chem.* **18**, 3288 (1979).
15. K. E. Newman and A. E. Merbach, *Inorg. Chem.* **19**, 2481 (1980).
16. T. W. Swaddle, *Inorg. Chem.* **19**, 3203 (1980).
17. D. A. Palmer, R. van Eldik, and H. Kelm, *Z. Anorg. Allg. Chem.* **468**, 77 (1980).
18. R. van Eldik, D. A. Palmer, and H. Kelm, *Inorg. Chem.* **18**, 572 (1979).
19. M. J. Blandamer, J. Burgess, and S. J. Hamshere, *Trans. Met. Chem.* **4**, 291 (1979).
20. Yu. M. Kukushkin and V. K. Krylov, *Zh. Obsch. Khim.* **49**, 2782 (1979); [*J. Gen. Chem. USSR* **49**, 2466 (1979)].
21. R. S. Omarova, A. V. Babkov and A. I. Busev, *Koord. Khim.*, **5**, 1085 (1979); [*Chem. Abs.* **91**, 113055u (1979)].
22. A. V. Babkov, R. S. Omarova, and A. I. Busev, *Koord. Khim.* **6**, 291 (1980); [*Chem. Abs.* **92**, 221559r (1980)].
23. L. I. Elding and A. -B. Gröning, *Inorg. Chim. Acta* **38**, 59 (1980).
24. S. Miya, K. Kashiwabara, and K. Saito, *Inorg. Chem.* **19**, 98 (1980).
25. M. J. Blandamer, J. Burgess, and S. J. Hamshere, *J. Chem. Soc. Dalton Trans.*, 1539 (1979).
26. R. Gosling and M. L. Tobe, *Inorg. Chim. Acta* **42**, 223 (1980).
27. B. P. Kennedy, R. Gosling, and M. L. Tobe, *Inorg. Chem.* **16**, 1744 (1977).
28. I. M. Al-Najjar, M. Green, S. J. S. Kerrison, and P. J. Sadler, *J. Chem. Res. (S)*, 206 (1979).
29. H. Motschi, C. Nussbaumer, P. S. Pregosin, F. Bachechi, and P. Mura, *Helv. Chim. Acta* **63**, 2071 (1980).
30. M. V. Cooper, P. J. Guerney, and M. Partlin, *J. Chem. Soc. Dalton Trans.*, 349 (1980).
31. G. Albertini, E. Bordignon, A. A. Orio, B. Pavoni, and H. B. Gray, *Inorg. Chem.* **18**, 1451 (1979).

32. V. I. Kasbanov, G. D. Mal'chikov, S. P. Kalenyuk, and T. K. Kasbanova, *Izv. Sib. Otd. Akad. Nauk SSSR Ser. Khim. Nauk,* (4), 14 (1980; [*Chem. Abs.* 93, 210902s (1980)].

33. O. N. Shumilo. V. A. Likholobov, and N. N. Bulgakov, *Izv. Sib. Otd. Akad. Nauk SSSR, Ser. Khim. Nauk,* (4), 19, 25 (1980); *Chem. Abs.* 93, 192814h (1980).

34. D. A. Palmer and H. Kelm, *Aust. J. Chem.* 32, 1415 (1979).

35. D. A. Palmer and H. Kelm, *Inorg. Chim. Acta* 39, 275 (1980).

36. M. J. Blandamer, J. Burgess, P. P. Duce, S. J. Hamshere, and J. J. Walker, *J. Chem. Soc. Dalton Trans.,* 1809 (1980).

37. M. Cusumano, G. Guglielmo, and V. Ricevuto, *J. Chem. Soc. Dalton Trans.,* 2044 (1980).

38. S. Matsumoto and S. Kawaguchi, *Bull. Chem. Soc. Japan* 53, 1577 (1980).

39. S. Balt and J. Meuldijk, *Z. Naturforsch. B: Anorg. Chem. Org. Chem.* 34B, 843, (1979).

40. M. U. Fayyaz and M. W. Grant, *Aust. J. Chem.* 32, 2159 (1979).

41. R. G. Pearson and D. A. Sweigart, *Inorg. Chem.* 9, 1167 (1970).

42. I. N. Marov, M. N. Vargaftik, V. K. Belyaeva, G. A. Evtikova, E. Hoyer, R. Kirmse, and W. Dietzsch, *Russ. J. Inorg. Chem.* 25, 101 (1980).

43. D. C. Olson and D. W. Margerum. *J. Am. Chem. Soc.* 85, 297 (1963).

44. M. Cusumano, *Inorg. Chem.* 18, 3612 (1979).

45. M. Cusumano, *J. Chem. Soc. Dalton Trans.,* 2456 (1980).

46. W. L. Darby and L. M. Vallarino, *Inorg. Chim. Acta* 36, 253 (1979).

47. B. I. Peshchevitskii and G. I. Shamovskaya, *Koord. Khim.* 6, 1657 (1980).

48. L. H. Skibsted, *Acta Chem. Scand.* A33, 113 (1979).

49. E. Jorgensen and J. Bjerrum, *Acta Chem. Scand.* 13, 2075 (1959).

50. N. W. Alcock, D. J. Benton, and P. Moore, *Trans. Faraday Soc.* 66, 2210 (1970).

51. R. A. Reinhardt and J. S. Coe, *Inorg. Chim. Acta* 3, 438 (1969).

52. R. D. Alexander and P. N. Holper, *Transition Met. Chem.* 5, 108 (1980).

53. M. J. Blandamer, J. Burgess, S. J. Hamshere, and P. Wellings, *Transition Met. Chem.* 4, 161 (1979).

54. C. F. Weik and F. Basolo, *Inorg. Chem.* 5, 576 (1966).

55. D. Chatterji and S. K. Podder, *Indian J. Chem.* 17A, 456 (1979).

56. Y. Ohtani, Y. Yamagishi, and M. Fujimoto, *Bull. Chem. Soc. Japan* 52, 1537, 2149 (1979).

57. F. H. Jumean, *J. Indian Chem. Soc.* 57, 312 (1980).

58. W. J. Louw and C. E. Hepner, *Inorg. Chem.* 19, 7 (1980).

59. H. Voss, K. J. Wannowius, and H. Elias, *Inorg. Chem.* 18, 1454 (1979).

60. H. Elias, U. Fröhn, A. von Irmer, and K. J. Wannowius, *Inorg. Chem.* 19, 869 (1980).

61. K. R. Adam, G. Anderegg, L. F. Lindoy, H. C. Lip, M. McPartlin, J. H. Rea, R. J. Smith, and P. A. Tasker, *Inorg. Chem.* 19, 2956 (1980).

62. M. Kodama, T. Yatsunami, and E. Kimura, *J. Chem. Soc. Dalton Trans.,* 1783 (1979).

63. L. S. W. L. Sokol, T. D. Fink, and D. B. Rorabacher, *Inorg. Chem.* 19, 1263 (1980).

64. M. Kodama and E. Kimura, *J. Chem. Soc. Dalton Trans.,* 2447 (1980).

65. G. K. Anderson and R. J. Cross, *Chem. Soc. Rev.* 9, 185 (1980).

66. R. Favez, R. Roulet, A. N. Pinkerton, and D. Schwarzenbach, *Inorg. Chem.* 19, 1356 (1980).

67. T. Uchiyama, T. Nakamura, T. Miwa, S. Kawaguchi, and S. Okeya, *Chem. Lett.,* 337 (1980).

68. A. Okumura, M. Kitani, Y. Toyomi, and N. Okazaki, *Bull. Chem. Soc. Japan* 53, 3143 (1980).

69. L. E. Nivorozhkin, L. E. Konstantinowskii, V. I. Minkin, M. S. Korobov, and I. Ya. Kvito, *Koord. Khim.* 6, 568 (1980); *Chem. Abs.* 93, 81138w (1980).

70. J. H. Coates, P. R. Collins, and S. F. Lincoln, *Aust. J. Chem.* 33, 1381 (1980).

71. H. L. Collier and E. Grimley, *Inorg. Chem.* 19, 511 (1980).

72. T. Ramasami and J. H. Espenson, *Inorg. Chem.* 19, 1523 (1980).

73. P. M. Burkinshaw, D. T. Dixon, and J. A. S. Howell, *J. Chem. Soc. Dalton Trans.,* 999 (1980).

References for Chapter 5

1. J. Burgess and P. Moore, in *Specialist Periodical Reports: Inorganic Reaction Mechanisms* (A. G. Sykes, ed.), Vol. 7, The Chemical Society, London (1981), Chap. 5.
2. G. A. Lawrance and D. R. Stranks, *Acc. Chem. Res.* **12**, 403 (1979).
3. e. g., F. K. Meyer, K. E. Newman, and A. E. Merbach, *J. Am. Chem. Soc.* **101**, 5588 (1979).
4. C. H. Langford, *Inorg. Chem.* **18**, 3288 (1979).
5. E. Buncel and H. Wilson, *Acc. Chem. Res.* **12**, 42 (1979).
6. M. J. Blandamer and J. Burgess, *Coord. Chem. Rev.* **31**, 93 (1980).
7. M. J. Blandamer and J. Burgess, *Pure Appl. Chem.* **51**, 2087 (1979).
8. J. W. Vaughn, *Synth. React. Inorg. Met. -Org. Chem.* **9**, 585 (1979).
9. D. A. Palmer and H. Kelm, *Coord. Chem. Rev.* **36**, 89 (1981).
10. J. I. Byington, R. D. Peters, and L. O. Spreer, *Inorg. Chem.* **18**, 3324 (1979).
11. M. J. Akhtar and L. O. Spreer, *Inorg. Chem.* **18**, 3327 (1979).
12. S. J. Wang and E. L. King, *Inorg. Chem.* **19**, 1506 (1980).
13. G. Rabai, G. Bazsa, and M. T. Beck, *Acta Chim. Acad. Sci. Hung.* **102**, 223 (1979).
14. D. E. Pennington, *J. Coord. Chem.* **10**, 135 (1980).
15. A. I. Kofi and E. Deutsch, *Inorg. Chem.* **19**, 1366 (1980).
16. P. Riccieri and E. Zinato, *Inorg. Chem.* **19**, 853 (1980).
17. T. R. Griffiths and D. C. Pugh, *Anal. Chim. Acta* **107**, 279 (1979).
18. V. Holba and O. Grancicova, *Chem. Abs.* **92**, 65339m (1980).
19. M. F. Amira, P. Carpenter, and C. B. Monk, *J. Chem. Soc. Dalton Trans.*, 1726, 1980.
20. M. Iida and H. Yamatera, *Bull. Chem. Soc. Jpn.* **54**, 441 (1981).
21. R. C. Thompson, P. Wieland, and E. H. Appelman, *Inorg. Chem.* **18**, 1974 (1979).
22. L. Moensted and O. Moensted, *Acta Chem. Scand. Ser A* **A34**, 259 (1980).
23. J. Casabo, J. Ribas, V. Cubas, G. Rodriguez, and J. Fernandez, *Inorg. Chim. Acta.* **36**, 183 (1979).
24. Y. N. Shevchenko, V. V. Satok, and N. K. Davidento, *Russ. J. Inorg. Chem.* **24**, 1475 (1979).
25. A. Heatherington, S. M. Oon, R. Vargas, and N. A. P. Kane-Maguire, *Inorg. Chim. Acta.* **44**, L279 (1980).
26. D. Malek and J. C. Chang, *J. Inorg. Nucl. Chem.* **42**, 1313 (1980).
27. J. C. Chang, J. E. Guerrero, and D. Malek, *J. Inorg. Nucl. Chem.* **42**, 1654 (1980).
28. S. Wajda and A. Szycik, *Chem. Abs.* **94**, 91110r (1981).
29. M. J. Saliby, P. S. Sheridan, and S. K. Madan, *Inorg. Chem.* **19**, 1291 (1980).
30. M. J. Saliby, D. West and S. K. Madan, *Inorg. Chem.* **20**, 723 (1981).
31. G. J. Samuels and J. H. Espenson, *Inorg. Chem.* **19**, 233 (1980); **18**, 2587 (1979).
32. M. C. Pohl and J. H. Espenson, *Inorg. Chem.* **19**, 235 (1980).
33. M. J. Saliby and S. K. Madan, *Inorg. Chem.* **20**, 289 (1981).
34. T. W. Kallen, M. J. Root, and K. A. Schroeder, *Inorg. Chem.* **18**, 3318 (1979).
35. K. Bauer, H. Elias, R. Gaubatz, and G. Lang, *Inorg. Chim. Acta* **36**, 55 (1979).
36. H. Elias, E. Horst, G. Lang, and M. Wüzt, *Inorg. Chim. Acta.* **44**, L119 (1980).
37. C. J. O'Connor and R. E. Ramage, *Aust. J. Chem.* **33**, 695 (1980).
38. C. J. O'Connor, E. J. Fendler, and J. H. Fendler, *J. Chem. Soc. Dalton Trans.*, 625 (1974).
39. S. Wajda and A. W. Szemik, *Pol. J. Chem.* **53**, 1425, 1707 (1979).
40. H. Kido, *Bull. Chem. Soc. Jpn.* **53**, 82 (1980).
41. K. Wieghardt, W. Schmidt, R. VanEldik, B. Nuber and J. Weiss, *Inorg. Chem.* **19**, 2922 (1980).
42. R. A. Holwerda and J. S. Petersen, *Inorg. Chem.* **19**, 1775 (1980).
43. S. C. Tyagi and A. A. Khan, *J. Inorg. Nucl. Chem.* **41**, 1447 (1979).

44. F. K. Meyer, A. R. Monnerat, K. E. Newman, and A. E. Merbach, *Inorg. Chem.* **21**, 774 (1982).
45. S. C. Tyagi and A. A. Khan, *J. Chem. Soc. Dalton Trans.*, 420 (1979).
46. J. N. Mandal and G. S. De, *Indian J. Chem. Sec. A* **17A**, 254 (1979).
47. J. N. Mandal and G. S. De, *Indian J. Chem. Sec. A* **19A**, 25 (1980).
48. R. P. Pantaler and I. V. Pulyaeva, *Chem. Abs.* **92**, 186488t (1980).
49. N. F. Kosenko, T. V. Mal'kova, K. B. Yatsimirskii, and I. I. Batishcheva, *Chem. Abs.* **94**, 198216g (1981).
50. P. J. Preece, H. B. Gray, and F. C. Anson, *Inorg. Chem.* **18**, 2593 (1979).
51. W. U. Malik, S. P. Srivastava, K. K. Thallam, and V. K. Gupta, *Chem. Abs.* **91**, 163647 (1979).
52. P. Kita, L. Chamarczuk, and R. Kniahnicka, *Pol. J. Chem.* **53**, 575 (1979).
53. M. Ferrer and A. G. Sykes, *Inorg. Chem.* **18**, 3345 (1979).
54. T. W. Kallen, W. C. Stevens, and D. G. Hammond, *Inorg. Chem.* **18**, 1358 (1979).
55. K. R. Ashley, J. G. Leipoldt, and V. K. Joshi, *Inorg. Chem.* **19**, 1608 (1980).
56. D. R. Prasad, T. Ramasami, D. Ramaswamy, and M. Santappa, *Inorg. Chem.* **19**, 3181 (1980).
57. H. Ogino, M. Shimura, and N. Tanaka, *Inorg. Chem.* **18**, 2497 (1979).
58. T. W. Kallen and R. E. Hamm, *Inorg. Chem.* **18**, 2151 (1979).
59. T. G. Dunne and J. M. Hurst, *Inorg. Chem.* **19**, 1152 (1980).
60. A. J. Miralles and A. Haim, *Inorg. Chem.* **19**, 1158 (1980).
61. H. Ogino and K. Tsukahara, *Inorg. Chem.* **19**, 255 (1980).
62. G. J. Samuels and J. H. Espenson, *Inorg. Chem.* **18**, 2587 (1979).
63. M. Sriram, T. Ramasami, and D. Ramaswamy, *Inorg. Chim. Acta* **36**, L433 (1979).
64. K. Angermann, R. Van Eldik, H. Kelm, and F. Wasgestian, *Inorg. Chem.* **20**, 955 (1981).
65. P. Riccieri and E. Zinato, *Inorg. Chem.* **19**, 3279 (1980).
66. R. R. Ruminski and W. F. Coleman, *Inorg. Chem.* **19**, 2185 (1980).
67. L. G. Vanquickenborne and A. Ceulemans, *Inorg. Chem.* **18**, 3475 (1979).
68. S. C. Pyke and R. G. Linck, *Inorg. Chem.* **19**, 2468 (1980).
69. A. D. Kirk, L. A. Frederick, and C. F. C. Wong, *Inorg. Chem.* **18**, 448 (1979).
70. A. D. Kirk and C. F. C. Wong, *Inorg. Chem.* **18**, 593 (1979).
71. A. D. Kirk and G. B. Porter, *Inorg. Chem.* **19**, 445 (1980).
72. A. D. Kirk, L. A. Frederick, and S. G. Glover, *J. Am. Chem. Soc.* **102**, 7120 (1980).
73. R. Sriram, M. Z. Hoffman, M. A. Tamieson, and N. Serpone, *J. Am. Chem. Soc.* **102**, 1754 (1980).
74. M. C. Cimolino, N. C. Shipley, and R. G. Linck, *Inorg. Chem.* **19**, 3291 (1980).
75. G. B. Porter and J. Can Houten, *Inorg. Chem.* **19**, 2903 (1980).
76. E. Zinato, P. Riccieri, and P. S. Sheridan, *Inorg. Chem.* **18**, 720 (1979).
77. C. J. O'Connor and A. L. Odell, *Aust. J. Chem.* **33**, 1129 (1980).
78. K. Uchida and Y. Takinami, *Bull. Chem. Soc. Jpn.* **53**, 3522 (1980).
79. P. Kita and A. Swinarski, *Z. Anorg. Allg. Chem.* **464**, 195 (1980).
80. P. Kita and R. Kostrzewska-Zeidler, *Pol. J. Chem.* **53**, 1183 (1979).
81. L. E. Metcalf and J. E. House, *J. Inorg. Nucl. Chem.* **42**, 961 (1980).
82. J. E. House and L. E. Metcalf, *Inorg. Nucl. Chem. Lett.* **16**, 49 (1980).
83. S. Mitra, T. Yoshikuni, A. Uehara, and R. Tsuchiya, *Bull. Chem. Soc. Jpn.* **52**, 2569 (1979).
84. S. Mitra, A. Uehara, and R. Tsuchiya, *Thermochim. Acta.* **34**, 189 (1979).
85. K. Nakano, Y. Narusawa, M. Tsuchiya and H. Moroi, *Chem. Abs.* **94**, 24191e (1981).
86. A. Okumura, M. Kitani, Y. Toyomi, and N. Okazaki, *Bull. Chem. Soc. Jpn.* **53**, 3143 (1980).
87. P. Hambright, J. McRae, P. E. Valk, A. J. Bearden, and B. A. Shipley, *J. Nucl. Med.* **16**, 478 (1975).

88. V. W. Eckelman and S. M. Levason, *Int. J. Appl. Radiat. Isot.* **28,** 67 (1977).
89. E. Ianovici, P. Lerch, Z. Proso, and A. G. Maddock, *J. Radioanal. Chem.* **46,** 11 (1978).
90. G. E. Boyd, *J. Chem. Educ.* **36,** 3 (1959); K. V. Kotegov, O. N. Pavlov, and V. P. Shvedov, *Adv. Inorg. Chem. Radiochem.* **11,** 1 (1968).
91. B. Noll, S. Seifert, and R. Münze, *Radiochem. Radioanal. Lett.* **43,** 215 (1980).
92. V. I. Spitsyn, A. F. Kuzina, and S. V. Kryuchkov, *Russ. J. Inorg. Chem.* **25,** 406 (1980).
93. R. W. Thomas, G. W. Estes, R. C. Elder, and E. Deutsch, *J. Am. Chem. Soc.* **101,** 4581 (1979).
94. H. Spies, B. Johannsen, R. Münze, and K. Unverfurth, *Radiochem. Radioanal. Lett.* **43,** 311 (1980).
95. N. Vanlić-Razumenić, B. Johannsen, H. Spies, R. Syhre, M. Kretschmar, and R. Berger, *Int. J. Appl. Radiat. Isot.* **30,** 661 (1979).
96. R. Münze, *Isotopenpraxis* **14,** 81 (1978).
97. M. J. Blandamer, J. Burgess, and R. I. Haines, *J. Chem. Soc. Dalton Trans.,* 607 (1980).
98. H. E. Toma and J. M. Malin, *Inorg. Chem.* **12,** 1039 (1973).
99. G. C. Pedrosa, N. L. Hernández, N. E. Katz, and M. Katz, *J. Chem. Soc. Dalton Trans.,* 2297 (1980).
100. N. E. Katz, M. E. G. Posse, and M. A. Martinez, *J. Inorg. Nucl. Chem.* **42,** 1782 (1980).
101. H. E. Toma, J. M. Martins, and E. Giesbrecht, *J. Chem. Soc. Dalton Trans.,* 1610 (1978).
102. M. L. Bowers, D. Kovacs, and R. E. Shepherd, *J. Am. Chem. Soc.* **99,** 6555 (1977).
103. D. H. Macartney and A. McAuley, *Inorg. Chem.* **18,** 2891 (1979).
104. J. -J. Jwo, P. L. Gans, and A. Haim, *J. Am. Chem. Soc.* **101,** 6189 (1979).
105. J. M. Malin and R. C. Koch, *Inorg. Chem.* **17,** 752 (1978).
106. A. Szecsy and A. Haim, *Inorg. Chim. Acta* **28,** 189 (1978).
107. R. Juretić, D. Pavlović, and S. Ašperger, *J. Chem. Soc. Dalton Trans.,* 2029 (1979).
108. M. Jakševac-Mikša, V. Hankonyi, and V. Karas-Gašparec, *Z. Phys. Chem. (Leipzig)* **261,**1041 (1980).
109. G. Davies and A. R. Garafalo, *Inorg. Chem.* **19,** 3542 (1980).
110. M. J. Blandamer, J. Burgess, and S. H. Morris, *J. Chem. Soc. Dalton Trans.,* 1717 (1974).
111. S. Raman, *J. Inorg. Nucl. Chem.* **40,** 1073 (1978).
112. A. D. Pethybridge and J. E. Prue, *Progr. Inorg. Chem.* **17,** 327 (1972).
113. N. E. Katz, M. A. Blesa, J. A. Olabe, and P. J. Aymonino, *J. Inorg. Nucl. Chem.* **42,** 581 (1980).
114. E. R. Gardner, F. M. Mekhail, and J. Burgess, *Int. J. Chem. Kinet.* **6,** 133 (1974).
115. J. Burgess and G. M. Burton, *Rev. Latinoamer. Quim.* **11,** 107 (1980).
116. V. L. Goedken, *J. Chem. Soc. Chem. Commun.,* 207 (1972).
117. J. Burgess, A. J. Duffield, D. R. Russell, and L. Sherry, to be published.
118. J. Burgess and R. H. Prince, *J. Chem. Soc. A,* 434 (1967); J. Burgess, *J. Chem. Soc. A,* 955 (1967).
119. J. Burgess and R. H. Prince, *J. Chem. Soc. A,* 2111 (1970).
120. R. Somuthevan, R. Renganathan, and P. Maruthamuthu, *Inorg. Chim. Acta* **45,** L165 (1980).
121. J. Burgess, *J. Chem. Soc. A,* 497 (1968).
122. E. Chaffee, I. I. Creaser, and J. O. Edwards, *Inorg. Nucl. Chem. Lett.* **7,** 1 (1971); J. O. Edwards, *Coord. Chem. Rev.* **8,** 87 (1972).
123. D. Soria, M. L. de Castro, and H. L. Chum, *Inorg. Chim. Acta* **42,** 121 (1980).
124. T. Fujiwara and Y. Yamamoto, *Inorg. Chem.* **19,** 1903 (1980).
125. T. Fujiwara and Y. Yamamoto, *Inorg. Nucl. Chem. Lett.* **15,** 397 (1979).
126. E. Iwamoto, T. Fujiwara, and Y. Yamamoto, *Inorg. Chim. Acta* **43,** 95 (1980).
127. G. A. Lawrance, D. R. Stranks, and S. Suvachittanont, *Inorg. Chem.* **18,** 82 (1979).

128. T. Asano and W. J. le Noble, *Chem. Rev.* **78,** 407 (1978); W. J. le Noble and H. Kelm, *Angew. Chem. Int. Ed. Eng.* **19,** 841 (1980).
129. D. R. Stranks, personal communication.
130. J. Burgess, A. J. Duffield, and R. Sherry, *J. Chem. Soc. Chem. Commun.,* 350 (1980).
131. J. M. Lucie, D. R. Stranks, and J. Burgess, *J. Chem. Soc. Dalton Trans.,* 245 (1975).
132. T. Fujiwara, K. Matsuda, and Y. Yamamoto, *Inorg. Nucl. Chem. Lett.* **16,** 301 (1980).
133. M. H. Abraham and A. F. Danil de Namor, *J. Chem. Soc. Faraday Trans. 1,* **72,** 955 (1976).
134. D. J. Farrington, J. G. Jones, and M. V. Twigg, *Inorg. Chim. Acta* **25,** L75 (1977).
135. R. D. Gillard, *Inorg. Chim. Acta* **37,** 103 (1979).
136. S. Raman, *Inorg. Chim. Acta* **40,** 273 (1980).
137. M. J. Blandamer, J. Burgess, J. G. Chambers, R. I. Haines, and H. E. Marshall, *J. Chem. Soc. Dalton Trans.,* 165 (1977).
138. M. J. Blandamer, J. Burgess, and A. J. Duffield, *J. Chem. Soc. Dalton Trans.,* 1 (1980).
139. M. J. Blandamer, J. Burgess, P. P. Duce, and R. I. Haines, *J. Chem. Soc. Dalton Trans.,* 2442 (1980).
140. R. D. Gillard, *Inorg. Chim. Acta* **11,** L21 (1974); *Coord. Chem. Rev.* **16,** 67 (1975).
141. J. Burgess, *Inorg. React. Mech.* **7,** 232, 235 (1981), and references therein.
142. R. D. Gillard, C. T. Hughes, W. S. Walters, and P. A. Williams, *J. Chem. Soc. Dalton Trans.,* 1769 (1979).
143. R. D. Gillard, R. P. Houghton, and J. N. Tucker, *J. Chem. Soc. Dalton Trans.,* 2102 (1980).
144. J. W. Bunting, *Adv. Heterocycl. Chem.* **25,** 1 (1979).
145. R. J. Wademan and P. A. Williams, *Transition Met. Chem.* **4,** 333 (1979).
146. R. D. Gillard, D. W. Knight, and P. A. Williams, *Transition Met. Chem.* **4,** 375 (1979).
147. L. I. Simandi, S. Nemeth, and E. Budozahonyi, *Inorg. Chim. Acta* **45,** L143 (1980).
148. F. T. T. Ng and P. M. Henry, *Can. J. Chem.* **58,** 1773 (1980).
149. J. Martinsen, M. Miller, D. Trojan, and D. A. Sweigart, *Inorg. Chem.* **19,** 2162 (1980).
150. D. E. Hamilton, T. J. Lewis, and N. K. Kildahl, *Inorg. Chem.* **18,** 3364 (1979).
151. C. J. Barbour, J. H. Cameron, and J. M. Winfield, *J. Chem. Soc. Dalton Trans.,* 2001 (1980).
152. D. H. Cuatecontzi and J. D. Miller, *Inorg. Chim. Acta* **38,** 157 (1980).
153. A. Drummond, J. F. Kay, J. H. Morris, and D. Reed, *J. Chem. Soc. Dalton Trans.,* 284 (1980).
154. C. R. Johnson, R. E. Shepherd, B. Marr, S. O'Donnell, and W. Dressick, *J. Am. Chem. Soc.* **102,** 6227 (1980).
155. P. Collings and G. Stedman, *Inorg. Chim. Acta* **41,** 1 (1980).
156. D. A. Sweigart and W. Fiske, in *NATO Advanced Study Institutes, Series C, Vol. 50: Techniques and Applications of Fast Reactions in Solution* (W. J. Gettins and E. Wyn-Jones, eds.), Reidel, Dordrecht (1979), p. 315. [*Chem. Abstr.* **92,** 11809w (1980)].
157. H. Kido and K. Saito, *Bull. Chem. Soc. Jpn.* **53,** 424 (1980).
158. A. Yeh and H. Taube, *Inorg. Chem.* **19,** 3740 (1980).
159. H. Lehmann, K. J. Schenk, G. Chapuis, and A. Ludi, *J. Am. Chem. Soc.* **101,** 6197 (1979).
160. C. -K. Poon, C. -M. Che, and Y. -P. Kan, *J. Chem. Soc. Dalton Trans.,* 128 (1980).
161. B. P. Sullivan, J. M. Calvert, and T. J. Meyer, *Inorg. Chem.* **19,** 1404 (1980).
162. J. D. Birchall, T. D. O'Donoghue, and J. R. Wood, *Inorg. Chim. Acta* **37,** L461 (1979).
163. N. Farrel and N. G. de Oliveira, *Inorg. Chim. Acta* **44,** L255 (1980).
164. R. A. Krause and K. Krause, *Inorg. Chem.* **19,** 2600 (1980).
165. D. A. Johnson and V. C. Dew, *Inorg. Chem.* **18,** 3273 (1979).
166. S. Kohata, N. Itoh, H. Kawaguchi, and A. Ohyoshi, *Bull. Chem. Soc. Jpn.* **52,** 2264 (1979).
167. J. A. A. Sagüés, R. D. Gillard, and P. A. Williams, *Inorg. Chim. Acta* **44,** L253 (1980).
168. W. M. Wallace and P. E. Hoggard, *Inorg. Chem.* **19,** 2141 (1980).

169. P. E. Hoggard and G. B. Porter, *J. Am. Chem. Soc.* **100**, 1457 (1978).
170. B. Durham, J. L. Walsh, C. L. Carter, and T. J. Meyer, *Inorg. Chem.* **19**, 860 (1980).
171. S. H. Peterson and J. N. Demas, *J. Am. Chem. Soc.* **101**, 6571 (1979).
172. E. Tfoumi and P. C. Ford, *Inorg. Chem.* **19**, 72 (1980).
173. T. Tsuihiji, T. Akiyama, and A. Sugimori, *Bull. Chem. Soc. Jpn.* **52**, 3451 (1979).
174. S. Bagger, *Acta Chem. Scand. Ser. A* **34**, 63 (1980).
175. M. T. Fairhurst and T. W Swaddle, *Inorg. Chem.* **18**, 3241 (1979).
176. M. E. Rerek and P. S. Sheridan, *Inorg. Chem.* **19**, 2646 (1980).
177. S. Sasaki, T. Uchida, and A. Ohyoshi, *Inorg. Chim. Acta* **45**, L269 (1980).
178. S. C. Pati and Y. Sriramula, *Z. Phys. Chem. (Leipzig)* **260**, 834 (1979).
179. E. I. Maslov and A. V. Goncharov, *Russ. J. Inorg. Chem.* **25**, 264 (1980).
180. J. Dehand and J. Rose, *Inorg. Chim. Acta* **37**, 249 (1979).
181. S. Sugimoto, *Radioisotopes* **28**, 669 (1979) [*Chem. Abstr.* **92**, 186536g (1980)].
182. M. M. Taqui Khan, M. Ahmeed, and A. Kumar, *Inorg. Chim. Acta* **43**, 137 (1980).
183. J. D. Buhr, J. R. Winkler, and H. Taube, *Inorg. Chem.* **19**, 2416 (1980).
184. N. M. Sinitsyn, A. A. Svetlov, and L. V. Bobrova, *Russ. J. Inorg. Chem.* **25**, 428 (1980).
185. U. Dietl and W. Preetz, *Z. Anorg. Allg. Chem.* **451**, 93 (1979).
186. A. K. Shukla, H. D. Zerbe, and W. Preetz, *Z. Anorg. Allg. Chem.* **468**, 39 (1980).
187. U. Dietl and W. Preetz, *Z. Anorg. Allg. Chem.* **456**, 81 (1979).
188. H. Müller, H. Scheible, and S. Martin, *Z. Anorg. Allg. Chem.* **462**, 18 (1980).
189. Foo Chuk Ha and D. A. House, *Inorg. Chim. Acta* **38**, 167 (1980).
190. A. C. Dash, M. S. Dash, and S. K. Mohapatra, *J. Chem. Res. (S)*, 354; *(M)*, 4531 (1979).
191. A. Yamada, T. Yoshikumi, Y. Kato, and N. Tanaka, *Bull. Chem. Soc. Jpn.* **53**, 942 (1980).
192. J. R. Flückiger, C. W. Schläpfer, and C. Couldwell, *Inorg. Chem.* **19**, 2493 (1980).
193. D. A. House, G. Hall, A. J. Matheson, W. T. Robinson, Foo Chuk Ha, and C. B. Knobler, *Inorg. Chim. Acta* **39**, 257 (1980).
194. N. Ise, T. Maruno, and T. Okubo, *Proc. R. Soc. London Ser. A,* **370A**, 485 (1980).
195. D. A. Palmer and H. Kelm, *Z. Anorg. Allg. Chem.* **450**, 50 (1979).
196. G. A. Lawrance, *Inorg. Chim. Acta* **45**, L275 (1980).
197. G. A. Lawrance and S. Suvachittanont, *Inorg. Chim. Acta* **44**, L61 (1980).
198. I. E. Konstantinova, E. A. Grigoryan, R. V. Kurnysheva, L. E. Semina, V. V. Slukina, I. P. Smirenkina, and L. M. Yusupova, *Kinet. Katal.* **21**, 378 (1980) [*Chem. Abstr.* **92**, 221613d (1980)].
199. M. Glavaš and W. L. Reynolds, *J. Chem. Soc. Dalton Trans.*, 1446 (1979).
200. O. Grancicova and V. Holba, *Chem. Abstr.* **92**, 65339m (1980).
201. C. Chatterjee and A. K. Basak, *Transition Met. Chem.* **5**, 212 (1980).
202. C. F. Wells, *J. Chem. Soc. Faraday Trans. 1* **69**, 984 (1973).
203. C. F. Wells, *J. Chem. Soc. Faraday Trans. 1* **70**, 694 (1974).
204. C. F. Wells, *J. Chem. Soc. Faraday Trans. 1* **72**, 601 (1976).
205. C. F. Wells, *J. Chem. Soc. Faraday Trans. 1* **74**, 1569 (1978).
206. C. F. Wells, *J. Chem. Soc. Faraday Trans. 1* **71**, 1868 (1975).
207. C. F. Wells, *J. Chem. Soc. Faraday Trans. 1* **77**, 1515 (1981).
208. C. F. Wells, *J. Chem. Soc. Faraday Trans. 1* **73**, 1851 (1977).
209. C. N. Elgy and C. F. Wells, *J. Chem. Soc. Dalton Trans.*, 2405 (1980).
210. M. J. Blandamer, J. Burgess, and P. P. Duce, unpublished observations.
211. M. F. Amira, P. Carpenter, and C. B. Monk, *J. Chem. Soc. Dalton Trans.*, 1726 (1980).
212. N. C. Naik and R. K. Nanda, *Indian J. Chem. Sec. A* **19**, 113 (1980).
213. R. van Eldik and G. M. Harris, *Inorg. Chem.* **19**, 880 (1980).
214. J. N. Cooper, J. D. McCoy, M. G. Katz, and E. Deutsch, *Inorg. Chem.* **19**, 2265 (1980).
215. J. N. Cooper, D. S. Buck, M. G. Katz, and E. Deutsch, *Inorg. Chem.* **19**, 3856 (1980).

216. R. C. Thompson, *Inorg. Chem.* **18,** 2379 (1979).
217. A. C. Dash and B. Mohanty, *Transition Met. Chem.* **5,** 183 (1980).
218. M. I. Khalil, N. Logan, and A. D. Harris, *J. Chem. Soc. Dalton Trans.,* 314 (1980).
219. L. Omelka and A. Tkac, *Collect. Czech. Chem. Commun.* **45,** 464 (1980).
220. J. A. Howard and S. B. Tong, *Can. J. Chem.* **58,** 1962 (1980).
221. J. Reinhold, R. Benedix, H. Zwanziger, and H. Hennig. *Z. Phys. Chem. (Leipzig)* **261,** 989 (1980).
222. T. Inoue and G. M. Harris, *Inorg. Chem.* **19,** 1091 (1980).
223. K. B. Nolan and A. A. Soudi, *J. Chem. Soc. Dalton Trans.,* 1419 (1979).
224. P. D. Ford and K. B. Nolan, *J. Chem. Res. (S),* 220; *(M),* 2654 (1979).
225. P. D. Ford and K. B. Nolan, *Inorg. Chim. Acta* **43,** 83 (1980).
226. J. MacB. Harrowfield, D. R. Jones, L. F. Lindoy, and A. M. Sargeson, *J. Am. Chem. Soc.* **102,** 7733 (1980).
227. P. Comba and W. Marty, *Helv. Chim. Acta* **63,** 693 (1980).
228. Y. Kitamura, *Bull. Chem. Soc. Jpn.* **52,** 3280 (1979).
229. T. Okubo, T. Maruno, and N. Ise, *Proc. R. Soc. London Ser. A* **370A,** 501 (1980).
230. M. V. Twigg, *Inorg. Chim. Acta* **24,** L84 (1977).
231. R. W. Hay, *Inorg. Chim. Acta* **45,** L83 (1980).
232. R. W. Hay and P. R. Norman, *J. Chem. Soc. Chem. Commun.,* 734 (1980).
233. M. L. Tobe, *Acc. Chem. Res.* **3,** 377 (1970).
234. D. A. House, P. R. Norman, and R. W. Hay, *Inorg. Chim. Acta* **45,** L117 (1980).
235. L. S. Bark, M. B. Davies, and M. C. Powell, *J. Inorg. Nucl. Chem.* **40,** 1661 (1978).
236. M. E. Farago and C. F. V. Mason, *J. Chem. Soc. A,* 3100 (1970).
237. C. F. V. Mason and M. E. Farago, *J. Inorg. Nucl. Chem.* **42,** 131 (1980).
238. A. S. Janardan, V. Kesavan, and D. S. Mahadevappa, *Curr. Sci.* **49,** 738 (1980) [*Chem. Abst.* **93,** 210922y (1980)].
239. A. S. Janardan, V. Kesavan, and D. S. Mahadevappa, *Aust. J. Chem.* **33,** 1485 (1980).
240. S. Balt, W. E. Rankema, and P. C. M. van Zijl, *Inorg. Chim. Acta* **45,** L241 (1980).
241. M. Iida and H. Yamatera, *Bull. Chem. Soc. Jpn.* **52,** 2290 (1979).
242. J. Burgess and A. J. Duffield, *J. Inorg. Nucl. Chem.* **42,** 1531 (1980).
243. M. Iida and H. Yamatera, *Chem. Lett.,* 693 (1980).
244. W. G. Jackson, G. A. Lawrance, and A. M. Sargeson, *Inorg. Chem.* **19,** 1001 (1980).
245. W. L. Reynolds, S. Hafezi, A. Kessler, and S. Holly, *Inorg. Chem.* **18,** 2860 (1979).
246. D. A. Buckingham, C. R. Clark, and W. S. Webley, *J. Chem. Soc. Dalton Trans.,* 2255 (1980).
247. S. Ašperger, *Inorg. Chim. Acta* **40,** X4 (1980).
248. B. Mohanty and A. C. Dash, *Indian J. Chem., Sec. A* **17,** 296 (1979).
249. A. C. Dash and M. S. Dash, *J. Coord. Chem.* **10,** 79 (1980).
250. A. C. Dash and B. Mohanty, *J. Inorg. Nucl. Chem.* **42,** 1161 (1980).
251. R. van Eldik and G. M. Harris, *Inorg. Chem.* **14,** 10 (1975).
252. R. van Eldik, *Inorg. Chim. Acta* **44,** L197 (1980).
253. R. Varadarajan and C. F. Wells, *J. Chem. Soc. Faraday Trans. 1* **76,** 2017 (1980).
254. H. Gamsjäger, G. A. K. Thompson, W. Sagmüller, and A. G. Sykes, *Inorg. Chem.* **19,** 997 (1980).
255. M. Pal and G. S. De, *Indian J. Chem. Sec. A* **17,** 36 (1979).
256. C. F. Wells, *J. Chem. Soc. Dalton Trans.,* 1494 (1980).
257. G. Biswas, S. Aditya, A. Bhattacharyya, and S. C. Lahiri, *J. Indian Chem. Soc.* **54,** 1137 (1977).
258. J. Burgess and R. I. Haines, *J. Chem. Engn. Data* **23,** 196 (1978).
259. F. M. van Meter and H. M. Neumann, *J. Am. Chem. Soc.* **98,** 1382, 1388 (1976).

260. D. K. Hazra and S. C. Lahiri, *Anal. Chim. Acta* **79**, 335 (1975).
261. J. Burgess and R. I. Haines, *Chem. and Ind.*, 289 (1980).
262. M. J. Blandamer, J. Burgess, R. Sherry, and R. I. Haines, *J. Chem. Soc. Chem. Commun.*, 353 (1980).
263. W. L. Reynolds and M. S. El-Nasr, *Inorg. Chem.* **18**, 2864 (1979).
264. W. L. Reynolds and M. S. El-Nasr, *Inorg. Chem.* **19**, 1006 (1980).
265. N. V. Shokhirev and P. V. Schastnev, *Koord. Khim.* **6**, 1177 (1980) [*Chem. Abst.* **93**, 210903t (1980)].
266. G. Schiavon and C. Paradisi, *Inorg. Chim. Acta* **41**, 99 (1980).
267. S. Tsuboyama, K. Tsuboyama, T. Sakurai, and J. Uzawa, *Inorg. Nucl. Chem. Lett.* **16**, 267 (1980).
268. R. K. Murmann and K. M. Rahmoeller, *Inorg. Chim. Acta* **42**, 53 (1980).
269. H. Kido, *Bull. Chem. Soc. Jpn.* **53**, 82 (1980).
270. H. Kido and K. Saito, *Bull. Chem. Soc. Jpn.* **52**, 3545 (1979).
271. S. Wajda and A. W. Szemik, *Pol. J. Chem.* **53**, 1425 (1979) [*Chem. Abst.* **91**, 182109b (1979)].
272. M. Kostanski and S. Magas, *Radiochim. Acta* **27**, 27 (1980).
273. F. Monacelli and A. A. G. Tomlinson, *Gazz. Chim. Ital.* **110**, 43 (1980).
274. V. Favaudon, M. Momenteau, and J. -M. Lhoste, *Inorg. Chem.* **18**, 2355 (1979).
275. G. A. Lawrance and S. Suvachittanont, *J. Coord. Chem.* **9**, 13 (1979).
276. G. A. Lawrance and S. Suvachittanont, *Aust. J. Chem.* **33**, 1649 (1980).
277. B. Chakravarty, P. K. Das, and A. K. Sil, *Inorg. Chim. Acta* **40**, X152 (1980).
278. Y. Fujii, K. Shiono, K. Ezuka, and T. Isago, *Bull. Chem. Soc. Jpn.* **53**, 3537 (1980).
279. A. Sault, F. Fry, and J. C. Bailar, *J. Inorg. Nucl. Chem.* **42**, 201 (1980).
280. C. J. Hilleary, T. F. Them, and R. E. Tapscott, *Inorg. Chem.* **19**, 102 (1980).
281. W. G. Jackson, G. A. Lawrance, P. A. Lay, and A. M. Sargeson, *Inorg. Chem.* **19**, 904 (1980).
282. A. W. Zanella and A. F. Fucaloro, *Inorg. Nucl. Chem. Lett.* **16**, 515 (1980).
283. H. Mäcke, V. Houlding, and A. W. Adamson, *J. Am. Chem. Soc.* **102**, 6888 (1980).
284. G. C. Levy, J. D. Cargioli, and F. A. L. Anet, *J. Am. Chem. Soc.* **95**, 1527 (1973).
285. L. R. Gahan and M. J. O'Connor, *Aust. J. Chem.* **32**, 1653 (1979).
286. G. A. Lawrance, S. Suvachittanont, D. R. Stranks, P. A. Tregloan, L. R. Gahan, and M. J. O'Connor, *J. Chem. Soc. Chem. Commun.*, 757 (1979).
287. G. A. Lawrance, M. J. O'Connor, S. Suvachittanont, D. R. Stranks, and P. A. Tregloan, *Inorg. Chem.* **19**, 3443 (1980).
288. R. van Eldik and G. M. Harris, *Inorg. Chem.* **19**, 3684 (1980).
289. R. W. Hay and B. Jeragh, *J. Chem. Soc. Dalton Trans.*, 1343 (1979).
290. R. W. Hay and B. Jeragh, *Transition Met. Chem.* **4**, 288 (1979).
291. R. W. Hay and B. Jeragh, *Transition Met. Chem.* **5**, 252 (1980).
292. K. E. Hyde, E. W. Hyde, J. Moryl, R. Baltus, and G. M. Harris, *Inorg. Chem.* **19**, 1603 (1980).
293. N. M. Samus' and A. V. Ablov. *Coord. Chem. Rev.* **28**, 177 (1979).
294. N. M. Samus', I. A. Fridman, and T. S. Luk'yanets, *Russ. J. Inorg. Chem.* **25**, 1507 (1980).
295. G. P. Syrtsova and Luong Ngoc The. *Koord. Khim.* **5**, 702 (1979) [*Chem. Abst.* **91**, 79475u (1979)].
296. G. P. Syrtsova and G. I. Shpakov, *Russ. J. Inorg. Chem.* **24**, 882 (1979).
297. V. E. Belevantsev. E. I. Evdokimova, and B. I. Peshchevitskii, *Russ. J. Inorg. Chem.* **24**, 1497 (1979).

298. V. N. Shafranskii, I. V. Dranka, and Yu. Ya. Kharitonov, *Russ. J. Inorg. Chem.* **24,** 1053 (1979).
299. J. A. Kargol, K. D. Lavin, R. W. Crecely, and J. L. Burmeister, *Inorg. Chem.* **19,** 1515 (1980).
300. N. Yoshida and M. Fujimoto, *Bull. Chem. Soc. Jpn.* **53,** 3526 (1980).
301. A. Bakač and J. H. Espenson, *Inorg. Chem.* **19,** 242 (1980).
302. K. L. Brown and R. K. Hessley, *Inorg. Chem.* **19,** 2410 (1980).
303. G. Cros, M. H. Darbieu, and J. P. Laurent, *Inorg. Nucl. Chem. Lett.* **16,** 349 (1980).
304. G. Tauzher, R. Dreos, G. Costa, and M. Green, *Inorg. Chem.* **19,** 3790 (1980).
305. T. Shibahara, H. Kuroya, and M. Mori, *Bull. Chem. Soc. Jpn.* **53,** 2834 (1980).
306. M. Ferrer, T. D. Hand, and A. G. Sykes, *J. Chem. Soc. Dalton Trans.,* 14 (1980).
307. H. Siebert, *Z. Anorg. Allg. Chem.* **463,** 155 (1980).
308. S. Fallab, M. Zehnder, and U. Thewalt, *Helv. Chim. Acta* **63,** 1491 (1980).
309. N. Al-Shatti, M. Ferrer, and A. G. Sykes, *J. Chem. Soc. Dalton Trans.,* 2533 (1980).
310. P. Natarajan and N. V. Raghavan, *J. Chem. Soc. Chem. Commun.,* 268 (1980).
311. M. Zehnder, U. Thewalt, and S. Fallab, *Helv. Chim. Acta* **62,** 2099 (1979).
312. K. Wieghardt, W. Schmidt, B. Nuber, and J. Weiss, *Chem. Ber.* **112,** 2220 (1979).
313. J. Phillips and A. Haim, *Inorg. Chem.* **19,** 1616 (1980).
314. Yu. P. Manuilov and G. A. Shagisultanova, *Russ. J. Inorg. Chem.* **25,** 1510 (1980).
315. A. L. Poznyak and V. V. Pansevich, *Russ. J. Inorg. Chem.* **24,** 399 (1979).
316. N. Sabbatini, R. Rossi, M. A. Scandola, and F. Scandola, *Inorg. Chem.* **18,** 2633 (1979).
317. C. Conti, F. Castelli, and L. S. Forster, *J. Phys. Chem.* **83,** 2371 (1979).
318. H. D. Wohlers, K. D. Van Tassel, B. A. Bowerman, and J. D. Petersen, *Inorg. Chem.* **19,** 2837 (1980).
319. A. L. Poznjak and V. I. Pawlowski, *Z. Anorg. Allg. Chem.* **465,** 159 (1980).
320. K. F. Purcell, S. F. Clark, and J. D. Petersen, *Inorg. Chem.* **19,** 2183 (1980).
321. U. Sakaguchi, K. Morito, and H. Yoneda, *Bull. Chem. Soc. Jpn.* **53,** 2821 (1980).
322. T. Ama, H. Kawaguchi, M. Kanekiyo, and T. Yasui, *Bull. Chem. Soc. Jpn.* **53,** 956 (1980).
323. C. Boreham and D. A. Buckingham, *Aust. J. Chem.* **33,** 27 (1980).
324. U. Sakaguchi, K. Morito, and H. Yoneda, *Inorg. Chim. Acta* **37,** 209 (1979).
325. M. Yamaguchi, S. Yamamatsu, T. Furusawa, S. Yano, M. Saburi, and S. Yoshikawa, *Inorg. Chem.* **19,** 2010 (1980).
326. R. Job, *Inorg. Chim. Acta* **40,** 59 (1980).
327. D. A. Buckingham and C. R. Clark, *J. Chem. Soc. Dalton Trans.,* 1757 (1979).
328. A. W. Zanella and P. C. Ford, *Inorg. Chem.* **14,** 700 (1975).
329. R. D. Cornelius, *Inorg. Chem.* **19,** 1286 (1980).
330. P. W. A. Hübner and R. M. Milburn, *Inorg. Chem.* **19,** 1267 (1980).
331. M. Z. Hoffman, D. W. Kimmel, and M. G. Simic, *Inorg. Chem.* **18,** 2479 (1979).
332. N. V. Brezniak and M. Z. Hoffman, *Inorg. Chem.* **18,** 2934 (1979).
333. J. P. Candlin and J. Halpern, *J. Am. Chem. Soc.* **85,** 2518 (1963).
334. J. P. Mittleman, J. N. Cooper, and E. A. Deutsch, *J. Chem. Soc. Chem. Commun.,* 733 (1980).
335. I. K. Adzamli and E. Deutsch, *Inorg. Chem.* **19,** 1366 (1980).
336. D. A. Palmer, *Aust. J. Chem.* **32,** 2589 (1979).
337. A. V. Belyaev, A. B. Venediktov, and B. I. Peshchevitskii, *Koord. Khim.* **5,** 1071 (1979) [*Chem. Abst.* **91,** 113054t (1979).
338. H. L. Chung and E. J. Bounsall, *Can. J. Chem.* **56,** 709 (1978).
339. R. W. Hay and P. R. Norman, *J. Chem. Soc. Dalton Trans.,* 1441 (1979).

340. A. J. Poë and C. P. J. Vuik, *Can J. Chem.* **53,** 1842 (1975); *J. Chem. Soc. Dalton Trans.,* 661 (1976).

341. A. J. Poë and C. P. J. Vuik, *Inorg. Chem.* **19,** 1771 (1980).

342. C. Chatterjee and A. K. Basak, *Bull. Chem. Soc. Jpn.* **52,** 2710 (1979).

343. K. R. Ashley, Shaw-Bey Shyu, and J. G. Leipoldt, *Inorg. Chem.* **19,** 1613 (1980).

344. L. H. Skibsted and P. C. Ford, *Acta Chem. Scand. Ser. A* **34,** 109 (1980).

345. S. Kohata, H. Kawaguchi, N. Itoh, and A. Ohyoshi, *Bull. Chem. Soc. Jpn.* **53,** 807 (1980).

346. A. Peloso, *J. Chem. Soc. Dalton Trans.,* 2033 (1979).

347. D. A. Palmer, R. van Eldik, H. Kelm, and G. M. Harris, *Inorg. Chem.* **19,** 1009 (1980).

348. R. van Eldik, D. A. Palmer, and G. M. Harris, *Inorg. Chem.* **19,** 3673 (1980).

349. R. van Eldik, D. A. Palmer, H. Kelm, and G. M. Harris, *Inorg. Chem.* **19,** 3679 (1980).

350. K. Wieghardt, W. Schmidt, R. van Eldik, B. Nuber, and J. Weiss, *Inorg. Chem.* **19,** 2922 (1980).

351. R. van Eldik, *Inorg. Chim. Acta* **42,** 49 (1980).

352. T. Ramasami and J. H. Espenson, *Inorg. Chem.* **19,** 1846 (1980).

353. C. Kutal and A. W. Adamson, *Inorg. Chem.* **12,** 1454 (1973).

354. L. H. Skibsted, D. Strauss, and P. C. Ford, *Inorg. Chem.* **18,** 3171 (1979).

355. A. F. Diaz, K. K. Kanazawa, and G. P. Gardini, *J. Chem. Soc. Chem. Commun.,* 635 (1979).

356. L. H. Skibsted and P. C. Ford, *J. Chem. Soc. Chem. Commun.,* 853 (1979).

357. L. H. Skibsted and P. C. Ford, *Inorg. Chem.* **19,** 1828 (1980).

358. S. F. Clark and J. D. Petersen, *Inorg. Chem.* **18,** 3394 (1979).

359. S. F. Clark and J. D. Petersen, *Inorg. Chem.* **19,** 2917 (1980).

360. J. D. Petersen and F. P. Jakse, *Inorg. Chem.* **18,** 1818 (1979).

361. A. Vogler and A. Kern, *Ber. Bunsenges. Phys. Chem.* **83,** 500 (1979).

362. V. I. Kravtsov, L. Ya. Smirnova, and G. P. Tsayun, *Radiokhimiya* **15,** 1737 (1979) [*Chem. Abst.* **92,** 65331c (1980)].

363. V. I. Kravtsov, L. Ya. Smirnova, and L. V. Lazareva, *Elektrokhimiya* **16,** 1586 (1980) [*Chem. Abst.* **93,** 226445f (1980)].

364. J. Szalma, V. I. Kravtsov, L. Ya. Smirnova, and G. P. Tsayun, *Magy. Kem. Foly.* **86,** 293 (1980) [*Chem. Abst.* **93,** 121171k (1980)].

365. M. Talebinasab-Sarvari, A. W. Zanella, and P. C. Ford, *Inorg. Chem.* **19,** 1835 (1980).

366. M. Talebinasab-Sarvari and P. C. Ford, *Inorg. Chem.* **19,** 2640 (1980).

367. B. Divisia, P. C. Ford, and R. J. Watts, *J. Am. Chem. Soc.* **102,** 7264 (1980).

368. J. Burgess, E. R. Gardner, and F. M. Mekhail, *J. Chem. Soc. Dalton Trans.,* 487 (1972).

369. F. Bozon-Verduraz and C. Leclere, *Bull. Soc. Chim. Fr. I,* 101 (1980).

370. B. Harrison, N. Logan, and A. D. Harris, *J. Chem. Soc. Dalton Trans.,* 2382 (1980).

371. M. Schmidt and G. G. Hoffmann, *Z. Anorg. Allg. Chem.* **452,** 112 (1979).

372. N. H. Agnew, T. G. Appleton, and J. R. Hall, *Inorg. Chim. Acta* **41,** 71 (1980).

373. N. H. Agnew, T. G. Appleton, and J. R. Hall, *Inorg. Chim. Acta* **41,** 85 (1980).

374. A. V. Nikolaev, T. K. Kazbanova, G. D. Mal'chikov, and V. I. Kazbanov, *Russ. J. Inorg. Chem.* **24,** 898 (1979).

375. A. V. Babkov, M. N. Kuznetsova, and V. S. Smurova, *Koord. Khim.* **6,** 1593 (1980) [*Chem. Abst.* **93,** 246229s (1980)].

376. E. W. Abel, M. Booth, and K. G. Orrell, *J. Chem. Soc. Dalton Trans.,* 1994 (1979).

377. E. W. Abel, A. R. Khan, K. Kite, K. G. Orrell, and V. Šik, *J. Chem. Soc. Dalton Trans.,* 1175 (1980).

378. E. W. Abel, M. Booth, and K. G. Orrell, *J. Chem. Soc. Dalton Trans.,* 1582 (1980) and references therein.

379. E. W. Abel, A. R. Khan, K. Kite , K. G. Orrell, and V. Šik, *J. Chem. Soc. Dalton Trans.*, 2208 (1980).
380. E. W. Abel, A. R. Khan, K. Kite, K. G. Orrell, and V. Šik, *J. Chem. Soc. Dalton Trans.*, 2220 (1980).

References for Chapter 6

1. T. J. Gilligan and G. Atkinson, *J. Phys. Chem.* **84**, 208 (1980).
2. Y. Ducommun, K. E. Newman, and A. E. Merbach, *Inorg. Chem.* **19**, 3696 (1980).
3. F. K. Meyer, K. E. Newman, and A. E. Merbach, *J. Am. Chem. Soc.* **101**, 5588 (1979).
4. Y. Ducommun, W. L. Earl, and A. E. Merbach, *Inorg. Chem.* **18**, 2754 (1979).
5. Y. Ducommun, K. E. Newman, and A. E. Merbach, *Helv. Chim. Acta* **62**, 2511 (1979).
6. F. K. Meyer, K. E. Newman, and A. E. Merbach, *Inorg. Chem.* **18**, 2142 (1979).
7. Y. Yano, M. T. Fairhurst, and T. W. Swaddle, *Inorg. Chem.* **19**, 3267 (1980).
8. C. H. Langford, *Inorg. Chem.* **18**, 3288 (1979).
9. K. E. Newman and A. E. Merbach, *Inorg. Chem.* **19**, 2481 (1980).
10. T. W. Swaddle, *Inorg. Chem.* **19**, 3203 (1980).
11. J. Frahm and H. H. Füldner, *Ber. Bunsenges. Phys. Chem.* **84**, 173 (1980).
12. J. Frahm, *Ber. Bunsenges. Phys. Chem.* **84**, 754 (1980).
13. J. C. Thomas, C. M. Frey, and J. E. Stuehr, *Inorg. Chem.* **19**, 501 (1980).
14. T. Takahashi and T. Koiso, *Bull. Chem. Soc. Japan* **53**, 3400 (1980).
15. P. K. Chattopadhyay and B. Kratochvil, *Inorg. Chem.* **18**, 2953 (1979).
16. M. Tanaka, *Inorg. Chem.* **19**, 3205 (1980).
17. L. S. W. L. Sokol, T. D. Fink, and D. B. Rorabacher, *Inorg. Chem.* **19**, 1263 (1980).
18. M. N. Tkaczuk and S. F. Lincoln, *Aust. J. Chem.* **32**, 1915 (1979).
19. M. N. Tkaczuk and S. F. Lincoln, *Aust. J. Chem.* **33**, 2621 (1980).
20. C. Ammann, P. Moore, A. E. Merbach, and C. H. McAteer, *Helv. Chim. Acta* **63**, 268 (1980).
21. A. E. Merbach, P. Moore, O. W. Howarth, and C. H. McAteer, *Inorg. Chim. Acta* **39**, 129 (1980).
22. D. L. Pisaniello, S. F. Lincoln, and E. H. Williams, *J. Chem. Soc. Dalton Trans.*, 1473 (1979).
23. D. L. Pisaniello and S. F. Lincoln, *J. Chem. Soc. Dalton Trans.*, 699 (1980).
24. H. Gamsjäger, G. A. K. Thompson, W. Sagmüller, and A. G. Sykes, *Inorg. Chem.* **19**, 997 (1980).
25. G. J. Honan, S. F. Lincoln, and E. H. Williams, *Aust. J. Chem.* **32**, 1851 (1979).
26. S. F. Lincoln, *Pure Appl. Chem.* **51**, 2059 (1979).
27. Y. Ikeda, S. Soya, H. Fukutomi, and H. Tomiyasu, *J. Inorg. Nucl. Chem.* **41**, 1333 (1979).
28. B. G. Cox, I. Schneider, and H. Schneider, *Ber. Bunsenges. Phys. Chem.* **84**, 470 (1980).
29. B. G. Cox, D. Knop, and H. Schneider, *J. Phys. Chem.* **84**, 320 (1980).
30. R. Gresser, D. W. Boyd, A. M. Albrecht-Gary, and J. P. Schwing, *J. Am. Chem. Soc.* **102**, 651 (1980).
31. J. C. Thomas, C. M. Frey, and J. E. Stuehr, *Inorg. Chem.* **19**, 505 (1980).
32. A. D. Pacheco, F. Creazzola de O., J. D. Medina, V. de Santis, J. L. Calderon, *Rev. Latinoam. Quim.* **10**, 31–34 (1979); *Chem. Abs.* **91**, 63279x (1979).
33. S. Harada, M. Kawasawa, and T. Yasunaga, *Bull. Chem. Soc. Japan* **53**, 2074 (1980).
34. G. Arcoleo, F. P. Cavasino, E. Di Dio, and C. Sbriziolo, *J. Chem. Soc. Dalton Trans.*, 41 (1980).

35. P. J. Nichols and M. W. Grant, *Aust. J. Chem.* **32,** 1679 (1979).
36. K. Ohashi and H. Freiser, *Anal. Chem.* **52,** 767 (1980).
37. E. Mentasti, E. Pelizzetti, F. Secco, and M. Venturini, *Inorg. Chem.* **18,** 2007 (1979).
38. E. Mentasti, F. Secco, and M. Venturini, *Inorg. Chem.* **19,** 3528 (1980).
39. V. C. Reinsborough and B. H. Robinson, *J. Chem. Soc. Faraday Trans. 1* **75,** 2395 (1979).
40. S. Diekmann and J. Frahm, *J. Chem. Soc. Faraday Trans. 1* **75,** 2199 (1979).
41. M. J. Blandamer, J. Burgess, R. Sherry, and R. I. Haines, *J. Chem. Soc. Chem. Commun.,* 353 (1980).
42. T. J. Kemp, P. Moore, and G. R. Quick, *J. Chem. Soc. Dalton Trans.,* 1377 (1979).
43. A. Ekstrom, L. F. Lindoy, and R. J. Smith, *J. Amer. Chem. Soc.* **101,** 4014 (1979).
44. A. Ekstrom, L. F. Lindoy, and R. J. Smith, *Inorg. Chem.* **19,** 724 (1980).
45. R. W. Hay and P. R. Norman, *Inorg. Chim. Acta* **45,** L139 (1980).
46. M. J. Bain-Ackerman and D. K. Lavallee, *Inorg. Chem.* **18,** 3358 (1979).
47. A. N. Thompson and M. Krishnamurthy, *J. Inorg. Nucl. Chem.* **41,** 1251 (1979).
48. J. Turay and P. Hambright, *J. Inorg. Nucl. Chem.* **41,** 1385 (1979).
49. Y. Matsushima and S. Sugata, *Chem. Pharm. Bull.* **27,** 3049–3053 (1979); *Chem. Abs.* **92,** 100094b (1980).
50. J. M. T. Raycheba and D. W. Margerum, *Inorg. Chem.* **19,** 497 (1980).
51. J. H. Ferguson, K. Kustin, and A. Phipps, *Inorg. Chim. Acta* **43,** 49 (1980).
52. B. Perlmutter-Hayman, F. Secco, E. Tapuhi, and M. Venturini, *J. Chem. Soc. Dalton Trans.,* 1124 (1980).
53. B. Perlmutter-Hayman and E. Tapuhi, *Inorg. Chem.* **18,** 2872 (1979).
54. B. Monzyk and A. L. Crumbliss, *J. Am. Chem. Soc.* **101,** 6203 (1979).
55. M. J. Hynes and B. D. O'Regan, *J. Chem. Soc. Dalton Trans.,* 7 (1980).
56. T. M. Che and K. Kustin, *Inorg. Chem.* **19,** 2275 (1980).
57. S. Harada, Y. Funaki, and T. Yasunaga, *J. Am. Chem. Soc.* **102,** 136 (1980).
58. M. Kodama and E. Kimura, *J. Chem. Soc. Dalton Trans.,* 327 (1980).
59. M. Kodama and E. Kimura, *Inorg. Chem.* **19,** 1871 (1980).
60. R. B. Jordan and B. E. Erno, *Inorg. Chem.* **18,** 2895 (1979).
61. R. W. Renfrew, P. Osvath, and D. C. Weatherburn, *Aust. J. Chem.* **33,** 45 (1980).
62. J. H. Coates, P. R. Collins, and S. F. Lincoln, *Aust. J. Chem.* **33,** 1381 (1980).
63. A. Kumar and P. Neta, *J. Am. Chem. Soc.* **102,** 7284 (1980).
64. I. J. Ostrich, G. Liu, H. W. Dodgen, and J. P. Hunt, *Inorg. Chem.* **19,** 619 (1980).
65. R. I. Haines and A. McAuley, *Inorg. Chem.* **19,** 719 (1980).
66. C. G. Ekström, L. Nilsson, I. A. Duncan, and I. Grenthe, *Inorg. Chim. Acta* **40,** 91 (1980).
67. D. E. Hamilton, T. J. Lewis, and N. K. Kildahl, *Inorg. Chem.* **18,** 3364 (1979).
68. S. Balt, J. Meuldijk, and W. E. Renkema, *Inorg. Chim. Acta* **43,** 173 (1980).
69. A. Yamagishi, T. Masui, and F. Watanabe, *J. Phys. Chem.* **84,** 34 (1980).
70. H. Vanni and A. E. Merbach, *Inorg. Chem.* **18,** 2758 (1979).
71. E. H. Curzon, N. Herron, and P. Moore, *J. Chem. Soc. Dalton Trans.,* 574 (1980).
72. M. U. Fayyaz and M. W. Grant, *Aust. J. Chem.* **32,** 2159 (1979).
73. H. Elias, U. Fröhn, A. von Irmer, and K. J. Wannowius, *Inorg. Chem.* **19,** 869 (1980).

References for Chapter 7

1. J. K. Kochi, *Organometallic Mechanisms and Catalysis: the Role of Reactive Intermediates in Organic Processes,* Academic Press, New York (1979).

2. J. P. Collman and L. S. Hegedus, *Principles and Applications of Organotransition Metal Chemistry*, University Science Books, Mill Valley, California (1980).
3. J. K. Kochi, *Pure Appl. Chem.* **52**, 571 (1980).
4. J. Halpern, *Pure Appl. Chem.* **51**, 2171 (1979).
5. J. Halpern, F. T. T. Ng, G. L. Rempel, *J. Am. Chem. Soc.* **101**, 7124 (1979).
6. K. L. Brown, *J. Am. Chem. Soc.* **101**, 6600 (1979).
7. J. H. Grate and G. N. Schrauzer, *J. Am. Chem. Soc.* **101**, 4601 (1979).
8. J. H. Espenson and D. M. Wang, *Inorg. Chem.* **18**, 2853 (1979).
9. J. I. Byington, R. D. Peters, and L. O. Spreer, *Inorg. Chem.* **18**, 3324 (1979); M. J. Akhtar and L. O. Spreer, *Inorg. Chem.* **18**, 3327 (1979).
10. M. E. Vol'pin, I. Y. Levitin, A. L. Sigan, J. Halpern, and G. M. Tom, *Inorg. Chim. Acta* **41**, 271 (1980).
11. W. N. Rogers, J. A. Page, and M. C. Baird, *Inorg. Chem. Acta* **37**, L539 (1979).
12. F. Ozawa, T. Ito, and A. Yamamoto, *J. Am. Chem. Soc.* **102**, 6457 (1980).
13. P. Foley, R. Di Cosimo, and G. M. Whitesides, *J. Am. Chem. Soc.* **102**, 6713 (1980).
14. L. M. Dyagileva, V. P. Mar'in, E. I. Tsyganova, and G. A. Razuvaev, *J. Organomet. Chem.* **175**, 63 (1979).
15. J. R. Chipperfield, J. C. R. Sneyd, and D. E. Webster, *J. Organomet. Chem.* **178**, 177 (1979).
16. J. P. Fawcett, A. J. Poë, and K. R. Sharma, *J. Chem. Soc. Dalton Trans.*, 1886 (1979).
17. R. A. Jackson and A. J. Poë, *Inorg. Chem.* **18**, 3331 (1979).
18. F. Ungváry and L. Markó, *J. Organomet. Chem.* **193**, 383 (1980).
19. R. W. Wegman and T. L. Brown, *J. Am. Chem. Soc.* **102**, 2494 (1980).
20. N. P. Forbus, R. Oteiza, S. G. Smith, and T. L. Brown, *J. Organomet. Chem.* **193**, C71 (1980).
21. J. A. Stone, R. M. Camicioli, and M. C. Baird, *Inorg. Chem.* **19**, 3128 (1980).
22. J. M. T. Raycheba and G. Geier, *Inorg. Chem.* **18**, 2486 (1979).
23. L. J. Kirschenbaum and D. Meyerstein, *Inorg. Chem.* **19**, 1373 (1980).
24. T. Ramasami and J. H. Espenson, *Inorg. Chem.* **19**, 1523 (1980).
25. H. E. Bryndza, E. R. Evitt, and R. G. Bergmann, *J. Am. Chem. Soc.* **102**, 4948 (1980).
26. J. F. Endicott, K. P. Balakrishnan, and C. -L. Wong, *J. Am. Chem. Soc.* **102**, 5519 (1980).
27. G. J. Samuels and J. H. Espenson, *Inorg. Chem.* **19**, 233 (1980).
28. S. Fakuzumi and J. K. Kochi, *J. Am. Chem. Soc.* **102**, 7290 (1980).

References for Chapter 8

1. J. K. Kochi, *Organometallic Mechanisms and Catalysis: the Role of Reactive Intermediates in Organic Processes*, Academic Press, New York (1979).
2. J. P. Collman and L. S. Hegedus, *Principles and Applications of Organotransition Metal Chemistry*, University Science Books, Mill Valley, California (1980).
3. M. J. Wovkulich and J. D. Atwood, *J. Organomet. Chem.* **184**, 77 (1979); M. J. Wovkulich, S. J. Feinberg, and J. D. Atwood, *Inorg. Chem.* **19**, 2608 (1980).
4. D. J. Darensbourg, *Inorg. Chem.* **18**, 2821 (1979).
5. M. J. Blandamer, J. Burgess, and A. J. Duffield, *J. Organomet. Chem.* **175**, 293 (1979); J. Burgess and A. J. Duffield, *J. Organomet. Chem.* **177**, 435 (1979).
6. R. P. Stewart, *Inorg. Chem.* **18**, 2083 (1979).
7. D. Sonnenberger and J. D. Atwood, *J. Am. Chem. Soc.* **102**, 3484 (1980).
8. S. Aime, G. Gervasio, R. Rossetti, and P. L. Stanghellini, *Inorg. Chim. Acta* **40**, 131 (1980).

9. G. Bellachioma, G. Reichenbach, and G. Cardaci, *J. Chem. Soc. Dalton Trans.*, 634 (1980).
10. J. R. Fox, W. C. Gladfelter, and G. L. Geoffroy, *Inorg. Chem.* **19**, 2574 (1980).
11. R. Huq, A. J. Poë, and S. Chawla, *Inorg. Chim. Acta* **38**, 121 (1980).
12. D. J. Darensbourg and M. J. Incorvia, *Inorg. Chem.* **19**, 2585 (1980).
13. E. R. Evitt and R. G. Bergman, *J. Am. Chem. Soc.* **102**, 7003 (1980).
14. D. W. Parker, M. Marsi, and J. A. Gladysz, *J. Organometal. Chem.* **194**, C1 (1980).
15. K. J. Asali and G. R. Dobson, *J. Organometal. Chem.* **179**, 169 (1979).
16. E. L. Muetterties, J. R. Bleeke, and A. C. Sievert, *J. Organometal. Chem.* **178**, 197 (1979).
17. P. M. Burkinshaw, D. T. Dixon, and J. A. S. Howell, *J. Chem. Soc. Dalton Trans.*, 999, 2237 (1980).
18. S. Miya, K. Kashiwabara, and K. Saito, *Inorg. Chem.* **19**, 98 (1980).
19. T. T. Tsou, J. C. Huttman, and J. K. Kochi, *Inorg. Chem.* **18**, 2311 (1979).
20. M. Gower and L. A. P. Kane-Maquire, *Inorg. Chim. Acta* **37**, 79 (1979).
21. R. H. Crabtree and G. G. Alatky, *Inorg. Chem.* **19**, 571 (1980).
22. H. -U. Blaser and J. Halpern, *J. Am. Chem. Soc.* **102**, 1684 (1980).
23. S. H. Goh and L. Y. Goh, *J. Chem. Soc. Dalton Trans.*, 1641 (1980).
24. G. S. Samuels and J. H. Espenson, *Inorg. Chem.* **18**, 2587 (1979).
25. T. Ramasami and J. H. Espenson, *Inorg. Chem.* **19**, 1846 (1980).
26. J. A. Labinger and J. A. Osborn, *Inorg. Chem.* **19**, 3230 (1980); J. A. Labinger, J. A. Osborn, and N. J. Coville, *Inorg. Chem.* **19**, 3236 (1980).
27. T. L. Hall, M. F. Lappert, and P. W. Lednor, *J. Chem. Soc. Dalton Trans.*, 1448 (1980).
28. T. A. B. M. Bolsman and J. A. Van Doorn, *J. Organomet. Chem.* **178**, 381 (1979).
29. G. Butler, C. Eaborn, and A. Pidcock, *J. Organomet. Chem.* **181**, 47 (1979).
30. E. L. Weinberg and M. C. Baird, *J. Organomet. Chem.* **179**, C61 (1979).
31. G. Smith and J. K. Kochi, *J. Organomet. Chem.* **198**, 199 (1980); see also T. T. Tsou and J. K. Kochi, *J. Am. Chem. Soc.* **101**, 7547 (1979).
32. R. G. Pearson and P. E. Figdore, *J. Am. Chem. Soc.* **102**, 1541 (1980).
33. J. A. Connor and P. I. Riley, *J. Chem. Soc. Dalton Trans.*, 1318 (1979).
34. T. T. Tsou and J. K. Kochi, *J. Am. Chem. Soc.* **101**, 6319 (1979).
35. A. Puxeddu, G. Costa, and N. Marsich, *J. Chem. Soc. Dalton Trans.*, 1489 (1980).
36. R. J. Klinger, K. Mochida, and J. K. Kochi, *J. Am. Chem. Soc.* **101**, 6626 (1979).
37. E. V. Blackbrun and D. D. Turner, *J. Am. Chem. Soc.* **102**, 692 (1980).
38. H. R. Rogers, G. L. Hill, Y. Fujiwara, R. J. Rogers, H. L. Mitchell, and G. M. Whitesides, *J. Am. Chem. Soc.* **102**, 217 (1980).
39. H. R. Rogers, J. Deutch, and G. M. Whitesides, *J. Am. Chem. Soc.* **102**, 226 (1980).
40. H. R. Rogers, R. J. Rogers, H. L. Mitchell, and G. M. Whitesides, *J. Am. Chem. Soc.* **102**, 231 (1980).
41. J. J. Barber and G. M. Whitesides, *J. Am. Chem. Soc.* **102**, 239 (1980).
42. G. B. Sergeev, V. V. Smirnov, and V. V. Zagorsky, *J. Organomet. Chem.* **201**, 9 (1980).
43. E. C. Ashby, *Pure Appl. Chem.* **52**, 545 (1980).
44. G. Kramer, J. Patterson, and A. J. Poë, *J. Chem. Soc. Dalton Trans.*, 1165 (1979).
45. G. Kramer, J. Patterson, A. Poë, and L. Ng, *Inorg. Chem.* **19**, 1161 (1980).
46. T. Kohara, T. Yamamoto, and A. Yamamato, *J. Organomet. Chem.* **192**, 265 (1980).
47. R. J. Al-Essa, R. J. Puddephatt, P. J. Thompson, and C. F. H. Tipper, *J. Am. Chem. Soc.* ·**102**, 7546 (1980).
48. R. J. Puddephatt, P. J. Thompson, and C. F. H. Tipper, *J. Organomet. Chem.* **177**, 403 (1979).
49. E. J. Kuhlmann and J. J. Alexander, *Coord. Chem. Rev.* **33**, 195 (1980).
50. C. P. Casey and D. M. Scheck, *J. Am. Chem. Soc.* **102**, 2723 (1980).
51. D. Saddei, H. J. Freund, and G. Hohlneicher, *J. Organomet. Chem.* **186**, 63 (1980).

52. H. Des Abbayes and A. Buloup, *J. Organomet. Chem.* **198,** C36 (1980).
53. S. B. Butts, S. H. Strauss, E. M. Holt, R. E. Stimson, N. W. Alcock, and D. F. Shriver, *J. Am. Chem. Soc.* **101,** 5864 (1979); **102,** 5093 (1980).
54. F. Correa, R. Nakamura, R. E. Stimson, R. C. Burwell, and D. F. Shriver, *J. Am. Chem. Soc.* **102,** 5112 (1980).
55. C. F. J. Barnard, J. A. Daniels, and R. J. Mawby, *J. Chem. Soc. Dalton Trans.,* 1331 (1979).
56. G. K. Anderson and R. J. Cross, *J. Chem. Soc. Dalton Trans.,* 1246 (1979).
57. G. K. Anderson and R. J. Cross, *J. Chem. Soc. Dalton Trans.,* 712 (1980).
58. R. J. Puddephatt and M. A. Stalteri, *J. Organomet. Chem.* **193,** C27 (1980).
59. R. G. Severson, T. W. Leugn, and A. Wojcicki, *Inorg. Chem.* **19,** 915–923 (1980).
60. H. Fischer, *J. Organomet. Chem.* **195,** 55 (1980).
61. H. Fischer, *J. Organomet. Chem.* **197,** 303 (1980).
62. R. Ugo, G. M. Zanderighi, A. Fusi, and D. Carreri, *J. Am. Chem. Soc.* **102,** 3745 (1980).

References for Chapter 9

1. (a) R. Romeo, D. Minniti, and S. Lanza, *Inorg. Chem.* **19,** 3663 (1980); (b) H. Kelm, W. J. Louw, and D. A. Palmer, *Inorg. Chem.* **19,** 843 (1980); (c) W. J. Louw, R. van Eldik, and H. Kelm, *Inorg. Chem.* **19,** 2878 (1980).
2. L. L. Costanzo, S. Giuffrida, and R. Romeo, *Inorg. Chim. Acta* **38,** 31 (1980).
3. H. Kurosawa and S. Numata, *J. Organomet. Chem.* **175,** 143 (1979).
4. T. Yoshida, D. L. Thorn, T. Okano, S. Otsuka, and J. A. Ibers, *J. Am. Chem. Soc.* **102,** 6451 (1980).
5. K. D. Tau, R. Uriarte, T. J. Mazanec, and D. W. Meek, *J. Am. Chem. Soc.* **101,** 6614 (1979).
6. T. J. Mazanec, K. D. Tau, and D. W. Meek, *Inorg. Chem.* **19,** 85 (1980).
7. D. J. Darensbourg and B. J. Baldwin, *J. Am. Chem. Soc.* **101,** 6447 (1979).
8. W. A. Schenk, *J. Organomet. Chem.* **184,** 195 (1980).
9. D. A. Clark, D. L. Jones, and R. J. Mawby, *J. Chem. Soc. Dalton Trans.,* 565 (1980).
10. T. Tsuihiji, T. Akiyama, and A. Sugimori, *Bull. Chem. Soc. Jpn.* **52,** 3451 (1979).
11. F. A. Van-Catledge, S. D. Ittel, C. A. Tolman, and J. P. Jesson, *J. Chem. Soc. Chem. Commun.,* 254 (1980).
12. H. W. Choi, R. M. Gavin, and E. L. Muetterties, *J. Chem. Soc. Chem. Commun.,* 1085 (1979).
13. S. S. Wreford, J. K. Kouba, J. F. Kirner, E. L. Muetterties, I. Tavanaiepour, and V. W. Day, *J. Am. Chem. Soc.* **102,** 1558 (1980).
14. U. Puttfarcken and D. Rehder, *J. Organomet Chem.* **185,** 219 (1980).
15. D. J. Bevan and R. J. Mawby, *J. Chem. Soc. Dalton Trans.,* 1904 (1980).
16. B. J. Brisdon, D. A. Edwards, K. E. Paddick, and M. G. B. Drew, *J. Chem. Soc. Dalton Trans.,* 1317 (1980).
17. J. R. Briggs, C. Crocker, W. S. McDonald, and B. L. Shaw, *J. Organomet. Chem.,* **181,** 213 (1979).
18. T. A. Albright, R. Hoffman, J. C. Thibeault, and D. L. Thorn, *J. Am. Chem. Soc.* **101,** 3801 (1979).
19. R. M. Pitzer and H. F. Schaefer, *J. Am. Chem. Soc.* **101,** 7176 (1979).
20. K. Kawakami and M. Okajima, *J. Inorg. Nucl. Chem.* **41,** 1501 (1979).
21. K. Sato, K. Kawakami, and T. Tanaka, *Inorg. Chem.* **18,** 1532 (1979).
22. D. L. Reger and C. J. Coleman, *Inorg. Chem.* **18,** 3270 (1979).

23. F.-W. Grevels, M. Lindermann, R. Benn, R. Goddard, and C. Krüger, *Z. Naturforsch.* **35b,** 1298 (1980).

24. B. C. Ward and J. L. Templeton, *J. Am. Chem. Soc.* **102,** 1532 (1980).

25. G. Erker and F. Rosenfeldt, *J. Organomet. Chem.* **188,** C1 (1980).

26. M. Brookhart, J. R. Tucker, T. C. Flood, and J. Jensen, *J. Am. Chem. Soc.* **102,** 1203 (1980).

27. P. K. Baker and N. G. Connelly, *J. Organomet. Chem.* **178,** C33 (1979).

28. P. V. Rinze and U. Müller, *Chem. Ber.* **112,** 1973 (1979).

29. F. A. Van-Catledge, S. D. Ittel, and J. P. Jesson, *J. Organomet. Chem.* **168,** C25 (1979).

30. A.-R. Al-Ohaly and J. F. Nixon, *J. Organomet. Chem.* **202,** 297 (1980).

31. M. Moll, H. Behrens, and W. Popp, *Z. Anorg. Allg. Chem.* **458,** 202 (1979).

32. M. Moll, H.-J. Seibold, and W. Popp, *J. Organomet Chem.* **191,** 193 (1980).

33. J.-M. Bassett, M. Green, J. A. K. Howard, and F. G. A. Stone, *J. Chem. Soc. Dalton Trans.,* 1779 (1980).

34. R. E. Davis and P. E. Riley, *Inorg. Chem.* **19,** 674 (1980).

35. J. L. Davidson, *J. Chem. Soc. Chem. Commun.,* 113 (1980).

36. S. Carter and J. N. Murrell, *J. Organomet. Chem.* **192,** 399 (1980).

37. P. Seiler and J. D. Dunitz, *Acta Crystallogr.* **B35,** 1068, 2000 (1979).

38. A. Haaland, *Acc. Chem. Res.* **12,** 415 (1979).

39. F. Takusagawa and T. F. Koetzle, *Acta Crystallogr.* **B35,** 1074 (1979).

40. A. Almenningen, A. Haaland, S. Samdal, J. Brunvoll, J. L. Robbins, and J. C. Smart, *J. Organomet. Chem.* **173,** 293 (1979).

41. S. Sorriso, *J. Organomet. Chem.* **179,** 205 (1979).

42. A. Eisenberg, A. Shaver, and T. Tsutsui, *J. Am. Chem. Soc.* **102,** 1416 (1980).

43. J. W. Faller, D. F. Ghodosh, and D. Katahira, *J. Organomet. Chem.* **187,** 227 (1980).

44. J. W. Faller and Y. Shvo, *J. Am. Chem. Soc.* **102,** 5396 (1980).

45. R. Benn and E. G. Hoffman, *J. Organomet Chem.* **193,** C33 (1980).

46. M. W. Schoonover, E. C. Baker, and R. Eisenberg, *J. Am. Chem. Soc.* **101,** 1880 (1979).

47. B. Henc, P. W. Jolly, R. Salz, G. Wilke, R. Benn, E. G. Hoffman, R. Mynott, G. Schroth, K. Seevogel, J. C. Sekutowski, and C. Krüger, *J. Organomet. Chem.* **191,** 425 (1980).

48. B. Henc, P. W. Jolly, R. Salz, S. Stobbe, G. Wilke, R. Benn, R. Mynott, K. Seevogel, R. Goddard, and C. Krüger, *J. Organomet. Chem.* **191,** 449 (1980).

49. H. Werner and H.-J. Kraus, *J. Chem. Soc. Chem. Commun.,* 814 (1979); *Angew. Chem. Int. Ed. Eng.* **18,** 948 (1979).

50. H. Werner and A. Kühn, *Angew. Chem. Int. Ed. Eng.* **18,** 416 (1979).

51. H.-O. Stühler, *Angew. Chem. Int. Ed. Eng.* **19,** 468 (1980).

52. R. Benn, B. Bogdanovic, P. Goettsch, and K. Schlichte, *Z. Naturforsch.* **35b,** 200 (1980).

53. G. Erker, J. Wicher, K. Engel, F. Rosenfeldt, W. Dietrich, and C. Krüger, *J. Am. Chem. Soc.* **102,** 6344 (1980).

54. J. S. Fredericksen, R. E. Graf, D. G. Gresham, and C. P. Lillya, *J. Am. Chem. Soc.* **101,** 3863 (1979).

55. J. A. Gibson and B. E. Mann, *J. Chem. Soc. Dalton Trans.,* 1021 (1979).

56. M. A. Bennett, T. W. Matheson, G. B. Robertson, A. K. Smith, and P. A. Tucker, *Inorg. Chem.* **19,** 1014 (1980).

57. C. P. Lewis, W. Kitching, A. Eisenstadt, and M. Brookhart, *J. Am. Chem. Soc.* **101,** 4896 (1979).

58. K. S. Chang and A. Storr, *Can. J. Chem.* **58,** 2278 (1980).

59. D. M. Heinekey and W. A. G. Graham, *J. Am. Chem. Soc.* **101,** 6115 (1979).

60. D. M. P. Mingos and C. R. Nurse, *J. Organomet. Chem.* **184,** 281 (1980).

61. B. E. Mann, A. Keasey, A. Sonoda, and P. M. Maitlis, *J. Chem. Soc. Dalton Trans.,* 338 (1979).

62. S. P. Solodovnikov, A. N. Nesmayanov, N. A. Vol'kenau, and L. S. Kotova, *J. Organomet. Chem.* **201,** 447 (1980).

63. H. Werner and R. Feser, *Angew. Chem. Int. Ed. Eng.* **18,** 157 (1979).

64. J. Müller, H. Menig, and P. V. Rinze, *J. Organomet. Chem.* **181,** 387 (1979).

65. S. D. Ittel, F. A. Van-Catledge, and J. P. Jesson, *J. Am. Chem. Soc.* **101,** 6905 (1979); R. L. Harlow, R. J. McKinney, and S. D. Ittel, *J. Am. Chem. Soc.* **101,** 7496 (1979); R. K. Brown, J. M. Williams, A. J. Schultz, G. D. Stucky, S. D. Ittel, and R. L. Harlow, *J. Am. Chem. Soc.* **102,** 981 (1980).

66. C. Crocker, R. J. Errington, W. S. McDonald, K. J. Odell, B. L. Shaw, and R. J. Goodfellow, *J. Chem. Soc. Chem. Commun.,* 498 (1979); C. Crocker, R. J. Errington, R. Markham, C. J. Moulton, K. J. Odell, and B. L. Shaw, *J. Am. Chem. Soc.* **102,** 4373 (1980).

67. W. Lamanna and M. Brookhart, *J. Am. Chem. Soc.* **102,** 3490 (1980).

68. S. G. Davies, H. Felkin, and O. Watts, *J. Chem. Soc. Chem. Commun.,* 1980, 159.

69. W. R. Roper, G. E. Taylor, J. M. Waters, and L. J. Wright, *J. Organomet. Chem.* **182,** C46 (1979).

70. M. A. Bennett and J. C. Jeffery, *Inorg. Chem.* **19,** 3763 (1980).

71. M. Brookhart, J. R. Tucker, and R. G. Husk, *J. Organomet. Chem.* **193,** C23 (1980).

72. M. D. Rausch, G. F. Westover, E. Mintz, G. M. Reisner, I. Bernal, A. Clearfield and J. M. Troup, *Inorg. Chem.* **18,** 2605 (1979).

73. P. Hofman and T. A. Albright, *Angew. Chem. Int. Ed. Eng.* **19,** 728 (1980).

74. A. L. Burrows, K. Hine, B. F. G. Johnson, J. Lewis, D. G. Parker, A. Poe, and E. J. S. Vichi, *J. Chem. Soc. Dalton Trans.,* 1135 (1980).

75. W. E. Carroll, M. Green, A. M. R. Galas, M. Murray, T. W. Turney, A. J. Welch, and P. Woodward, *J. Chem. Soc. Dalton Trans.,* 80 (1980).

76. J. A. S. Howell and P. Mathur, *J. Organomet. Chem.* **174,** 335 (1979); J. A. S. Howell and A. J. Rowan, *J. Chem. Soc. Dalton Trans.,* 503 (1980).

77. C. E. Sumner, P. E. Riley, R. E. Davis, and R. Pettit, *J. Am. Chem. Soc.* **102,** 1752 (1980).

78. N. M. Boag, M. Green, R. M. Mills, G. N. Pain, F. G. A. Stone, and P. Woodward, *J. Chem. Soc. Chem. Commun.,* 1171 (1980).

79. H. H. Wang and L. H. Pignolet, *Inorg. Chem.* **19,** 1470 (1980).

80. N. W. Alcock, O. W. Howarth, P. Moore, and G. E. Morris, *J. Chem. Soc. Chem. Commun,* 1160 (1979).

81. M. D. Curtis, K. R. Han, and W. M. Butler, *Inorg. Chem.* **19,** 2096 (1980).

82. J. A. Beck, S. A. R. Knox, G. H. Riding, G. E. Taylor, and M. J. Winter, *J. Organomet. Chem.* **202,** C49 (1980).

83. K. I. Gell, G. M. Williams, and J. Schwartz, *J. Chem. Soc. Chem. Commun.,* 550 (1980).

84. S. Aime, L. Milone, E. Sappa, A. Tiripicchio, and M. Tiripicchio Camellini, *J. Chem. Soc. Dalton Trans.,* 1155 (1979).

85. A. F. Dyke, S. A. R. Knox, P. J. Naish, and G. E. Taylor, *J. Chem. Soc. Chem. Commun.,* 409 (1980).

86. L. Messerle and M. D. Curtis, *J. Am. Chem. Soc.* **102,** 7789 (1980).

87. M. H. Chisholm and I. P. Rothwell, *J. Am. Chem. Soc.* **102,** 5950 (1980).

88. C. Brown, B. T. Heaton, A. D. C. Towl, P. Chini, A. Fumagalli, and G. Longoni, *J. Organomet. Chem.* **181,** 233 (1979).

89. M. J. Mays and P. D. Gavens, *J. Chem. Soc. Dalton Trans.,* 911 (1980).

90. S. Aime, L. Milone, R. Rossetti, and P. L. Stanghellini, *J. Chem. Soc. Dalton Trans.,* 46 (1980).

91. P. A. Dimas, E. N. Duesler, R. J. Lawson, and J. R. Shapley, *J. Am. Chem. Soc.* **102,** 7787 (1980).

92. W. A. Herrmann, J. Plank, E. Guggolz, and M. L. Ziegler, *Angew. Chem. Int. Ed.* **19**, 651 (1980).
93. B. T. Heaton, L. Longhetti, L. Garleschelli, U. Sartorelli, *J. Organomet. Chem.* **192**, 431 (1980).
94. A. Fumagalli, T. F. Koetzle, F. Takusagawa, P. Chini, S. Martinengo, and B. T. Heaton, *J. Am. Chem. Soc.* **102**, 1740 (1980).
95. B. T. Heaton, C. Brown, D. O. Smith, L. Strona, R. J. Goodfellow, P. Chini, and S. Martinengo, *J. Am. Chem. Soc.* **102**, 6175 (1980).
96. G. Longoni, B. T. Heaton, and P. Chini, *J. Chem. Soc. Dalton Trans.*, 1537 (1980).
97. W. A. Herrmann, M. L. Ziegler, K. Weidenhammer, and H. Biersack, *Angew. Chem.* **91**, 1026 (1979).
98. L. N. Lewis and K. G. Caulton, *Inorg. Chem.* **19**, 3201 (1980).
99. M. R. Churchill, R. A. Lashewycz, J. R. Shapley, and S. I. Richter, *Inorg. Chem.* **19**, 1277 (1980).
100. B. F. G. Johnson, J. Lewis, A. G. Orpen, P. R. Raithby, and G. Suess, *J. Organomet. Chem.* **173**, 187 (1979).
101. S. C. Brown, J. Evans, and L. E. Smart, *J. Chem. Soc. Chem. Commun.*, 1021 (1980).
102. S. A. R. Knox, R. J. McKinney, V. Riera, F. G. A. Stone, and A. C. Szary, *J. Chem. Soc. Dalton Trans.*, 1801 (1979); J. A. K. Howard, R. F. D. Stansfield, and P. Woodward, *J. Chem. Soc. Dalton Trans.*, 1812 (1979).

References for Chapter 10

1. H. Fischer, *J. Organomet. Chem.* **197**, 303 (1980).
2. H. Fischer and K. H. Dötz, *Chem. Ber.* **113**, 193 (1980).
3. A. G. Constable and J. A. Gladysz, *J. Organomet. Chem.* **202**, C21 (1980).
4. N. Calderon, J. P. Lawrence, and E. A. Ofstead, *Adv. Organomet. Chem.* **17**, 449 (1979).
5. T. J. Katz and S. J. Lee, *J. Am. Chem. Soc.* **102**, 422 (1980).
6. T. Masuda, N. Sasaki, and T. Higashimura, *Macromolecules* **8**, 717 (1975).
7. E. L. Weinberg and M. C. Baird, *J. Organomet. Chem.* **179**, C61 (1979).
8. Yu. A. Alexandrov, S. A. Lebedev, N. V. Kuznetsova, and G. A. Razuvaev, *J. Organomet. Chem.* **177**, 91 (1979).
9. K. Yasufuku, A. Hamada, K. Aoki, and H. Yamazaki, *J. Am. Chem. Soc.* **102**, 4363 (1980).
10. I. M. Al-Najjar and M. Green, *J. Chem. Soc. Dalton Trans.*, 1651 (1979).
11. B. S. Tovrog, F. Mares, and S. E. Diamond, *J. Am. Chem. Soc.* **102**, 6616 (1980).
12. J. Halpern and R. A. Jewsbury, *J. Organomet. Chem.* **181**, 223 (1979).
13. L. Cosslett and L. A. P. Kane-Maguire, *J. Organomet. Chem.* **178**, C17 (1979).
14. P. J. Lennon, A. Rosan, M. Rosenblum, J. Tancrede, and P. Waterman, *J. Am. Chem. Soc.* **102**, 7033 (1980).
15. (a) B. M. Trost, L. Weber, P. E. Strege, T. J. Fullerton, and T. J. Dietsche, *J. Am. Chem. Soc.* **100**, 3416 (1978); (b) B. M. Trost and E. Keinan, *J. Am. Chem. Soc.* **100**, 7779 (1978), and references cited therein.
16. B. Åkermark, J.-E. Bäckvall, A. Löwenborg, and K. Zetterberg, *J. Organomet. Chem.* **166**, C33 (1979), and references cited therein.
17. J.-E. Bäckvall, R. E. Nordberg, E. E. Björkman, and C. Moberg, *J. Chem. Soc. Chem. Commun.*, 943 (1980).
18. B. M. Trost, T. R. Verhoeven, and J. M. Fortunak, *Tetrahedron Lett.*, 2301 (1979).
19. D. A. Clark, D. L. Jones, and R. J. Mawby, *J. Chem. Soc. Dalton Trans.*, 565 (1980).

20. G. Cardaci, *J. Organomet. Chem.* **202,** C81 (1980).
21. S. Sakaki, M. Nishikawa, and A. Ohyoshi, *J. Am. Chem. Soc.* **102,** 4062 (1980).
22. A. J. Pearson, *Acc. Chem. Res.* **13,** 463 (1980).
23. A. L. Burrows, B. F. G. Johnson, J. Lewis, and D. G. Parker, *J. Organomet. Chem.* **194,** C11 (1980).
24. A. L. Burrows, K. Hine, B. F. G. Johnson, J. Lewis, D. G. Parker, A. Poë, and E. J. S. Vichi, *J. Chem. Soc. Dalton Trans.,* 1135 (1980).
25. D. A. Brown, W. K. Glass, and F. M. Hussein, *J. Organomet. Chem.* **186,** C58 (1980).
26. R. S. Bayoud, E. R. Biehl, and P. C. Reeves, *J. Organomet. Chem.* **174,** 297 (1979).
27. L. A. P. Kane-Maguire, T. Odiaka, S. Turgoose, and P. A. Williams, *J. Organomet. Chem.* **188,** C5 (1980).
28. G. R. John and L. A. P. Kane-Maguire, *J. Chem. Soc. Dalton Trans.,* 1196 (1979).
29. T. I. Odiaka and L. A. P. Kane-Maguire, *Inorg. Chim. Acta* **37,** 85 (1979).
30. M. Gower, G. R. John, L. A. P. Kane-Maguire, T. I. Odiaka, and A. Salzer, *J. Chem. Soc. Dalton Trans.,* 2003 (1979).
31. L. F. Kelly, A. S. Narula, and A. J. Birch, *Tetrahedron Lett.,* 4107 (1979).
32. A. J. Birch, W. D. Raverty, and G. R. Stephenson, *Tetrahedron Lett.,* 197 (1980).
33. A. J. Birch, W. D. Raverty, and G. R. Stephenson, *J. Chem. Soc. Chem. Commun.,* 857 (1980).
34. P. J. Domaille, S. D. Ittel, J. P. Jesson, and D. A. Sweigart, *J. Organomet. Chem.* **202,** 191 (1980).
35. D. A. Sweigart, *J. Chem. Soc. Chem. Commun.,* 1159 (1980).
36. A. C. Knipe, S. J. McGuinness, and W. E. Watts, *J. Chem. Soc., Chem. Commun.,* 842 (1979).
37. M. F. Semmelhack, H. T. Hall, R. Farina, M. Yoshifuji, G. Clark, T. Bargar, K. Hirotsu, and J. Clardy, *J. Am. Chem. Soc.* **101,** 3535 (1979).
38. T. A. Albright and B. K. Carpenter, *Inorg. Chem.* **19,** 3092 (1980).
39. D. W. Clack and L. A. P. Kane-Maguire, *J. Organomet. Chem.* **174,** 199 (1979).
40. A. Solladie-Cavallo and E. Tsamo, *J. Organomet. Chem.* **172,** 165 (1979).
41. P. L. Pauson, *J. Organomet. Chem.* **200,** 207 (1980).
42. J. W. Faller, *Inorg. Chem.* **19,** 2859 (1980).
43. J. G. Atton, L. A. Hassan, and L. A. P. Kane-Maguire, *Inorg. Chim. Acta* **41,** 245 (1980).
44. C. A. Bunton, N. Carrasco, F. Davoudzadeh, and W. E. Watts, *J. Chem. Soc. Perkin Trans. 2,* 1520 (1980).
45. C. A. Bunton, N. Carrasco, and W. E. Watts, *J. Chem. Soc. Perkin Trans. 2,* 1267 (1979).
46. N. Cully and W. E. Watts, *J. Organomet. Chem.* **182,** 99 (1979).
47. K. L. Brown and A. W. Awtrey, *J. Organomet. Chem.* **195,** 113 (1980).
48. K. J. Asali and G. R. Dobson, *J. Organomet. Chem.* **179,** 169 (1979).
49. M. Rosenblum and P. S. Waterman, *J. Organomet. Chem.* **187,** 267 (1980).
50. Yu. T. Struchkov, V. G. Andrianov, V. N. Setkina, N. K. Baranetskaya, V. I. Losilkina, and D. N. Kursanov, *J. Organomet. Chem.* **182,** 213 (1979).
51. G. Kramer, J. Patterson, and A. Poë, *J. Chem. Soc. Dalton Trans.,* 1165 (1979).
52. G. Kramer, J. Patterson, A. Poë, and L. Ng, *Inorg. Chem.* **19,** 1161 (1980).

References for Chapter 11

1. F. G. A. Stone and R. West, *Advances in Organometallic Chemistry,* Vol. 18, Academic Press, London (1980).

2. R. J. Kazlauskas and M. S. Wrighton, *J. Am. Chem. Soc.* **102,** 1727 (1980).
3. C. L. Reichel and M. S. Wrighton, *J. Am. Chem. Soc.* **101,** 6769 (1979).
4. K. J. Ivin, J. J. Rooney, C. D. Stewart, M. L. H. Green, and R. Mahtab, *J. Chem. Soc. Chem. Commun.,* 604 (1978).
5. E. R. Evitt and R. G. Bergman, *J. Am. Chem. Soc.* **101,** 3973 (1980).
6. E. R. Evitt and R. G. Bergman, *J. Am. Chem. Soc.* **102,** 7003 (1980).
7. A. S. C. Chan and J. Halpern, *J. Am. Chem. Soc.* **102,** 838 (1980).
8. W. D. Jones and R. G. Bergman, *J. Am. Chem. Soc.* **101,** 5447 (1979).
9. D. L. Thorn, *J. Am. Chem. Soc.* **102,** 7109 (1980).
10. R. C. Brady and R. Petit, *J. Am. Chem. Soc.* **102,** 618 (1980).
11. J. S. Bradley, G. B. Ansell, and E. W. Hill, *J. Am. Chem. Soc.* **101,** 7419 (1979).
12. K. Whitmire and D. F. Shriver, *J. Am. Chem. Soc.* **102,** 1456 (1980).
13. H. W. Walker, C. T. Kresge, P. C. Ford, and R. G. Pearson, *J. Am. Chem. Soc.* **101,** 7428 (1979).
14. M. B. Mooiman and J. M. Pratt, *J. Chem. Soc. Chem. Commun.,* 33 (1981).
15. Lyon Symposium on Metathesis, *J. Mol. Cat.* **8,** (1980).
16. H. Rudler, *J. Mol. Cat.* **8,** 53 (1980).
17. A. F. Dyke, S. A. R. Knox, P. J. Naish, and G. E. Taylor, *J. Chem. Soc. Chem. Commun.,* 803 (1980).
18. T. J. Katz, S. J. Lee, M. Nair, and E. B. Savage, *J. Am. Chem. Soc.* **102,** 7940 (1980).
19. T. J. Katz and S. J. Lee, *J. Am. Chem. Soc.* **102,** 422 (1980).
20. C. P. Casey and A. J. Shusterman, *J. Mol. Cat.* **8,** 1 (1980).
21. E. L. Muetterties and E. Band, *J. Am. Chem. Soc.* **102,** 6572 (1980).
22. J. R. M. Kress, M. J. M. Russell, M. G. Wesolek, and J. A. Osborn, *J. Chem. Soc. Chem. Commun.,* 431 (1980).
23. A. K. Rappe and W. A. Goddard, *J. Am. Chem. Soc.* **102,** 5114 (1980).
24. G. A. Rupprecht, L. W. Messerle, J. D. Fellmann, and R. R. Schrock, *J. Am. Chem. Soc.* **102,** 6236 (1980).
25. R. R. Schrock, S. Rocklage, J. Wengrovius, G. Rupprecht, and J. Fellman, *J. Mol. Cat.* **8,** 73 (1980).
26. J. H. Wengrovius, R. R. Schrock, M. R. Churchill, J. R. Missert and W. J. Youngs, *J. Am. Chem. Soc.* **102,** 4515 (1980).
27. S. M. Rocklage, J. D. Fellmann, G. A. Rupprecht, L. W. Messerle, and R. R. Schrock, *J. Amer. Chem. Soc.* **103,** 1440 (1981).
28. G. J. A. Adam, S. G. Davies, I. C. A. Ford, M. Ephritikhine, P. F. Todd, and M. L. H. Green, *J. Mol. Cat.* **8,** 15 (1980).
29. F. N. Tebbe, G. W. Parshall, and D. W. Ovenall, *J. Am. Chem. Soc.* **101,** 5074 (1979).
30. T. R. Howard, J. B. Lee, and R. H. Grubbes, *J. Am. Chem. Soc.* **102,** 6877 (1980).
31. U. Klabunde, F. N. Tebbe, G. W. Parshall, and R. L. Harlow, *J. Mol. Cat.* **8,** 37 (1980).
32. T. J. Katz, E. B. Savage, C. J. Lee, and M. Nair, *J. Am. Chem. Soc.* **102,** 7942 (1980).
33. L. Bencze, K. J. Ivin, and J. J. Rooney, *J. Chem. Soc. Chem. Commun.,* 830 (1980).
34. M. L. Econte and J. M. Bassett, *J. Am. Chem. Soc.* **101,** 7296 (1979).
35. T. J. Katz, S. J. Lee, and M. A. Shippey, *J. Mol. Cat.* **8,** 219 (1980).
36. K. J. Ivin, G. L. Apienis, J. J. Rooney, and C. D. Stewart, *J. Mol. Cat.* **8,** 203 (1980).
37. B. R. James, *Homogeneous Hydrogenation,* Wiley–Interscience, Chichester, England (1973).
38. J. Cole, R. Ramage, K. Cann, and R. Pettit, *J. Am. Chem. Soc.* **102,** 6182 (1980).
39. E. Samuel, *J. Organometal. Chem.* **198,** C65 (1980).
40. S. Komiya and A. Yamamoto, *J. Mol. Cat.* **5,** 279 (1979).
41. R. Crabtree, *Ace. Chem. Res.* **12,** 331 (1979).

42. M. S. Delaney, C. B. Knobler, and M. F. Hawthorne, *J. Chem. Soc. Chem. Commun.*, 849 (1980).
43. J. A. Davies, F. R. Hartley, and S. G. Murray, *J. Chem. Soc. Dalton Trans.*, 2246 (1980).
44. C. U. Pittman, R. C. Ryan, J. McGee, and J. P. O'Connor, *J. Organometal Chem.* **178**, C43 (1979).
45. M. Airoldi, G. Deganello, G. Dia, and G. Gennard, *J. Organometal. Chem.* **187**, 391 (1980).
46. K. Kawakami, T. Mizoroki, and A. Ozaki, *J. Mol. Cat.* **5**, 175 (1979.
47. R. G. Nuzzo, D. Feitler, and G. M. Whitesides, *J. Am. Chem. Soc.* **102**, 3683 (1980).
48. S. Vastag, B. Heil, and L. Marko, *J. Mol. Cat.* **5**, 189 (1979).
49. R. A. Sanchez-Delgado and O. L. De Ocaoa, *J. Organometal. Chem.* **202**, 427 (1980).
50. M. Bianchi, G. Menchi, F. Francalanci, F. Piacenti, V. Matteoli, P. Fradiani, and C. Botteghi, *J. Organometal. Chem.* **188**, 109 (1980).
51. R. A. Grey, G. P. Pez, A. Wallo, and J. Corsi, *J. Chem. Soc. Chem. Commun.*, 783 (1980).
52. T. Yoshida, T. Okano, and S. Otsuka, *J. Chem. Soc. Chem. Commun.*, 870 (1979).
53. R. C. Ryan, G. M. Wilemon, M. P. Dalsanto, and C. U. Pittman, *J. Mol. Cat.* **5**, 319 (1979).
54. B. R. James and D. K. W. Wang, *Can. J. Chem.* **58**, 245 (1980).
55. A. S. C. Chan, J. J. Pluth, and J. Halpern, *J. Am. Chem. Soc.* **102**, 5952 (1980).
56. J. M. Brown, P. A. Chaloner, G. Descotes, R. Glaser, D. Lafont, and D. Sinoll, *J. Chem. Soc. Chem. Commun.*, 611 (1979).
57. J. M. Brown and P. A. Chaloner, *J. Chem. Soc. Chem. Commun.*, 613 (1979).
58. J. M. Brown and P. A. Chaloner, *J. Am. Chem. Soc.* **102**, 3040 (1980).
59. J. M. Brown and P. A. Chaloner, *J. Chem. Soc. Chem. Commun.*, 344 (1980).
60. W. C. Christopfel and B. D. Vineyard, *J. Am. Chem. Soc.* **101**, 4406 (1979).
61. J. M. Brown and D. Parker, *J. Chem. Soc. Chem. Commun.*, 342 (1980).
62. K. Onuma, T. Ito, and A. Nakamura, *Chem. Lett.* 905 (1979).
63. H. Brunner and W. Pieronczyk, *Angew. Chem. Int. Eng. Ed.* **18**, 620 (1979).
64. M. Fiorini and G. M. Giongo, *J. Mol. Cat.* **7**, 411 (1980).
65. A. Miyashita, A. Yasuda, H. Taraya, K. Toriumi, T. Ito, T. Souchi, and R. Noyori, *J. Am. Chem. Soc.* **102**, 7932 (1980).
66. W. R. Cullen, F. W. B. Einstein, C. H. Huang, A. C. Willis, and E-S Yeh, *J. Am. Chem. Soc.* **102**, 988 (1980).
67. R. Dabard, G. Jaoven, G. Simonnehux, M. Cais, D. H. Kohn, A. Lapid, and D. Tatarsky, *J. Organometal. Chem.* **184**, 91 (1979).
68. E. Cesarotti, R. Ugo, and H. B. Kagan, *Angew. Chem. Int. Eng. Ed.* **18**, 779 (1979).
69. B. Heil, S. Toros, J. Bakos, and L. Marko, *J. Organometal. Chem.* **175**, 229 (1979).
70. H. M. Feder and J. Halpern, *J. Am. Chem. Soc.* **97**, 7186 (1975).
71. E. L. Muetterties and J. R. Bleeke, *Acc. Chem. Res.* **12**, 324 (1979).
72. R. A. Grey, G. P. Pez, and A. Wallo, *J. Am. Chem. Soc.* **102**, 5948 (1980).
73. C. Fragale, M. Gargano, and M. Rossi, *J. Mol. Cat.* **5**, 65 (1979).
74. R. Spogliarich, G. Zassinovich, G. Mestroni, and M. Graziani, *J. Organometal. Chem.* **198**, 81 (1980).
75. R. Spogliarich, G. Zassinovich, G. Mestroni, and M. Graziani, *J. Organometal. Chem.* **179**, C45 (1979).
76. G. Mestroni, G. Zassinovich, A. Camus, and F. Martinelli, *J. Organometal. Chem.* **198**, 87 (1980).
77. A. Dobson, D. S. Moore, S. D. Robinson, M. B. Hursthouse, and L. New, *J. Organometal. Chem.* **177**, C8 (1979).
78. R. H. Crabtree, J. M. Mihelcic, and J. M. Quirk, *J. Am. Chem. Soc.* **101**, 7738 (1979).
79. M. Bianchi, U. Matteoli, G. Menchi, P. Frediani, S. Pratesi, and F. Piacenti, *J. Organometal. Chem.* **198**, 73 (1980).

80. J. Cook, J. E. Hamlin, A. Nutton, and P. M. Maitlis, *J. Chem. Soc. Chem. Commun.*, 144 (1980).
81. T. Hayashi, M. Tanaka, Y. Ikeda, and I. Ogata, *Bull. Chem. Soc. Jpn.* **52**, 2605 (1979).
82. C. Botteghi, M. Branca, and A. Saba, *J. Organometal. Chem.* **184**, C17 (1980).
83. Y. Watanabe, T. Mitsudo, Y. Yasunori, J. Kikuchi, and Y. Takegami, *Bull. Chem. Soc. Jpn.* **52**, 2735 (1979).
84. D. A. van Bezard, G. Consiglio, F. Morondini, and P. Pino, *J. Mol. Cat.*, 431 (1980).
85. D. Forster, *J. Chem. Soc. Dalton Trans.*, 1639 (1979).
86. G. Cavinato and L. Tonilo, *J. Mol. Cat.* **6**, 111 (1979).
87. T. F. Murray and J. R. Norton, *J. Am. Chem. Soc.* **101**, 4107 (1979).
88. G. Cometti and G. P. Chiusoli, *J. Organometal. Chem.* **181**, C14 (1979).
89. J. E. Hallgrew and R. O. Matthews, *J. Organometal. Chem.* **175**, 135 (1979).
90. R. B. King, A. D. King, and M. Z. Iqbal, *J. Am. Chem. Soc.* **101**, 4893 (1979).
91. K. Kurtev, D. Ribola, R. A. Jones, D. J. Cole-Hamilton, and G. Wilkinson, *J. Chem. Soc. Dalton Trans.*, 55 (1980).
92. W. Keim, M. Berger, and J. Schlupp, *J. Catal* **61**, 359 (1980).
93. J. S. Bradley, *J. Am. Chem. Soc.* **101**, 7421 (1979).
94. K. S. Wong and J. A. Labinger, *J. Am. Chem. Soc.* **102**, 3652 (1980).
95. J. A. Marsella and K. G. Caulton, *J. Am. Chem. Soc.* **102**, 1747 (1980).
96. B. D. Dombeck, *J. Am. Chem. Soc.* **102**, 6857 (1980).
97. H. Dumas, J. Levisalles, and H. Rudler, *J. Organometal Chem.* **177**, 239 (1979).
98. H. Dumas, J. Levisalles, and H. Rudler, *J. Organometal. Chem.* **187**, 405 (1980).
99. B. D. Dombek, *J. Am. Chem. Soc.* **101**, 6466 (1979).
100. K. Murata, A. Matsuda, K-I Bando, and Y. Sugi, *J. Chem. Soc. Chem. Commun.*, 785 (1979).
101. C. Ungermann, V. Landis, S. A. Moya, H. Cohen, H. Walker, and R. G. Pearson, *J. Am. Chem. Soc.* **101**, 5922 (1979).
102. A. D. King, R. B. King, and D. B. Yang, *J. Chem. Soc. Chem. Commun.*, 529 (1980).
103. E. C. Baker, D. E. Hendriksen, and R. Eisenberg, *J. Am. Chem. Soc.* **103**, 1030 (1981).
104. A. D. King, R. B. King, and D. B. Yang, *J. Am. Chem. Soc.* **102**, 1028 (1980).
105. A. D. King, R. B. King, and D. B. Yang, *J. Am. Chem. Soc.* **103**, 2699 (1981).
106. P. Giannoccard, G. Vasapollo, and A. Sacco, *J. Chem. Soc. Chem. Commun.*, 1136 (1980).
107. T. Yoshida, T. Okano, and S. Otsuka, *J. Am. Chem. Soc.* **102**, 5966 (1980).
108. C. P. Kubiak and R. Eisenberg, *J. Am. Chem. Soc.* **102**, 3637 (1980).
109. D. J. Cole-Hamilton, *J. Chem. Soc. Chem. Commun.*, 1213 (1980).
110. J. Halpern and R. A. Jewsbury, *J. Organomet. Chem.* **181**, 223 (1979).
111. H. Mimoun, R. Charpentier, A. Mitschler, J. Fischer, and R. Weiss, *J. Am. Chem. Soc.* **102**, 1047 (1980).
112. S. Bhaduri, L. Casella, R. Ugo, P. R. Raithby, C. Zuccaro, and M. B. Hursthouse, *J. Chem. Soc. Dalton Trans.*, 1624 (1979).
113. M. T. Atlay, L. R. Gahan, K. Kite, K. Moss, and G. Read, *J. Mol. Cat.* **7**, 31 (1980).
114. K. M. Nicholas, *J. Organomet. Chem.* **188**, C10 (1980).
115. M. Kubota, K. J. Evans, C. A. Koerntgen, and J. C. Masters, *J. Mol. Cat.* **7**, 481 (1980).
116. B. J. Tovrog, F. Mares, and S. E. Diamond, *J. Am. Chem. Soc.* **102**, 6616 (1980).
117. T. Okada and Y. Kamiya, *Bull. Chem. Soc. Jpn.* **52**, 3321 (1979).
118. D. M. Roundhill, M. K. Dickson, N. S. Dixit, and B. P. Sudha-Dixit, *J. Am. Chem. Soc.* **102**, 5538 (1980).
119. T. Hosokawa, T. Uno, S. Inui, and S-I Murahashi, *J. Am. Chem. Soc.* **103**, 2318 (1981).

Author Index

The page on which an author is cited is given first, followed by the reference number(s) in parantheses.

Freed, R.J., 86 (151)
Freeman, F., 45 (47)
Freiser, H., 200 (36)
Frey, C.M., 195 (13), 198 (31)
Freund, H.J., 243 (51)
Fridman, I.A., 172 (294)
Fröhn, U., 99 (60), 208 (73)
Frolov, K.M., 46 (49), 53 (88), 53 (90)
Fry, F., 168 (279)
Fucaloro, A.F., 169 (282)
Fuhrhop, J.H., 29 (80)
Fujihira, M., 81 (107)
Fujii, Y., 168 (278)
Fujimoto, M., 99 (56), 173 (300)
Fujiwara, T., 135 (124–126), 137 (132)
Fujiwara, Y., 240 (38)
Fukutomi, H., 197, 206 (27)
Fukuzumi, S., 10 (19), 222 (28)
Füldner, H.H., 194 (11)
Fullerton, T.J., 279, 282 (15a)
Fumagalli, A., 268 (88), 269 (94)
Funaki, Y., 203 (57)
Furusawa, T., 178 (325)
Fusi, A., 247 (62)

Gafney, H.D., 27, 33 (62)
Gahan, L.R., 169 (285), 170 (286), 315 (113)
Gamsjäger, H., 161 (254), 197 (24)
Galas, A.M.R., 265 (75)
Gans, P.L., 129, 130, 131 (104)
Garafalo, A.R., 132 (109)
Garber, A.R., 67 (9)
Gardini, G.P., 187 (355)
Gardner, E.R., 134 (114), 190 (368)
Gargano, M., 307 (73)
Garleschelli, L., 269 (93)
Gaubatz, R., 115, 120 (35)
Gauss, P.L., 27 (59–60)
Gavens, P.D., 269 (89)
Gavin, R.M., 251 (12)
Geanangel, R.A., 67 (8)
Gebert, E., 84 (135)
Geier, G., 218 (22)
Gell, K.I., 266, 267 (83)
Gennard, G., 303 (45)
Geoffroy, G.L., 226 (10)
Gervasio, G., 226 (8)
Geselowitz, D.A., 34 (118)
Ghodosh, D.F., 257 (43)
Giannoccard., P., 313 (106)
Gibian, M.J., 81 (105)
Gibson, J.A., 259 (55)
Giesbrecht, E., 129 (101)
Gillard, R.D., 137 (135), 138 (140) (142–143) (146), 142 (167), 156 (140)
Gilligan, T.J., 193 (1)
Giongo, G.M., 306 (64)

Giordano, R., 52, 53 (78)
Giro, G., 31 (102)
Giuffrida, S., 250 (2)
Giurgiu, M., 41 (30)
Gjerde, H.B., 28 (72)
Gladfelter, W.C., 226 (10)
Gladysz, J.A., 228 (14), 275 (3)
Glaser, R.V., 305 (56)
Glass, W.K., 283 (25)
Glavas, M., 149, 164 (199)
Glennon, C.S., 25 (47)
Glidewell, C., 83 (127)
Glover, S.G., 125 (72)
Goddard, R., 253 (23), 257, 258 (48)
Goddard, W.A., 298 (23)
Godfrey, A.F., 48 (64)
Goedken, V.L., 134 (116)
Goettsch, P., 258 (52)
Goh, L.Y., 231, 233 (23)
Goh, S.H., 231, 233 (23)
Goldkorn, T., 36 (137)
Goncharov, A.V., 144 (179)
Goodfellow, R.J., 262 (66), 270 (95)
Gordon, S., 53 (89)
Gosling, R., 91 (26), 92 (27)
Gosney, A.P., 73 (52)
Gould, E.S., 24 (42), 25 (44–45), 57 (112)
Gower, M., 230 (20)
Gowenlock, B.G., 73 (54)
Gower, M., 286 (30)
Graf, R.E., 259 (54)
Graham, W.A.G., 259, 260 (59)
Grancicova, O., 109 (18), 149 (200)
Granoth, I., 78 (84)
Grant, M.W., 95 (40), 200 (35), 207 (72)
Grate, J.H., 212 (7)
Grätzel, M., 31 (104–105)
Gray, H.B., 31 (100–101), 34 (122), 35 (123–124), 36 (141), 88 (9), 92 (31), 119 (50)
Graziani, M., 307 (74–75)
Green, M., 92 (28), 173 (304), 254, 255 (33), 265 (75) (78), 277 (10)
Green, M.L.H., 296 (4), 300 (28)
Grenthe, I., 206 (66)
Gresham, D.G., 259 (54)
Gresser, R., 198 (30)
Grevels, F.-W., 253 (23)
Grey, R.A., 303 (51), 307 (72)
Griannotti, C., 33 (110)
Grifiths, T.R., 108 (17)
Grigoryan, E.A., 148 (198)
Grimley, E., 101 (71)
Gröning, A.-B., 90 (23)
Grubes, R.A., 300 (30)
Guerin, C., 69 (18)
Guerney, P.J., 92 (30)

Guerrero, J.E., 110, 111 (27)
Guggolz, E., 269 (92)
Guglielmo, G., 95 (37)
Gupta, B.G.B., 72 (38)
Gupta, K.K.S., 39 (21)
Gupta, V.K., 86 (156), 120 (51)
Gupta, Y.K., 55 (103), 78 (91)
Gutman, V., 31 (97)

Haaland, A., 256 (38) (40)
Habashi, E., 28 (71)
Habon, I., 86 (154)
Haddad, S., 86 (148) (150)
Hafezi, S., 160 (245)
Haim, A., 10 (17), 26 (57), 27 (17) (58) (60), 33 (112), 56 (106), 123 (60), 129 (104), 130 (104) (106), 131 (104) (106), 175 (313)
Haines, R.I., 42 (34), 55 (101), 82 (118), 128 (97), 137 (137), 138 (139), 159 (97) (137), 162 (258), 163 (261), 164 (262), 201 (41), 206 (65)
Halko, D.J., 81 (109)
Hall, G., 148, 159 (193)
Hall, H.T., 289 (37)
Hall, J.R., 191 (372–373)
Hall, T.L., 234 (27)
Hallett, G., 73 (55) (58)
Hallgrew, J.E., 310 (89)
Halpern, J., 23 (29), 58 (116), 59 (124), 179 (333), 211 (4), 212 (5), 214 (10), 231, 232 (22), 278 (12), 296 (7), 304 (55), 315 (110)
Hamada, A., 276 (9)
Hambright, P., 28 (73), 127, 128 (87), 202 (48)
Hamilton, D.E., 139 (150), 207 (67)
Hamlin, J.E., 308 (80)
Hamm, R.E., 122 (58)
Hammond, D.G., 120 (54)
Hammond, G.S., 34 (117)
Hamshere, S.J., 90 (19), 91 (25), 93 (36), 98 (53)
Han, K.R., 266 (81)
Hand, T.D., 24 (34), 25 (47), 174 (306)
Hankonyi, V., 131 (108)
Happer, D.A.R., 70 (27)
Harada, S., 198 (33), 203 (57)
Harger, M.J.P., 78 (87)
Harlow, R.L., 261 (65), 300 (31)
Harris, A.D., 153 (218), 190 (370)
Harris, G.M., 60 (128), 152 (213), 154 (222), 161 (213) (251), 170 (222) (288), 172 (213) (292), 183 (347), 185 (348–349)
Harris, P.J., 79 (95)

Nagarajan, K., 45 (44)
Nagase, S., 67 (13)
Naik, N.C., 152 (212)
Nair, M., 298 (18), 300 (32)
Naish, P.J., 266, 267 (85), 297 (17)
Nakadaira, Y., 69 (24)
Nakamura, A., 306 (62)
Nakamura, R., 245 (54)
Nakamura, Y., 100 (67)
Nakano, K., 26 (52), 127 (85)
Nanda, R.K., 152 (212)
Nanni, E.J., 81 (105–106)
Napoleone, V., 73 (44)
Narang, S.C., 72 (38)
Narula, A.S., 286 (31)
Narusawa, Y., 26 (52), 127 (85)
Nascu, C., 54 (96)
Nasielski, J., 34 (115)
Natarajan, P., 23 (31), 33 (111), 175 (310)
Natile, G., 54 (94)
Navon, G., 4 (4)
Nazer, A.F.M., 38 (20), 46 (48)
Nazer, M., 75 (65)
Needham, J., 83 (127)
Nemeth, S., 138 (147)
Nesmayanov, A.N., 260 (62)
Neta, P., 54 (92), 206 (63)
Netzel, T.L., 31 (107)
Neubecker, T.A., 29 (74), 59 (125)
Neumann, H.M., 162 (259)
New, L., 307 (77)
Newman, K.E., 90 (15), 103 (3), 118 (44), 193 (2), 194 (3), (5–6) (9)
Ng., F.T.T., 139 (148), 212 (5)
Ng., L., 241 (45), 293 (52)
Nicholas, K.M., 315 (114)
Nichols, P.J., 200 (35)
Nickel, U., 86 (159)
Nicolaidou, P., 73 (51)
Nikolaev, A.V., 191 (374)
Nikonova, L.A., 58 (114)
Nilsson, L., 206 (66)
Nisbet, M.P., 76 (77)
Nishikawa, M., 281 (21)
Niswander, R.H., 24 (36)
Nivorozhkin, L.E., 101 (69)
Nixon, J.F., 79 (93–94), 254, 255 (30)
Nolan, K.B., 154 (223–225), 158 (223)
Noll, B., 128 (91)
Nord, G., 41, 42 (33)
Nordberg, R.E., 280, 281 (17)
Norman, P.R., 156 (232) (234), 180 (339), 201 (45)
Norton, J.R., 310 (87)
Noszticzius, Z., 86 (153)
Novakov, T., 83 (124)
Novice, M.H., 69 (20)

Noyau, A., 86 (145)
Noyes, R.M., 37 (4), 86 (144) (152)
Noyori, R., 306 (65)
Nuber, B., 117 (41), 175 (312), 186 (350)
Numata, S., 250 (3)
Nurse, C.R., 260 (60)
Nussbaumer, C., 92 (29)
Nutter, D.E., 75 (68)
Nutton, A., 308 (80)
Nuzzo, R.G., 303 (47)

Ocaoa, O.L., 303 (49)
O'Connor, C.J., 116 (37–38), 126 (77)
O'Connor, J.P., 302 (44)
O'Connor, M.J., 169 (285), 170 (286–287)
Odell, A.L., 126 (77)
Odell, K.J., 262 (66)
Odiaka, T.I., 285 (29), 286 (30), 288 (27)
Odom, J.D., 67 (9–10)
O'Donnell, S., 140 (154)
O'Donoghue, T.D., 142 (162)
Ofstead, E.A., 275 (4)
Ogata, I., 308 (81)
Ogatta, Y., 85 (139)
Ogino, H., 26 (55), 122 (57), 123 (61)
Ohashi, K., 200 (36)
Ohashi, O., 79 (93)
Ohno, A., 48 (61)
Ohno, T., 55 (104)
Ohta, Y., 30 (91)
Ohtani, Y., 99 (56)
Ohyoshi, A., 142 (166), 144 (177), 182 (345), 281 (21)
Ojo, J.F., 55 (100)
Oka, S., 48 (61)
Okada, T., 315 (117)
Okajima, M., 253 (20)
Okamoto, M., 67 (14)
Okamoto, T., 48 (61)
Okano, T., 250 (4), 303 (52), 314 (107)
Okazaki, N., 80 (99), 101 (68), 127 (86)
Okeya, S., 100 (67)
Okpanachi, M.I., 36 (140)
Okubo, T., 148, 159 (194), 155 (229)
Okumura, A., 80 (99), 101 (68), 127 (86)
Okumura, M., 31 (107)
Olabe, M., 133 (113)
Olah, G.A., 72 (38)
Olcott, R.J., 88 (9)
Olson, D.C., 96 (43)
Olubuyide, O., 55 (100)
Omarova, R.S., 89 (21–22)
Omelka, L., 153 (219)

Onimura, Y., 80 (102)
Onuma, K., 306 (62)
Onwood, D.P., 74 (64)
Oon, S.M., 110, 111, (25)
Orbán, M., 86 (146) (152) (154)
O'Regan, B.D., 203 (55)
Orhanović, M., 24 (41)
Orio, A.A., 92 (31)
Orpen, A.G., 271 (100)
Orrell, K.G., 191 (376), 192 (377–380)
Osa, T., 81 (107)
Osborn, J.A., 234 (26), 298 (22)
Ostrich, I.J., 206 (64)
Osvath, P., 205 (61)
Oteiza, R., 217 (20)
Otsuka, 250 (4), 303 (52), 314 (107)
Outram, J.R., 73 (48)
Ovenall, D.W., 300 (29)
Owens, G.D., 29 (79), 38 (12)
Ozaki, A., 303 (46)
Ozawa, F., 215 (12)

Pacheco, A.D., 198 (32)
Paddick, K.E., 251 (16)
Page, J.A., 214 (11)
Page, M.I., 73 (52)
Pain, G.N., 265 (78)
Pakulski, M., 78 (85)
Pal, M., 162 (255)
Pal, S., 86 (156)
Palmer, D.A., 88 (10–12), 90 (13) (17–18), 93 (34–35), 105 (9), 148, 161 (195), 180 (336), 183 (347), 185 (348–349), 250 (1b)
Panda, R., 79 (92)
Pandey, D.P., 48 (62)
Panigrahi, G.P., 79 (92)
Pansevich, V.V., 170 (315)
Pantaler, R.P., 118 (48)
Paradisi, C., 165 (266)
Parish, R.V., 69 (25)
Parker, D., 305, 306 (61)
Parker, D.G., 264 (74), 282 (23), 283 (24)
Parker, D.W., 228 (14)
Parshall, G.W., 37 (2), 300 (29) (31)
Partlin, M., 92 (30)
Pasto, D.J., 75 (66)
Pathak, A., 45 (45)
Pati, S.C., 144 (178)
Patterson, J., 241 (45), 293 (51–52)
Pauson, P.L., 290 (41)
Pavlov, O.N., 127 (90)
Pavlovic, D., 130, 133 (107)
Pavoni, B., 92 (31)
Pawlowski, V.I., 176 (319)
Peacock, R.D., 29 (82), 34 (119)
Pearson, A.J., 282 (22)

General Subject Index